Essentials of Enzymology

Essentials of Enzymology

Edited by **John Herald**

SYRAWOOD
PUBLISHING HOUSE
New York

Published by Syrawood Publishing House,
750 Third Avenue, 9th Floor,
New York, NY 10017, USA
www.syrawoodpublishinghouse.com

Essentials of Enzymology
Edited by John Herald

International Standard Book Number: 978-1-68286-228-5 (Hardback)

Printed in China.

Contents

Preface

Enzymology deals with in-depth study and analysis of enzymes and is crucial for the understanding of many physiological processes. The aim of this book is to provide an understanding of the multiple aspects of enzymology through discussions on topics like metabolism, enzyme kinetics, industrial applications, etc. A number of latest researches have been included to keep the readers up-to-date with the global progress in this area of study. This book is an essential guide for both researchers and students who wish to delve deeper into the scientific study of enzymes.

The information shared in this book is based on empirical researches made by veterans in this field of study. The elaborative information provided in this book will help the readers further their scope of knowledge leading to advancements in this field.

Finally, I would like to thank my fellow researchers who gave constructive feedback and my family members who supported me at every step of my research.

Editor

A study on the combined administration of chloroquine and insulin effect on serum transferases and phosphatases activity during increased dietary fat and calcium consumption

Ajani E. O.*, Salau B. A., Ezima E. N. , Oyefuga O. H. and Osineye P. O

Department of Biochemistry, Faculty of Basic Medical Sciences, Olabisi Onabanjo University, Remo Campus, Ikenne, Ogun State, Nigeria.

Previous studies reported on relationship between chloroquine administration, insulin and glucose homeostasis, but the role of dietary fat and the combined effect of the drugs on liver function have not been exploited. The present study was set up to investigate the effect of combined administration of insulin and chloroquine on serum transferase and phosphates activity during increased dietary fat and calcium consumption. Thirty (30) adult male albino rats randomly assigned into five (5) groups were used in the study. Group A was placed on normal diet (normal control); B (test control) C, D, and E were placed on an increased dietary fat and calcium. Drugs were administered as follows: group C, insulin (100 µg/kg body weight once daily); group D, chloroquine (Imarsel Chemical Co. Ltd., Chinon, Hungary, 15 mg/kg body weight) thrice weekly over a period of 12 weeks, and group E, chloroquine combined with insulin. All rats were sacrificed after 15 weeks treatment period. Blood was withdrawn and activities of the enzymes were determined. Combined administration of the drugs increased serum alanine amino transferase, aspartate transaminase and alkaline phosphate activity and decreased acid phosphatase activity. The result suggests that combined administration of insulin and chloroquine may result in hepatic injury particularly in individual whose dietary regimen includes increased fat and calcium.

Key words: Chloroquine, insulin, liver function, transaminases, phosphatases, dietary calcium, fat.

INTRODUCTION

Calcium is a key component in all cells for maintenance of proper structure of membranes and organelles. It is also a pivotal regulator for a wide variety of cellular functions as a major second messenger from plasma receptors (Newmark, 2000; Newmark et al., 1984). The ubiquitous role of calcium and its binding proteins in the regulation of cell function is an established principle of cellular physiology (Ramussen et al., 1984; Chung, 1980). Calcium is recognized to be a kind of universal second messenger within cells, transforming a variety of signals into appropriate cellular actions and involving a broad range of processes ranging from blood clotting, neuromuscular transmission, secretion and cell division (Robert, 1990). Calcium is regarded so powerful in this regard that the intracellular free calcium ion levels in a resting cell must be kept extremely low (four to six order of magnitude below the level in extracellular fluid). Because of this central importance of calcium for fundamental cell functions, most organisms have evolved a variety of mechanisms for dealing with excess or deficiencies of calcium (more often than not it is the excess) (Heaney et al., 1977). Calcium toxicity takes two forms: hypercalciuria and hypercalcemia.

Chloroquine is a drug that is widely used in West Africa for the treatment of malaria fever. It has been reported to be concentrated in the liver and many other tissues following its administration (Adelusi and Salako, 1982). In toxic doses, it is known to cause appreciable cellular damage to liver, kidney and heart muscle (deGroot et al., 1981; Ngaha, 1982). It is an acidophilic weak drug that will accumulate in the acid environment of the endosome,

*Corresponding author. E-mail: immanbisi@yahoo.com.

raising its pH. This may interfere with the dehydration of the internalized insulin receptor complex. It has a wide distribution in the body with a particular affinity for melanin-containing tissue and may affect the function of tissues such as the eye and the uterus (Anderson et al., 1980). A direct interaction of chloroquine with insulin receptor, resulting in a reduced rate of dissociating insulin from the receptor has been reported (Adelusi and Salako, 1982). This may increase the biological half-life of the activated receptor and may prolong the action of the insulin (Smith et al., 1987).

The release of insulin is calcium dependent and appears to be associated with changes in the plasma membrane potential. It is believed that the mechanism of insulin release is that an increase in blood glucose increases the influx of calcium ions into the B cells leading to membrane depolarization and subsequent release of contents of the granules (Bevan et al., 1997). Although it is usual to associate insulin with regulation of blood sugar, it is well known that insulin affects fat and protein metabolism almost as much as it does for carbohydrate metabolism. Indeed in diabetic patients, abnormalities of fat metabolism result in such complications as acidosis and atherosclerosis which could be the cause of death (Frank et al., 1994).

Of particular interest to this study are reports linking chloroquine to insulin and glucose homeostasis. In an earlier study, we reported that administration of insulin and chloroquine resulted in reduction of serum glucose level and an improved kidney function (Ajani et al., 2004). The liver plays essential role in drug metabolism and this possibly explains the wide variety of drug-induced liver lesions that are produced and their relative frequency. The measurement of the activities of various enzymes in tissues and body fluids provided a significant and well known aid in disease investigation, diagnosis, assault on the organs/tissues and to a reasonable extent the toxicity of the drug (Yakubu et al., 2003).

The present study aimed at investigating the activities of the phosphatases and transaminases enzymes, (which are important in diagnosing liver functions) when chloroquine and insulin are co-administered to subject whose dietary regimen consist of increased fat and calcium.

METHODOLOGY

Experimental subjects

The study was carried out with thirty (30) adult male albino rats of the Sprague- Dawley strain weighing 165 ± 18 g. The rats were randomly assigned into 5 groups and housed in plastic cages with wood-chip bedding in a humidity controlled room with 12 h light/dark cycle. The animals were kept in separate cages labeled A - E. They were all allowed free access to their respective diets and clean drinking water on which they were stabilized for three (3) weeks.

Dietary regime

Two (2) separate diets were used in the study. The feeds were composed according to Olowokere et al. (1994). The control diet consist of 57% carbohydrate; 28% protein; 7.0% fats; 1.5% vitamins and 5.95% salt mix (consisting of 21.7% calcium). Rats in group labeled A were placed on this diet. The test diet on the other hand consists of 46.7% carbohydrate; 24.5% protein; 21.3% fats; 1.5% vitamins and 5.95% salt mix (consisting of 51.97% calcium). Rats in groups labeled B, C, D and E were placed on this diet. Each group was maintained on their respective diet *ad libitum* for 15 weeks.

Administration of drugs

Insulin (100 µg/Kg body weight) was administered intramuscularly as a single dose daily into rats in groups C and E. Chloroquine phosphate (Imarsel Chemical Co. Ltd. Chinon, Hungary; 15 mg/Kg body weight) dissolved in 0.2 ml of sterile saline was administered subcutaneously into rats in groups D thrice weekly over a period of 12 weeks. All treatments were made to commenced 3 weeks after adaptation to respective diet. At the end of the treatment period, all rats were fasted overnight and anaesthetized by keeping in an enclosed container containing diethylether. Blood was then withdrawn by cardiac puncture.

Enzyme assays

Serum acid phosphatase (EC. 3.1.3.2) and alkaline phosphatase (EC. 3.1.3.1) were determined using carbonate acetate buffer by measuring the 4-nitrophenol liberated from 4-nitrophenyl phosphate at 400 nm (Wright et al., 1972a; Wright et al., 1972b). Protein concentration was measured by Biuret method (Wright et al., 1972b; Plummer, 1978). Serum alanine aminotransferase (GPT) was determined by monitoring the concentration of pyruvate hydrazone formed with 2,4-dinitrophenylhydrazine (King, 1978). Aspartate amino transferase (GOT) activity was measured by monitoring the concentration of oxaloacetate hydrazone formed with 2, 4-dinitrophenyl hydrazine (Reitman and Frankel, 1957).

Statistical analysis

Results were expressed as mean ± SEM. n values were the same (5) for both control and test animals. The data were analyzed using paired Student's t-test, choosing a p value of < 0.05 as the level of significance.

RESULTS

Results for the serum phosphatase activities are presented in Table 1. Increased dietary fat and calcium signifycantly increased ($p < 0.05$) the serum acid phosphatase activity but has no significant ($p > 0.05$) effect on the serum alkaline phosphatase activity. Separate administration of chloroquine and insulin apparently does not alter the phosphatase activity. However, combined administration of insulin and chloroquine significantly reduced acid phosphatase activity. The observed activity for the acid phosphatase was not significantly different from the normal control value and the activity observed for the enzyme in other test groups, that is, groups D and E. The observed activity for the serum alkaline phosphatase, following combined administration of insulin and chloroquine was significantly higher than the activity of the enzyme in all other groups. Table 2 illustrates the effect

Table 1. Effect of treatment on phosphatase activity.

Group/ treatment	Enzyme activity (nM/min/mg protein)	
	Acid phosphatase	Alkaline phosphatase
A [Control diet]	0.17 ± 0.06[a]	0.26 ± 0.03[a]
B [Test diet, test control]	0.31 ± 0.04[b]	0.25 ± 0.08[a]
C [Test diet+ insulin]	0.19 ± 0.11[ab]	0.29 ± 0.02[a]
D [Test diet + chloroquine]	0.21 ± 0.08[ab]	0.28 ± 0.06[a]
E [Test diet + chloroquine + insulin]	0.19 ± 0.04[a]	0.36 ± 0.03[b]

Note:
- Values are mean ± SEM
- Number of subjects (n) = 5
- All values in the same column with different superscripts are significantly different from each other ($p < 0.05$).

Table 2. Effect of treatment on transaminase activity.

Group/ treatment	Enzyme activity (U/L)	
	Alanine amino transferase	Aspartate transaminase
A [Control diet]	10.21 ± 0.60[a]	10.72 ± 0.03[a]
B [Test diet, test control]	10.25 ± 0.30[a]	10.69 ± 0.21[a]
C [Test diet+ insulin]	30.65 ± 0.52[b]	40.01 ± 0.08[b]
D [Test diet + chloroquine]	30.45 ± 0.23[b]	40.18 ± 0.14[b]
E [Test diet + chloroquine + insulin]	43.61 ± 0.32[c]	50.16 ± 1.01[c]

Note:
- Values are mean ± SEM
- Number of subjects (n) = 5
- All values in the same column with different superscripts are significantly different from each other ($p < 0.05$).

of administration of the drugs on the serum transaminase activity. Increased dietary fat and calcium do not signify-cantly altered transaminase activity in the serum. Separate administration of chloroquine and insulin, both increase significantly alanine amino transferase and aspartate tranaminase activity ($p < 0.05$) when compared with either the normal control or the test control value. The activity obtained for the insulin administered group and the chloroquine administered group were however not different from each other. Combined administration of the drugs significantly raised ($p < 0.05$) the transaminase activity. The values obtained for the alanine amino transferase was significantly higher than that of all other groups, whereas the observed value for the aspartate amino transaminase though significantly higher ($p < 0.05$) than the control value was not significantly different from the values obtained for the separate administration of the drugs ($p > 0.05$).

DISCUSSION

Normally, enzymes will not be found in the serum except there is damage to one or more organs or tissues of the body. Therefore, enzymes from diseased tissues or organs and from drug assault and other xenobiotics may become manifested in the serum resulting in increased activity since the increased might be a consequence of leakage from damaged tissue. Calcium toxicity has been associated with calcium deposition in the kidney and to subsequent renal failure (Heaney, 1982). There has however, been no report sighted on its hepatotoxicity. The report from this study thus suggest that high calcium diet is largely tolerated and may not potentate liver damage. Calcium has been described as an extremely safe nutrient with a wide margin of safety between repletion and toxicity except for unusual combination circumstances (Johnston et al., 1950).

Alkaline phosphatase is a marker enzyme for the plasma membrane and endoplasmic reticulum (Akanji and Niumanse, 1987). It is often employed to access the integrity of plasma membrane (Khan et al., 1985). Similarly, a rise in serum alanine amino transferase and aspartate transaminase activities has been described as the most important determinant in liver diseases (King, 1978). The increased in alanine amino transferase, aspartate transaminase and alkaline phosphatase activity following combined administration of both chloroquine and insulin observed in this study could indicate a hepatocellular damage. In our opinion, the increase in the activity of these enzymes observed here could be attributable to the loss of membrane components into the extracellular fluid. Chloroquine in toxic doses has been reported to cause appreciable cellular damage to liver, kidney and heart muscle (Kukoyi et al., 2000). Coupled with this is the fact that chloroquine interacts with insulin from the receptor (Kukoyi et al., 2000), hence prolonging the action of insulin. This interaction of chloroquine and insulin may be the predisposing factor in the hepatocellular damage suggested by the observed changes in the marker enzyme's activity. Though chloroquine administration independently increases the activity of these enzymes, the increase was more significant with combined administration. The values obtained for the serum alanine amino transferase and aspartate amino transaminase in the groups studied were within the normal range for rat (10-45U/L, for AST and 10-35U/L for ALT), (Karen, 2001) except in group E where the values obtained for the combined administration of the drugs were noticed to be above normal reference range (43.61 ± 0.32 U/L for alanine amino transferase and 50.16 ± 1.01 U/L for aspartate transaminase). In one study, Khan et al. (1985) noted that chloroquine injection caused a marked accumulation of both insulin and its receptor distribution.

The observed decrease in activity of acid phosphatase following combined administration of the drugs, may suggest the possibility of the drugs playing an important role in cancer chemotherapy (though not monitored here). An increase in plasma acid phosphatase activity has been observed in men with metastasis from prostate carcinoma (Bevan et al., 1997). Again, dietary calcium has been reported to play a role in the promotion of mammary cancer caused by dietary fat (Newmark et al., 1984).

Conclusion

The result of this study indicates that combined administration of chloroquine and insulin may result in hepatic injury. Based on the findings reported in this study, we thus opined that the role of combined administration of chloroquine and insulin in cancer particularly prostate cancer need to be further investigated.

REFERENCES

Adelusi SA, Salako LA (1982). Tissue and blood concentration of chloroquine following chronic administration in the rat. J. Pharm. Pharmacol. 34: 733-735.

Ajani EO, Salau BA, Fagbohun TR, Ogun AO (2004). Combined administration of chloroquine and insulin effect on some biochemical parameters in rats placed on high fat and calcium diet. Afr. J. Med. 33: 365-369.

Akanji MA, Niumanse SE (1987). Alkaline phosphatase activities following repeated suramin administration in some rat tissues in relation to their functions. Pharmacol and Toxicol. 61:182-183.

Anderson A, Oilson S, Tjalve H (1980). Chloroquine inhibits the insulin production of isolated islets. Biochem. Pharmacol. 29:1729-1735.

Bevan AP, Krook A, Tikerpae J, Siddle K, Smith G (1997). Chloroquine extends the lifetime of the activated insulin receptor complex in endosomes. J. Biol. Chem. 24: 26833-26840.

Chung WY (1980). Calmodulin plays a pivotal role in cellular regulation. Science 207: 19-27.

deGroot PQ, Eiferink RQ, Hollemans M, Khand M, Tager JM (1981). Activation of B galactosidase in cultured human skin fibroblast". Exp. Cell. Res. 136: 327-333.

Frank MT, Colin SC, Peter C, Catherine SB, Judith LB, Allan MR (1994). Effects of insulin- like growth factors on protein and energy metabolism in tumor bearing rats. Biochem. J. 301:769-775.

Heaney RP (1982). Calcium nutrition and bone health in the elderly. Am. J. Clin. Nutr. 36: 986-1013.

Heaney RP, Reker RR, Saville PD (1977). Calcium balance and calcium requirement in middle aged women. Am. J. Clin. Nutr. 30: 12603-12611.

Hgaha EO (1982. Some biochemical changes in the rat during repeated chloroquine administration. Toxicol. Letter 10: 145-149.

Johnston FA, Sclaphott D, McMillan TC (1950). Calcium retained from one level of intake by six girls. J. Nutr. 41: 137-147.

Khan W, Savole S, Khan RJ, Posner BI (1985). Insulin and insulin receptor uptake into rat liver: Chloroquine action on receptor recycling. Diabetes 34: 1025-1030.

King J. (1978). Aspartate amino transaminase and alanine amino transferase in "clinical Biochemistry: Principles and Methods" Curtis HC and Marc R (eds). pp 1148-1158.

Kukoyi BI, Thomas KD, Alemji GA (2000). Chloroquine- induced inhibition of hepatic tryptophan dioxygenase activity in rats. Afr. J. Med. Pharm. Sci. 1:7 5-77.

Newmark H C (1989). Calcium in cellular function. Triamgle (Suppl.) 9-13.

Newmark HC, Wargovich MJ, Bruce WR (1984). Colon cancer and dietary fat, phosphate and calcium: A hypothesis. J. Natl. Cancer Inst. 72: 1323-1325.

Olowookere JO (1984). Bioenergetics of Kwashiokor and obesity. An overview of Kwashiokor and obesity. Triumph Book Publisher, Ijebu-Ode. pp. 32-36.

Plummer DT (1978). In "An introduction to practical Biochemistry" 2nd ed. McGraw- Hill. London pp. 144-145.

Ramussen H, Barreh PQ (1984). Calcium messenger system: an integrated view. Physiol. Rev. 64: 938-984.

Reitman GA, Frankel M (1957). Analytical tests for GPT and GOT. Am. J. Clin. Path. 5: 28-32.

Robert PH (1990). Calcium metabolism. Basic Clin. Pharm. 4:28-34.

Smith GD, Amos T, Mahler R, Peters TJ (1987). Effects of chloroquine on insulin and glucose homeostasis in normal subjects and patients with non insulin- dependent diabetes mellitus. Brit. Med. J. 294: 465-467.

Wright PJ, Leathwood PD, Plummer DT (1972a) Enzymes in rat urine: alkaline phosphatase. Enzymologia 42: 312-327.

Wright PJ, Leathwood PD, Plummer DT (1972b) Enzymes in rat urine: acid phosphatase. Enzymologia 42: 459-468.

Yakubu MT, Salau IO, Muhammed NO (2003). Phospahatase activities in selected rat tissues following repeated administration of rantidine. NJMB 18: 21-24.

2

Higher tolerance of a novel lipase from *Aspergillus flavus* to the presence of free fatty acids at lipid/water interface

Faouzi Laachari[1], Fatimazahra El Bergadi[1], Moulay Sadiki[1], Adel Sayari[3], Wifak Bahafid[1], Soumya Elabed[2], Iraqui Mohammed[1] and Saad Koraichi Ibnsouda[1]

[1]Laboratoire de Biotechnologie Microbienne. Université Sidi Mohamed Ben Abdellah, B.P. 2202 – Route d'Imouzzer, FES, Morocco.
[2]Université Sidi Mohamed Ben Abdellah, Centre Universitaire Régional d'Interface, Fès, Morocco.
[3]Laboratoire de biochimie et de génie enzymatique des lipases, ENIS, BPW 3038 Sfax, Tunisia.

The main objective of this work was to identify novel lipases of industrial interest. In this paper, *Aspergillus flavus* lipase (AFL) was isolated from the traditional tannery of Fez city in Morocco; it kept its stability even in the presence of high concentrations of detergent from 0 to 10 mM sodium deoxycholate (NaDC). Bile salts showed no inhibitory effect on the lipolytic activity, whereas the calcium salts showed a stimulating action on the lipase activity. Unlike most of the lipases which were quickly denatured at the lipid/water interface, the accumulation of free fatty acids at the oil/water interface did not affect the activity of the enzyme which effectively hydrolyzed the emulsified olive oil even in the absence of bile salts. Furthermore, AFL was more active on long chain triacylglycerols than on short chain triacylglycerols. This study allowed us to prove that AFL had the interfacial activation phenomenon. A 3D structure model of AFL was built and we have concluded that the ratio hydrophobic surface/hydrophilic surface was 51% versus 50%; it could be responsible for a higher tolerance to the presence of free fatty acids at lipid/water interface.

Key words: *Aspergillus flavus* lipase (AFL), detergent, interfacial activation, free fatty acids, model.

INTRODUCTION

Triacylglycerol acyl hydrolase (EC.3.1.1.3), lipases belong to the carboxylic ester hydrolases family. The physiological role of lipases is to hydrolyze triglycerides to diglycerides, monoglycerides, glycerol and fatty acids (Mats and Karl, 1994). These enzymes exist in all living organisms; they have the ability to achieve synthesis reactions such as esterification (reaction between an acid and an alcohol), the transesterification (ester andalcohol), interesterification (ester and ester), and in transfer reactions acetyl group of an ester. Lipolytic enzymes are perfectly soluble in water.

They are responsible for the hydrolysis of lipids which act on insoluble lipid substrates in water (Bénarouche et al., 2014). Lipids are formed by an aliphatic backbone, which are cyclic or polycyclic, that constitute the hydrophobic portion, which can be fixed polar groups and form the hydrophilic portion. Lipids include fats, oils, waxes and certain substances that are related (sterols, steroids, terpenes, etc).

The study of lipases has contributed to the development of the interfacial enzymology, which catalysis occurs in a heterogeneous medium at oil-water interface. The biochemical properties of these enzymes depend much on the quality of the interface and certain conventional parameters such as pH or ionic strength (Bénarouche et al., 2014). Lipases and esterases are important biocatalysts and particularly suitable for industrial applications, as they are very stable and active in organic solvents (Schmidt et al., 2004).

MATERIALS AND METHODS

Chemicals

Tributyrin (99%; puriss) and benzamidine were from Fluka (Buchs, Switzerland); tripropionin (99%, GC) was from Jansen (Pantin, France); phosphatidylcholine, sodium deoxycholic acid (NaDC), Tween 20, yeast extract and ethylene diamine tetraacetic acid (EDTA) were from Sigma Chemical (St. Louis, USA); gum arabic was from Mayaud Baker LTD (Dagenham, United Kingdom); acrylamide and electrophoresis grade were from BDH (Poole, United Kingdom); PVDF membrane was purchased from Applied Biosystems (Roissy, France); casein peptone was from Merck (Darmstadt, Germany); and pH-stat was from Metrohm (Switzerland).

Lipase activity determination

The lipase activity was measured titrimetrically at pH 8.5 and 40°C with a pH-stat under standard conditions using TC_4 (tributyrin) (0.25 ml) in 30 ml of 2.5 mM Tris–HCl pH 8.2, 2 mM $CaCl_2$, 2 mM NaDC (sodium deoxycholate) or olive oil emulsion (10 ml in 20 ml of 2.5 mM Tris–HCl pH 8.2, 2 mM $CaCl_2$, 4 mM NaDC); the olive oil emulsion was obtained by mixing (in a Waring blender) 10 ml of olive oil in 90 ml of 10% GA (gum Arabic) as substrate (Sarda and Desnuelle, 1958). Some lipase assays were performed in the presence of bile salts. Assays were carried out in 30 ml of 2.5 mM Tris–HCl buffer pH 7.0 containing 0.1 M NaCl. Standard conditions for measuring enzyme activity at increasing esters concentrations have been described previously (Stöcklein, 1993). When measuring AFL activity in the absence of $CaCl_2$, we added EDTA to the lipolytic system. Lipolytic activity was expressed as units. One unit corresponds to 1 µmol of fatty acid released per minute.

pH-stat titrimetric assay

Titration method pH-stat of the liberated fatty acids from triacylglycerol continuously stirred, and coupled to an automatic burette, recorder and connected to a circulating water bath. This technique allows for automatical titration of the fatty acids released by addition of 0.1 N sodium hydroxide, maintaining the reaction medium at a constant pH during the hydrolysis of triacylglycerol emulsion by a lipase, and permits in varying the assay conditions, including substrates and bile salts. The pH chosen is generally appropriate in the optimum pH of the enzyme studied (Taylor, 1985). The enzyme activity is expressed in International Units (1U = 1µmole of fatty acid released / min).

Determination of protein concentration

Protein concentration was determined as described by Bradford (1976) using BSA ($E^{1\%}{}_{1cm} = 6.7$).

Procedure of AFL purification

1000 ml of culture medium, obtained after 72 h of cultivation, were centrifuged for 15 min at 8500 rpm and filtered to remove the cells. The supernatant containing extracellular lipase was used as the crude enzyme preparation.

Ammonium sulphate precipitation

The cell-free culture supernatant was precipitated using solid ammonium sulphate to 70% saturation. The pellet obtained after centrifugation (30 min at 8500 rpm) was dissolved in 10 ml of buffer A (20 mM sodium acetate pH 5. 4, 20 mM NaCl, and 2 mM benzamidine) and within 5 min, insoluble material was removed by centrifugation at 13,000 rpm.

Heat treatment

The supernatant obtained (10 ml) was incubated for 30 min at 70°C and in 5 min; insoluble material was removed by centrifugation at 13,000 rpm. Filtration on Sephadex G-100: The enzyme solution (10 ml) was applied to a Sephadex G-100 column (3 ×100 cm) previously equilibrated in buffer A at a rate of 30 ml/h. The fractions containing the lipase activity (eluted at a void volume) were pooled.

Cation exchange chromatography

Active fractions eluted from Sephadex G-100 column were poured into a mono-S Sepharose cation exchanger equilibrated in buffer A. The column (2 ×30 cm) was rinsed with 400 ml of the same buffer. No lipase activity was detected in the washing flow. Adsorbed material was eluted with a linear NaCl gradient (300 ml of 20 to 500 mM in buffer A) at a rate of 45 ml/h. AFL activity was eluted between 170 and 220 mM NaCl. The fractions containing the lipase activity were pooled and concentrated.

Effect of free fatty acids

The lipase activity was measured according to various substrate (TC_4, olive oil) in assigned concentrations ranging from 0-40 mM. The Michaelis-Menten (KMapp) and the maximum velocity (Vmax) for the reaction were calculated by Lineweaver-Burk (Taylor, 1985).

Effect of detergents

The lipase activity was measured using tributyrin and olive oil as a substrate in the presence of increasing concentrations of NaDC

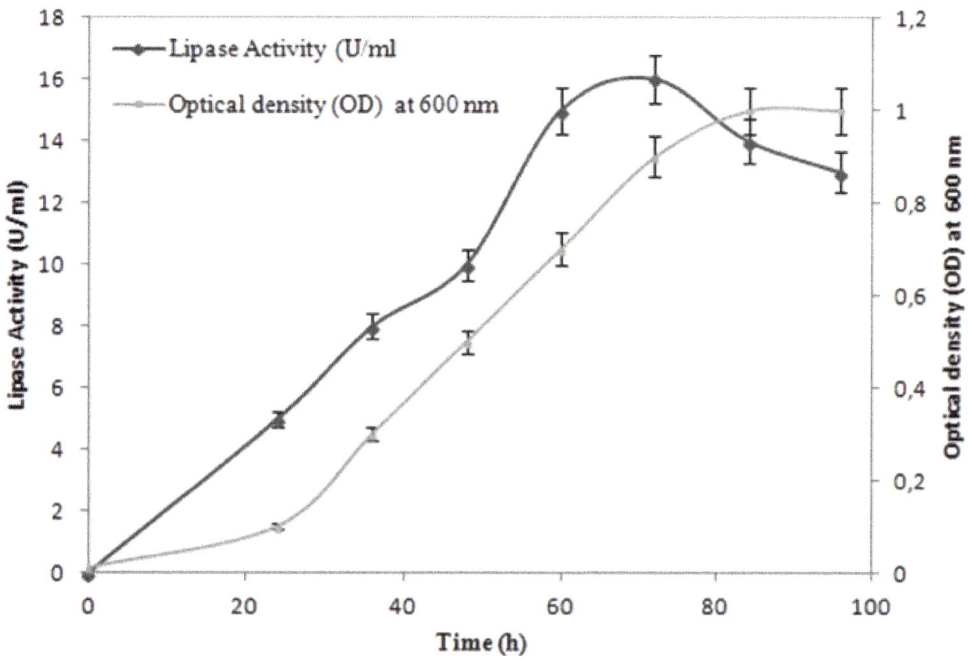

Figure 1. The time courses of lipase production. The culture was carried out at 28°C in shaking at 150 rpm. Bars correspond to standard deviation

ranging from 0 to 10 mM, under optimum conditions of pH and temperature (Taylor, 1985).

Effect of calcium

The lipase activity was measured using tributyrin and olive oil as a substrate in the presence of increasing concentrations of calcium from 0 to 6 mM and under optimum conditions of pH and temperature. In the absence of calcium, the lipase activity is measured in the presence of 10 mM EDTA or EGTA (Taylor, 1985).

3D structure prediction

The AFL structure was modelled using the 3D coordinates of the closed form of the Aspergillus lipases (AL) (PDB code 3DXL_A). The method of minimization (Gromos96) (located in the server Swiss-PDB-Viewer) was used with the version of force field GROMOS43B1. This method enabled the evaluation of minimizing energy of the three dimensional structure and the distortion compensation. The geometry quality of the model was checked using PROCHECK program (Laskowski et al., 1993).

RESULTS AND DISCUSSION

Production of lipase

The maximal production of AF lipase with inoculum size of 3 × 10^7 cells/ml was 16 U/mL, which was obtained by incubating 1 ml of the enzyme with olive oil emulsion (10 ml in 20 ml of 2.5 mM Tris-HCl pH 8.2, 2 mM $CaCl_2$, 4 mM NaDC) as substrate, The lipase activity was measured titrimetrically at pH 8.5 and 40°C with a pH-stat. The time course of lipase production was followed at 28°C with cell growth (Figure 1). The lipase activity was observed to start after incubation and reached the maximum (16 U/mL) at the end of the exponential phase corresponding to 72 h of cultivation.

Purification of AFL

The AFL was purified according to the procedure described. The protein elution profile which was obtained at the final step of the purification was shown in (Figure 2A).The purified lipase was homogeneous when tested by the Coomassie blue staining in SDS-PAGE (Figure 2). This figure shows that one band was revealed for AFL with a molecular mass of 55 kDa. The specific activity of the pure lipase reached 4300 U/mg using olive oil emulsified with gum Arabic as substrate at pH 8.5 and 40°C. Under the same conditions, a specific activity of 3400 U/mg was obtained when using TC4 as substrate.

Kinetic studies of AFL

Lipase hydrolysis emulsified triglycerides in the presence of bile salts. Some microbial lipases like *Rhizopus oryzae* lipase (Ben Salah et al., 1994) may lose its enzymatic activity, when TC_4 or olive oil is used as substrate in the

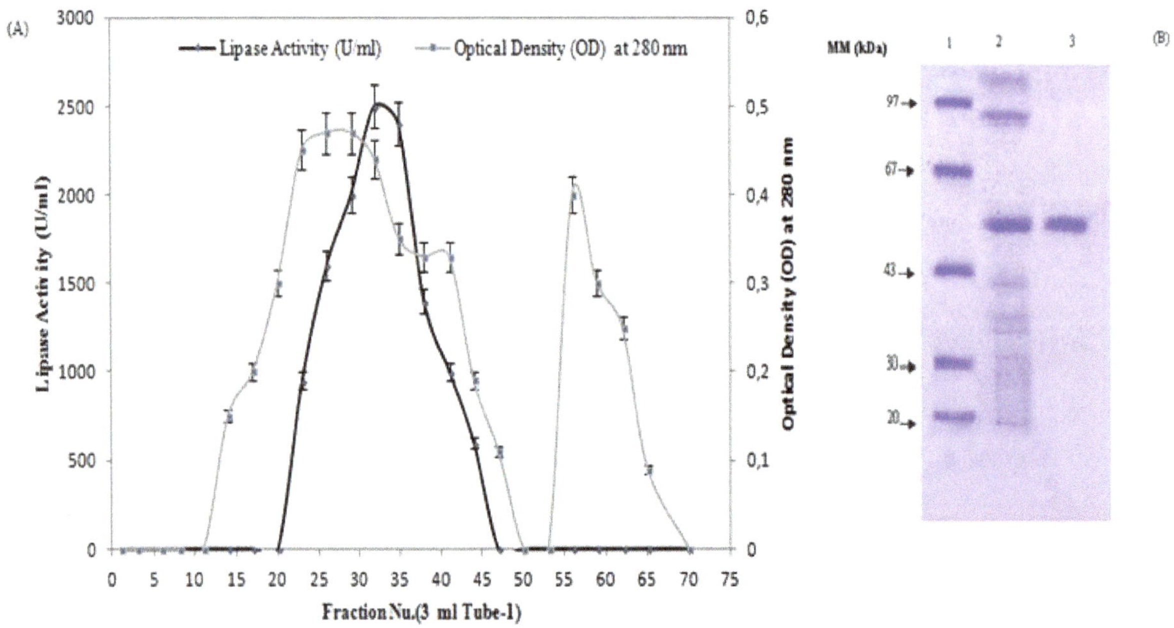

Figure 2. (A) Chromatography of AFL on Sephacryl S-200. The colomn (3 × 100 cm) was equilibrated with buffer A (20 mM sodium acetate pH 5.4, 20 mM NaCl, 1 mM benzamidine). The elution of lipase was performed with the same buffer at a rate of 26 ml/h. Lipolytic activity was measured under standard conditions at pH 8.5 and 40°C using a pH-stat. (B) SDS/PAGE (15%). Lane 1, molecular mass markers (Pharmacia); lane 2, characterisation of the AFL obtained after Mono-S chromatography chromatography; lane 3, 20 µg of purified AFL. Bars correspond to standard deviation.

absence of amphipathic reagent; the high energy existing at lipid/water interface is responsible for their irreversible denaturation. The enzyme denaturation cannot be reflected in the number of disulfide bridges, but behave very differently at interfaces. AFL was able to hydrolyze the TC_4 or the olive oil emulsion alone (Figure 3). The kinetics of substrate hydrolysis remained linear for more than 20 min, accordingly; AFL probably presented a three-dimensional structure allowing it to hydrolyse its substrate efficiently and without any denaturation at high interfacial energy. Also, it tolerated the presence of long-chain free fatty acids, at the olive oil/water interface without any addition of amphipathic reagent (NaDC, Triton X-100); a difference between Aspergillus lipase and other microbial lipases which had a strong preference for short-chain substrates (Simons et al., 1996; Laachari et al., 2013; Sayari et al., 2001).

Effect of calcium on AFL activity

Metal cations, particularly Ca^{2+}, plays important role in influencing the structure and function of lipases (Shangguan et al., 2011). The activity of lipases may depend on the presence of Ca^{2+} ions, like the staphylococcal lipases (Elkhattabi et al., 2003). The effect of various Ca^{2+} concentrations on the rate of hydrolysis of AFL was studied. A specific activity of 4300 U/mg was measured in the presence of 10 mM of chelator such as EDTA or EGTA, when using olive oil emulsion as substrate. In the absence of chelators, the specific activity of AFL reached 4900 U/mg at 2 mM $CaCl_2$ (Figure 4). The enzymatic activity of AFL was stimulated by Ca^{2+}. The lipases from *P. Glumae* and *S. hyicus* (Elkhattabi et al., 2003; Tiesinga et al., 2007) contained a Ca^{2+} binding site which was formed by two conserved aspartic acid residues near the active-site, dramatically enhanced the activities of these enzymes.

Effect of detergents

To verify whether AFL was capable to hydrolyze triglycerides in the presence of surfactants such as bile salts, we measured the rate of hydrolysis of TC_4 and emulsified olive oil in the presence of various NaDC concentrations, from the results as shown in Figure 5, we notice that the NaDC has no inhibitory effect on activity of lipolytic enzymes even at a high concentration (10 mM). This result confirmed that unlike many lipases, such as *Staphylococcus xylosus* lipase, the maximal activity was reached in presence of 2 mM of NaDC using tributyrin as substrate. In presence of 4mM of NaDC, only 40% of residual activity was detected (Mosbah, et al., 2005). AFL

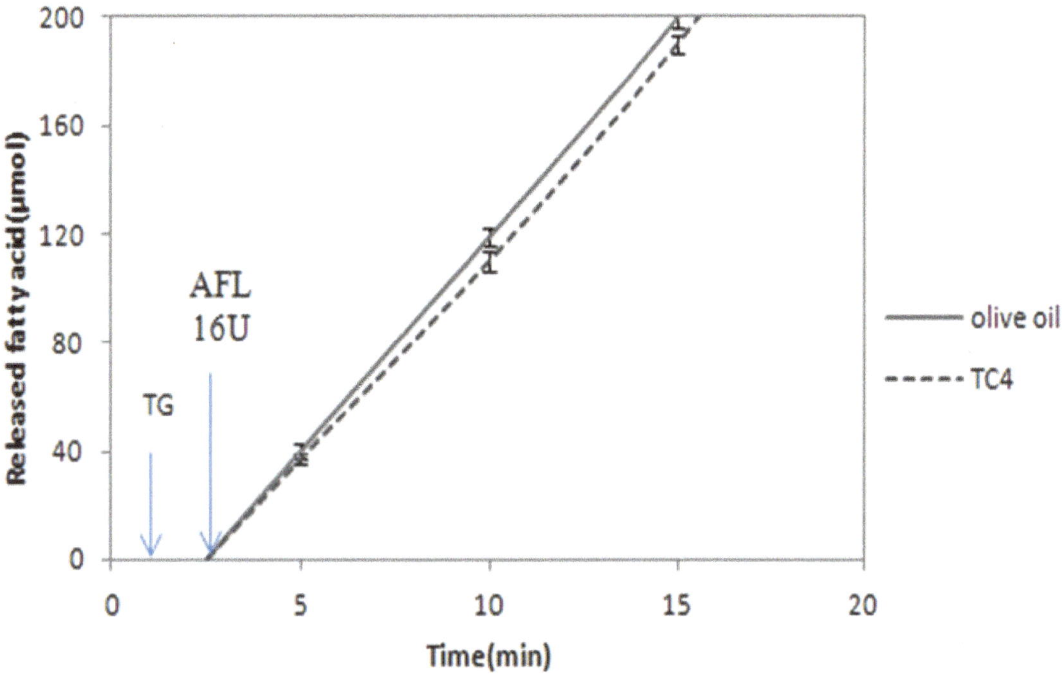

Figure 3. Kinetic of hydrolysis of olive oil emulsions or tributyrin by AFL (16 U). Lipolytic activity was followed at pH 8.5 and 40°C in the absence of bile salts under standard conditions at pH 8.5 and 40°C using a pH-stat.

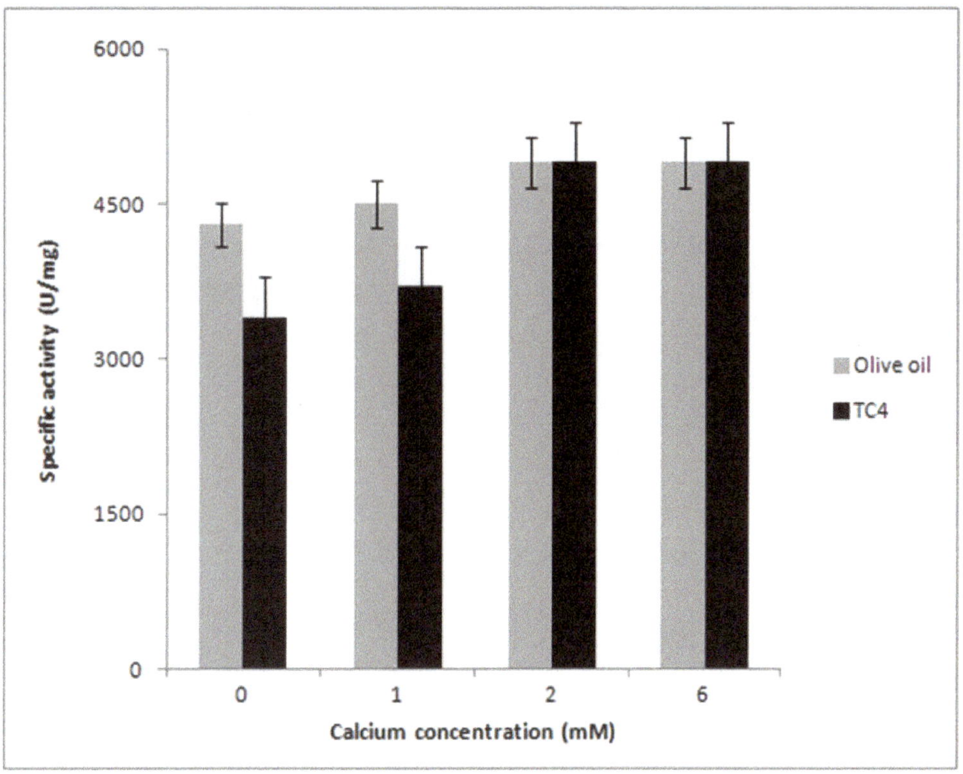

Figure 4. Effect of increasing concentrations of calcium on the rate of hydrolysis of tributyrin and olive oil emulsion by AFL measured in the presence of 10 mM EDTA or EGTA under standard conditions at pH 8.5 and 40°C using a pH-stat.

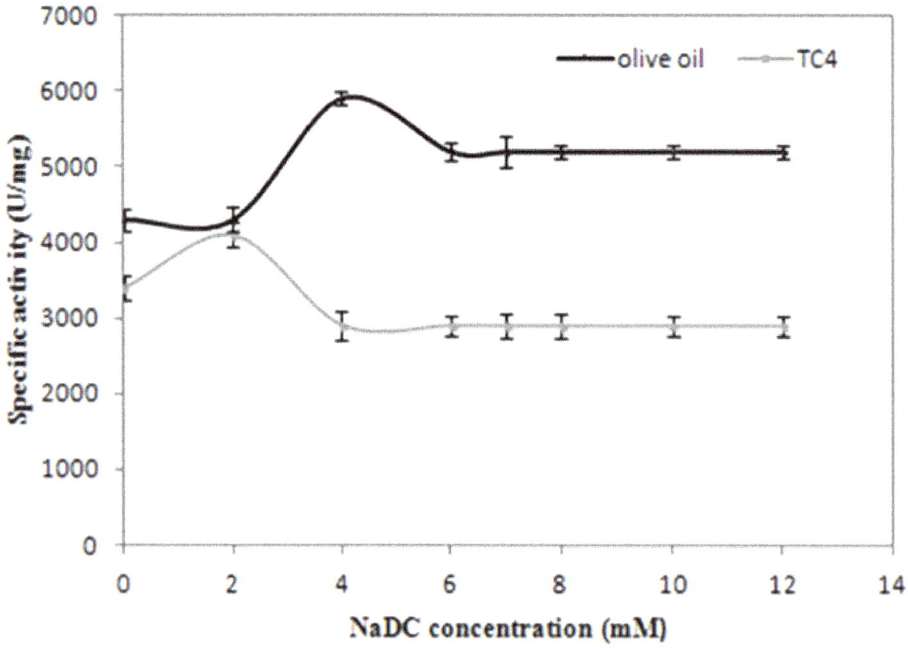

Figure 5. Effect of increasing concentration of NaDC on the rate of hydrolysis of tributyrin and olive oil emulsion by AFL. Lipolytic activity was measured under standard conditions at pH 8.5 and 40°C using a pH-stat.

Alignments Download GenPept Graphics Distance tree of results Multiple alignment

Description	Max score	Total score	Query cover	E value	Ident	Accession
lipase/serine esterase [Aspergillus oryzae RIB40]	64.7	64.7	100%	1e-10	100%	XP_001823770.2
lipase/serine esterase, putative [Aspergillus flavus NRRL3357]	64.7	64.7	100%	1e-10	100%	XP_002380732.1
lipase/serine esterase [Ajellomyces dermatitidis SLH14081]	61.7	61.7	100%	1e-09	94%	XP_002620495.1
lipase/serine esterase [Aspergillus fumigatus Af293]	61.3	61.3	100%	1e-09	94%	XP_748741.1
lipase/serine esterase, putative [Neosartorya fischeri NRRL 181]	61.3	61.3	100%	1e-09	94%	XP_001259182.1

Figure 6. Alignment of the amino acid sequence of AFL using the program Blast.

was able to reach its substrate even in the presence of certain active agents surface, such as bile salts. Similar results were obtained with SSL (lipase from *Staphylococcus simulans*) (Sayari et al., 2001). Similarly, SSL was not inhibited by anionic detergents such as NaDC (Simons et al., 1997). Therefore, it can be inferred that probably AFL had a higher penetrating power than those of other microbial lipases that enabled it to hydrolyze the olive oil or TC_4 in the presence of bile salts.

3D structure model of AFL

The research of homologous with the AFL sequence was made in the database using the BLASTp (Basic Local Alignment Search Tool protein) program by multiple alignment and the results are shown in Figure 6. Sequence analysis allowed us to reveal a 100% homology with *A. flavus* NRRL3357 (33.5 kDa), *A. oryzae* RIB40 and *Ajellomyces dermatitidis* SLH14081. In order

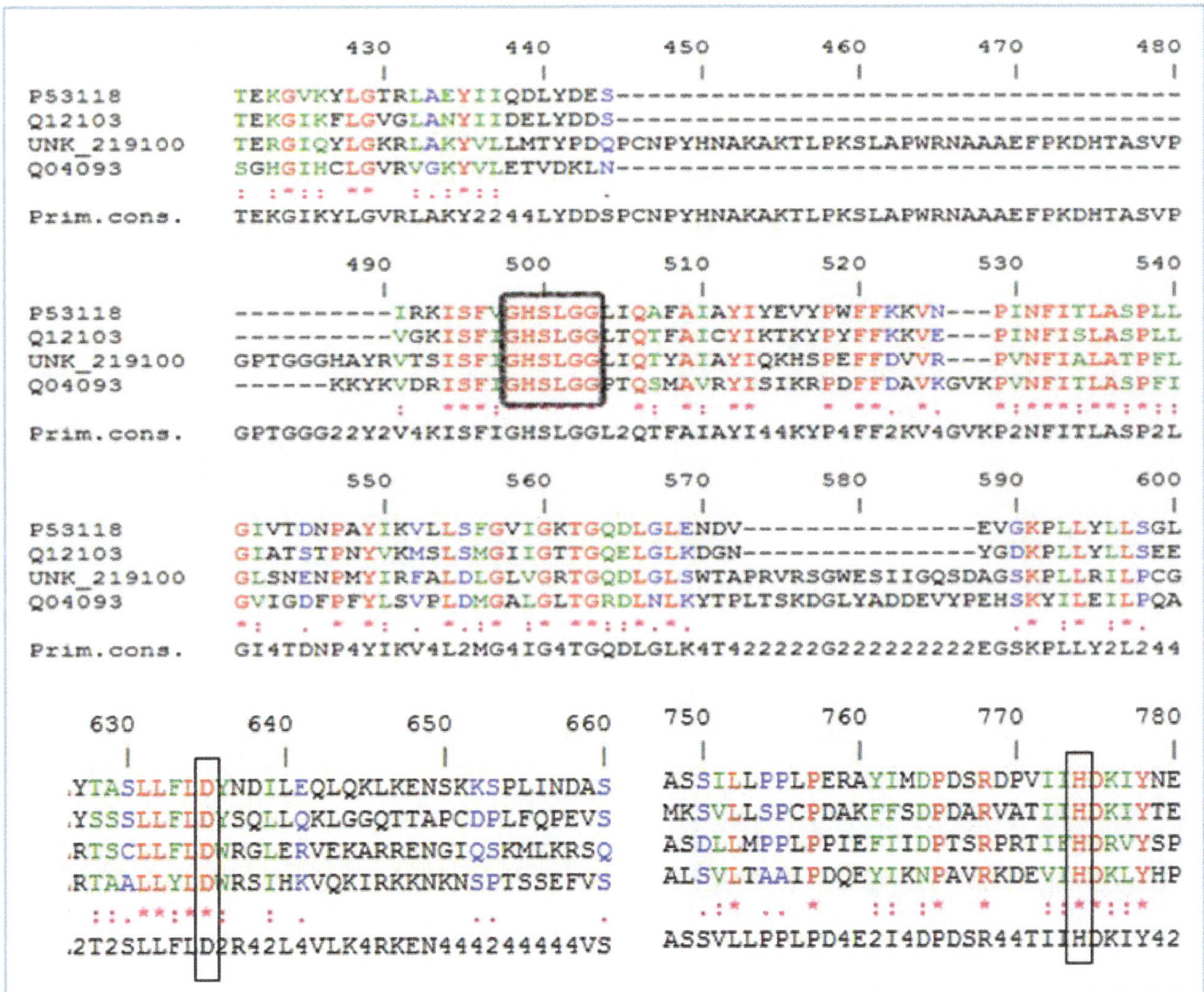

Figure 7. The multiple sequence alignment of AFL (UNK_219100) with lipase sequences of Aspergillus genus: chain (P53118), chain (Q12103) and chain (Q04093) with the additional sequence (cons prim). The catalytic triad is framed and the catalytic serine500 is present in the conserved pentapeptide G-X-S-X-G, Asp635 and His774.

to analyze sequences in databases NPSA (network protein sequence analysis) that allowed the comparison of protein sequences, a sequence alignment of the N-terminal portion of AFL was completed. From Figure 7, we observe that the catalytic triad of AFL constituted Ser500, Asp635 and His774. The three amino acids were located on the C-terminal side of β core lamina. We also noticed that the Ser500 was a part of the pentapeptide Gly-X-Ser-X-Gly and located at the apex of the nucleophilic elbow between strand β5 and helix α5. The alignment of 3 sequences structures with the reference sequence allowed us to identify the location of the catalytic triad of AFL (Figure 7). Indeed, the catalytic triad (Ser-His-Asp) was a characteristic structure of a well known serine proteases (Brumlik and Buckley, 1996). It was also observed in the catalytic sites in several lipases with a carboxylate residue of aspartic acid or glutamic acid. These amino acids were generally determined by chemical modification studies (Ruiz et al., 2007) or by site directed mutagenesis (Hyun-Ju et al., 2000). The model of the *Bacillus stearothermophilus* lipase P1 constructed by the use of the basis of secondary structure predictions, showed an organization in fold and this enzyme α/β-hydrolase was identified by the catalytic triad of Ser-113, Asp-317 and His-358, in close proximity to each other which played a key role in the catalytic mechanism (Sinchaikul et al., 2001).

In order to create a sample of the closed form of the *A. flavus* lipase, we used the automatic modeling by the Swiss-Model server (http://www.expasy.org/spdvb). The model of AFL was a subject of several cycles of energy minimization using Gromos96 installed on the server

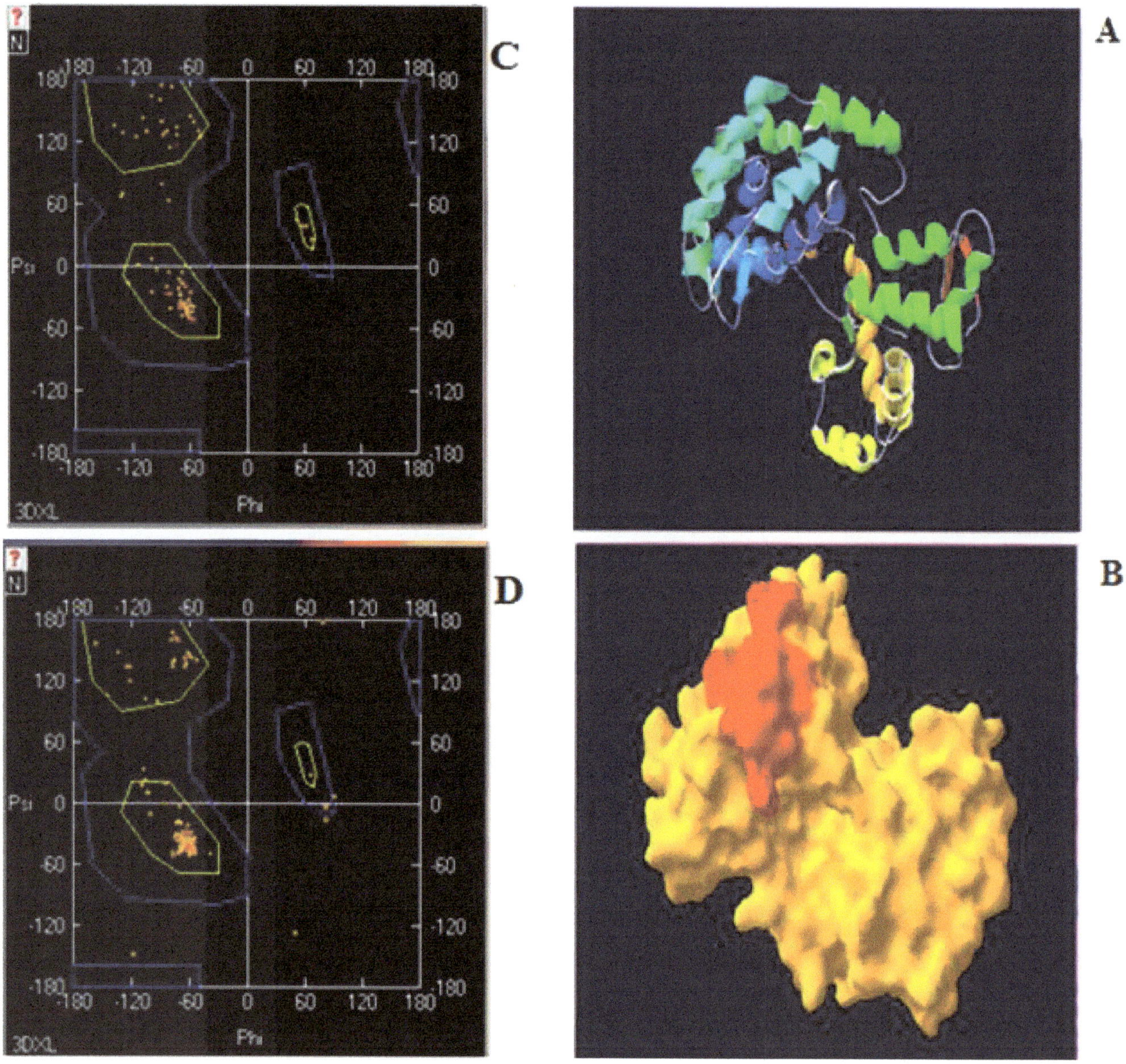

Figure 8. Modeling of a closed model of the AFL (A) and the surface structure (B). The catalytic triad is colored in red (B). Hydrophobic amino acids, belonging to the polar N-terminal domain are located in the Ramachandran plot (C). The hydrophobic residues of the AFL involved in increasing the accessible surface interacting with an interface is indicated (D).

Swiss PDB. The superposition of the two closed structures gave an average standard deviation (rmsd) equal to 1.46 Å. The stereochemical quality of the closed model of AFL and statistical analysis of the distribution of amino acids in the Ramachandran plot were also tested by the program PROCHECK (Laskowski et al., 1993). According to the Ramachandran plot (Figure 8), we found that 99 and 97% of non-polar and polar amino acids respectively, were located in suitable areas (Figure 8).

The crystal structures of *Bacillus sp.* H257 lipase in its free form at 1.2 was found complexed with phenylmethyl-sulfonyl fluoride at 1.8 Å. The catalytic residues were buried at the bottom of a long chain ~ 22 Å, from the surface to the active site, the bottom of the binding pocket of the substrate had more polar character with contributions from catalytic residue Ser97, His226, Met98 and Phe29, built the oxyanion hole, which stabilized the tetrahedral intermediate formed during the hydrolysis reaction (Rengachari et al., 2012; Laachari et al., 2014). Furthermore, the study of the total surface accessible(N-terminal domains) of the closed form of AFL, showed that the ratio hydrophobic surface/hydrophilic surface was

51% versus 50%. This could explain the fact that AFL tolerates accumulation of long chain fatty acids in the lipid/water interface and had a linear kinetics over 15 min, during the hydrolysis of an emulsion of oil olive.

The structure of the lipase from *A. niger* was composed of a core domain (residues 1-82 and 97-269), showing the typical characteristics of the α/β hydrolase and a lid domain (residues 83-96), with a simple pattern of loop-helix-loop. The orientations and the positions of residues of the catalytic triad (Ser145-Asp198-His260) followed the movement of the lid (Liu et al., 2013). These observations allowed us to suggest that the structure of the lipase from *A. flavus* conformed to all the characteristics of the lipases in the open conformation.

Conclusion

The interest of microbial lipases in biotechnological applications has taken a meteoric rise in recent years. Therefore, the industry requires new enzymes, which meet the criteria of use, particularly in terms of thermostability. The *A. flavus* lipase kept their stability even in the presence of high concentrations of detergent (NaDC). Calcium salts showed a stimulating action of the lipase activity. Molecular modeling of the 3D structure was also carried out and the results have allowed us to demonstrate that AFL tolerates the accumulation of long chain fatty acids in the lipid / water interface. These results show that AFL has biochemical properties attractive for various industrial applications.

Conflict of interests

The authors did not declare any conflict of interest.

REFERENCES

Bénarouche A, Point V, Carrière F, Cavalier JF (2014). Using the reversible inhibition of gastric lipase by Orlistat for investigating simultaneously lipase adsorption and substrate hydrolysis at the lipid-water interface. Biochimie. 101:221-31.

Bradford MM (1976). A Rapid and Sensitive Method for the Quantitation of Microgram Quantities of Protein Utilizing the Principle of Protein-Dye Binding. Anal. Biochem. J. 72:248-254.

Brumlik MJ, Buckley JT (1996). Identification of the catalytic triad of the lipase/acyltransferase from Aeromonas hydrophila. J. Bacteriol. 178:2060-2064.

Elkhattabi M, van Geldr P, Bitter W, Tommassen J (2003). Role of the calcium ion and the disulfide bond in the Burkholderia glumae lipase, Mol. Catal. B. J. 22:329-338.

Hyun-Ju K, Amada K, Haruki M, Morikawa M, Kanaya S (2000). Identification of the histidine and aspartic acid residues essential for enzymatic activity of a family I.3 lipase by site directed mutagenesis. FEBS. Lett. 483:139-142.

Laachari F, Elabed S, Sayari A, Mohammed I, Harchali E, Ibnsouda SK (2013). Biochemical characterization of a thermoactive and thermostable lipase from a newly isolated Trichosporon coremiiforme strain. Afr. J. Biotechnol. 12(28):4503-4511.

Laachari F, El Bergadi F, Elabed S, Sayari A, Mohammed I, Ibnsouda SK (2014). Biochemical Study of the Effect of Free Fatty Acids in the Interface on a Novel Lipase from Bacillus pumilus Strain. World. Appl. Sci. J. 32(10):2070-2075.

Laskowski RA, MacArthur MW, Moss DS, Thornton JM (1993). PROCHECK: A program to check the stereochemical quality of protein structures. Appl. Cryst. J. 26:283-291.

Liu L, Lan D, Wang Q, Gao C, Li Z, Yang B, (2013). "bridge-like" structure responsible for the substrate selectivity of mono- and diacylglycerol lipase from Aspergillus oryzae. J. Mol. Catal. B: Enzym. 97:144-149.

Mats M, Karl H (1994). Kinetics of triglyceride lipase. In: Wooley P. & Petersen S.B., eds. Lipases. Cambridge, UK: Cambridge University Press. pp. 159-180.

Mosbah H, Sayari A, Bezzine S, Gargouri Y (2005). Expression, purification, and characterization of His-tagged Staphylococcus xylosus lipase wild-type and its mutant Asp 290 Ala. Protein. Express. Purif. 47:516-523.

Rengachari S, Bezerra GA, Riegler-Berket L, Gruber CC, Sturm C, Taschler U, Boeszoermenyi A, Dreveny I, Zimmermann R, Gruber K, Oberer M (2012). The structure of monoacylglycerol lipase from Bacillus sp. H257 reveals unexpected conservation of the cap architecture between bacterial and human enzymes. Biochim. Biophys. Acta. 1821(7):1012-21.

Ruiz C, Falcocchio S, Pastor FI, Saso L, Diaz P (2007). Helicobacter pylori EstV: identification, cloning and characterization of the first lipase isolated from an epsilon-proteobacterium. Appl. Environ. Microbiol. 73(8):2423-31.

Sarda L, Desnuelle P (1958). Action de la lipase pancréatique sur les esters en émulsion. J. Biochim. Biophys. Acta. 30:513-521.

Sayari A, Agrebi N, Jaoua S, Gargouri Y (2001). Biochemical and molecular characterization of Staphylococcus simulans lipase. Biochimie. J. 83:863-871.

Schmidt BM, Baumann M, Henke E, Konarzycka-Bessler M, Bornscheuer UT (2004). Directed evolution of Lipases and Esterases. Methods in enzymology. 388:199-207.

Shangguan JJ, Liu YQ, Wang FJ, Zhao J, Fan LQ, Li SX, Xu JH (2011). Expression and Characterization of a Novel Lipase from Aspergillus fumigatus with High Specific Activity. Appl. Biochem. Biotechnol. J. 165(3-4):949-62.

Simons JW, Adams H, Cox RC, Dekker N, Gotz F, Slotboom AJ, Verheij HM (1996). The lipase from Staphylococcus aureus. Expression in Escherichia coli, large-scale purification and comparison of substrate specificity to Staphylococcus hyicus lipase. Biochem. Eur. J. 242:760-769.

Stöcklein W, Sztajer H, Menge U, Schmid RD (1993). Purification and properties of lipase from Penicillium expansum. Biochim. Biophys. Acta. 1168:181-189.

Sinchaikul S, Sookkheo B, Phutrakul S, Wu YT, Pan FM, Chen ST (2001). Structural Modeling and Characterization of a Thermostable Lipase from Bacillus stearothermophilus P1. Biochem. Biophys. Res. Commun. 18;283(4):868-75.

Taylor F (1985). Flow-through pH-stat method for lipase activity. Anal. Biochem. 148(1):149-153.

Tiesinga JJ, van Pouderoyen G, Nardini M, Ransac S, Dijkstra BW (2007). Structural Basis of Phospholipase Activity of Staphylococcus hyicus lipase. Mol. Biol. J. 371:447-456.

Purification and characterization of the α-amylase isolated from *Penicillium camemberti* PL21

Tahar Nouadri*, Zahia Meraihi, Djekrif-Dakhmouche Shahrazed and Bennamoun Leila

Research Laboratory in Enzymatic Engineering, Department of Biochemistry and Microbiology, Mentouri- University, Constantine- Algeria.

The amylase family of enzymes is a great significance due to its wide area of potential application. α-amylase from *Penicillium camemberti PL21* obtained from I. N. A. (The Technological Laboratory of Agriculture National Institute Paris- France) and using orange waste as substrates was produced under optimum conditions, after 168 h of incubation and subjected for purification and characterization. The enzyme was purified by ammonium sulfate precipitation, dialysis, sephadex G-100 and DEAE-Sepharose CL-6B column chromatography. A trial for the purification of α-amylase resulted in an enzyme with specific activity of (154.2 units/ml/mg protein) with (38.5 folds) purification .The α-amylase activity increased by enzyme concentration rise. The optimum substrate concentration for soluble starch was 1 % (w/v) while the optimum incubation temperature was 30ºC. The purified α-amylase enzyme had a maximum activity at pH 6 and the *K*m value for soluble starch was 0.92 mg/ml. Analyses of this enzyme for molecular mass was carried out by SDS-PAGE electrophoresis, which revealed one band 60.5 Kda.

Key words: *Penicillium camemberti PL21*, α-amylase, purification, SDS-PAGE electrophoresis.

INTRODUCTION

Alpha amylase (E.C 3.2.1.1) catalyses the hydrolysis of α -D-(1, 4) glycosidic linkage in starch components and related carbohydrates it is a key enzyme in the production of starch derivatives and also used in desiring fabrics, in the baking industry, pharmaceuticals and detergents. Submerged fermentation has been used for the production of industrially important enzymes because of the ease of handling and greater control of environmental factors such as temperature and pH. The utilization of agro-industrial residues, including orange waste powder as the substrate for the fermentation (Diomi et al., 2008; Oleivera et al., 2006) has growing interests as they are inexpensive energy rich resources and also eliminates large-scale accumulation of the biomass (Pandey et al., 2000; Henning et al., 2006; Ramachandran et al., 2007; Djekrif-Dakhmouche et al., 2006). To meet the growing demands in the industry it is necessary to improve the performance of the system and thus increase the yield without increasing the cost of production. α-amylase purification has largely been limited to a few species of fungi (AbouZeid et al., 1997; Khoo et al., 1994). Nevertheless, (Amirul et al., 1996) produced alpha-glucosidase, α-amylase and two forms of glucoamylase from *Aspergillus Niger* grown on a liquid medium containing raw tapioca starch as the carbon source.

On the other hand, as bacterial α-amylase has generally been produced from the strains belonging to genus *Bacillus* (Asgher et al., 2007,Hema et al., 2006) several attempts have been made at their purification and characterization, from both *mesophilic* and *thermophilic* strains (Pandey et al., 2000 and Kathiresan et al.,2006). An extra-cellular α-amylase produced by: *Penicillium chrysogenum* in solid state fermentation (Bel en orgaz et al., 2006, Bilal and Figen, 2007, Neuro et al., 1993). *P. olsonii* (Afifi et al., 2008) and *P. camemberti* (Isabelle et al., 2006; Nathalie et al., 1993) was partially purified using acetone and ammonium sulfate with of 60% of saturation, and 43.7% recovery of enzyme with 6.2-fold purification was recorded (Malhotra et al., 2002, Amirul et al., 1996) purified α-amylase, alpha-glucosidase, and

*Corresponding author: E-mail: nouadri2000@yahoo.fr.

glucoamylase to homogeneity by ammonium sulfate precipitation, ion-exchange and two cycles of gel filtration chromatography. In addition to that, an extra cellular α-amylase produced by *Lactobacillus fermentum* was purified by glycogen precipitation and ion exchange chromatography. The purification was approximately 28-fold with 27% yield (Talamond et al., 2002). The aim of this present work was to investigate the purification and partial characterization of the α-amylase enzyme which produced by *Penicillium camemberti PL 21*.

MATERIALS AND METHODS

Microorganism

Penicillium camemberti strain PL21 was provided by I.N.A (The technological laboratory of Agriculture National Institute Paris-France). Stock culture of *P. camemberti PL21* was stored on potato dextrose agar (PDA) at 4℃.

α-Amylase production

α-amylase from *P. camemberti PL 21* was produced in a 2-l fermenter (New Brunswick Scientific-USA) at 22℃, agitation 200 rpm, pH 5 and an aeration of 0.50 l per min .The culture medium (1.5 L) was prepared to include powder orange waste peel (20% w/v), and enriched by: yeast extracts 5 g/l, the corn-steep-liquor 20%, calcium chloride ($CaCl_2$) 0.125 mg/l, starch 1% and glycerol 10%, KH_2PO_4, 0.2 mg/l $MgCl_2{\cdot}6H_2O$. After 168 h of incubation, the cultures were kept in initial pH and temperature was controlled, (Fogarty et al., 1990). The extract was clarified by centrifugation at 10,000 rpm at 4℃ for 15 min. The clear supernatant was used for enzyme activity measurements (Diomi et al., 2008).

α-Amylase assay

Amylase activity was determined by measuring the release of reducing sugar from soluble starch. The reaction mixture contained 0.5 ml of enzyme and 4.5 ml of 0.1 M phosphate buffer (pH 6.0) added to 2 ml of soluble starch (1%). After incubation at 30℃ for 30 min in a shaking water bath, the reaction was stopped by the addition 2 ml of 3-5-Dinitrosalicylic acid method (Bernfeeld, 1955). The tubes were kept in boiling water for 5 min to develop the color and cooled. The absorbance was read at 540 nm after making up the volume to 10 ml. Maltose was used to construct a standard curve. One unit of enzyme activity is defined as the amount of enzyme releasing 1 μmol maltose per min under the assay conditions Specific activity is expressed as amylase activity per mg of protein.

Protein assay

Protein was measured by the method of (Lowry et al., 1951) with bovine serum albumin (BSA) as standard. The concentration of protein during purification studies was calculated from the standard curve. The eluted fractions from the chromatographic separations were monitored at 280 nm.

Confirmation of amylase production

Amylase production by *P. camemberti PL 21* was confirmed on starch agar plates containing peptone 5 g, yeast extract 1.5 g, soluble starch 2.0 g, agar 15 g, a media composed with Na_2HPO_4 4 g, KH_2PO_4 2 g, NaCl 0.5 g, $MgSO_4$ 0.24 g, $CaCl_2$ 0.01 g, and (Burhan et al., 2003). The initial pH was adjusted with NaOH to pH 6.0 after autoclaving. Culture was maintained for 72 h at 30℃ and pH 6.0. Amylase production was detected after flooding the plates with iodine solution (Hols et al., 1994), (Burhan et al., 2008, Saxena et al., 2007).

Enzyme purification

The α-amylase from *P. camemberti PL 21 was* purified by a combination of $(NH_4)_2SO_4$ precipitation, dialysis, gel filtration sephadex G-100 and DEAE Sepharose-CL-6B, and the clear supernatant was subjected to the following purification steps:

(1) After 168 h cultivation of *P. camemberti PL 21* in a 2-l fermenter containing 1.5 l of the amylase-producing medium, the fermentation broth was centrifuged at 10,000 rpm for 30 min to remove cells. The supernatant was then brought to 30-50% saturation with ammonium sulfate. The precipitate was removed by centrifugation at 12,000 rpm for 30 min, and then ammonium sulfate was added to the supernatant to 50 - 60% saturation.

(2) The precipitate protein was dissolved in 0.1M phosphate buffer, pH 6.0 and dialyzed against the same buffer. Precipitation, dialysis and centrifugation were carried out at 0 - 4℃, to remove the remaining salt. Specific activity of enzyme was estimated in the dialyzed fraction.

(3) The desalted enzyme solution was collected dissolved in phosphate buffer 0.1M; pH 6.0 and fractionated through Sephadex G-100 column (1.6 x 36 cm) (Frac-100.Pharmacia –Fine Chemicals) previously equilibrated with the same buffer. Elution was carried out with the respective buffer at 1ml/min. The protein content of each fraction was determined at 280 nm. The eluted enzymatically active fractions were pooled and used as the purified enzyme.

The purified enzyme was applied to a DEAE Sepharose CL-6B column chromatography (2.6 × 20 cm) which had been previously equilibrated with 0.1 M phosphate buffer (pH 6.0) containing 0.1 mM $CaCl_2$ and was then eluted at a flow rate of 1 ml/min.

SDS-polyacrylamide gel electrophoresis (SDS-PAGE) and molecular weight determination

The molecular mass of the purified enzyme was determined by sodium dodecyl sulfate polyacrylamide gel electrophoresis SDS-PAGE (10 % polyacrylamide) (Laemmli, 1970) using BIO-RAD MINI PROTEAN apparatus. Phosphorylase b (97.0 Kda,), BSA (66.0 Kda, Carbonic anhydrase (30.0 Kda), Soybean trypsin inhibitor (20.1 Kda), α-Lactabumin (14.4 Kda) were used as molecular mass standards (molecular weight marker kit S7 were from Sigma Chemical Co., USA.).The electrophoresis was carried at 80 volt for 2 h. The gel was stained with 0.25% Coomassie Brilliant Blue R-250 and distained by washing overnight with a mixture of acetic acid-methyl alcohol-water (5:5:1 v/v).

Amylolytic activity in polyacrylamide gel electrophoresis (PAGE)

Non-denaturing polyacrylamide gel electrophoresis (PAGE) was performed in 10% gels according to Davis (1964). After running, the gel was incubated with 0.1M phosphate buffer (pH 6.0) for 30 min and immediately immersed in 1% starch solution. The gel was

Table 1. Summary of purification procedures for α-amylase produced from *P. camemberti PL21.*

Purification step	Enzyme volume (ml)	Total activity (unit/ml)	Total protein (mg/ml)	Specific activity (unit/ml/mg protein)	Purificatio n (x- fold)	Recovery (%)
Culture supernatant	500	3998.5	997.80	4.0	1.0	100
Ammonium $(NH_4)_2SO_4$ Fractionation	10	2024.7	36.50	55.4	13.8	50.6
Dialysis	10	1980.5	30.30	65.3	16.3	49.5
Sephadex G-100 chromatography	5	1386.2	13.30	104.2	26.0	34.6
DEAE-Sepharose CL-6B column chromatography	5	925.20	6.30	154.2	38.5	23.1

incubated for 20 min and amylolytic activity determined with a mixture 0.3% of KI and 0.1% I_2, and fixed with (3 %) acetic acid solution until the appearance of the bands with amylolytic activity (Silva et al.,2009). The gel was stained with 0.15% Coomassie Brilliant Blue R-250 and distained by washing overnight with a mixture of acetic acid-methyl alcohol-water (5:5:1 v/v). (Damodara et al., 2005)

pH optimum

The relative amylase activity was measured under the standard assay conditions, using (1% w/v) soluble starch; enzyme was incubated at room temperature 30℃ for 30 minutes at various pH (3-10) in 0.1M acetate and phosphate buffer (McTigue et al., 1995).Relative activities at the test pH were calculated assuming the maximum activity observed as 100% (during the experiment).

Temperature optimum

Enzyme samples were incubated for 30 minutes at various temperatures between 10 and 90℃ in 0.1M phosphate buffer pH 6.0 (Lo et al., 2001).The temperature at which maximum activity was observed was taken as 100% and relative activities at different temperatures calculated.

Thermal stability of enzyme

Thermostability of the amylase was performed by enzyme samples at optimum pH for 30 min pre-incubating at temperatures 30,40,50,60 and 70℃. The remaining activity was determined by incubating enzymes at optimum temperature for 30 min (Egas et al., 1998).

Kinetic determinations

Initial rates of starch hydrolysis were determined at various substrate concentrations (0.25 to 2 mg/ml). The kinetic constants Km and Vmax were estimated by the method of Lineweaver-Burk.

Effect of metal ions on specific activity

The purified enzyme was incubated for 10 min with different concentrations of each of the following substances (Table 2), all the metal ions added as chloride salts, then the relative activity was assayed.

RESULTS AND DISCUSSION

Purification of α-amylase

The results of the purification of α-amylase produced by the strain *Penicillium camemberti PL 21* is summarized in (Table 1). The first step in purification was conventional $(NH_4)_2SO_4$ fractionation. The fraction (50 - 60%) saturation of ammonium sulfate revealed maximum enzyme specific activity (55.4 unit/ml/mg protein). The α-amylase subjected to dialysis against resulted in specific activity of (65.3 units/ml/mg protein) , then it was further purified by Sephadex G-100 in order to desalt the solution (Xu and yan, 2007), remove the low molecular weight proteins This has induced the specific activity up to (104.2 units/ml/mg protein).

The purified enzyme obtained with the DEAE-Sepharose CL-6B column chromatography (Figure 1) exhibited 23.1% of the total initial activity and there was a 38.5 fold increase in specific activity (154.2 units/ml/mg protein) when compared with the crude culture filtrate.

Properties of the purified α-amylase

The α-amylase isolated from *P. camemberti PL 21* was approximately (38 folds) pure with a yield of 23%. As shown in Figure 2, the purified enzyme has migrated as a single protein band in both native PAGE and SDS-PAGE (10% gel), which indicates its homogeneity. The molecular mass of the α-amylase isolated from *P. camemberti PL 21* was estimated to be 60.5 kDa from its mobility relative to those of standard proteins on SDS-PAGE, indicating that the purified enzyme is a monomer. The enzyme containing amylolytic activity was applied in non denaturing polyacrylamide gel electrophoresis (PAGE) (Figure 2C); a clear-zone forming band indicated the presence of the active α-amylase in the native PAGE (Figure 2D). The purified enzyme could be stored at 0.5 M phosphate buffer, pH 6, at 4℃, for more than 3 months without any apparent loss of activity. However, repeated freezing and thawing led to considerable loss of its activity (Iiori et al., 1997; Lin et al., 1998; Burhan, 2008;

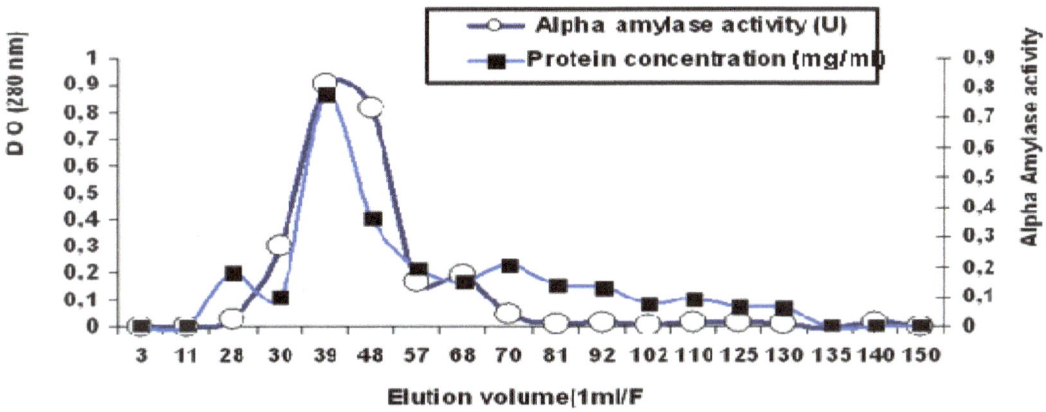

Figure 1. Purification profile of α-amylase produced by *P. camemberti* PL21.

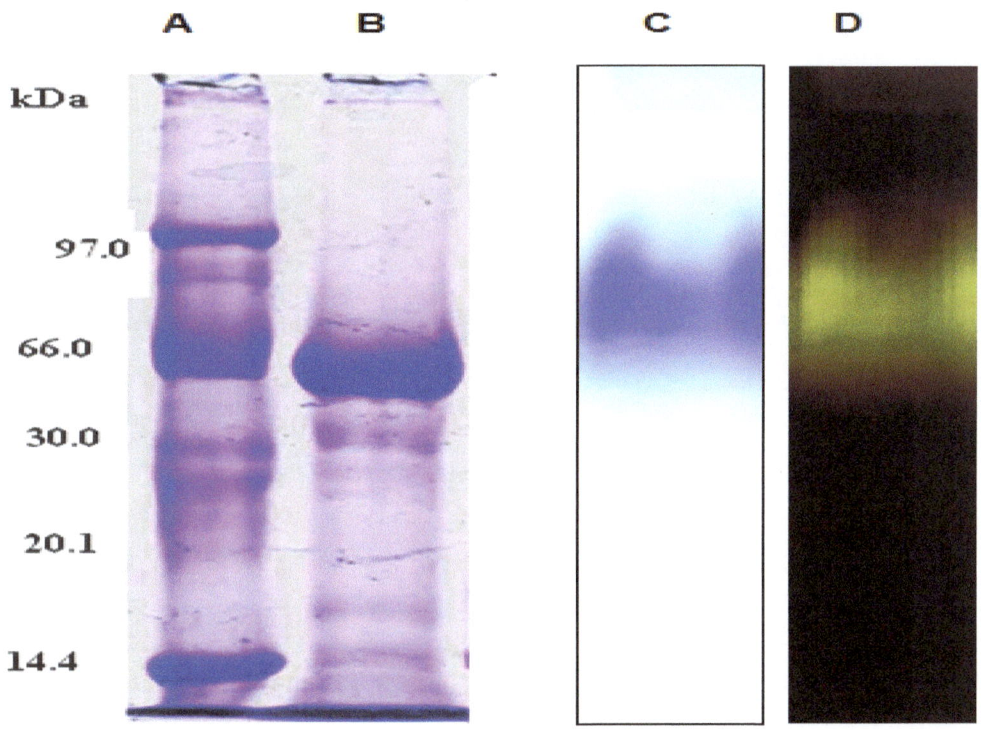

Figure 2. Polyacrylamide gel electrophoresis of the purified α-amylase from *P. Camemberti PL21.* SDS-PAGE, Lane (A); Molecular mass standard proteins: Phosphorylase B (97.0 kDa), SA (66.0 KDa), Carbonic anhydrase (30.0 KDa), soybean trypsin inhibitor (20.1 KDa) ,α-ctabumin (14,4 KD), Lane (B); the purified α-amylase from DEAE-Sepharose CL-6B column chromatography. Native-PAGE Lane (C); amylolytic activity on PAGE contained 1% starch. The protein of the purified α-amylase stained by coomassie Brilliant blue R-250, Lane (D); The purified α-amylase stained by KI/I_2 solution. Electrophoresis conditions: 80 V/2h.

Biljana et al., 2007; Yang et al., 2004).

Influence of pH on activity

The influence of pH on enzymatic activity is represented in (Figure 3). The α-amylase showed optimal activity (80% of the maximal activity 153 U) from pH 5.0 to 6.0 and this result was similar to the values reported for the most bacterial (Afifi et al., 2008, Mazen et al., 2002) and fungal (Ray et al., 2001) amylases.

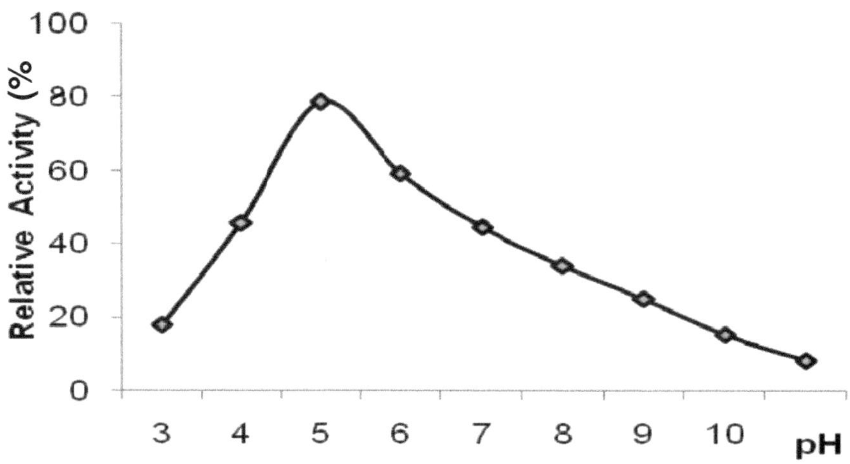

Figure 3. Effect of pH values on the specific activity of the purified alpha amylase from *P. camemberti PL 21.*

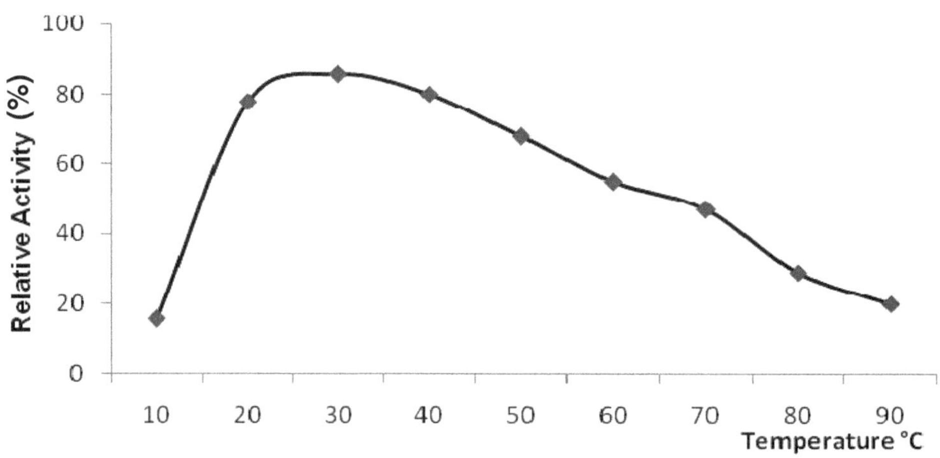

Figure 4. Effect of different temperatures on the specific activity of purified alpha amylase *P. camemberti PL 21.*

Effect of temperature and thermal stability on activity

The enzyme activity was determined at different temperatures (10 - 90°C) ,in order to estimate the optimum temperature. The enzyme has reflected a broad temperature range between (20 - 60°C) and the optimum temperature was observed around 30°C (Figure 4). The results accumulated on the thermal stability of this enzyme are represented in Figure 5. Apparently the activity of α-amylase was affected largely by exposing to temperature above 60°C. The enzyme lost about 50% of its activity at temperature ranging between 50 - 60°C for two hours (Afifi et al., 2008).

Kinetic determinations

With respect of the effect of different starch's concentrations on the enzyme activity, the results in Figure 6 showed that the activity of α-amylase increased with the increase of starch up to 1.00 mg/ml followed by slight decrease in the activity. From the Lineweaver- Burk plot of the reciprocal of initial velocities and substrate concentrations (Figure 7), The K_m and V_{max} values were 0.92 mg/ml and 38.5 µmole/min at 30°C and pH 6.0 with 0.1 M phosphate buffer. This is equal to Km value of α-amylase produced from *Bacillus sphaericus* (Al-Qodah et al., 2007).

Influence of different metal ions on activity

The study has showed a different influence of various metal ions on α-amylase activity (Table 2) summarized the results. Induction of the enzyme activity by Ca^{2+} (2mM) which seemed to be in accordance with what was reported in the literature concerning α- amylase from *A.*

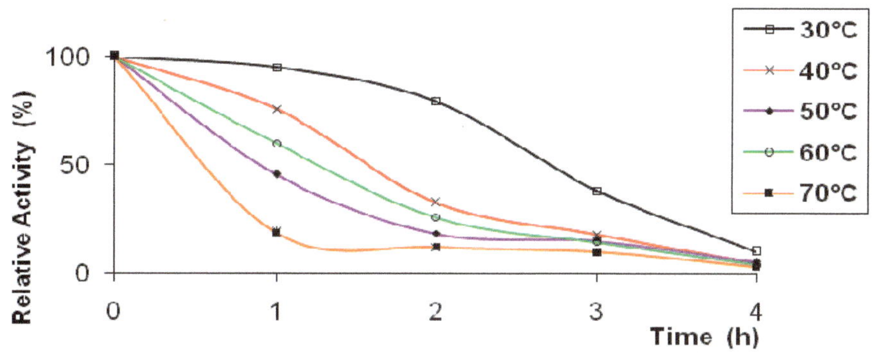

Figure 5. Thermal stability of the purified alpha amylase from *P. camemberti PL 21.*

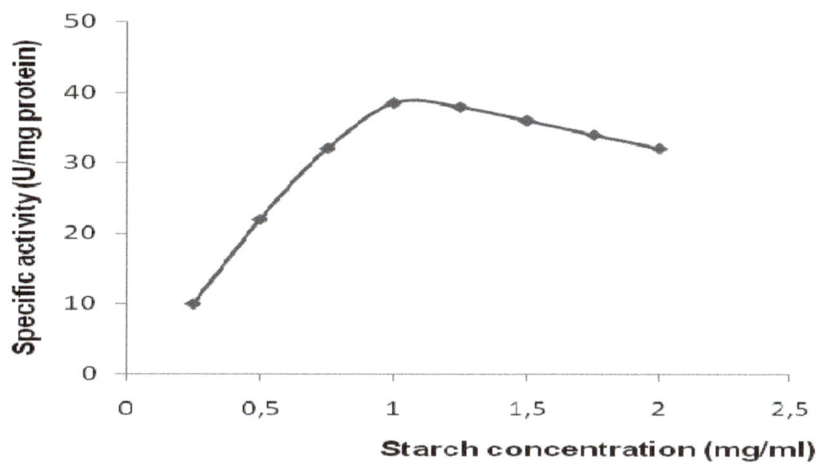

Figure 6. Effect of starch concentrations on the specific activities of purified α-amylase from *P. camemberti PL 21.*

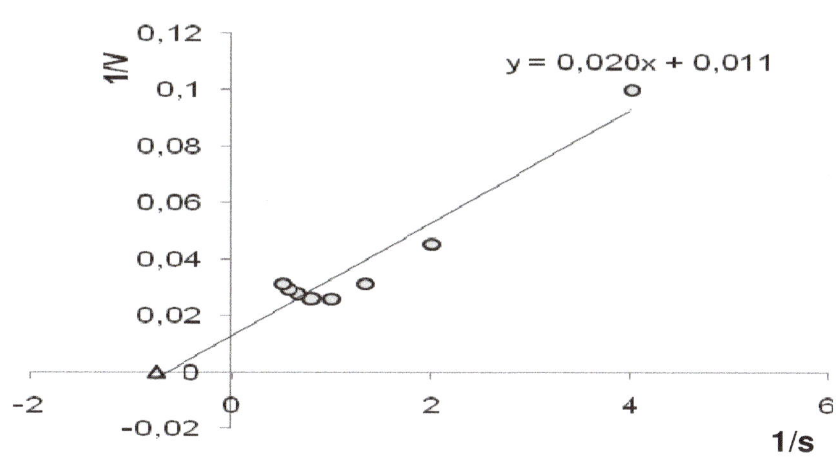

Figure 7. Lineweaver-Burk plot of the reciprocal of initial velocities and starch concentration.

oryzae EI212 (Bose and Das, 1996). However this induction was retraced when (1 mM) of Ca^{2+} was used. On the other hand, Mg^{2+} has decreased α- amylase activity by 10 % through Mg^{2+} ions were reported to inhibit slightly α- amylase activity (Chao-Hsun et al., 2004). Inhibition of α- amylase activity by Hg^{2+}, Ag^{+}

Table 2. Effect of some metal ions on the specific activity of α-amylase produced from *P. camemberti PL21*.

Metal ions	Concentration (mM)	Relative activity (%)
None	-	100
Ca^{2+}	2.0	108
Mg^{+}	1.0	90
Hg^{2+}	0.5	15
Ag^{+}	0.5	20
Cu^{2+}	1.0	75

(0.5 mM), Cu^{2+} (1 mM), respectively, which indicated by the involvement of the sulfhydryl groups in the catalytic activity of α- amylase.

Conclusion

The *P. camemberti PL21* extra-cellular α-amylase was purified by selective fractionation with ammonium sulfate, followed by two chromatography steps, which included gel filtration (Sephadex G-100 and Sepharose-CL-6B). The purification factor was (38.5-folds) and a recovery was (23.1%). α-amylase maximum activity was (154.2 U/ml.mg protein). The molecular mass was estimated to be around 60.5 Kda. The enzyme was active with pH and temperature optima at 6.0 and 30℃ similar with the values reported for α-amylase from *Penicillium expansum* and *Aspergillus nedulans*. The α-amylase was inhibited by (0.5 Mm) $Hg2^{+}$ and Ag^{+} up to (85 and 80%) respectively. *P.camemberti PL21* is different from other fungal α-amylase several respects. In Cloning and characterization of the α-amylase gene from *P. camemberti PL21* will help to elucidate the structure and the regulation of this enzyme.

REFERENCES

AbouZeid AM (1997). Production, purification and characterization of an extracellular α-amylase enzyme isolated from *Aspergillus flavus*. Microbios. 89(358): 55-66.

Afifi AF, Kamel EM, Khalil AA, Fouaad E, Fazxi M, Housery M (2008). Purification and characterization of α-amylase from *Penicillium olsonii* under the effect some Antioxidant Vitamins, Global J. Biotechnol. Biochem. 3(1): 14-21.

Al-Qodah Z, Daghstani H, Geopel P, Lafi W (2007). Determination of kinetic parameters of α-amylase producing thermophile *Bacillus sphaericus, Center* of Environmental Studies, The Hashemite University Zarqa, 13115, P.O. Box 330127 Jordan. Afri. J. Biotechnol. 6: 699-706.

Amirul AA, Khoo SL, Nazalan MN, Razip MS, Azizan MN (1996). Purification and properties of two forms of glucoamylase from *Aspergillus niger*. Folia Microbial (Praha).41(2): 165-174.

Asgher M, Javaid M, Asad S, Rahman U, Legg RL (2007). A thermostable a-amylase from a moderately *thermophilic Bacillus subtilis* strain for starch processing; J. Food Engineering. 79: 950-955.

Bel´en O, Juliana K, Ana M, Pedregosa I, Monistrol F, Fernando L, Carmen S (2006). Bacterial biofilm removal using fungal enzymes, Enzym. Microbial Technol. 40: 51-56.

Bernfeeld P (1955). Amylases A and B methods Enzymology, 1: 149-158.

Bilal B, Figen E (2007). Production and Properties of alpha amylase from *Penicillium chrysogenum* and its Application in Starch Hydrolysis, Preparative Biochem. Biotechnol. 35: 169-178.

Biljana D, Nataša B, Vera N, Jelisaveta I, Zoran V (2007). Purification and properties of midgut α-amylase isolated from *Morimus funereus (Coleoptera: Cerambycidae) larvae*, University of Belgrade, Studentski trg. 3: 12-16.

Bose K, Das D (1996) .Thermostable alpha-amylase production using *B. licheniformis NRRLB 14368*. Indian J. Exp. Biol. 34(12): 1279-1282.

Burhan A (2008). Highly thermostable, thermophilic, alkaline, SDS and chelator resistant amylase from a *thermophilic Bacillus sp.* isolate A3-15, Bioresour. Technol. 99: 3071-3076.

Burhan A, Unaldi N, Coral G, Colak O, Aygan A, Gu¨lnaz O (2003). Enzymatic properties of a novel thermostable, thermophilic, alkaline and chelator resistant amylase from an *alkaliphilic Bacillus sp. Isolat ANT-6*. Process Biochem. 38: 1397-1403.

Chao-Hsun Y, Wen-Hsiung L (2004). Purification and properties of a maltotriose-producing Alpha amylase from *Thermobifida fusca* ,Institute of Microbiology and Biochemistry, National Taiwan University; Enzym. Microbial Technol. 35: 254-59.

Damodara RM, Ratnam BV, Purnima A, Ayyanna C (2005). Affinity chromatography of α-amylase from *Bacillus licheniformis*. Department of Pediatrics, Division of Infectious Diseases, Johns Hopkins University School of Medicine, 720.

Davis BJ (1964). Disc electrophoresis. II. Method and application to human serum proteins", Annals of the New York Academy Sciences, 121: 404-427.

Diomi M, Elisavet K, Paul C (2008). Fungal multienzyme production on industrial by-products of the citrus-processing industry, Biotechnology Laboratory, School of Chemical Engineering, National Technical University of Athens. Technology. 99: 2373–2383.

Djekrif-Dakhmouche S, Gheribi-Aoulmi Z, Meraihi Z, Bennamoun L (2006). Application of a statistical design to the optimization of culture medium for a-amylase production by *Aspergillus niger ATCC 16404* grown on orange waste powder. J. Food Eng. 73: 190-197.

Egas MCV, da Costa MS, Cowan DA, Pires EMV (1998). Extracellular a-amylase from *Thermus filiformis Ork A2*: purification and biochemical characterization. Extremophiles .2: 23-32.

Fogarty WM, Kelly CT (1990). Microbial Enzymes and Biotechnology .Applied Science, London, New York.71- 132.

Hema A, Ujjval T, Kamlesh P (2006). Alpha Amylase Production by *Bacillus cereus* MTCC 1305 Using Solid-State Fermentation. Department of Biosciences, Patel University, Vidyanagar. 388 -120.

Henning J, Lisbeth O (2006). Production of cellulases by *Penicillium brasilianum* IBT 20888. Enzyme and Microbial Technology. (38): 381-390.

Hols P, Ferain T, Garmyn D (1994). Use of expression secretion signals and vector free stable chromosomal integration in engineering of *Lactobacillus plantarum* for α-amylase and levanase expression. App. Ennviron. Microbial. 60 : 1401-1403.

Iiori MO, Amund O, Omidiji O (1997). Purification and properties of an alpha-amylase produced by a cassava-fermenting strain of *Micrococcus lueus*. Folia Microbiology (Praha). 42(5): 445-449.

Isabelle K, Pascal M, Gilles F, Catherine V, Alain D, Jean-François C, Pascale C (2006). Impact of solid medium composition on the conidiation in *Penicillium camemberti, INRA*. UMR .1082.Journal de Microbiologie, INRA. Dijon . France. 2: 45-53.

Kathiresan K, Manivannana S (2006). α-Amylase production by *Penicillium fellutanum* isolated from mangrove rhizosphere soil, Afri. J. Biotechnol. 5(10): 829-832.

Khoo SL, Nazalan MN, Razip MS, Azizan MN (1994). Purification and properties of two forms of glucoamylase from *Aspergillus niger.*Folia Microbiol. (Praha). 41(2): 65-174.

Laemmli UK (1970). Cleavage of structural protein during the assembly of the head of bacteriophage T4. Nature. 227: 680-685.

Lin LL, Chyau C, Hsu WH (1998). Production and properties of a taw-starch-degrading amylase from the *thermophilic* and *alkaliphilic Bacillus sp. TS-23.* Nature.6: 68-74.

Lo HF, Lin LL, Chen HL, Hsu HH, Chang CT (2001). Enzymatic properties of a SDS-resistant Bacillus sp. TS-23 a-amylase produced by recombinant *Escherichia coli.* Process Biochem. 36: 743-750.

Lowry OH, Rosebrough NJ, Farr A, Land Randall RJ (1951). Protein measurement with the folin phenol reagent. J. Biol. Chem. 193: 265-275.

Malhotra R, Noorwez SM, Satyanarayana T (2002). Production and partial characterization of thermostable and calcium-independent alpha-amylase of an extreme thermophile *Bacillus thermooleovorans* NP54. Lett. Appl. Microbial. 31(5): 378-384.

Mazen A, Abdeltif A, Yves P (2002). Carbon and nitrogen substrates consumption, ammonia release and proton transfer in relation with growth of *Geotrichum candidum* and *Penicillium camemberti* on a solid medium, J. Biotechnol. .95: 99-108.

McTigue MA, Kelly CT, Doyle EM, Fogarty WM (1995). The alkaline amylase of the *alkalophilic Bacillus sp. IMD 370.* Enzym. Microbial Technol. 17: 570-573.

Nathalie J, Jean-Marc B (1993). Comparaison of Volatile Flavor Compounds Produced by Ten Strains of *Penicillium camemberti anda* Horn, Université de Bourgogne Ecole Nationale Supérieure de Biologie Appliquée et Alimentation, Laboratoire de Biotechnologie. 21000 Dijon, France. 148-157.

Nuero OM, Alfonso C, Del A, Fand Reyes F (1993). Study of beta 1-3-glucanase activity during autolysis of *Aspergillus nidulans* by HPLC ion exchange chromatography. Lett. Appl. Microbiol.17: 04-108.

Oliveira Luciana A, Porto Ana LF, Tambourgi Elias B (2006). Production of xylanase and protease by *Penicillium janthinellum* CRC 87M-115 from different agricultural wastes, Bioresour. Technol. 97(6): 862-867.

Pandey A, Nigam P, Soccol CR, Soccol VT, Singh D, Mohan R (2000). Advances in microbial amylases. Biotechnol. Appl. Biochem. 31: 135-152.

Ramachandran S, Singh S, Larroche C, Soccol CR, Pandey A (2007). Oil cakes and biotechnological applications. A review. Bioresour. Technol. 98(10): 200-209.

Ray Rina R (2001). Production of alpha-amylase and Xylanase by an Alkalophilic *Penicillium griseoroseum* RR-99, Acta, Microbiol. Polonica. 50(3): 305-9.

Saxena KR, Dutt K, Agarwal L, Nayyar P (2007). A highly and thermostable alkaline amylase from a *Bacillus sp.* PN5. Bioresour. Technol. 98: 260–265.

Silva TM (2009). Use of Cassava Peel as Carbon Source for Production of Amylolytic Enzymes by *Aspergillus niveus* Int. J. Food Eng. Pp. 5-5.

Talamond P, Desseaux V, Moreau Y, Santimone M, Marseille-Mouren G (2002). Isolation, characterization and inhibition by acarbose of the alpha-amylase from *Lactobacillus* fermentation: comparison with *Lb.Manihotivorans* and *Lb.Plantarum* amylases. Comp Biochem. Physiol. Biochem. Mol Biol. 133(3): 351-60.

Xu Dong L, Yan X (2007). A novel raw starch digesting a-amylase from a newly isolated *Bacillus sp. YX-1*; Purification and characterization Key; Laboratory of Industrial Biotechnology of Ministry of Education and School of Biotechnology, Southern Yangtze University, Wuxi 214036, PR China. Appl. Environ. Microbiol. 10(2): 7229-7235.

Yang J, Kang S, Park S, Yoon H, Kim M, Heu S, Ryu S (2004). Characterization of a novel amylolytic enzyme encoded by a gene from a soil-derived metagenomic library. Appl. Environ. Microbiol. 70(12): 7229-7235.

4

Effect of electromagnetic field on acetylcholinesterase activity: *In vitro* study

Ezzatollah Fathi[1] and Raheleh Farahzadi[2]*

[1]Department of Clinical Sciences, Faculty of Veterinary Medicine, University of Tabriz, Tabriz-Iran.
[2]Department of Clinical Biochemistry, School of Medical Sciences, Tarbiat Modarres University, Tehran-Iran.

The *in vitro* effects of extremely low frequency (ELF)-electromagnetic field (EMF) (frequencies of 50, 100, 217 Hz, induction 300 µT) on the structure and activity of acetylcholinesterase (AChE) was investigated. AChE activity was measured with UV-vis spectroscopy. Circular dichroism spectroscopy and fluorescence spectroscopy were used for investigating the secondary and tertiary structure of enzyme. The results demonstrated that the ELF-EMF decreased enzyme activity and the main inactivation took place in 217 Hz ($p<0.05$). It can be concluded that increasing the frequencies of ELF-EMF caused an increase in the helicity of AChE; therefore, under ELF-EMF, the α-helix structure can induce the secondary structure of AChE. Also, the intrinsic fluorescence decreased by increasing the frequency of field. In conclusion, according to changes observed in the secondary and tertiary structure of enzyme, it is proposed that these fields are able to affect the structure and dynamics of the active site gorge of AChE.

Key words: Acetylcholinestrase, circular dichroism spectroscopy, extremely low frequency electromagnetic field, fluorescence spectroscopy, UV-visible spectroscopy.

INTRODUCTION

The effects of extremely low frequency electromagnetic fields (ELF-EMFs) on biological systems like different enzymes, nucleic acids, cell proliferation etc. have been investigated (Volpe, 2003; Zwirska-Korczala, 2005). Small effects of ELF-EMFs on the activities of different enzymes such as antioxidative enzymes and acid phosphatase have been already reported (Zwirska-Korczala, 2005; Prashanth et al., 2009). As ELF-EMFs were found to affect the activity of membrane enzymes such as Na, K-ATPase or cytochrome oxidase and adenylate kinase (Ravera et al., 2004; Morelli et al., 2005), other studies have demonstrated that ELF-EMF with the frequency of 75 Hz and amplitudes above a 250 µT threshold decreased the activity of three membrane-bound blood cell- or synaptosome-enzymes: Alkaline phosphatase, phosphoglycerate kinase and acetylcholinesterase (AChE) up to 54 to 61% (Morelli et al., 2005). The effects of these fields on human body depend on amperage, frequency and exposure duration (Seyyedi et al., 2007). Therefore, this hypothesis could be checked that ELF-EMF affects the activity of AChE, which occurs with high specific activity in the brain and nervous tissues. AChE is one of the most important enzymes in nervous system and a little amount in the liver (Parveen et al., 2004), placenta, as well as the surface of erythrocytes membrane (Dvir et al., 2004).

AChE is found in multiple forms. The predominant forms of this enzyme are Globular (monomers, G1; dimmers, G2; and tetramers, G4) and asymmetric forms (one, A4; two, A8; or three, A12, tetramers linked to a collagen-like tail) (Massoulie and Bon; 1982). Mammalian brain AChE consists almost exclusively of globular forms, principally the tetramer (Grassi et al., 1982; Massoulie, 2002). This enzyme is a glycoprotein, a member of serine hydrolase enzymes and has four conserved amino acids

*Corresponding author. E-mail: farahzadi@ibb.ut.ac.ir.

including Serine, Histidin, Aspartate and Glutamate in its own active site and causes the termination of signal transduction in the somatic nervous systems with the degradation of neurotransmitter, acetylcholine (ACh) (Axelsen et al., 1994). It is, therefore, remarkable that the active site of this enzyme is found at the bottom of a deep and narrow gorge (Axelsen et al., 1994). One study demonstrated that this gorge provides a mechanism that somehow facilitates the rapid action of AChE (Barteri et al., 2005).

The aim of this study was the most basic in this contex: to determine the influence of ELF-EMF on the activity and structure of acetylcholinesterase. The reason for this choice was both the availability of highly purified enzyme and tremendous increase in the use of electrical devices. Furthermore, AChE's enzymatic activity can easily be tested "*in vitro*" preparations, and thus, it provides a suitable model system to study the influence of EMF emission on cholinergic activity.

MATERIALS AND METHODS

In this study, acetylcholinesterase [(EC 3.1.1.7 from Electrophorus electricus (electric eel)] were supplied by sigma (St. Louis, MO, USA) and another chemicals obtained from Sigma and Merck chemical companies (Germany). All experimental procedures were repeated for three times.

Enzyme assay

AChE activity was determined at 25°C by the spectrophotometeric method of Ellman et al. (1961), based on the following colorimetric reactions:

$$\text{Acetylthiocholine iodide} \xrightarrow{\text{AChE}} \text{Thiocholine + Acetate}$$

$$\text{Thiocholine + Dithiobisnitrobenzoic acid (DTNB)} \longrightarrow \text{5-thio-2-nitro-benzoat (yellow)}$$

The reaction mixture containing the 3 μM AChE with the 12.5×10^{-6} M acetylthiocholine iodide (ACh iodide) was kept in the electromagnetic fields 50, 100, 217 Hz, 300 μT for 30 min at 4°C and immediately after the DTNB was added to the reaction mixture the enzymatic activity was measured for 10 min in the absence of the field. Control sample were run in the same experimental condition as used previously but in the absence of field. The principle of the method is the measurement of the rate of production of thiocholine as acetylthiocholine iodide is hydrolyzed. This is accomplished by the continuous reaction of the thiol with DTNB (Ellman reagent) to produce the yellow anion of 5-thio-2-nitro-benzoat. The rate of color production was measured at 412 nm for 10 min with a Cary Spectrophotometer, 100 Bio-model, with jacketed cell holders. All the experiments were run in phosphate buffer (0.1 M) at pH 7.4 in conventional thermostated quartz cell to maintain the temperature at 20±0.1°C. The initial velocity of enzyme was calculated using an absorption velocity of enzyme calculated using an absorption coefficient, $\varepsilon = 1.36\times10^4\ M^{-1}\ cm^{-1}$ and one unit of enzymatic initial velocity is given as μmol/min (Hekmat et al., 2008). The general schemes of enzymic reaction can be summarized as follows:

$$\text{Enz + AcCh} \underset{K_2}{\overset{K_1}{\rightleftharpoons}} \text{(EnzAcCh)} \xrightarrow{K_3} \text{EnzAc + Ch}$$

$$\text{EnzAc} + H_2O \xrightarrow{K_4} \text{Enz + AcOH}$$

Where: Enz is the enzyme, AcCh is the substrate (ACh), and Ch and AcOH are the products (choline and acetic acid, respectively). The rate is calculated from the equations as follows:

$$\text{Rate } (\mu\text{mol /min}) = \frac{\Delta \text{ absorbance/min}}{1.36\times10^4}$$

Circular dichroism (CD) measurements

Circular dichroism spectroscopy is a powerful tool for revealing any conformational change to the protein structure (Sreerama and Woody, 2004; Hekmat et al., 2008). The secondary structure estimation of protein was studied using the Aviv circular dichroism spectrometer, Model 215 (Lakewood, NJ, USA). The spectro-photometer was equipped with a temperature programmer for controlling the rate of temperature changes. The CD spectrum of AChE was recorded from 200 to 250 nm using quartz cell, 0.1 cm path length, with a resolution of 0.2 nm, scan speed of 20 nm min^{-1}, time constant of 2.0 s, 10 nm band width and sensitivity of 20 m^{o}. Protein solutions were prepared in the same buffer as used for the enzymatic assay. Protein solutions of 0.4 mg/ml were used to obtain the spectra with incubation in the presence and absence of ELF-EMF at 4°C.

Intrinsic fluorescence measurements

Fluorescence emission spectra were measured on a Hitachi spectrofluorometer (MPF-4 model), equipped with a thermostatically controlled cuvette compartment. The intrinsic emission of 3 μM protein was seen at the excitation wavelength of 280 nm. The conditions of incubation were the same as those used for CD studies.

Extremely low frequency electromagnetic fields (ELF-EMFs)

ELF-EMF was generated by a parallel set of Helmholtz solenoids with 500 turns of 0.7 mm coated copper wire. Each solenoid diameter was 27 cm. The coils were driven by 50, 100 and 217 Hz power through a variable transformer and generated a magnetic flux density of 300 μT. The samples were placed in the center half way between the plains of coils to receive a uniform field for 30 min. Attention was paid to eliminate all the metal elements around the field.

Statistical analysis

Statistical analysis was performed by SPSS software using one way variance analysis ANOVA (Ver. 16, IBM Corporation, USA). Data were reported as mean ± standard deviation at significance level of $p < 0.05$.

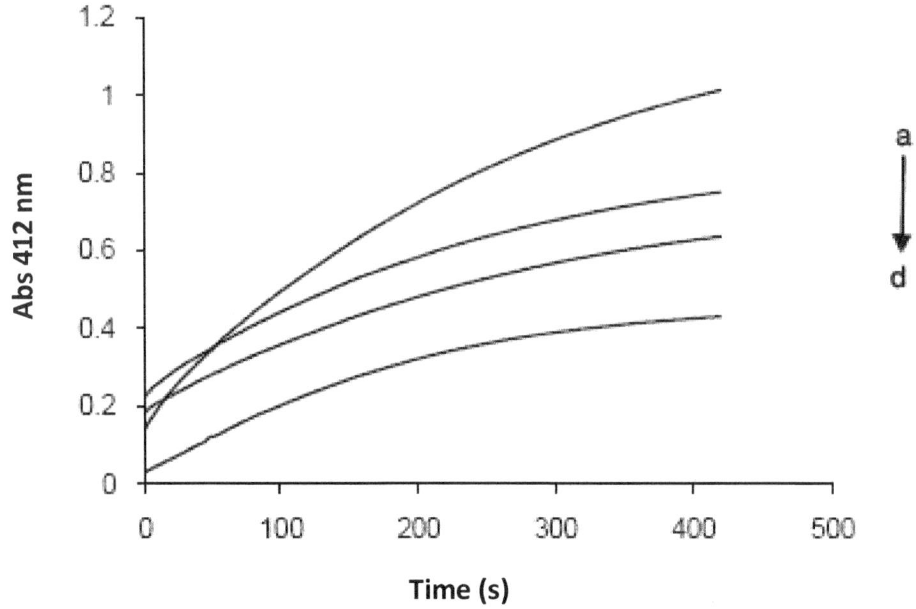

Figure 1. Kinetic plots of acetylcholinesterase after incubation for 30 min at 4°C, and different frequencies with 300 μT field intensity. a: without Electromagnetic field, b: 50 Hz, c: 100 Hz and d: 217 Hz. The mean±SD of three measurements for a, b, c and d were 0.37±0.03, 0.18±0.02, 0.11±0.01 and 0.05±0.01, respectively.

Table 1. Average rates of hydrolysis of ACh iodide.

Samples	Rate (μmol /min)
AChE without ELF-EMF	1.07 ± 0.34
AChE under 50 Hz, 300 μT	0.89 ± 0.06
AChE under 100 Hz, 300 μT	0.54 ± 0.04
AChE under 217 Hz, 300 μT	0.31 ± 0.02*

Each value represents the mean ± SD of three measurements. *Indicates $p<0.05$.

RESULTS

Enzyme activity measurements

Kinetic plot of AChE in the presence and absence of ELF-EMF were recorded, as described in Figure 1. When the enzymatic activity was measured after the exposure to the field, a dramatic decrease in enzymatic activity under the 217Hz was found ($p<0.05$). According to Table 1, with increasing the frequency of ELF-EMF the average rates of hydrolysis of ACh iodide was reduced, indicating decline in AChE activity.

Circular dichroism studies

Far-UV CD spectra (190 to 260 nm) are used for the determination of a protein secondary structure. The peptide bond is the main absorbing group. During the present analysis, in order to consider the effect of different frequencies of ELF-EMF, on the conformational changes of AChE, Far-UV CD technique was used and the spectra can be observed in Figure 2. Table 2 shows the content of the secondary structure of AChE in the presence and absence of ELF-EMF. Figure 2 demonstrates the increase of helicity of AChE under 50, 100 and 217 Hz (300 μT). Within the wavelength region of 205 to 250 nm, the CD spectrum of a protein gives information about its conformation in relation to the secondary structure. The ELF-EMF increase both of these bands, clearly indicating the increase of helical content in the protein in the presence of 50, 100 and 217 Hz (300 μT).

Fluorescence studies

Intrinsic Triptophan (Trp) fluorescence was used to

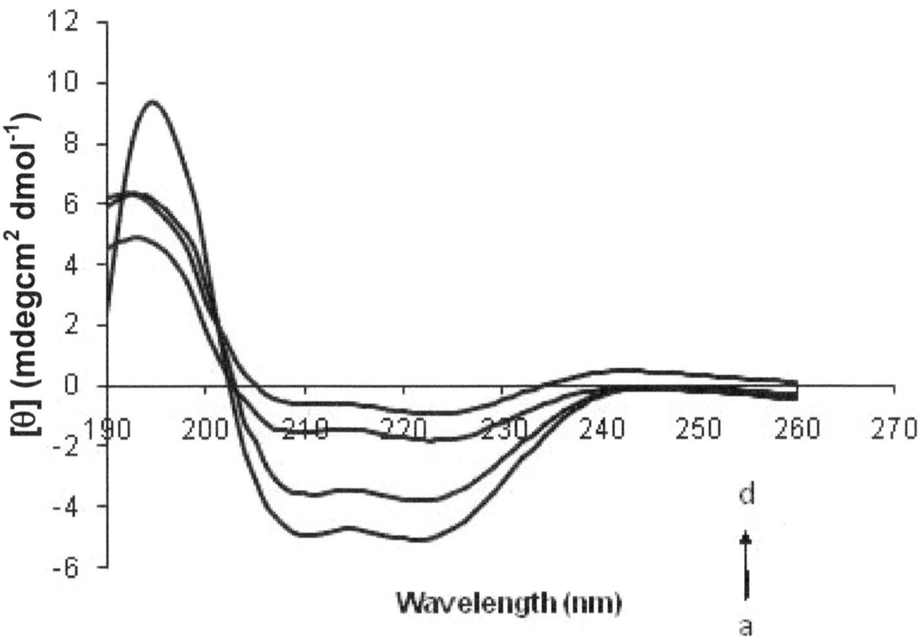

Figure 2. Far-UV-CD spectra of Acetylcholinesterase after incubation for 30 min at 4°C, and different frequencies with 300 μT field intensity.a: 217 Hz, b: 100 Hz, c: 50 Hz and d: without electromagnetic field.

Table 2. Content of the secondary structure of the AChE in the presence and absence of ELF-EMF using the Aviv program.

Samples	α-Helix (%)	β-Sheet (%)	Random coil (%)
AChE without ELF-EMF	49.8	14.3	33
AChE under 50 Hz, 300 μT	55.3	13.2	19.4
AChE under 100 Hz, 300 μT	77.8	11	11.2
AChE under 217 Hz, 300 μT	81.5	9.4	9.1

monitor changes in the protein conformation after incubation for 30 min at 4°C in the presence of 50, 100 and 217 Hz, 300 μT (Figure 3). Graphs a to d illustrate the intrinsic fluorescence emission spectra of AChE in the absence and presence of EMFs 50, 100 and 217 Hz (300 μT), respectively. These graphs demonstrate a decrease in the intrinsic fluorescence by increasing the frequency of field. This indicates that with increasing the frequencies of ELF-EMF, the tryptophan residues are gradually expose to more hydrophilic environment.

DISCUSSION

Exposures to many EMF frequencies are increasing significantly as technology advances unabated and new applications are found. While the enormous benefits of using electricity in everyday life and health care are unquestioned, during the past 20 years the general public has become increasingly concerned about potential adverse health effects of exposure to electric and magnetic fields at ELF (Volpe, 2003). The effects of electromagnetic fields on living microorganisms have been investigated for a long time. The initial effect of electromagnetic field is to acute the key biochemical processes in different metabolic pathway (Ebadi et al., 2006). Various studies showed that electromagnetic fields affect some biochemical processes and result in changing some serum biochemical parameters and enzymes (Ames et al., 2005; Yaghmaei et al., 2009). Different mechanisms proposed for EMF action on biological systems include exchange of electrons, ions and dipoles (Ebadi et al., 2006). The effect of ELF-EMF on triton X-100 solublized acetylcholinesterase has been reported previously.

The full activity of enzyme is rapidly restored when the enzyme is removed from the field (Morelli et al., 2005), however, this study did not confirm such an effecting.

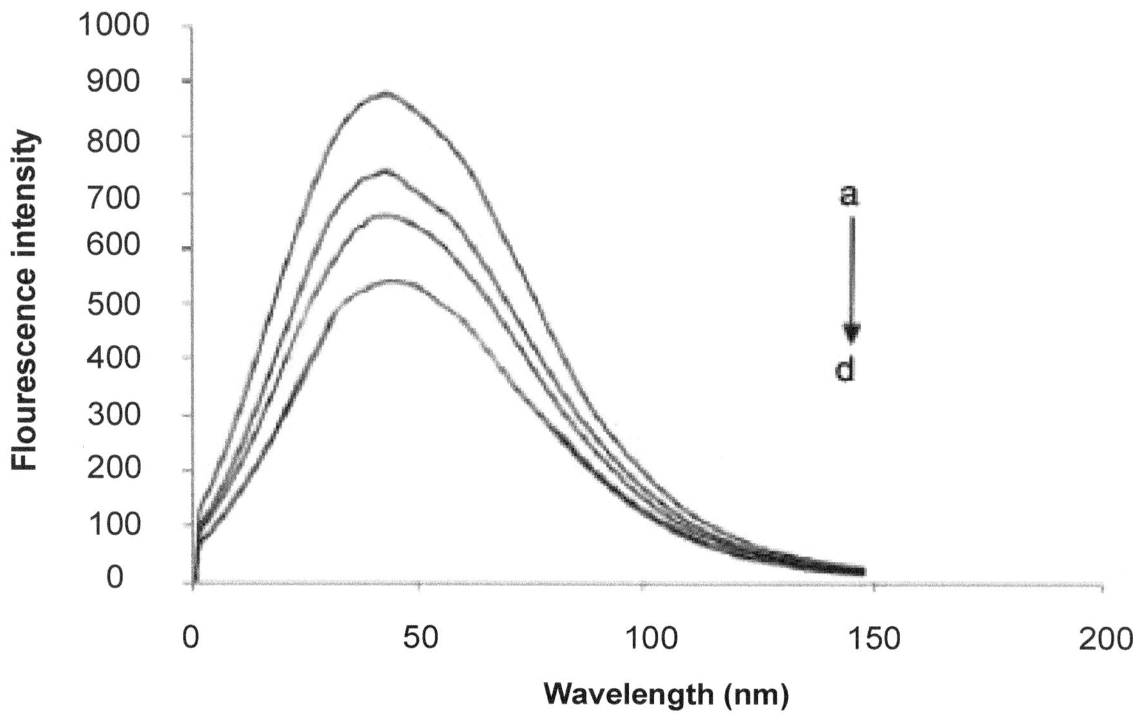

Figure 3. Intrinsic fluorescence spectra of acetylcholinesterase after incubation for 30 min at 4°C, and different frequencies with 300 μT field intensity. a: without electromagnetic field, b: 50 Hz, c: 100 Hz and d: 217 Hz.

This research and previous researchers strongly suggested that the field action is mediated by the structure which is crucial in determining the conditions of the enzyme inactivation (Ravera et al., 2004). The effects might be related to structural changes in the secondary and tertiary structure of the enzyme. Another previous study demonstrates that cellular phone emissions affect the structural and biochemical characteristics of acetylcholinesterase (Barteri et al., 2005). In the present study, UV-vis spectroscopy, CD and fluorescence techniques have been used to monitor the changes of structure of AChE induced by ELF-EMF. Activity studies (Figure 1) showed that different frequencies of EMF have an effect on the activity of AChE. Statistical evaluations elucidate significant differences in the 217 Hz, in which changes in the structure occurs leading to the enzyme inactivation ($p<0.05$; Table 1). CD has proved to be a ideal technique to monitor conformational changes in proteins, which can occur as a result of changes in experimental parameters such as pH, temperature, binding of ligands and so on (Kelly and price, 2000). Far-UV CD studies of AChE at different frequencies of ELF-EMF showed significant effect of EMF on the secondary structure of the protein. The native protein has alpha/beta structure (Axelsen et al., 1994) while under ELF-EMF the content of α-helix structure was increased (Figure 2 and Table 2) then, it can be concluded that with increasing the frequency of ELF-EMF, the α-helix structure can induce in the secondary structure of AChE and consequently prevent the access of the substrate (ACh iodide) to the enzyme active site. Taken together, the conclusion is reached that different frequencies of EMF can bring about changes in enzyme structure, which in turn can affect AChE interaction with the substrate (ACh iodide), and its activity. EMF-induced changes were observed in intrinsic fluorescence intensity of the Trp residues in AChE: with increasing the frequency of ELF-EMF, Trp fluorescence decreased. Combining the results from UV-vis, Far-UV CD and intrinsic fluorescence analysis, we conclude that the structure of AChE is altered by ELF-EMF.

Peculiar characteristic of acetylcholinesterase is the presence of a large negative potential near the active site (gorge), due to the presence of negatively charged residues located at the entrance, midway down and near the gorge base. This charged group distribution leads to an electrostatic field, whose potential was calculated solving the Poisson-Boltzman equation by means of the finite difference method (Warwicker and Watson, 1982) implemented in the Delphi algorithm (Gilson et al., 1988). Cholinesterase have a rather strong first moment of 800-1800 Debye roughly aligned along the gorge axis, so that a positive charged substrate will be drawn to the active site by its electrostatic field, creating a selective and efficient substrate-binding site interaction. Current theories and models on the effects of ELF-EMFs on enzyme activities refer to changes of motion of ions such as calcium at the active site (Edmonds, 1993) or alterations in the binding parameters of ligand-receptor (Bianco and Chiabrera, 1992; Moggia et al., 1997).

However, the decrease in AChE activity in the prescence of ELF field, dose not seems to depend on changes of motion of ions such as calcium or magnesium. On the contrary, the effect of the field could be on the membrane organization and structure which would limit in some way the binding of the substrate to the active site of the enzyme. Neuropathological studies have been reported that Alzheimer's diseases (AD) were associated with deficiency in the brain neurotransmitter, ACh (Tabet, 2006). The inhibition of AChE enzyme, which catalyzes the breakdown of ACh, may be one of the most realistic approaches to the symptomatic treatment of AD (Pangestuti and Kim, 2010). Therefore, we suggest that the ELF-EMF could be useful in treatment of AD disease.

Conclusions

In conclusion, our results suggested that ELF-EMF application affects the AchE activity. The effects might be related to structural changes in the secondary and tertiary structure of the enzyme.

ACKNOWLEDGEMENT

The present study was supported by a research grant from the University of Tabriz.

REFERENCES

Ames BD, Yamasak WE, Lee FD (2005). Dose generalized anxiety disorder predict coronary heart disease risk factors in dependently of major depressive disorder. Affect Disord., 88: 87-91.

Axelsen PH, Harel M, Silman I, Sussman JL (1994). Structure and dynamics of the active site gorge of acetylcholinesterase. Synergistic use of molecular dynamics simulation and X-ray crystallography. Protein Sci., 3: 188-197.

Barteri M, Pala A, Rotella S (2005). Structural and kinetic effects of mobile phone microwaves on acetylcholinesterase activity. Biophys. Chem., 113: 245-253.

Bianco B, Chiabrera A (1992). From the Langevin-Lorentz to the Zeeman model of electromagnetic effects on ligand-receptor binding. Bioelectrochem. Bioenerg., 28: 355-365.

Dvir H, Harel M, Bon S, Silman I (2004). The synaptic acetylcholinesterase tetramer assembles around a polyproline II helix. EMBO J., 23: 4394-4405.

Ebadi A, Sedghi GH, Zare S, Hayatgeibi H, Alivandi S (2006). Biological effect of power frequency magnetic fields on serum biochemical parameters in Guinea pigs. Pak. J. Biol. Sci., 6: 1083-1087.

Edmonds DT (1993). Larmor precession as a mechanism for the detection of static and alternating magnetic fields. Bioelectrochem. Bioenerg., 30: 3-12.

Ellman GL, Curtney KD, Andres V, Featherstone RM (1961). A new and rapid colorimetric determination of acetylcholinesterase activity. Biochem. Pharm., 7: 88-95.

Gilson MK, Sharp KA, Homig BH (1988). Calculating electrostatic interactions in biomolecules: Method and error assessment. J. Comput. Chem., 9: 327-335.

Grassi J, Vigny M, Massoulie J (1982). Molecular forms of acetylcholinesterase in bovine caudate nucleus and superior cervical ganglion: solubility properties and hydrophobic character. J. Neurochem., 38: 457-469.

Hekmat A, Saboury AA, Divsalar A, Khanmohammadi M (2008). Conformational and structural changes of choline oxidase from Alcaligenes species by changing pH Values. Bull. Korean Chem. Soc., 29: 1510-1518.

Kelly SM, Price NC (2000). The use of circular dichroism in the investigation of protein structure and function, Curr. Prot. Pept. Sci., 1: 349-384.

Massoulie J, Bon S (1982). The molecular forms of cholinesterase in vertebrates. Annu. Rev. Neurosci., 5: 57-106.

Massoulie J (2002). The origin of the molecular diversity and functional anchoring of cholinesterase. Neurosignals. 11: 130-143.

Moggia E, Chiabrera A, Bianco B (1997). Fokker-Planck analysis of the Langevin-Lorentz equation: Application to ligand-receptor binding under electromagnetic exposure. J. Appl. Phys., 82: 4669-4677.

Morelli A, Ravera S, Panfoli I, Pepe IM (2005). Effects of extremely low frequency electromagnetic fields on membrane-associated enzymes. Arch. Biochem. Biophys., 441: 191-198.

Pangestuti R, Kim SK (2010). Neuroprotective properties of chitosan and its derivatives. Mar Drugs. 8: 2117-2128.

Parveen M, Kumar S, Singh P (2004). Kinetic Analysis of the in vivo Inhibition of Liver AchE in Air Breathing Fish Clarias batrachus (Linnaeus, 1758). J. Fish. Aquat. Sci., 21: 143-144.

Prashanth KS, Chouhan TRS , Nadiger S (2009). Effect of 50 Hz electromagnetic fields on acid phosphatase activity. Afr. J. Biochem. Res., 3: 060-065

Ravera S, Repaci E, Morelli A, Pepe IM, Botter R, Beruto D (2004). Electromagnetic field of extremely low frequency decreased adenylate kinase activity in retinal rod outer segment membranes. Bioelectrochemistry. 63(1-2): 317-320.

Seyyedi SS, Dadras MS, Tavirani MR, Mozdarani H, Toossi P, Zali AR (2007). Proteomic analysis in human fibroblasts by continuous exposure to extremely low frequency electromagnetic field. Pak. J. Biol. Sci., 22: 4108-4112.

Sreerama N, Woody RW (2004). On the analysis of membrane protein circular dichroism spectra. Protein Sci., 13: 100-112.

Tabet N (2006). Acetylcholinesterase inhibitors for Alzheimer's disease: Anti-inflammatories in acetylcholine clothing! Age and Ageing (AGE). 35: 336-338.

Volpe P (2003). Interactions of zero-frequency and oscillating magnetic fields with biostructures and biosystems, Photochem. Photobiol. Sci., 2: 637-648.

Warwicker J, Watson HC (1982). Calculation of the electric potential in the active site cleft due to alpha-helix dipoles. J. Mol. Biol., 157: 671-679.

Zwirska-Korczala K (2005). Effect of extremely low frequency electromagnetic fields on cell proliferation, antioxidative enzyme activities and lipid peroxidation in 3T3-L1 predipocytes- an *in vitro* study. J. Physiol. Pharmacol., 56: 101-108.

Yaghmaei P, Parivar k, Doranian D, Hashemi M, Torkaman F (2009). Study on the effect of extremely low frequency electromagnetic fields on some blood serum's lipoproteins, liver enzymes and P 448/P 450 cytochrome enzyme system in NMRI mice. Journal of Paramedical Sciences (JPS).1: 1735-7802.

5

Kinetics studies of protease in fermenting locust beans (*Parkia biglobosa*) and melon seed (*Citrullus vulgaris*)

C. Egwim Evans[1]*, J. Yisa[2] and P.O. Egwim[3]

[1]Biochemistry Department, Federal University of Technology, P.M.B 65, Minna, Niger State, Nigeria.
[2]Chemistry Department, Federal University of Technology, P.M.B 65, Minna, Niger State, Nigeria.
[3]Science Laboratory Technology, Federal Polytechnic, P.M.B 55, Bida, Niger State, Nigeria.

Melon seed (*Citrullus vulgaris*) and locust bean (*Parkia biglobosa*) were separately subjected to a natural solid – substrate fermentation for 5 days. The peak of protease (P.I) production from the fermenting melon seeds was 48 h while that of the protease (P.II) from fermenting locust bean was 96 h. Both P.I and P.II demonstrated maximum activity at 40°C. The peak of activity for P.I was observed between pH 6.6 to 7.6, whereas for P.II, the optimum pH was 5.8. The Michaelis – Menten's constant (K_M) as well as the maximum velocity (V_{Max}) of activity were 4.8 x 10^{-2} M and 3.43 x 10^{-2}/s, and 5.8 x 10^{-2} M and 5.1x10^{-2}/s for both P.I and P.II respectively. The work concludes that fermenting melon and locust bean seeds may be a cheap and alternative source of proteases for industrial processes.

Key words: Kinetics, protease, natural fermentation, locust bean, melon seed, alternative sources.

INTRODUCTION

Proteases are a group of enzymes that catalyze the cleavage of peptide bonds in proteins, thereby causing the disassembly of protein molecules. They differ in their ability to hydrolyze various peptide bonds as each type of protease has a specific kind of peptide bond it cleaves.

They can be obtained from animal, plant and microbial sources some examples include, trypsin, chymotrypsin (animal pancrease), pepsin (gastric stomach), bormelain (pineapple), papain (papaya latex), ficin (fiscus latex), substilisin (*Bacillus subtilis* collagenase (clostridia), fungal proteases, viral proteases and other bacteria proteases are also kwon (Rao et al., 1998).

Proteases execute a large variety of complex physiological functions; their importance in conducting essential metabolic and regulatory functions as evident from their occurrence in all forms of living organisms. Some of the functions includes activation of zymogens and blood coagulation (Shimomura et al., 1995), cell growth and migration (Liaw et al., 1999), morphogenesis, tissue arrangement (Zahedi, 2001) inflammation (Holzhausen et al., 2005). Proteases also catalyse important proteolytic steps in tumor invasions or infection cycle of a number of pathogenic micro-organism and vessels, a 'quality' that makes proteases a valuable target for new pharmaceuticals, they also aid metastasis (Camera et al., 2004). Pozo et al. (2004) reported that proteases can be involveed in fungal development and have been related to pathogenic or bio-control of pests and diseases in stored grains where fungal infections are major problems.

Besides the numerous physiological functions of proteases they have a variety of applications in the food and detergents industry (Phadatare et al., 1993). Fungal neutral proteases supplement the action of plant, animal and bacterial proteases in reducing bitterness of food protein hydrolysate. Fungal alkaline proteases are also used in food protein modification (Impoolsup et al., 1981). In view of the recent trend of developing environmentally friendly technologies, proteases are seen to have extensive application in leather treatment (tanning) (Kamini et al., 2004) and in several bioremediation processes and in pharmaceutical industries for debridement of wounds (Roa et al., 1998).

The inability of plant and animal proteases to meet current world demand has led to an increased interest in microbial proteases both for basic understanding of enzyme mechanism and industrial applications (Rao et al., 1998). Microorganisms represent an excellence source of enzymes owing to their broad biochemical diversity, limited spaces required for their cultivation and their susceptibility to genetic manipulation. Protreases are ubiquitous

*Corresponding author. E-mail: evanschidi@gmail.com.

in microorganisms and except for a few carboxypeptidases, the majority of these enzymes are endopeptidases (or proteinases) which are predominantly extracellular and are isolated in active forms from the culture filtrates of the appropriate organism (Matsubara and Feder, 1971). Some of these organisms are natural associates of Nigerian fermented foods. For instance, *B. subtilis* and *Aspergillus spp* (Odunfa, 1981) are common isolates of dadawa production. However dadawa production can be from locust beans (Achi, 1992) or from melon seed (Jideani and Okeke, 1991) depending on the locality. Kuddus et al. (2005) have shown that *Bacillus spp* and *Staphylococcus spp* are bacteria isolates associated with fermenting foods. Such extracellular proteases from fermenting foods, particularly from the production of dadawa, a popular food seasoning in Nigeria, have not been studied. It is on this wise that the present work is designed to study the characteristics of extracellular proteases produced in the process of fermenting locust beans and melon seeds.

MATERIALS AND METHODS

Source of material and chemicals

African locust bean and melon seeds were obtained from the local market in Bida, Niger State, Nigeria, while most of the chemicals used were of the Analar grade from BDH Chemicals Limited. Poole, England.

Fermentation of seeds

The fermentation of locust bean and melon seeds was carried out according to the method reviewed by Achi, 2005.

Production of dadawa

Raw African locust bean was boiled for 12 h and further soaked in the boiling water another 12 h (preferably overnight). Excess water was drained off and the seeds were dehulled by marching in a large wooden mortar and further removal of the seed coat was achieved by rubbing the cotyledons between the palms of the hand. The cotyledons were again cooked for another 6 h, drained off and the cotyledons were then spread into trays, covered with jute sacks and fermented for 5 days to produce dadawa

Crude enzyme extraction

Locust bean and melon seeds (500 g each) were allowed to ferment as described above and samples were taken from the fermenting seeds at 24 h interval 5 days. Samples (1 g) collected in each case were suspended in 5 ml of pre-chilled 0.02 M phosphate buffer (pH 7.0) and the mixture was blended in warren blender and stirred, the resulting homogenate was centrifuged at 10000 X g for 10 min. The resulting supernatant constitutes the crude enzyme extract which was used for subsequent analysis. Proteases from fermenting locust bean and melon seeds are designated as P_1 and P_{11} respectively.

Assay for protease activity

Proteolytic activity was determined by monitoring the rate degradation of protein (1% albumin) by a modification of the biurette method according to Henry et al. (1974). 1 unit of enzyme activity was defined as the amount of enzyme required to degrade a unit of albumin (mMol/s) in the reaction mixture condition.

Proteases activity levels were monitored at 24 h intervals as fermentation progressed from 0 - 120 h, enzyme for subsequent analysis was harvested at the highest enzyme yield.

Determination of kinetic parameters

The Michaelis-Menten's constant (K_M) and the maximum attainable velocity (V_{Max}) were determined by investigating the effect of substrate concentration on enzyme activity. Enzyme activity was determined at different substrate concentrations [s], 0.1 – 1 mg/ml, the Lineweaver-Burk plot (1/V vs 1/[S], where V is the reaction velocity) was then constructed, and from this graph, the K_M and V_{Max} were determined for both P_I and P_{II}.

Effect of temperature on proteases activity

Reaction mixture was incubated at different temperatures (T) viz: 15, 20, 30, 35, 40, 45, 50 and 55^{0}C and the optimum temperature was determined from the graph of enzyme activity Vs temperature (T).

Effect of pH on proteases activity

Proteases activity was determination at various pH values, 0.02 M phosphate buffer at pH values ranging between 5.0 and 7.8 were used to prepare crude enzyme extract as earlier described and the proteases activity of each enzyme extract was assayed. From the result obtained a graph of enzyme activity Vs pH was plotted, and the pH optima for P_I and P_{II} were subsequently determined.
The results are mean of three determinations.

RESULT AND DISCUSSION

Fermentation

Protease activity profile with fermentation time is shown in Figure 1 and reflects that maximum yield of P_I is at 48 h, while P_{II} production was highest at 96hrs. P_I activity increased rapidly between 24 and 48 h and began to decrease afterwards, while for P_{II}, activity increased rapidly between 48 and 98 h and declined progressively afterwards. This observation suggests a period of maximum growth of fermenting organisms and protease yield in fermenting melon seeds and locust bean within 2 days and 4 days, respectively. This agrees with the report of Jidean et al. (1989), who reported that good aroma development is enhanced between 2 – 5 days of fermenting locust bean and melon seed at 37 and 40°C

Optimal temperature and pH

Temperature and pH-activity profiles of protease from fermenting locust bean and melon seed are shoen in Figures 2 and 3. The result (Figure 2) of the effect of temperature on the proleolytic activity on albumin (1%) indicated that both P_I and P_{II} demonstrated maximum activity at 40°C under the assay conditions. The same range of

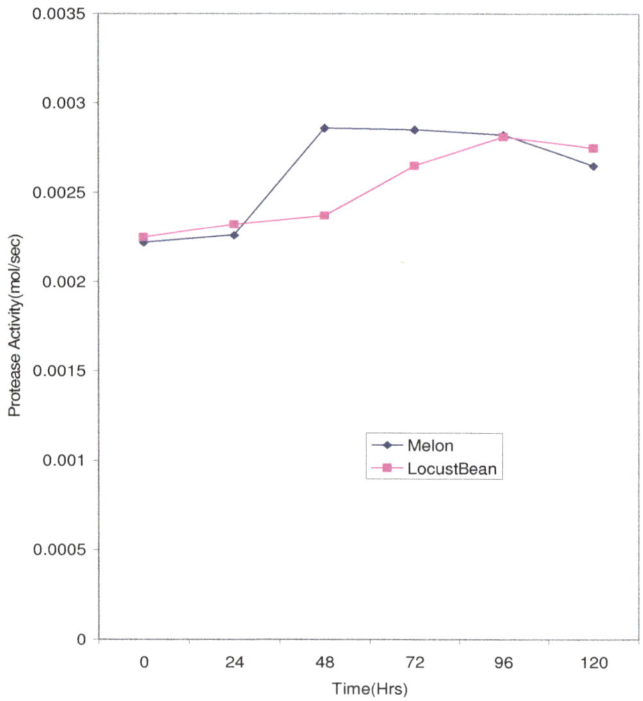

Figure 1. Protease activity profile with fermentation time.

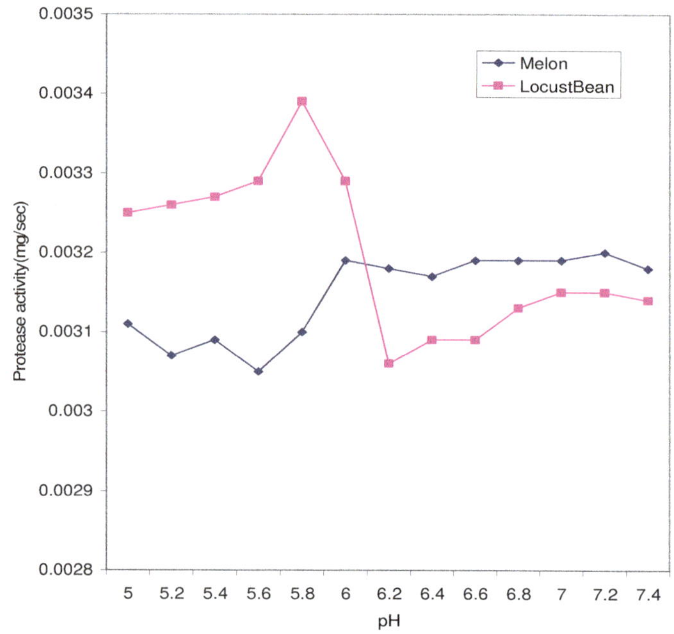

Figure 3. pH activity profile of fermenting melon and locust bean seed.

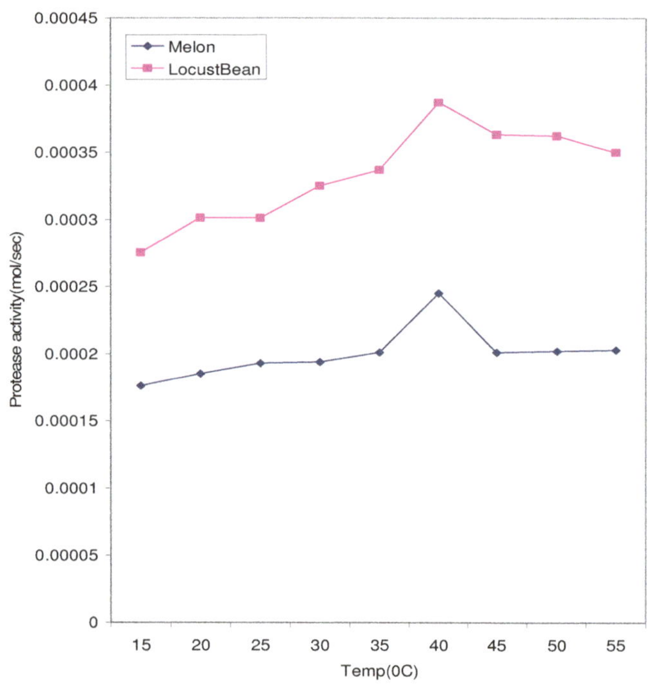

Figure 2. Temperature protease activity in fermenting melon and locust beans seed

temperature has been reported for protease activity isolated from *Bacillus lecheniformis* (Al-sheri and Mustafa, 2004). This result also suggests that protease from fermenting melon seeds and locust bean may be applied in in industrial processes with mild temperature as in leather industry (Kamini et al., 2004).

Crude protease from fermenting locust bean and melon seed showed maximum activity at pH 5.8 and 6 - 7.2 respectively (Figure 3). The fact that the activity of protease from fermenting melon was almost constant between pH 6.6 and 7.6 and that a single peak of activity at pH 5.8 was observed for crude protease from fermenting locust bean is an indication that a single acid protease and neutral were involved in each case. Though crude protease from fermenting melon seeds exhibited a relatively broad pH optima, it is suspected that different groups of peptide bonds were possibly hydrolysed in the substrate, moreover, the rate of cleavage of any particular susceptible bond is dependent upon the charge on the substrate molecule and hence, on the pH of the medium (Martin and Jonsson, 1965). That the crude protease from fermenting melon seeds and locust bean shared the same temperature optimum and different pH optima indicates that these protease could have different industrial applications or may be applied to different stages of an industrial process. This agrees with the report of Kuddus et al., (2005), who have shown that protease isolated from different organisms having different temperature and pH optima could have different industrial and biotechnological applications.

Enzyme kinetics

On examining the effect of substrate concentration on proteolytic activity, crude protease from fermenting melon seed and locust bean were found to obey the Michaelis-

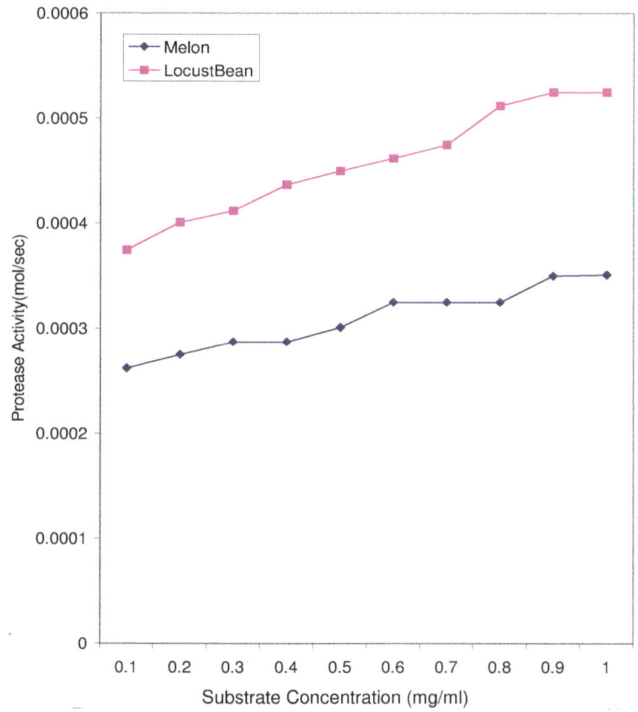

Figure 4. Substrate activity profile of protease from fermenting melon and locust bean seed.

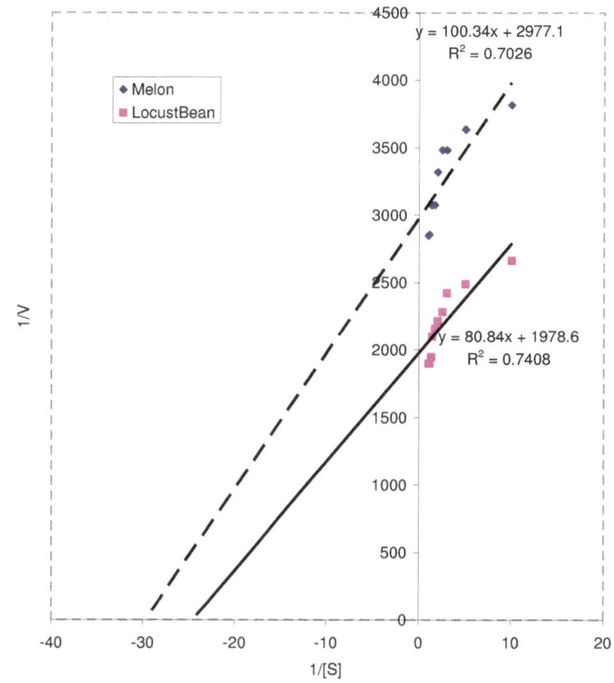

Figure 5. Line weaver -burk plot of protease from fermenting melon and locust bean seed.

Menten kinetics (Figure 4).

From the Linweaver-burk plot (Figure 5), the K_M of crude protease from fermenting locust bean and melon seed were 5.8 X 10^{-2} M and 4.8 X 10^{-2} M while the V_{Max} were 5.1 X 10^{-2} s^{-1} and 3.43 X 10^{-2} s^{-1} respectively, this shows that the enzymes exhibited considerable proteolytic activity on albumin.

However, protease from fermenting locust bean has a higher affinity for the substrate and higher proteolytic activity than protease from melon seed. This therefore suggest that crude protease from fermenting locust bean (P_1) may have more industrial applications particularly in egg processing.

The present work concludes that fermenting locust bean and melon seed may be an alternative and cheap source of industrial protease. However, further work may be required to find the activity of the crude and pure enzymes on other substrates.

REFERENCE

Achi OK (2005). The potential of upgrading traditional fermented foods through biotechnology. African J. Biotechnol. 4(5): 375 – 380.

Al-sheri MA, Mustafa SV (2004). properties of protease produced by *Bacillus licheniformis.* Bio. Sci. 7(9): 1631 – 1635.

Camerer E, Quazi AA, Duong DN, Carnelissea I, Advincular R, Coughlin SR (2004). Plateletes, protease-activated receptors and fribrinogen in hematogenous metastasis. J. Am. Society Haematol. (blood). 104: 397 – 401.

Henry RJ, Marmion OP, PHA (1974). Clinical principle and techniques, 2nd edition London harper and Rov. p. 25

Holzhausen M, Spolidorio LC. vergndle N (2005). Role of protease-activated receptor-2 in inflammation and its possible implications as putative mediator of periodontitis. Mem. Inst. Oswaldo Cruz, Rio de Janciso 100 (Suppl. 1): 177 – 180

Impoolsup A. Bhumiratana A, Flegel TW (1981). Isolation of alkaline and neutral proteases from *Aspergillus flavus Var columnaris*, a soy source koji mold. Appl. Environ. Microbiol. 42 (4): 619 – 628.

Jideani ERJ, Adegboye DP (1989). Effects of fermentation on the instruct status of locust bean. Nig. J. Sci. 5(3):19-22

Jideani IAO, Okeke CR (1991). Comparative study of microorganisms and sensory attrinutes of condiments from the fermentation of different seeds. Plant Food Hum. Nutr. 41:27 – 34.

Kamini NR, Hemachander C, Mala GSJ, pavanakrishnan R (2004). Microbial enzyme technology as an alternative to conventional chemicals in leather industry. Financial times information limited – asia intelligence wire. February 3, 2004.

Kuddus M, Ramteke PW, Mishra R (2005). Isolation and partial characterization of protease producing bacteria from food samples. Federation of European Biochemical Society. 272 (SI). A2 – O39P.

Liaw L, Crawford HC (1999). functions of the extracellular matrix and matrix degradation protease during tumor progression. Brazillian J. Med. Bio. Res. 32:805 – 812.

Martin SM, Jonsson AG (1965). An extracellular protease from *Aspergillus Fumigatus.* Canadian J. Biochem. 43: 1745 – 1753.

Matsubara H, Feder J (1971). in: Boyer, P.D. (ed) the enzymes 3rd edition academic press inc. New York, N.Y IV: 721 – 795.

Odunfa SA (1981). Microorganisms association with fermentation of African Locust Beans during preparation. J. Plant Foods 25:245-250.

Phadatare SV, Srinivasan MC, Deshpande VV (1993). High activity alkaline protease from Conidiobolus coronatus (NCL 86. 8. 20); enzyme

production and compatibility with commercial detergents. Enzyme Microb. Technol. 15 : 72-76.

Pozo M J, Back J M, Garcia J M, Kenerly C M (2004). Functional analysis of tvspl, a serine protease – encoding gene in bio-control agent, Trichoderma virens. Fungal Genet. Biol. 41 : 336-348.

Rao M B, Tanksale A M, Ghalge M S, Deshpande W (1998). Molecular and Biotechnological aspects of microbial proteases. Microbiol. Mol. Biol. Rev. 62 : 597 – 635.

Shimomura T, Miyazawa K, Komiyama Y, Hiraoka H, Naka D, Marimoto Y, Kitamura N (1995). Activation of hepatocyte growth factor activator. Eur. J. Biochem. 229: 257 -261.

Zahedi R (2000). Analysis of a protease invaded in mammary development. Lawrence Berkeley National Laboratory. www. Cbcrp.org/research/page Grant.asp? grant_ id = 2137

Characterization of partially purified cysteine protease inhibitor from *Tetracarpidium conophorum* (African walnut)

Obayomi, A.[1], Adeola, S. A.[1], Bankole, H. A.[1] and Raimi, O. G.[1,2]

[1]Department of Biochemistry, Lagos State University, Ojo, Lagos, Nigeria.
[2]Protein phosphorylation unit, College of Life Sciences, University of Dundee, Scotland, United Kingdom.

Cysteine protease inhibitors (CPIs) have been known to be present in a variety of seeds of plants, and have been intensively studied as useful tools for potential utilization in pharmacology. This study reports the isolation of CPI from *Tetracarpidium conophorum* by 65% ammonium sulphate saturation, followed by ion exchange chromatography; further purification was by gel filtration chromatography. The molecular weight of the partially purified protein inhibitor was analyzed by SDS-PAGE to be approximately 20 kDa. The inhibitor had an optimum pH and temperature of 8.0 and 40°C, respectively. The inhibitor competitively inhibited papain with the same V_{max} = 71.17×10^3 µmol/min, K_m = 166 µM, and K_i = 53.63 µM. Divalent metal ions such as, Mg^{2+}, Pb^{2+}, Mn^{2+}, Cu^{2+}, Co^{2+}, and Zn^{2+} had significant effect on inhibitory activity of CPI at concentration as low as 1 mM. Cysteine protease inhibitor of *T. conophorum* investigated in this study could serve as a template in biotechnology of herbal medicine to arrest the negative modulatory interactions of cysteine proteases in clinical pathogenic expressions.

Key words: *Tetracarpidium conophorum,* cysteine protease inhibitor, papain, purification, characterization.

INTRODUCTION

The interaction between cysteine proteases and their inhibitors is a requisite in preventing unwanted, potentially destructive proteolysis, which can be utilized in chemotherapy. Cysteine proteases (CPs) are widely distributed among living organisms, and are responsible for many biochemical processes occurring in living organisms, such as production of nutrients for cell growth and proliferation in *Rhizopus oligosoporus* (Lin et al., 2011). They play multi-faceted roles in plants, virtually in every aspect of their physiology and development, such as growth, senescence and apoptosis, in the accumu-lation and mobilization of storage proteins such as in seeds. They are also involved in signaling pathways triggered by protease-activated receptors (PARs), which are critical mediators of haemostasis, thrombosis, biotic and abiotic stresses, and have been implicated in cancer progressions, making this receptor class an important drug target (Chen et al., 2005). The *Carica papaya* enzyme papain (EC 3.4.22.2) was the first known proteolytic enzyme and its digestive properties were already being utilized in the

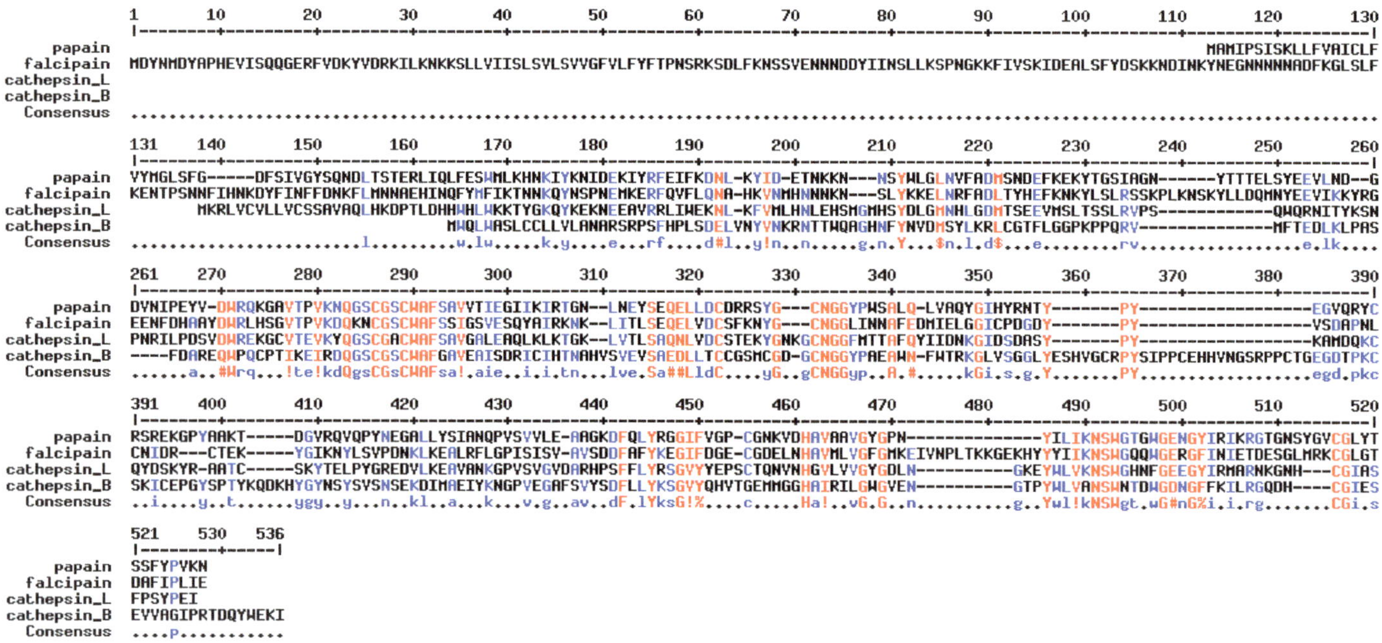

Figure 1. Sequence alignment of structurally equivalent residues of some C1 family of cysteine proteases. Coloured red are residues that are conserved; blue are similar residues and black are residues that are different in all the groups.

19th century (Redzynia et al., 2009). Since then papain has been used as a model enzyme in many studies and is a founding member of a large C1 family of papain-like cysteine proteases (Rawlings et al., 2008), while calpain (EC 3.4.22.52) constitute the C2 family of calcium-dependent non-lysosomal cysteine proteases expressed ubiquitously in mammals and many other organisms (Ohno et al., 1984). C1 family of proteases are evolutionarily old and are found in both prokaryotes and eukaryotes and show activities that are indispensable for the organism (Redzynia et al., 2009). Some of the mammalian cysteine proteases are evolutionarily closely related to papain and hence belong to this family for example cathepsins B, H, L, S and K (Figure 1). These enzymes function in every cell as components of the lysosomal degradation system taking part in the turnover of proteins as well as participate in a number of proteolytic cleavages, activating pro-hormone and regulation of antigens (Redzynia et al., 2009).

Obligatorily, the activities of these enzymes need to be strictly regulated and controlled (Corrion et al., 2010) , by cysteine protease inhibitors. Activities of proteases are generally regulated by protease inhibitors, which are usually proteins with domains that enter or block a protease active site to prevent substrate access (Figure 2). A control mechanism of proteases involves interaction of the active enzymes with proteins that inhibit their activities. These proteins have been isolated and characterized from a large number of organisms, including plants, animals and microorganisms (Outchkourov et al., 2003). C1 family of cysteine proteases are in equilibrium with protein inhibitors belonging to the cystatin family (Hartmann et al., 1997; Rawlings and Barrett, 1990; Lustigman et al., 1992; Rawlings et al., 2008). Most cystatins such as human cystatin B are single domain proteins of 100-120 residues with a characteristic wedge-like epitope consisting of the N-terminus and two β-hairpin loops which blocks the active site cleft of the target enzyme hence inhibiting the activity in a reversible manner (Abrahamson et al., 1986; Bode et al., 1988) (Figure 2).

The African walnut is an annual agricultural product found abundantly in Nigeria (Ayoola et al., 2011b). The plant, which belongs to the family of *Euphorbiaceae*, is a temperate fruit and nut crop, with a high fat content hence Walnuts are rich source of energy and contain health benefiting nutrients, minerals, antioxidants and vitamins that are essential for optimum health (Ayoola et al., 2011a). Walnuts are considered to be an herb in traditional Chinese medicine (Ganiyu and Mofoluso, 2004). The leaves, bark, root and fruit of *Plukenetia conophora* are considered to be medicinal (Enitan et al., 2014). The anti-microbial potential of *P. conophora* extracts and fractions against wide spectrum of bacteria (including *Staphylococci*, *Clostridia*, *Escherichia* and *Pseudomonas*) and some fungi like *Aspergillus niger* and *Candida albican* have been demonstrated (Ajaiyeoba and Fadare, 2006). The ethanol: water extract of the dried leaves had the best antioxidant activity; the broad range of antioxidant activity of this extract indicates the potential of the plant as a source of natural antioxidants or nutraceuticals with potential application to reduce oxidetive stress and consequent health benefits (Amaeze et

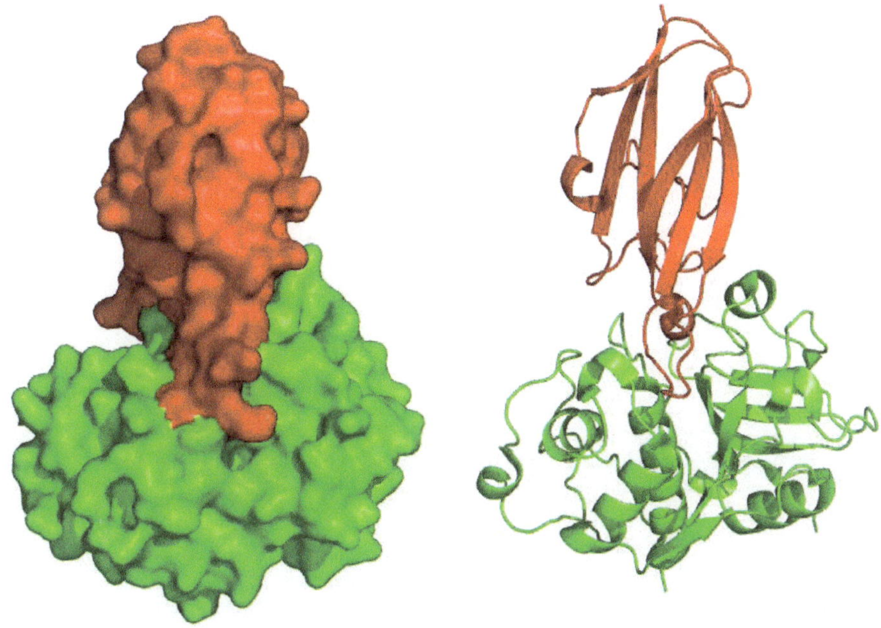

Figure 2. Surface and cartoon representations of the crystal structure of Chagasin-papain complex. The chagasin molecule is coloured in red and papain in green. The view is along the catalytic cleft of papain and corresponds to the standard orientation used for cysteine proteases (Redzynia et al., 2009; pdb: 3E1Z).

al., 2011). Antibacterial activity of methanolic leaf extract of *P. conophora* Mull. arg. against selected bacteria isolated from urinary tract infection has also been demonstrated (Enitan et al., 2014). The aim of the study was to isolate, purify and characterize cysteine protease inhibitor (CPI)from *Tetracarpidium conophorum* seeds.

MATERIALS AND METHODS

Sample collection and preparation

T. conophorum seeds and leaves were obtained at Ketu market in Lagos State, Southwestern Nigeria. The leaves were deposited at the herbarium of the Botany Department, Faculty of Science, Lagos State University, Ojo Lagos State, Nigeria for proper identification and authentication. 1000 g of *T. conophorum* seeds were de-shelled, dried and ground to powder with a grinding machine. The powdered sample was then defatted with n-hexane using the soxhlet apparatus, according to the method of Franz von Soxhlet (1879).

Extraction and isolation of cysteine protease inhibitor

Cysteine protease inhibitor was isolated as described by Benjakul et al., 1998. 40 g of the defatted sample was extracted in 500 ml ice-cold 100 mM phosphate buffer pH 7.2 containing 130 mM NaCl and 0.1% β-mercaptoethanol, stirred thoroughly for 1 h. The mixture was filtered using clean white piece of cloth. The filtrate was centrifuged at 7000 *g* for 10 min to collect cell debris. Cysteine protease inhibitor was then isolated from the resulting supernatant by ammonium sulphate precipitation according to the method described by Englard and Seifter (1990). Ammonium sulphate required to precipitate the protein was optimized by adding varying concentrations, (35, 55, 65, 75 and 90%) to the crude extract independently.

After each precipitation the precipitate was collected by centrifugation at 7000 *g* for 10 min. The precipitate was then re-dissolved in a small volume of buffer and dialyzed overnight against 100 mm Tris buffer pH 7.8 that was changed every 6 h.

Protein determination

Protein concentration was determined by Bradford (Bradford, 1976) using bovine serum albumin (BSA) as standard. The absorbance was read at 595 nm.

Inhibitory activity

The inhibitory activity of cysteine protease inhibitor on papain was monitored according to the modified method of Murachi, 1970, using casein as substrate. Papain (6 mg in 100 mM Tris-HCl buffer, pH 7.8, 0.5 mM cysteine, 0.2 mM EDTA) and the inhibitor extract (50 µl) were pre- incubated at 37°C for 15 min. The reaction mixture (150 µl) was then added to tubes containing 1.0 ml of 0.5% casein (casein prepared in 100 mM Tris-HCl pH 7.8 containing 0.2 mM EDTA and 0.5 mM cysteine) at 37°C. The assay was incubated for 30 min at 37°C and the reaction terminated by the addition of 1.0 ml of 5% trichloroacetic acid (TCA). The absorbance of the supernatant was measured at 280 nm after 30 min. One unit of inhibitor activity was defined as the decrease by one unit of absorbance of tricholoroacetic acid-soluble casein hydrolysis product liberated by protease action at 280 nm at 37°C in a given assay volume. Percentage Inhibition was determined as shown below:

Table 1. Purification table of cysteine protease inhibitor (CPI) from *T. conophorum*.

Sample	Total protein (mg)	Inhibitory activity (units)	Specific inhibitory activity (units/mg)	Protein yield (%)	Activity yield (%)	Purification fold
Crude extract	363	4000	11.02	100	100	1
Ammonium sulphate Fraction (65%)	126	1500	11.9	34.7	37.5	1.1
Ion Exchange chromatography (DEAE-cellulose)	46	3500	76.1	12.67	87.5	6.9
Gel filtration (Sephadex G-100)	26	3330	128.1	7.16	83.25	11.61

The purification fold of cysteine protease inhibitor from ammonium sulphate precipitation, ion exchange chromatography and gel filtration chromatography were 1.1, 6.9 and 11.6 respectively.

$$\% \text{ Inhibition by each tube} = \frac{\text{Abs of Std} - \text{Abs of Sample}}{\text{Abs of Std}} \times 100$$

Protein purification

The crude protein after dialysis was purified by ion exchange chromatography according to the method of Rossomando (1990), followed by size exclusion chromatography. 3.0 ml of the dialysate (with highest % inhibitory activity against papain) was then loaded on DEAE-cellulose column previously equilibrated with 100 mM Tris-HCl buffer pH 7.8. 5 ml fractions were collected into 60 test tubes using an increasing linear gradient of NaCl concentration from 0 to 0.3 M in the same buffer (Tris-HCl). Total protein and inhibitor activity were carried out on each fraction as earlier described. Fractions with highest inhibitory activity were pooled together concentrated and loaded on sephadex G-100 column previously equilibrated with 100 mM Tris-HCl buffer pH 7.8 and eluted using the same buffer. Fractions with highest inhibitory activity were pooled together and analyzed further by SDS-PAGE for molecular weight estimation.

Mechanism of inhibition

Mode of inhibition of the purified CPI was carried out as described for inhibitory assay above but now varying the concentration of casein. This was carried out both in the absence and presence of the inhibitor, respectively.

Optimum temperature

The optimum temperature for the inhibitor was determined by incubating the reaction mixtures at varying temperatures ranging from 10 - 100°C with 10-unit increase.

Optimum pH

Estimation of the pH optimum was carried out using Tris-HCl buffer with varying pH (6-11) differently in the reaction mixtures. Other steps were as described for inhibitory assay.

Effect of metals on inhibitory activity

Effect of different metal ions on protease inhibitory activity was carried out by incubating the protease inhibitor with different concentrations of various metals ions for 30 min followed by measuring inhibitory activity as described above. The metals that were investigated included Mg^{2+},Mn^{2+}, Zn^{2+}, Co^{2+}, Cu^{2+},and Pb^{2+}. Each with concentration between 1 to 10 mM.

RESULTS AND DISCUSSION

Plant CPIs have been researched extensively, but this study however is the first at attempting to characterize cysteine protease inhibitor from *T. conophorum*. *T. conophorum* is a temperate fruit and nut crop which possess an elevated level of fat content. The seed is an annual agricultural product, which is found abundantly in Nigeria (Ayoola et al., 2011).

CPI was isolated by ammonium sulphate precipitation, using different concentrations of ammonium sulphate (35, 55, 65, 75 and 90%) and maximal inhibitory activity was observed at 65% saturation and this was used in further studies. The increase in hinbitory activity observed at 65% might be due to an increase in concentration of the salt ions having a greater charge density, thus increasing its ionic strength through the decline in solvent molecules being used to solvate the salt ions for optimum saturation, stressing the role of hydrophobic aggregation of proteins in molecular interaction.

Bijina et al. (2011), reported 30 - 60% ammonium sulphate saturation for *Moringa oleifera* protein. The protein was separated from the salt ions by dialysis, which allows ammonium sulphate salts and small metabolites to diffuse across a semi-permeable membrane. This step of purification yielded a low purification fold close to the crude protein, which suggests the presence of impurities (Table 1). Further purification of cysteine protease inhibitor by chromatographic methods, using DEAE cellulose an anion exchanger, yielded multiple peaks with a single peak of maximum inhibitory activity and a slight increase in purification fold (Table 1).

The single peak with the maximum inhibitory activity obtained after ion exchange chromatography was purified further by gel filtration yielded multiple peaks with two

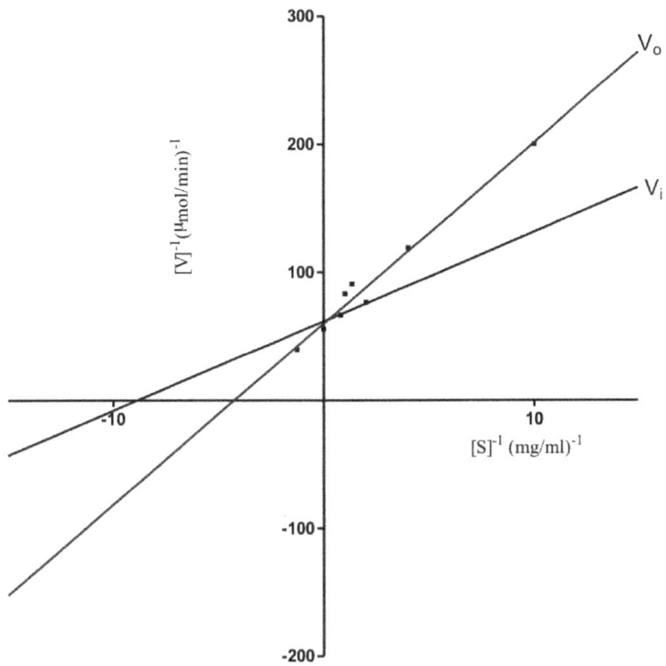

Figure 3. Line weaver-Burk plot of purified CPI. In the absence and presence of purified inhibitor extract, the reciprocal of the varying amount of substrate concentration [1/S] used and the reciprocal of the absorbance (1/V) at 280 nm were plotted against each other. The double reciprocal plot shows that inhibition is competitive having the same V_{max} of 71.17×10^3 µmol/min, different K_m =166 µM and K_i = 53.63 µM. $[V]^{-1}$ = with inhibitor and $[V_0]^{-1}$ = without inhibitor.

peaks having maximum inhibitory activity. This suggests that a splitting of the polypeptide chain might have occurred or there might be two different cysteine protease inhibitors present. The gel filtration chromatography using sephadex G-100, resulted in the highest purification fold of 11.61 and protein yield of 83.3% (Table 1). Bijina (2006), in his work, purified cysteine protease inhibitor from *Moringa oleifera,* using ammonium sulphate precipitation (30 - 90%) and ion exchange chromatography, and obtained purification fold of 1.5 and 2.5%, respectively. This could be accounted for by the fact that each purification step reduced the level of impurities associated with the inhibitor, which suggests that the fold can be increased by repetitive purification steps using advanced purification methods.

CPI from *T. conophorum,* demonstrated a competitive mode of inhibition on papain (Figure 3) with the same V_{max} = 71.17×10^3 µmol/min, K_m = 166 µM, and K_i = 53.63 µM. The result obtained was contrary to that reported by Bijina (2006), who reported the mechanism of protease inhibition of *Moringa oleifera* as uncompetitive. Kinetic studies between chagasin and papain (Figure 2) revealed the interaction to be a very strong reversible interaction with a K_i of 36 pM with chagasin blocking the active site cleft of papain (Redzynia et al., 2009) (Figure 2).

The protein inhibitor has an optimum pH of 8.0 (Figure 4) and observed to be inactive at extreme acidic and alkaline pHs. This indicates that extreme pH conditions could alter the electrostatic interactions between charged amino acids such as aspartate, lysine, arginine, and glutamate, thereby disrupting the structure of the cysteine protease inhibitor, making it loose its activity partially or completely. The optimum pH of cysteine protease inhibitors isolated in this study is in accordance with those reported for other isolated plants cysteine protease inhibitor such as *Dimorphandra mollis* seeds (pH 8.0) (Mello et al., 2002) and *M. oleifera* leaves (pH 7.0) (Bijina, 2006). *T. conophorum* CPI has a temperature optimum of 40°C (Figure 5). Most plant's cysteine protease inhibitors are active from this temperature up to 50°C (Bijina, 2006). However, the inhibitory activity declined at temperatures lower than its optimum temperature, and was totally inactive at temperatures above the optimum temperature.

The probable precursor for the modulating activities of cellular actino-regulatory proteases is a consequence of thermal influence on violently disrupted bonds of cysteine protease inhibitor from elevated kinetic energy (Bein and Simons, 2000). This result shows that cysteine protease inhibitor might undergo thermal denaturation when exposed to temperatures higher than its optimum temperature.

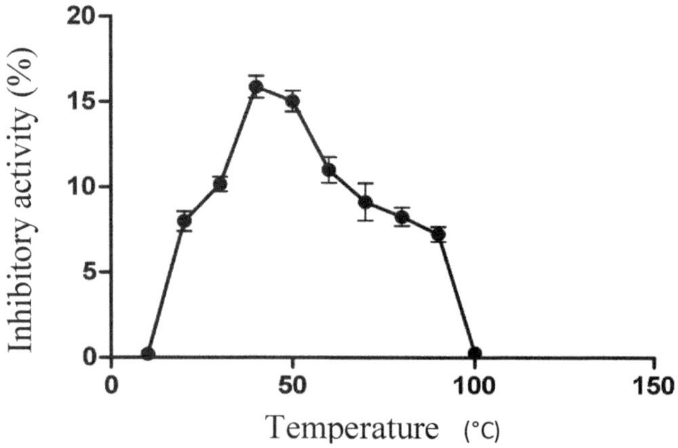

Figure 4. Optimum temperature of crude cysteine protease inhibitor. The result shows the optimal temperature of the cysteine protease inhibitor is 40°C. Cysteine protease inhibitor was almost inactive at low temperature, however as the temperature increases, the inhibitory activity increased gradually but decreased at temperatures above 40°C.

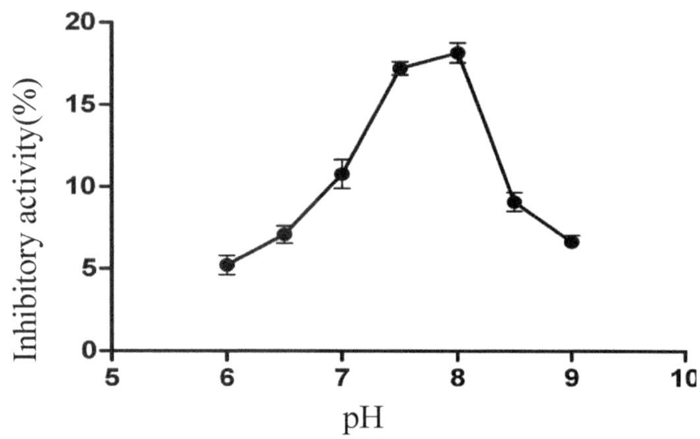

Figure 5. Optimum pH of purified cysteine protease inhibitor. The result shows the optimum pH of purified cysteine protease inhibitor at pH 8.0. Cysteine protease inhibitor was almost inactive at low pH values, however, as the pH increases, the inhibitory activity increased gradually but decreased at above pH 8.0.

This result is in agreement with those reported for other isolated plants cysteine protease inhibitor such as *Dimorphandra mollis* seeds (40°C) (Mello et al., 2002) and *M. oleifera* leaves (40°C) (Bijina, 2006).

Trace metal ions such as Co^{2+} and Zn^{2+} at low concentrations reduced the activity of CPI with a residual activity of 58 and 88%, respectively (Figure 6). However, when present in high concentrations, they are toxic and can denature proteins. Heavy metal ions such as Cu^{2+}, Mg^{2+}, Pb^{2+} and Mn^{2+} at 1 mm decreased the residual inhibitory activity of cysteine protease inhibitor up to 47.5, 51, 55 and 61.4%, respectively, compared to that of control. This suggest that these heavy metal ions interacted with the sulfhydryl groups of the cysteine protease inhibitor, thereby deactivating the inhibitor due to its sulphur-philic nature. The sulfhydryl group is respon-sible for the stabilization of tertiary and quaternary structure ofproteins. As a result, there is a conformational change in the three-dimensional structure of the protein and is denatured, thereby inhibiting their activity (Greenwood et al., 2002).

The heavy metal ions functioned in a concentration dependent manner, as their concentration increases, their inhibitory effect increases. The result obtained is in agreement with that of Jack et al. (2004), which confirmed that some divalent metal ions could play an important role in sustaining the structural integrity of protease inhibitors, such as Zn^{2+} and Mg^{2+}, that maintains the secondary and tertiary structure of cysteine protease inhibitor. The side chain carboxylates of glutamate and aspartate residues can however, participate in binding of divalent cations to metalloproteins, thereby resulting in amino acid modifications of the structure leading to a decrease in inhibitory activity, whereby the inhibitor is unable to bind to the active site of the enzyme or substrate to exhibit inhibitory activity. Most metalloproteases require zinc, but some use cobalt (Szeto et al., 2008).

The purified CPI was further analyzed by SDS-PAGE to determine its molecular weight. The molecular weight of cysteine protease inhibitor from *T. conophorum* was estimated to be 20 kDa (Figure 7). This result indicates that *T. conophorum* CPI is composed of a single polypeptide chain. Martinez-Vicente et al. (2005), reported a molecular weight of 23.1 kDa, for cysteine protease inhibitor obtained from *Fragaria x ananassa*. Cysteine protease inhibitor from black gram (*Vigna mungo Hepper)* and rice bean (*Vigna umbellata Thunb),* have molecular weight of 12 kDa by Tricine-SDS-PAGE (Benjakul and Visessanguan, 2000). Most of these protein inhibitors are small molecules with relative molecular masses ranging from 5 - 25 kDa (Singh and Rao, 2002; Lustigman et al., 1992).

The inhibition of papain and not trypsin, a serine protease (data not shown) reveals the specificity of this inhibitor against cysteine protease, which could be explored in biotechnology for the *in-vitro* synthesis and development of target drugs for pathogenesis of atherosclerosis, dementia and cancer. The inhibition of calpain has been shown to prevent NMDA-induced cell death and β-amyloid-induced synaptic dysfunction in hippocampal slice cultures (Nimmrich et al., 2010).

Conclusion

Cysteine protease inhibitor was isolated and partially purified from *T. conophorum*. The protein inhibitor was demonstrated to inhibit papain a representative of the C1 family of cysteine proteases in a competitive manner. The biological, specificity and physicochemical properties exhibited by this inhibitor certainly indicate its likely suitability for application in biotechnology as pharmaceutical agents in the treatment of some pathological conditions such as

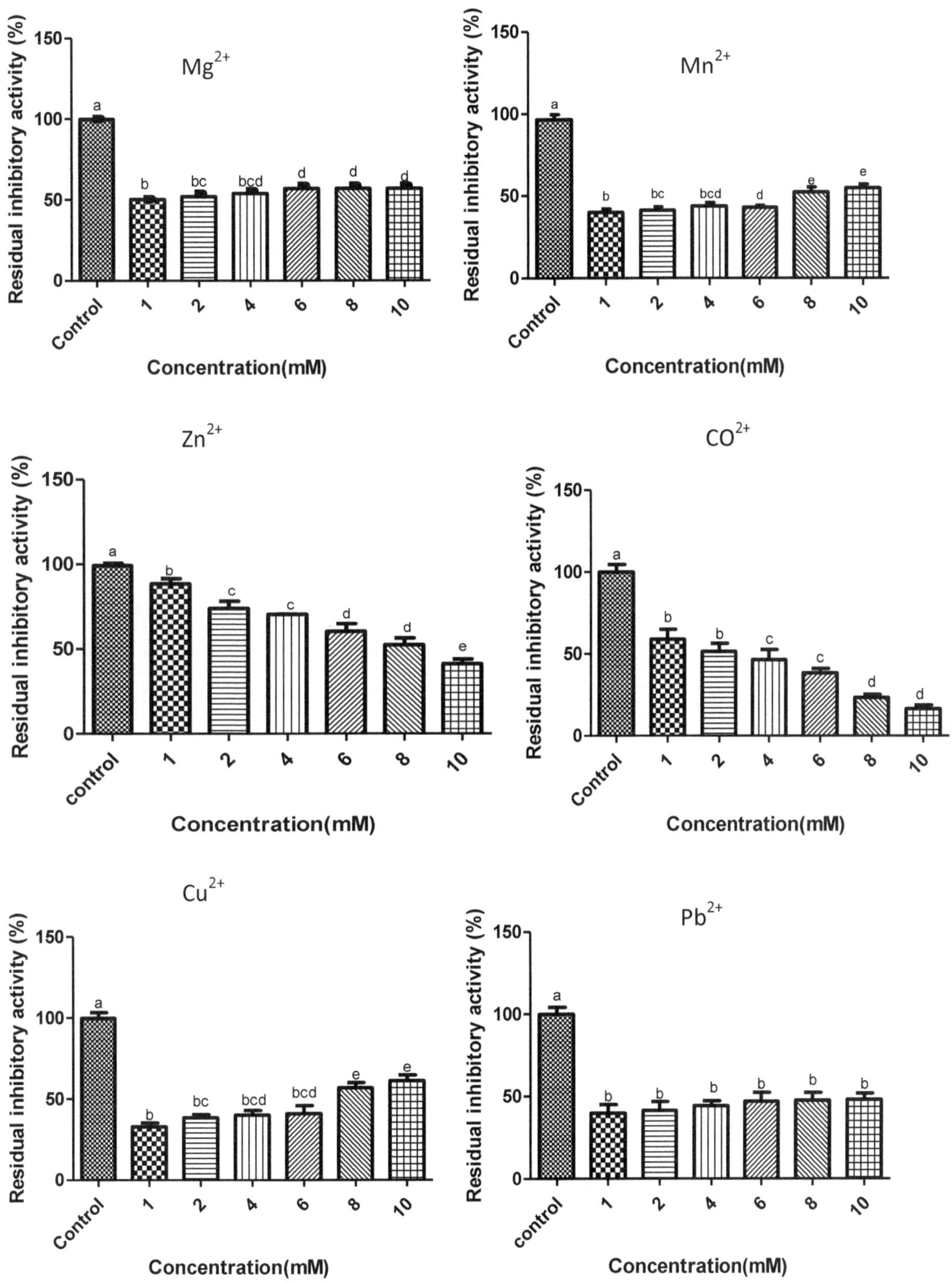

Figure 6. Effect of metal on CPI activity. The bars with the same superscripts are not statistically different (P<0.05).

cancer, atherosclerosis, and neurodegenerative diseases. However, further research is still required in determining its structure in other to elucidate its function and absolute specificity.

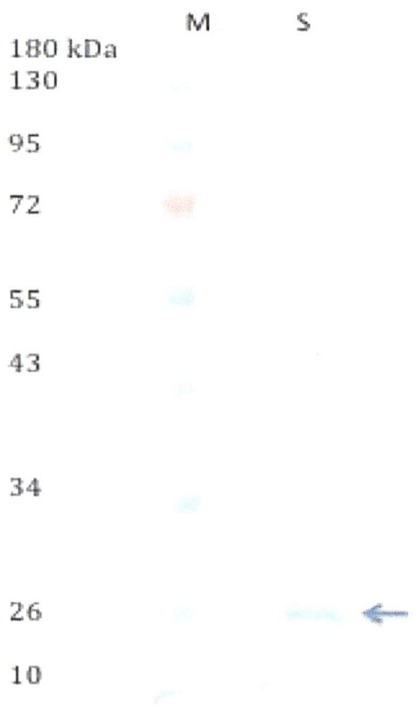

Figure 7. SDS-PAGE analysis of purified cysteine protease inhibitor. The gel pattern of the cysteine protease inhibitor fraction when subjected to SDS-PAGE yielded a single polypeptide band with a molecular weight of 20 kDa. M=marker; S=sample.

Conflict of interests

The authors did not declare any conflict of interest.

REFERENCES

Abrahamson M, Barrett AJ, Salvesen G, Grubb A (1986). Isolation of six cysteine proteinase inhibitors from human urine. Their physicochemical and enzyme kinetic properties and concentrations in biological fluids. J. Biol. Chem. 261:1282-1289.

Ajaiyeoba EO, Fadare DA (2006). Antimicrobial potential of extracts and fractions of the African walnut- Tetracarpidium conophorum. Afr. J. Biotechnol. 5(22):2322-2325.

Amaeze OU, Ayoola GA, Sofidiya MO, Adepoju-Bello AA, Adegoke, AO, Coker HAB (2011). Evaluation of Antioxidant Activity of Tetracarpidium conophorum (M¨ull. Arg) Hutch and Dalziel Leaves. Oxid. Med. Cell. Longev. Article ID 976701, 7pp

Ayoola PB, Adeyeye A, Onawumi, OO, Faboya OOP (2011a). Phytochemical and Nutrient Evaluation of Tetracarpidium Conophorum (Nigerian Walnut) Root. IJRRAS 7(2):197-202.

Ayoola PB, Onawumi OO, Faboya OOP (2011b). Chemical evaluation and nutritive values of Tetracarpidium conophorum (Nigerian walnut) seeds. JPBMS 11(15):1-5.

Bein K,Simons M (2000). Thrombospondin type 1 repeats interact with matrix metalloproteinase 2. Regulation of metalloproteinase activity. J. Biol. Chem. 275:32167-32173.

Benjakul S, Visessanguan W (2000). Pig plasma protein: Potential use as proteinase inhibitor for surimi manufacture; inhibitory activity and the active components. J. Sci. Food Agric. 80, 1351-1356. JSFA647>3.0.CO;2-I

Benjakul S, Visessanguan W, An H (1998). Screening and isolation of cysteine proteinase inhibitors from Thai legume seeds. Prince of Songkla University, Thailand. Unpublished data.

Bijina, B., 2006. Isolation, purification and characterization of protease inhibitor from Moringa oleifera. PhD D thesis, University of Science and Technology, Cochin.

Bijina B, Chellappan S, Krishna JG, Basheer SM, Elyas KK, Bahkali AH Chandrasekaran M (2011). Protease inhibitor from Moringa oleifera with potential for use as therapeutic drug and as seafood preservative. Saudi J. Biol. Sci. 18:273-281.

Bode W, Engh R, Musil D, Thiele U, Huber R, Karshikov A, Brzin J, Turk V (1988). The 2.0 Å X-ray crystal structure of chicken egg white cystatin and its possible mode of interaction with cysteine proteinases. EMBO J. 7:2593-2599.

Bradford MM (1976). Rapid and sensitive method for the quantitation of microgram quantities of protein utilizing the principle of protein-dye binding". Anal. Biochem. 72:248-254.

Chen XF, Zhang HT, Qi QY, Sun MM, Tao LY (2005). Expression of e-cadherin and nm23 is associated with the clinicopathological factors of human non-small cell lung cancer in china. Lung Cancer 48:69-76.

Corrion K, Scoffier S, Gernigon C, Cury F, d'Arripe-Longueville F (2010). Development and factorial validity of a moral disengagement in sport short scale. L'Encephale 36:495-503.

Englard S,Seifter S (1990). Precipitation techniques. Methods Enzymol. 182:285-300.

Enitan SS, Olley M, Uhunmwangho SE, Akele YR, Ehiaghe AF, Enitan CB (2014). Antibacterial Activity of Methanolic Leaf Extract of Plukenetia conophora Mull. arg. against Selected Bacteria Isolated from Urinary Tract Infection. Int. J. Microbiol. Appl. 1(1):1-10.

Ganiyu O, Mofoluso EM (2004). Nutritional evaluation of some Nigerian wild seeds. Nahrung 48:85-87.

Greenwood IA, Leblanc N, Gordienko DV, Large WA (2002). Modulation of icl(ca) in vascular smooth muscle cells by oxidizing and cysteine-reactive reagents. Pflugers Archiv: Eur. J. Physiol. 443:473-482.

Hartmann S, Kyewski B, Sonnenburg B, Lucius R (1997). A filarial cysteine protease inhibitor down-regulates T cell proliferation and enhances interleukin-10 production. Eur. J. Immunol. 27:2253-2260.

Jack NL, Cate NM, Rishipal RB, Hiba AB (2004). Inhibition of matrix metalloproteinase-I activity by the soybean Bowman-Birk inhibitor. Biotechnol. Lett. 26:901-905.

Lin WL, Dickson DW, Sahara N (2011). Immunoelectron microscopic and biochemical studies of caspase-cleaved tau in a mouse model of tauopathy. J. Neuropathol. Exp. Neurol. 70:779-787.

Lustigman S, Brotman B, Huima T, Prince AM, McKerrow JH (1992). Molecular cloning and characterization of onchocystatin, a cysteine proteinase inhibitor of Onchocerca volvulus. J. Biol. Chem. 267: 17339-17346.

Martinez-Vicente M,Sovak G,Maria Cuervo A (2005). Protein degradation and aging. Exp. Gerontol. 40:622-633.

Mello GC, Oliva MLV, Sumikava IT, Machado OLT, Marangoni S,Matos DGG (2000). Purification and characterization of new trypsin inhibitor from Dimorphandra mollis. J. Protein Chem. 20:625-632.

Murachi T (1970). Bromelain enzymes. Methods Enzymol. 19:274-283.

Nimmrich V,Reymann KG,Strassburger M,Schöder UH,Gross G,Hahn A, Schoemaker H, Wicke K,Möller A (2010). Inhibition of calpain prevents NMDA-induced cell death and b-amyloid-induced synaptic dysfunction in hippocampal slice cultures. Br. J. Pharmacol. 159(7): 1523-1531.

Ohno S, Emori Y, Imajoh S, Kawasaki H, Kisaragi M, Suzuki K (1984).

Evolutionary origin of a calcium-dependent protease by fusion of genes for a thiol protease and a calcium-binding protein?. Nature 312 (5994):566-570.

Outchkourov NS, Rogelj B, Strukelj B, Jongsma MA (2003). Expression of sea anemone equistatin in potato. Effects of plant proteases on heterologous protein production. Plant Physiol. 133:379-390.

Rawlings ND, Barrett AJ (1990). Evolution of proteins of the cystatin superfamily. J. Mol. Evol. 30:60-71.

Rawlings ND, Morton FR, Kok CY, Kong J, Barrett AJ (2008). MEROPS: the peptidase database. Nucleic Acids Res. 36:320–325.

Redzynia I, Ljunggren A, Bujacz A, AbrahamsonM,Jaskolski M, Bujacz G (2009). Crystal structure of the parasite inhibitor chagasin in complex with papain allows identification of structural requirements for broad reactivity and specificity determinants for target proteases. FEBS J. 276:793-806.

Rossomando EF (1990). Ion-exchange chromatography. Methods Enzymol. 182:309-317

Singh RR,Rao AG (2002). Reductive unfolding and oxidative refolding of a Bowman-Birk inhibitor from horsegram seeds (Dolichos bifloru): Evidence for 'hyperreactive' disulfide bonds and rate-limiting nature of disulfide isomerization in folding. Biochem. Biophys. Acta 1597:280-291.

Soxhlet F (1879). Die gewichtsanalytische Bestimmung des Milchfettes, Polytechnisches J. (Dingler's) 232: 461.

7

Maternal oxidative stress and enzymatic antioxidant status in premature rupture of membranes

Jafar Maleki[1], Ladan Haghighi[2], Mohsen Korani[1], Soudabeh Fallah[1*], Mohsen Firozrai[1] and Morteza Seifi[1]

[1]Department of Biochemistry, Tehran University of Medical Sciences, Hemmat High Way, Tehran, Iran.
[2]Tehran University of Medical Sciences, Shahid Akbarabadi Hospital, Tehran, Iran.

Premature rupture of the fetal membrane is a major cause of preterm birth and it is associated with infant morbidity. To examine the relationship between maternal oxidative stress (OS) and antioxidant statue and premature rupture of membranes (PROM). 30 pregnant women with PROM and 30 normal pregnant women were compared for maternal malonaldihyde concentration, erythrocyte glutathione (GSH) concentration, glutathione peroxidase (G-PX) and catalase (CAT) activities, using a cross sectional study and spectrophotometer method. Unpaired t test and Chi-square test were used for comparison of variables between controls and patient groups. A significant reduction was found in fthe GSH concentration of the PROM subjects when compared to the controls (6.40±0.43 µmol/g of Hb versus 7.28±0.40 µmol/g of Hb, P <0.001). The G-PX and CAT activities of the PROM population were significantly lower than controls (35±9.41 versus 44.17±11.92 U/g Hb, P=0.01 and 149.83±31.86 U/g of Hb versus 172.07±47.14U/g of Hb, P<0.01, respectively). The maximal rate of oxidation (Vmax) was shorter in the PROM group when compared to the control group (0.0068±0.00003 versus 0.008±0.00003, OD245 nm/min, P=0.01), whereas the maximal accumulation of absorbing products (ODmax) of the PROM subjects was higher than in the controls (0.73±0.013 versus 0.68±0.017 OD245 nm, P=0.01). The lag phase, reflecting resistance of serum lipids to oxidation, was different in the PROM group when compared to the control group (47.90±2.13 versus 53.53±2.54 min, P=0.01). The maternal malondialdehyde (MAD) percentage of the PROM group was higher than controls (75.62±3.73% versus 63.80±9.32%, P=0.001). Present study revealed that PROM is associated with decreased maternal erythrocyte antioxidant activities. As opposed to term labor, PROM is associated with increased maternal systemic OSwhen compared to normal pregnant women. The role for OSin preterm PROM warrants further studies.

Key words: Oxidative stress, antioxidant, premature, rubture, membrane.

INTRODUCTION

Premature rupture of the fetal membrane (PROM) is a major cause of preterm birth, accounting for 30 to 40% of all preterm births (Nourse and Steer, 1997) and its association with infant morbidity and mortality (Nourse and Steer, 1997; Woods, 2001). PROM results initially from damage to collagen in the chorioamnion leading to a tear in the membrane (Nourse and Steer, 1997). Multiple epidemiological and clinical factors are considered to be promoters of PROM (Ramkumar and Stephen, 2007). Numerous risk factors are associated with PROM. There appears to be no single etiology of preterm PROM

*Corresponding author. E-mail: s-fallah@tums.ac.ir or soudabeh.fallah@yahoo.com.

(Tanya et al., 2006). It has been shown that issue-damaging molecules called reactive oxygen species (ROS) are capable of damaging collagen in the chorioamnion that could lead to PPROM. This hypothesis was supported by epidemiological studies linking clinical conditions known to produce ROS or reduce antioxidant protection to Preterm premature rupture of membranes (PPROM), by *in vitro* studies in which membrane segments exposed to ROS exhibited tissue alterations consistent with PPROM, and by clinical studies showing that chorioamnion and amniotic fluid samples obtained from PPROM patients exhibit excessive collagen degradation (Nourse and Steer, 1997). It has been shown that (Ofer et al., 2007) oxygen free radicals, which are by products of inflammatory cells, increase the activity of matrix metalloproteinases in human fetal membranes. They suggested that oxygen free radicals, which are by products of inflammatory cells, increase the activity of matrix metalloproteinases in human fetal membranes.

This may cause degradation of the extracellular matrix, leading to rupture of the amniotic membranes. PROM is associated with elevated activity metalloproteinase in the amniotic fluid and amniochorionic membranes to which OSmay contribute (Nourse and Steer, 1997). It has also been shown (Fainaru et al., 2002) that active labor is associated with oxidative stress. However it is not known whether OStriggers labor or is an epiphenomenon and a byproduct of the labor process. The damage which is inflicted to tissues by reactive oxygen species can be widespread. ROS are capable of initiating lipid peroxidation, increasing intracellular free Ca^{2+}, damaging deoxyribonucleic acid (DNA), releasing destructive catalytic enzymes and damaging cell membranes (Nourse and Steer, 1997). Clinical conditions, in addition to infection, which frequently are linked to PROM fit with an ROS-induced mechanism of tissue damage. (Nourse and Steer, 1997). Lipid peroxidation is normal phenomenon that occurs continuously at low level in all animals. These peroxidation reactions are in part toxic to cells and cell membranes; however they are normally controlled by countervailing biological mechanisms (Akihito et al., 2006). Antioxidant enzymes G-PX, CAT, superoxidase dismutase (SOD) and GSH appear to be the key antioxidants that provide cell protection against oxidant agents generated in different conditions.

The GPx and SOD enzymes acting as free radical scavengers limit the effects of oxidant molecules in tissues and oxidative injury. These enzymes work together to eliminate active oxygen species and small deviations in their physiological concentrations may have an adverse effect on the resistance of cellular lipids, proteins and DNA to oxidative damage (Fallah et al., 2011). The role of antioxidants to protect the chorioamnion from ROS damage has been demonstrated in one *in-vitro* study. The result of a study conducted by Woods (2001) showed that a prospective, randomized blinded trial of antioxidant therapy during pregnancy is needed to evaluate this approach for the prevention of PPROM (Nourse and Steer, 1997). Normally a balance exists between production and elimination of ROS. OS occurs when prooxidants exceed antioxidants and the balance between antioxidant defenses is disturbed (Singh et al., 1998). SOD, G-Px, CAT and GSH constitute a team of antioxidant which provide defense against free radical mediated injury. The present study was conducted to assess the erythrocyte antioxidant enzymes G-Px, CAT activities and GSH concentration in PROM. In this study we tested "*in vivo*" the hypothesis that the antioxidant reduction and OS elevation associated with preterm PROM. In addition to measurement of the malonaldihyde levels as an OS index, we used an assay to determine serum lipid oxidizibility in pregnant women experiencing PROM and compared to pregnant women with intact membranes for investigation whether PROM is associated with OS.

PATIENTS AND METHODS

The study group consisted of 30 pregnant women aged 30±2.1 between 24 to 34 gestational weeks with PROM and 30 women aged 30±3.01 with normal pregnancy and intact membranes who had been admitted to the delivery unit of Shahid Akbar Abadi hospital of Southern Tehran because of a threatening preterm delivery with intact fetal membranes. Women in both groups were non-smokers, and were not taking any medications aside from iron supplementation, they were in good general health with no history of prior adverse pregnancy outcomes including preterm labor and delivery or premature rupture of the membranes, and known date of last menstrual period; more than 16 weeks of gestation at the beginning of the study. Women in both groups were not experiencing uterine contractions and did not show clinical or laboratory signs of chorioamnionitis (that is, uterine tenderness, systemic fever, fetal tachycardia or elevated white blood cell count). The groups were matched for age and gestational age and were of similar height and weight (Not shown). Gestational age was determined by the best obstetric estimate based on a combination of the last menstrual period dating and the earliest available ultrasonographic examination.

Exclusion criteria were as follows: Chronic diseases, (hypertension, diabetes, renal or cardiac diseases), genital tract anomalies of the mother, genetic or anatomical defects of the fetus, previous preterm deliveries. PROM diagnosis was identified by estimation the value of insulin-like growth factor-binding protein-1 (IGFBP–1) in cervical secretion in women with symptoms of preterm delivery and to investigate correlation of this test to the Bishop score in prediction of the preterm delivery. The concentration of IGFBP-1 in cervical mucus was estimated by an Actim Partus test (Medix Biochemica, Finland) according manufacturer's instructions. The test was considered positive if the concentration of IGFBP-1 in cervical mucus was higher than 10 µg/L. All experimental procedures were approved by the Human

Table 1. Antioxidant activities in women with PROM and women with normal pregnancy.

	Normal pregnancy	PROM	P
CAT (κ/g of Hb)	172.07±47.14	149.83±31.86	< 0.001
G-Px (U/g of Hb)	44.17±11.92	35.93±9.41	< 0.005
GSH (μmol/g of Hb)	7.28±0.40	6.40±0.43	< 0.001

Values are means±sd.

Table 2. Kinetic parameters of copper induced peroxidation in women with PROM and women with normal pregnancy.

	P	PROM (n=30)	Normal pregnancy (n=30)
Lag preceding oxidation (min)	0.01	47.90±2.13	53.53±2.54
t max (min)	0.01	90.67±3.48	97.80±3.13
V max (OD 245 /min)	0.01	0.0068±3×10-4	0.008±10-4
OD max (OD 245 nm)	0.01	0.73±0.013	0.68±0.017
OD initial (245 nm)	0.01	1.38±0.019	1.33±0.019

Values are mean±sd.

Ethics Committee of the Iran Health Ministry. Venous blood was drawn from each woman and serum was prepared, frozen immediately and stored at -80° C. Blood drawn from women in the PROM group was obtained no longer than 6 h after the onset of membrane rupture. The Hemoglobine (Hb) concentration was determined in 10-fold diluted hemolysate (by the cyanmethemoglubine) with Drabkin's reagent (200 mg $FeCN_6k_3$ + 50 mg CNk /Lit).

The erythrocyte antioxidant enzymes G-PX and CAT activities were assayed as described by Pagila and Valentin (1967) and Aebi (1984), respectively. The enzyme activities were reported as unit per gram of Hb. For G-PX activity, the decrease in nicotinamide adenine dinucleotide phosphate-oxidase (NADPH) absorbance was followed at 340 nm using a spectrophotometer (Cecil CE 9050 Modle, Germany). CAT activity was determined at 25°C by decrease in absorbance of HO at 240 nm Aebi (1984). The erythrocyte GSH concentration was measured spectrophotometrically by the method of Buteler et al. (1963). GSH concentration was measured in 5- sulphosalycilic acid-deproteinized samples. The rate of 2-nitro-5-thiobenzoic acid formation, which is proportional to the GSH present, is followed at 412 nm. Samples were assayed rapidly to minimize GSH oxidation (Beutler et al., 1963).

For OS assay, kinetics of copper-induced oxidation of serum lipids *ex vivo* for evaluating of oxidizibility of blood lipids (Ofer et al., 2007). The kinetic profiles were analyzed in terms of the lag preceding oxidation reflecting the resistance of serum lipids to oxidation, the maximal rate of accumulation of absorbing products (Vmax) as computed from the first derivative of the time course of absorption and the maximal accumulation of absorbing products (OD max) (Ofer et al., 2007). A method based on measurement of the oxidation products such as MAD as a plasma index of lipid peroxidation (Purnima et al., 2006) was also used for OS measurement. The absorbance of the sample was determined at 535 nm (using a spectrophotometer Ultrosec 300, Pharmacia Biotech England double-beam) against a blank that contained all the reagents minus the serum. The MAD concentration of the samples could be calculated using an extinction coefficient of $1.56\times10^5\ M^{-1}\ cm^{-1}$ (Purnima et al., 2006).

Statistical analysis

Statistical analysis was performed using the Statistical Package for Social Sciences (SPSS Inc., Chicago, IL, USA). The Kolmogorov-Smirnov test was used to assess the normality of data distribution. Normally distributed variables were presented as mean and standard deviation (±SD). The t test for independent samples was used for comparison of continuous variables, as appropriate. The difference of proportions between COC users and non-COC users was assessed by $\chi 2$ test. A p value ≤0.05 was considered significant.

RESULTS

Results of present investigation are shown in Tables 1 to 3. As it shown in Table 1 the erythrocyte G-PX activity of the PROM population was decreased significantly (-18%, P<0.005) as compared to controls. The mean G-PX activity of the PROM and normal pregnant groups was 35.93±9.41 and 44.17±11.92 U/g of Hb, respectively. The results of same comparison for erythrocyte CAT activity of the controls (172.07±47.14 U/g of Hb) and PROM group (149.83±31.86 U/ g of Hb) revealed a significant decrease (-20%, P≤0.001) in the PROM subjects. The erythrocyte GSH level of PROM subjects (6.40±0.43 μmol/g of Hb) was significantly lower (-12%, P≤0.001) than that in the normal pregnant group (7.28±0.40 μmol/g of Hb).

As shown in Table 2 the maximal rate of oxidation (Vmax) was shorter in the PROM group when compared

Table 3. Malonaldihyde concentration in women with PROM and women with normal pregnancy.

	Normal pregnancy (n=30)	PROM (n=30)	P
MDA-I(nmol/gofHb)	395.47±89.84	561.63±76.54	0.001
MDA-C(nmol/gof Hb)	257.73±85.94	427.17±76.54	0.001
MDA%	63.80±9.32	75.62±3.73	0.001

Values are means ± sd. MAD-I = MAD in the presence of CAT inhibitor (sodium azide as CAT inhibitor) MAD-C = MAD in the absence of CAT inhibitor. MAD%= the maximum of released MAD.

to the control group (0.0068±0.00003 versus 0.008±0.00003, OD245 nm/min, P=0.01), whereas the maximal accumulation of absorbing products (ODmax) of the PROM subjects was higher than in the controls (0.73±0.013 versus 0.68±0.017 OD245 nm, P=0.01). The lag phase, reflecting resistance of serum lipids to oxidation, was different in the PROM group when compared to the control group (47.90±2.13 versus 53.53±2.54 min, P=0.01). The maternal MAD percentage (Table 3) of the PROM group was higher than in the controls (75.62±3.73% versus 63.80±9.32%, P=0.001). All the data were analyzed by "SSP software" (version 16, for Windows). Continuous variables were compared by the two-tailed t test and two-tailed X^2. P<0.05 was considered significant.

DISCUSSION

The rupture of the fetal membranes before the onset of regular uterine contraction at preterm labor takes place in some 8 to 10% of pregnancies. This can result in an increased rate of maternal and fetal infection (Bryant-Greenwood and Millar, 2000). The results of present study demonstrated that the maternal antioxidant levels in erythrocyte decrease significantly in PROM subjects as compared to the normal pregnant population. Several studies have attempted to evaluate the antioxidants of maternal healthy pregnant and PROM pregnant women (Akihito et al., 2006). There are conflicting reports regarding to antioxidants levels changes throughout the gestation. G-PX enzyme, a component of antioxidant system, is decreased during pregnancy (Akihito et al., 2006). However, Akihito et al. (2006) showed an elevation in SOD activity throughout the normal pregnancy. In a study, Purnima et al. (2006) revealed that the maternal GSH concentration at PROM labor was significantly lower than those at normal labor (Purnima et al., 2006; Longini et al., 2007). It has been seen that lipid peroxidation products are more in case of preterm labor. GSH/G-PX system is very important in catabolizing H2O2 (Purnima et al., 2006).

They demonstrated that the G-PX activity of patients with preterm labor was higher than that in the controls. A rise in the G-PX activity was reported in the case of preterm labor patients which may be predominant cause of GSH depletion (Purnima et al., 2006). They also showed that the SOD and CAT activities of preterm labor patients were lower than those in the control group significantly (Purnima et al., 2006). Support for the concept that ROS are involved in the pathogenesis of PROM is derived from several lines of investigation. Generation of tissue-damaging molecules, called ROS, may impair the physical integrity (elasticity and strength) of amniotic epithelium and collagen in the amnion and chorion, thus resulting in PROM (James et al., 2001). Results of the present investigation are shown in Table 1. A significant decrease was found in the erythrocyte GSH concentration of PROM subjects as compared to the normal pregnant group. The mean erythrocyte G-PX and CAT activities in PROM group were lower than those in controls. The protective antioxidant mechanisms are complex and multifactorial. The susceptibility of cells to OS is a function of overall balance between the degree of the OS and oxidant defense capability (Akihito et al., 2006).

It is suggested that the decrease in antioxidant activities probably occur in response to an increase in the OS due to an elevation in the maternal lipid peroxidation levels, which might be resulting in PROM. Present study, which examined maternal lipid peroxidation in the healthy pregnant and PROM subjects, demonstrated that the MAD concentration of PROM group was significantly higher than those in the normal pregnant. These findings make it likely that the uncontrolled lipid peroxidation caused by ROS, which are produced in the consequences of the tissue reoxygenation that may occur during pregnancy, which caused PROM. It seems that in PROM, a possible OS is responsible for the induction of protective mechanisms through an increased consumption of antioxidants. Under physiological conditions, antioxidant defense systems have evolved to counterbalance their toxic actions by limiting the amount of lipid peroxides that can be formed (Martin et al., 2008). It is suggested that the PROM probably caused by oscillated oxygenation of both maternal and fetal tissue during pregnancy. Furthermore, it has also been confirmed that the ischemic-reperfusion in human and other species lead to production of free radicals (Akihito et al., 2006) and increased levels of ammonia may have toxic effects by generating free radicals (Fainarno et al., 2007; Loverro et al., 1997). The results of the present study indicate a disturbed antioxidant balance in case of PROM.

Conclusion

In this study we defined that the OS in pregnancy may enhance the risks of PROM. We indicated that women experiencing PROM are subject to OS. We propose that when an OS develops early in pregnancy, intrauterine growth restriction or PROM or both may occur depending on its entity and length. A topic for future prospective clinical trials is whether dietary supplementation with antioxidants may protect the fetal membranes and decrease the risk of PROM and preterm delivery As opposed to term labor, PROM is associated with increased maternal systemic OS when compared to normal pregnant women. The role for OS in preterm PROM warrants further studies. Our results also revealed a significant decrease in antioxidant enzymes (GP-X and CAT) activities and GSH concentration of PROM group when compared to the controls. It is proposed that an imbalance existed between production and elimination of ROS in the PROM subjects and suggested the need for further study.

Study limitations

A few limitations of the present study deserve comment. First a common limitation of this study was small sample size and these observations must be confirmed in a larger sample of patients with more analysis works. Other postulated limitation involving antioxidants and OS levels, variability between individuals in the diet feeding and cellular utilization and endogenous synthesis of them, the associations observed may still to some degree reflect the effect of varying metabolic responses to dietary fat and environmental situation. The other point might be considered in interpretation of these results as the role of genetic variations in adipose tissue fatty acid percentages, which we have not studied in the current project and might have affected our results. Our study is also limited by the non-representative nature of our study sample, because the pregnant subjects of southern Tehran who referred to Shahid Akbarabadi hospital have low social and economic status. This fact together with the promising results of experimental studies suggests the possibility that antioxidant replacement might become a new pharmacological approach to protect probably elevation of lipid peroxidation, antioxidant reduction and resultant PROM of pregnant women. Therefore further studies are required to elucidate mechanism underpinning PROM risk associated with decrease antioxidant and OS increase. Further, this observation interplay between genetic and environmental factors must be thoroughly considered in order to evaluate the etiological role of OS and antioxidants in PROM.

ACKNOWLEDGEMENTS

The authors wish to thank the Research Deputy of Tehran University of Medical Sciences for the funding of the project as well as Shahid Akbarabadi Hospital in southern Tehran for providing the samples. We acknowledge that because of relatively small number of subjects.

REFERENCES

Aebi H (1984). Catalaze invitro. Methods Enzymol., 105: 121-126.

Akihito NA, Astuko OY, Hirobumi KO, Hirobumi AS, Akishige YO, Tatsuo KO, Tsutomu AR (2006). Changes in maternal lipidperoxidation levels and antioxidant enzymatic activities before and after delivery. J. Nippon Med. Sch., 67: 434-437.

Beutler E, Duron O, Kelly BM (1963). Improved method for the determination of blood glutathione. J. Lab. Clin. Med., 61: 882-888.

Bryant-Greenwood GD, Millar LK (2000). Human fetal membranes: their preterm premature rupture. Biol. Reprod., 63(6): 1576-1579.

Fainarno A, Pinchuk I, Lichtenberg D, Lessing JB (2007). Serum lipid oxidizibility in term premature rupture of the membranes. Eur. J. Obstet. Gynecol. Reprod. Biol., 131(1): 28-31.

Fainaru O, Almog B, Pinchuk I, Kupferminc MJ, Lichtenberg D, Many A (2002). Active labour is associated with increased oxidizibility of serum lipids ex vivo. Bjog, 109(8): 938-941.

Fallah S, Valinejad SF, Firoozrai M (2011). Effect of contraceptive pills on the activity status of the antioxidant enzymes glutathione peroxidase and superoxide dismutase in healthy subjects. Contraception, 83: 385-389.

James R, Woods JR, Plessinger MA, Richard KM (2001). Vitamins C and E: Missing links in preventing preterm premature rupture of membranes. Am. J. Obstet. Gynecol., 185(1): 5-10.

Longini M, Perrone S, Vezzosi P, Marzocchi B, Kenanidis A, Centini G, Rosignoli L, Buonocore G (2007). Association between oxidative stress in pregnancy and preterm premature rupture of membranes. Clin. Biochem., 40(11): 793-797.

Loverro G, Greoo P, Capuano F, Carone D, Carmio G, Sevaggi L (1997). Liopoperoxidation and antioxidant enzymes activity in pregnancy complicated with hypertension. Eur. J. Obstet. Gynecol. Rerod. Biol., 70: 123-127.

Martin DR, Eleanor CK, John MW, Michael JD (2008). Oxidative damage to extracellular matrix and its role in human pathologies. Free Radic. Biol. Med., 44: 1973-2001.

Nourse CB, Steer PA (1997). Perinatal outcome following conservative management of mid-trimester pre-labor rupture of the membranes. J. Paediatr. Child Health, 33: 125-130.

Ofer F, Almog R, Pinchuk I, Lichtenberg D, Lessing JB, Kupferminc MJ (2007). Serum lipid oxidizibility in term premature rupture of the membranes. Eur. J. Obstet. Gynecol. Reprod. Biol., 131: 28-31.

Pagila D, Valentin W (1967). Studies on the quqntitative and quantitative characterization of erythrocyte glutathione peroxidase. J. Lab. Clin. Med., 70: 158-69.

Purnima DS, Tripti S, Maheshwar RS (2006). Oxidative stress and

antioxidants in preterm labor. Biomed. Res., 17(1): 41-43.
Ramkumar M, Stephen J (2007). Infection and the role of inflammation in preterm premature rupture of the membranes. Best Pract. Res. Clin. Obstet. Gynaecol., 21(3): 467-478.
Singh SK, Tandon A, Kumari S, Ravi RN, Ray GN, Batra S (1998). Changes in Anti-oxidant enzymes and lipid peroxidation in Hyaline membrane Disease. Indian J. pediatr., 65(4): 609-614.
Tanya M, Medina M, Ashley H (2006). Preterm Premature Rupture of Membranes: Diagnosis and Management. Am. Fam. Physician, 73(4): 659-664.
Woods JR (2001). Pathobiology oxidant stress, angiogenesis and neoplasia Reactive. Oxygen Species and Preterm Premature Rupture of Membranes—A Review. Placenta, 22(15): S38-S4.

Effects of sound waves on the enzyme activity of rice-*koji*

Noriaki Saigusa , Seika Imayama and Yuji Teramoto

Department of Applied Microbial Technology, Faculty of Biotechnology and Life Science, Sojo University; 4-22-1 Ikeda, Nishi Ward, Kumamoto 860-0082, Japan.

Rice-*koji* is an important agent for the saccharification of fermented food production in Japan. In *koji*-making, temperature, humidity, light and oxygen are important environmental factors. However, there is no report of *koji*-making controlled by sound wave. In this study, focusing on the enzyme activity of rice-*koji*, we confirmed the relationship between enzyme activity and frequency of sound waves. Here, the enzyme activities of rice-*koji* cultured with no sound and sound waves at seven different frequencies (1.0, 2.5, 6.3, 8.0, 10.0, 12.5 and 16.0 kHz) were compared. Using two types of polished rice from different production areas, the activities were investigated. In this study, we investigated the enzyme activity of four types of rice-*koji* and found frequencies where the enzyme activity was significantly different to activity with no sound. In particular, the frequency of significant difference was investigated to confirm both rice. In the results, due to the irradiation frequency of 6.3 kHz, glucoamylase activity of both rice-*koji* decreased significantly ($P<0.01$) compared to the silent control. Further, due to the irradiation frequency of 1.0 kHz, acid protease activity confirmed a significant difference between the silent control and a frequency of 1.0 kHz ($P<0.01$). That is, as the activity of one of rice-*koji* increased, that of another rice-*koji* decreased. However, the activity of α-amylase and acid carboxypeptidase was not affected at any frequency. These results suggest that sound waves, particularly at frequencies of 1.0 and 6.3 kHz, have marked effects on the specific enzymatic activity of rice-*koji*.

Key words: Frequency of sound waves, sound wave, rice-*koji*, enzyme activity, *koji*-making.

INTRODUCTION

In the production of fermented food in the Orient region, rice-*koji*, which is prepared by culturing fungal conidia on cooked rice, is an important saccharifying agent. During the production of rice-*koji*, temperature, humidity and oxygen are key factors affecting quality (Murakami, 1988). Yanagiuchi et al. (1993) reported that the supply of carbon dioxide affected the fatty acid composition, color of *koji* and components of *koji* extract. On the other hand, Iemura et al. (1996) reported that *koji*-making sealed in *toko* period, that is, incubation period, was higher in the activities of α-amylase and glucoamylase, while, lower in the activities of acid protease and acid carboxypeptidase than *koji*-making not sealed up in *toko* period. Furthermore, it was recently confirmed that the germination of conidia is affected by light and the possibility of using light to control the production of rice-*koji* was demonstrated (Hatakeyama and Kitamoto, 2008). Yan et al. (2014) reported the optimum conditions for glucoamylase, α-amylase and acid protease of *Aspergillus oryzae koji*. Glucoamylase and α-amylase activity could reach maximum under the following conditions; inoculation temperature at 35°C, 0.25% inoculation quality, culture humidity at 75%, culture temperature at 37°C and 56 h culture time. On the other hand, acid protease could reach maximum when the culture temperature was 33°C. Zhang and Gao (2011) reported that the optimum condition for amino peptidase activity in wheat bran *koji* making were water ratio 100%, raw material cooking for 21 min and *koji* making for 44 h. However, there are no reports for utilization of sound waves during the rice-*koji* making process.

In Japan, sound waves have been used in various processes, including the aging of spirits, particularly *shochu*, cultivation of fruit and vegetables and even the breeding of cattle. For *shochu*, acoustic aging of black sugar-*shochu* has been reported. For vegetables, growth promotion has been reported. For livestock, relaxation effects have been reported. However, there are no reports as paper. It is only been empirically applied. The reason for this is that elucidation of scientific mechanism of sound effects is thought to be difficult. The mechanism is not understood although it was been suggested that sound waves could affect the rate of proton transfer in proteins (Barnes et al., 1985). Furthermore, it was reported that sound waves at a frequency of 2.0 kHz have been shown to affect the sugar composition of plants (Satoh and Tanisu, 2013). Based on these findings, we speculated that the production of rice-*koji* could be optimized using sound waves and thus attempted to determine the critical sound wave frequencies that would specifically impact the enzyme activity of rice-*koji*.

MATERIALS AND METHODS

Materials

Rice-*koji* was prepared using commercially purchased polished rice (*Koshihikari*) produced in Niigata and Miyagi Prefectures, Japan. Hereafter, the former was expressed as N-rice, the latter was expressed M-rice. Tane-*koji* (*Aspergillus oryzae*) was purchased from Kawachi Genichiro Shoten Co., Ltd., Kagoshima, Japan.

Rice-*koji* making

Rice-*koji* was prepared according to the method of Saigusa and Ohba (2007). Briefly polished rice (200 g) was soaked in water at 15°C for 20 min and the water was then drained over a 2 h period at 15°C. The rice was then steamed for 40 min in a pot-type steamer, removed, and allowed to stand until the temperature decreased to 40°C. Next, 0.2 g of Tane-*koji* spores was inoculated onto the steamed rice, which was then mixed to uniformly disperse the spores. Except for the steaming process, the cultivation procedure was performed under sterile conditions. After inoculation, the steamed rice was transferred to a Petri dish of 152 mm in diameter and compacted with a spoon. A piece of filter paper was inserted inside the cover, and the plate was incubated at 30°C for 20 h with no sound waves. A further 20 h to irradiate sound wave, rice-*koji* was moved to a tray which covered gauze and containing water in bottom. Incubation was performed in incubator built in speaker that is connected to the audio generator MINIRATOR MR2 (NIT Japan Co., Ltd, Tokyo, Japan) and instrument for amplification. The rice-*koji* was incubated in the presence of sound waves with frequencies ranging from 0 to 16.0 kHz. The sound power level of audio generator was keep -50 dBv. To keep moisture, making-*koji* was performed in plastic bag. The effects of sound wave frequency on the enzyme activity of glucoamylase, α-amylase, acid protease, and acid carboxypeptidase as main enzymes of rice-*koji* was then investigated. Rice -*koji* was prepared one each time. After extraction of enzymes from three places at random from both rice-*koji*, enzyme activity was measured.

Enzyme activity of rice-*koji*

Glucoamylase

Glucoamylase produces glucose from starch or oligosaccharides. Glucoamylase activity was measured with G2-β-PNP (4-nitrophenyl-β-maltoside) and PNPG (4-nitrophenyl-α-glucoside) as substrates using a commercially available kit (Kikkoman Co., Ltd., Chiba, Japan).

α-Amylase

α-Amylase produces oligosaccharides from starch. The α-amylase activity was measured with N3-G5-β-CNP (2-chloro-4-nitrophenyl 65-azido-65-deoxy-β-maltopentaoside) as a substrate using a commercially available kit (Kikkoman Co., Ltd.).

Acid carboxypeptidase

Acid carboxypeptidase produces amino acids from proteins and oligopeptides. The acid carboxypeptidase activity was measured with Cbz-Tyr-Ala (carbobenzoxy-L-tyrosine-L-alanine) as a substrate using a commercially available kit (Kikkoman Co., Ltd.).

Acid protease

Acid protease produces oligosaccharides from proteins. The acid protease activity was measured with casein as a substrate according to the method described by the National Tax Administration (1993).

Moisture content

The moisture content of steamed rice was measured using an FD-600 Infrared Moisture Determination Balance (Kyushu Kett Co. Ltd., Fukuoka, Japan).

RESULTS AND DISCUSSION

In this study, we aimed to prove that sound waves are

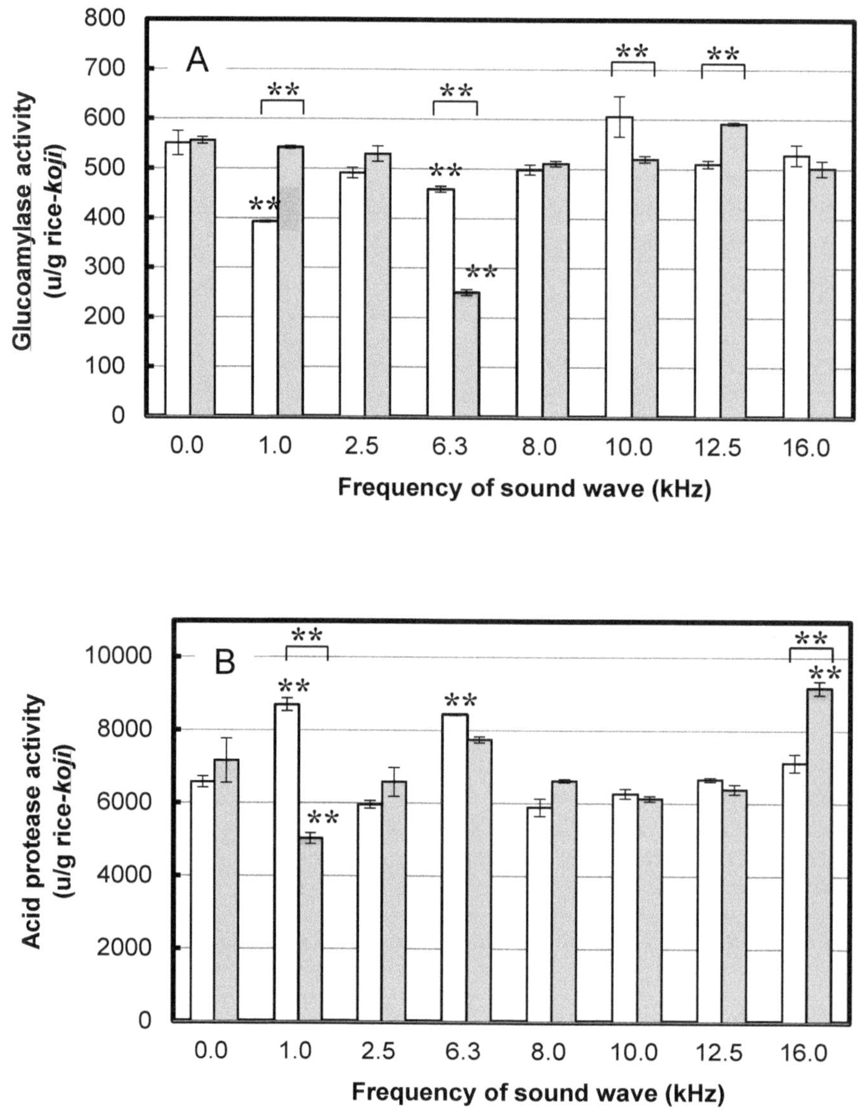

Figure 1. Effects of frequency of sound wave on enzyme activity of rice-*koji*. **A.** Glucoamylase activity. **B.** Acid protease activity. White bar showed N-rice, and black bar showed M-rice. Symbol (**) shows the significant difference (P<0.01). Especially, symbol (**) shown on the bar, indicated that there was significant difference between sound and no sound (0 kHz) condition (Scheffe's F test). Only the frequency of the indicated symbol (**) on the bar in both white and black, shows the frequency that affect the enzyme activity of rice-*koji*. Each value are presented as the mean ± S.D. (n=3).

one of the important environmental factors influencing the enzyme activities of rice-*koji*. To identify the critical frequencies affecting enzyme activity, rice-*koji* was produced in acoustic culture conditions using one of eight frequencies and the enzyme activities of the resulting rice-*koji* were compared.

For glucoamylase activity, significant difference was observed between no sound and a frequency of 6.3 kHz (Scheffe's F test). Also, glucoamylase activity of both rice-*koji* decreased significantly (P<0.01) (Figure 1A). On the other hand, for acid protease activity, significant difference was observed between no sound and a frequency of 1.0 kHz (Scheffe's F test). That is, as the activity of one of the rice-*koji* increased, another rice-*koji* decreased significantly (P<0.01) (Figure 1B). For α-amylase and acid carboxypeptidase activities, there were no frequencies at which significant differences were observed in both rice-*koji* (Scheffe's F test) (Figure 2A, B). In this experiment, three statistical analyses were carried as multiple comparison test. Results of significant different test in glucoamylase and acid protease activity were same results gotten for three methods, Scheffe's F

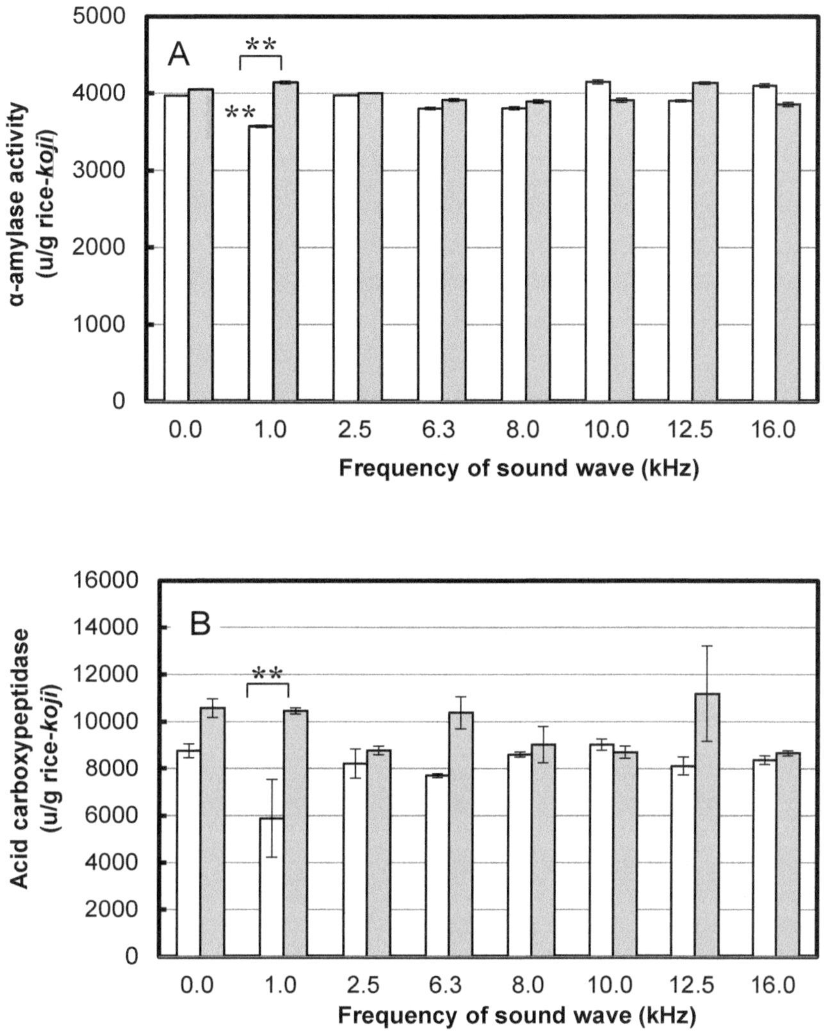

Figure 2. Effects of frequency of sound wave on enzyme activity of rice-*koji*. **A**. α-Amylase activity. **B.** Acid carboxypeptidase activity. White bar showed N-rice, and black bar showed M-rice. Symbol (**) showed the significant difference (P<0.01). Especially, symbol (**) shown on the bar, indicated that there was significant difference between sound and no sound (0 kHz) condition (Scheffe's F test). Only the frequency of the indicated symbol (**) on the bar in both white and black, shows the frequency that affect the enzyme activity of rice-*koji*. Each value are presented as the mean ± S.D. (n=3).

test, Tukey and Bonferroni/Dunn. Furthermore, from these analyses, we were unable to determine the cause and/or mechanism of the impact of frequency on rice-*koji* enzymes. To date, no studies have examined the effects of frequency on rice-*koji* enzymes. However, our findings do suggest that sound waves may be an important environmental factor influencing the rice-*koji* making process. In addition, we identified two sound wave frequencies (1.0 and 6.3 kHz) that might affect the enzyme activity of rice-*koji*.

We also observed that the rice-*koji* made from the two different sources responded differently to frequency of 1.0, 6.3 kHz. To determine the cause for this difference, the moisture content of rice-*koji* made from Niigata and Miyagi Prefectures was compared. Results show that the moisture content of cooked rice produced in Niigata Prefecture was 33.8%, while that of rice produced in Miyagi Prefecture was 40.7% (Table 1). Considering from the view point of vibration, it may be in close relationship between water and sound wave. In generally, the moisture content of steamed rice used in *sake* brewing is 27-32%. Furthermore, the more moisture content of steamed rice, enzyme activity decreases (Okazaki et al., 1979). However, in this study, it was confirmed that even when water content is high, enzyme activity increased significantly (P<0.01), by irradiating sound wave such as

Table 1. Water absorption and moisture content of rice used for rice-*koji*.

Rice	Water absorption rate (%) *1	Water absorption rate (%) *2	Moisture content (%) *3
N-rice	29.8	35.4	33.8
M-rice	35.0	40.2	40.7

*1 Absorption rate of water in raw rice. Calculation was performed as followed. [(weight of rice after drain – weight of rice before soak) / weight of rice before soak]×100. *2 Absorption rate of water in steamed rice. Calculation was performed as followed. [(weight of rice after steam – weight of rice before soak) / weight of rice before soak]×100. *3 Moisture content of rice after steam. Analysis method was described in the text. Both rice were produced in Sep. 2009 and purchased commercially in May 2010.

in glucoamylase (1.0, 12.5 kHz), acid protease (16.0 kHz), α-Amylase (1.0 kHz) and acid carboxypeptidase (1.0 kHz). Mishiro et al. (2000) reported that glucoamylase activity tends to increase according to the integration temperature of *ginjo koji*, that is starter for specially brewed Japanese sake. Okazaki et al. (1979) also reported the relationship between enzyme production and temperature. In that paper, they reported that α-Amylase and glucoamylase were produced well in optimum temperature of *koji* growth (37.5°C), while acid carboxypeptidase and acid protease were produced well under 35°C. In this study, in spite of constant temperature during *koji*-making, the difference of enzyme activity was observed by irradiation sound wave of various frequencies.

From these results, a possibility was suggested that a particular frequency might affect the enzyme production of rice-*koji*. In particular, with regard to 6.3 kHz, reduction of glucoamylase activity in the two types of rice-*koji* was also confirmed. Although the execution of repeat experiments is necessary, by the frequency of 6.3 kHz, a possibility of inhibition of the enzyme production or production of the active inhibitor is consider. Although the mechanism is not currently known about the effects of waves, overall impact of various factors, such as acoustic conditions, humidity, temperature and rice components (from the difference of water absorption rate), are considered.

As a result, enzyme activity does not depend at all on the high and low frequency. But, it was confirmed that each feature rice-*koji* was made with each frequency. From these results, the possibility was suggested that sound wave irradiation was one of the important environment condition for rice-*koji* making.

To further optimize the use of sound waves during the production of rice-*koji*, the underlying mechanism of frequency on enzyme production will need to be conclusively determined.

Conflict of interests

The authors did not declare any conflict of interest.

ACKNOWLEDGEMENT

The authors are grateful to Tsutoku Engineering Co., Ltd. for the assembly of the acoustic device.

REFERENCES

Barnes C, Evans JA, Lewis TJ (1985). Ultrasonic absorption of bovine serum albumin solutions in frequency range 200 kHz-1 kHz. J. Acoust Soc. Am. 78:6-11.

Hatakeyama R, Kitamoto K (2008). Koji mold Aspergillus oryzae response to light. J. Brew. Soc. Jpn 103:525-531.

Iemura Y, Katoh Y, Kataoka K (1996). Effects of koji-making sealed up in toko period or not sealed up in toko period on growth of mycelia and enzyme production. J. Brew. Soc. Jpn 91:439-445.

Mishiro Y, Katube T, Iwamoto M, Aoki N (2000). Effects of koji-making temperature and time on enzyme activity in ginjo-koji. J. Brew. Soc. Jpn 95:74-76.

Murakami H (1988). Koji gaku 2th ed. J. Brew. Soc. Jpn 317-319.

Okazaki N, Takeuchi K, Sugama S (1979). Effects of koji-making conditions on cell growth and enzyme production. J. Brew. Soc. Jpn 74:683-686.

Saigusa N, Ohba R (2007). Effects of koji production and saccharification time on the antioxidant activity of amazake. Food Sci. Tech. Res. 13:162-165.

Satoh Y, Tanitsu J (2013). Effects of sound wave on plant. Kagaku to Seibutsu, Nihon Nougei Kagaku Kai. 53:196-197.

The Method of the National Tax Administration (1993). J. Brew. Soc. Jpn. 4th ed. Tokyo, Japan.

Yan L, Jian GY, Ying YM, Hui BL, Qiu L, Hechuan W (2014). Optimization of conditions for glucoamylase, α-Amylase and acidic protease production by Aspergillus oryzae koji. J. Chem. Pharm. Res. 6:360-364.

Yanagiuchi T, Fukuda K, Kiyokawa Y, Wakai Y (1993). Influences of carbon dioxide on the making process of sake koji – Influences of various on the making process of sake koji (part 1). J. Brew. Soc. Jpn 88:475-481.

Zhang Y, Gao H (2011). Optimization of peptidase production conditions in koji making. Afr. J. Agric. Res. 6:4349-4356.

The distribution of cyanide detoxifying enzymes (rhodanese and 3- mercaptopyruvate sulphurtransferase) in different species of the family Cichlidae (*Tilapia zillii, Sarotherodon galilaeus* and *Oreochromis niloticus*)

Okonji Raphael Emuebie[1]*, Popoola Michael Olaoluwa[2], Kuku Adenike[1] and Aladesanmi Omolara Titilayo[1]

[1]Department of Biochemistry, Obafemi Awolowo University, Ile-Ife, Osun State, Nigeria.
[2]Department of Zoology, Obafemi Awolowo University, Ile-Ife, Osun State, Nigeria.

The cyanide metabolising enzymes (rhodanese and mercaptopyruvate sulphurtransferase) were estimated in different tissues of three species of tilapia (*Tilapia zilli, Sarotherodon galileus,* and *Oreochromis niloticus*) from two locations (Aiba and Osinmo reservoirs, both in the South-Western region of Nigeria). The enzyme activities were carried out by measuring the amount of thiocyanate produced by the two enzymes using specific substrates in each case. There was no significant difference in the activity of rhodanese in the tissues of the tilapia species from Aiba River. Mercaptopyruvate sulphurtransferase (3-MST) presented a different pattern of distribution with significant difference in the gut of the tilapia species. The study showed the activities of two cyanide detoxifying enzymes (rhodanese and mercaptopyruvate sulphurtransferase) in the different tilapia species indicating the presence of strong cyanide detoxifying mechanisms.

Key words: Rhodanese, mercaptopyruvate sulphurtransferase, cyanide, tilapia, detoxification.

INTRODUCTION

Cyanide is a highly toxic compound that is readily absorbed and causes death by preventing the use of oxygen by tissues (Egekeza and Oehme, 1980). This toxicant is widespread in the environment. Many naturally occurring substances as well as industrial products contain cyanide (Egekeza and Oehme, 1980). More than 2,000 species of plants are known to contain cyanogenic glycosides (Vennesland et al., 1982). It has been reported that ingestion of cyanogenic glycosides in forage crops can result in the death of grazing animals (Keeler et al., 1978). Many studies report the death of birds from cyanide poisoning through several routes, including exposure to cyanide salts or ingestion of cyanogenic plants (Wiemeyer et al., 1986). Numerous accidental spills of sodium cyanide or potassium cyanide into rivers and streams have resulted in massive kills of fishes, amphibians, aquatic insects, and aquatic vegetation (Leduc, 1984).

Two major enzymes (rhodanese: EC. 2.8.1.1., thiosulphate: cyanide sulphurtransferase and mercaptopyruvate sulphurtransferase: EC. 2.8.1.2) are involved in cyanide metabolism. Rhodanese is a ubiquitous enzyme that is known to be responsible for the biotransformation of cyanide to thiocyanate using thiosulphate as the donor substrate (Westley, 1981). The enzyme is well characterised among the cyanide detoxifying enzymes. The liver has always been considered to be the major source of rhodanese and is believed to be the major site of cyanide detoxification (Marrs and Ballantyne, 1987). However, Aminlari and Gilanpour (1991) and Aminlari and Shahbazi (1994) have shown that different parts of the

*Corresponding author. E-mail: okonjire@yahoo.co.uk, reokonji@oauife.edu.ng.

stomach in sheep and cattle and the proventriculus in chickens contain greater rhodanese activity than liver.

Unlike rhodanese, MST has not being fully characterised. The enzyme is also widely distributed in prokaryotes and eukaryotes and it is located mainly in the cytosol compared to rhodanese which is a mitochondrial enzyme (Jarabak and Westley, 1974; Jarabak and Westley, 1978; Westley, 1981; Nagahara et al., 1995). In the case of MST, mercaptopyruvate was found to be the donor substrate (Westley, 1980). Recent works have shown that different genes codes for the two cyanide detoxifying enzymes with speculations that they have the same origin (Nagahara et al., 1995; Nagahara et al., 1996). Reports have also shown that the pattern of distribution of these two enzymes might affect the physiological fate of cyanide (Ali et al., 2000). Also, the activity of the enzymes in a particular tissue/organ may reflect the ability of that tissue/organ to detoxify cyanide (Ali et al., 2000).

We report here the pattern of distribution of rhodanese and 3-MST in different internal sections of the tilapia species obtained at different locations. The results might indicate the parts of the tilapia species that are significantly involved in cyanide metabolism.

MATERIALS AND METHODS

Materials

Potassium cyanide, sodium thiosulphate, boric acid, sodium borate and mercaptoethanol were all purchased from Sigma Aldrich Chemicals, USA. All reagents used were of analytical grades.

Methods

Collection of samples

Different tilapia species were collected between April and August 2009, during a fishing tour of two reservoirs (Aiba and Osinmo reservoirs, both in the South-Western region of Nigeria). The fish species were collected alive from their normal environment and stored in an ice container before transporting to the laboratory where they were stored at temperature below 0°C until ready for use.

The fish species were *Tilapia zilli, Sarotherodon galileus* and *Oreochromis niloticus*. They were identified at the Fish Culture Laboratory, Department of Zoology, Obafemi Awolowo University, Ile-Ife, Nigeria.

Preparation of tissue extract

Prior to extraction, the tilapia species were dissected and the various tissues of interest (liver, guts and gills) were removed. These were later kept in the refrigerator until required. Tissue extracts were prepared by homogenising 10 g (w/v) of each tissue in 3 volume of homogenisation buffer (phosphate buffer, pH 7.2). The suspensions were centrifuged for 20 min at 4,000 rpm in a Microfield Centrifuge Model 800 D. The supernatants were used as the source of enzyme.

Protein and enzyme assays

Bradford method (1976) was used to measure the protein concentration of the enzyme using bovine serum albumin (BSA) as standard. Rhodanese was assayed by the method of Agboola and Okonji (2004). The reaction mixture contained 50 mM sodium thiosulphate, 50 mM potassium cyanide, 0.25 mM borate buffer, pH 9.4 and 10 µl of enzyme solution in a final volume of 1.0 ml. The reaction was carried out for 1 min at 37°C and stopped by adding 0.5ml 15% formaldehyde and 1.5 ml of Sorbo reagent (which is made up of ferric nitrate solution containing 0.025 g $Fe(NO_3)_3$,9 H_2O in 0.74 ml water and 0.26 ml concentrated nitric acid). Absorbance was measured at 460 nm. The unit of enzyme activity was defined as micromoles thiocyanate formed per minute at 37°C and pH 9.2.

Mercaptopyruvate sulphurtransferase activity was measured according to the modified method of Taniguchi and Kimura (1974). The reaction mixture in a final volume of 1 ml contained 0.38 M of Tris HCl buffer, pH 7.8, 0.5 M potassium cyanide, 0.3 M mercaptoethanol and 30 µl of enzyme solution. Absorbance was also measured at 460 nm. The unit of enzyme activity was defined as micromoles mercaptocyanate formed per minute at 37°C and pH 7.8

Statistical analysis

The results are presented as means ± SD. Data were analyzed by one-way ANOVA by using SAS/PC soft ware to examine whether there was any statistical difference among groups. Duncan multiple range test was used for paired comparisons. A *P* value less than 0.05 was considered statistically significant.

RESULTS

The activities of the two enzymes are presented in Tables 1-4. In most of the fishes, there were no significant differences in the values, though, the highest enzyme activities (rhodanese and 3-MST) were found in the livers of the three species of tilapia fish. Except for the livers of *S. galileus and T. zilli* where the 3-MST activity were lower compared to the activities found in the guts and gills of the fishes.

DISCUSSION

Fish and aquatic invertebrates are particularly sensitive to cyanide exposure. Free cyanide was reported to be the primary toxic agent in the aquatic environment. Environmentally relevant exposures to cyanide ions can cause stress, increase in mortality and place an appreciable metabolic load on fishes and other aquatic organisms (Eisler, 1991). Sulphurtransferases are widely distributed enzymes of prokaryotes and eukaryotes (Nakamura et al., 2000; Bordo and Bork, 2002). The enzymes catalyze the transfer of sulphane sulphur from a donor molecule, such as thiosulphate or 3-mercaptopyruvate, to a nucleophilic acceptor, such as cyanide or mercaptoethanol. Sulphurtransferases may also play a part in the management of the cytotoxicity of reactive oxygen species in aerobic tissues (Nandi et al., 2000). However the natural

Table 1. Mean rhodanese activity of tilapia species from Aiba reservoir.

Fish species	Liver	Gut	Gills
Oreochromis niloticus	0.132 ± 0.01 A	0.117 ± 0.04 A	0.113 ± 0.02 A
Sarotherodon galileus	0.298 ± 0.07 A	0.228 ± 0.05 A	0.118 ± 0.04 A
Tilapia zilli	0.146 ± 0.05 A	0.151 ± 0.01 A	O.127 ± 0.03 A
P- value	< 0.165	< 0.431	< 0.411

Values are means ±SD (n=15); Results are mean of five determinations.

Table 2. Mean 3-MST activity of tilapia species from Aiba reservoir.

Fish species	Liver	Gut	Gills
Oreochromis niloticus	0.157 ± 0.02 A	0.135 ± 0.02 A	0.140 ± 0.00 A
Sarotherodon galileus	0.100 ± 0.01 A	0.082 ± 0.01 B	0.075 ± 0.01 B
Tilapia zilli	0.196 ± 0.02 A	0.127 ± 0.01 A	0.144 ± 0.01 A
P-value	< 0.083	< 0.144	< 0.251

Values are means ±SD (n=15); Results are mean of five determinations

Table 3. Mean rhodanese activity of tilapia species from Osinmo reservoir.

Fish species	Liver	Gut	Gills
Oreochromis niloticus	0.199 ± 0.03 A	0.168 ±0.05 A	0.770 ± 0.06 A
Sarotherodon galileaus	0.326 ±0.03 A	0.222 ± 0.02 A	0.172 ± 0.08 A
Tilapia zilli	0.161 ± 0.04 A	0.503 ± 0.04 A	0.134 ± 0.07 A
P- value	< 0.112	< 0.218	< 0.156

Values are means ± SD (n = 15); Results are mean of five determinations.

Table 4. Mean 3-MST activity of tilapia species from Osinmo reservoir.

Fish species	Liver	Gut	Gills
Oreochromis niloticus	0.311 ± 0.00 A	0.145 ± 0.00 A	0.132 ± 0.02 A
Sarotherodon galileus	0.087 ± 0.01 B	0.136 ± 0.00 A	0.087 ± 0.01 B
Tilapia zilli	0.080 ± 0.01 B	0.139 ± 0.01 A	0.154 ± 0.03 A
P-value	< 0.491	< 0.622	<0.225

Values are means ±SD (n=15); Results are mean of five determinations.

sulphane donors and acceptors and the physiological functions of most Sulphurtransferases remain uncertain. The activities around the two rivers (Aiba and Osinmo reservoirs) have been described as those that enhance increase in environmental pollution especially the use of fertilizers and other chemicals by farmers, effluent from industries and poor sewage disposal by the locals. Tilapia fish (a herbivorous), ingest a wide variety of natural food organisms, including plankton, succulent green leaves, benthic organisms, aquatic invertebrates, larval fish, detritus and decomposing organic matter. The fish however, obtain substantial nutritional benefit from plant materials, some of which have been reported to be cyanogenic.

In this work, cyanide detoxifying enzymes (rhodanese and 3-MST) were estimated in some tissues of different tilapia species. There was no significant difference in the activity of rhodanese in the tissues of the three species of tilapia from Aiba reservoirs (Table1). In Table 2, there were slight variations in the distribution of 3-MST in the tilapia species from Aiba reservoirs. However there was significant correlation in the activity of 3-MST in the gut and gills of *S. galileus* as compared to the other two species.

The distribution pattern of rhodanese in the different species of tilapia in Osinmo reservoirs was similar to that of Aiba reservoirs (Table 3 and 4). Though no specific study of the subcellular distribution has been made. This distribution pattern may be attributed to the function of rhodanese in the detoxification of cyanide as this could be the primary function of rhodanese in these species. The activity of rhodanese in the gut of *T. zillii* was found to be very high with no significant difference as compared with the other two species. The activity of rhodanese in the gills of *Oreochromis* was highest compared to the other two species (Table 3). Kenneth (1991) had shown that gills are multifunctional organs responsible for respiration, osmoregulation and acid base balance. Therefore, the high activity of rhodanese in the gill of the tilapia species could be physiological, since the organ (gill) is involved in respiration, a function that is prone to cyanide attack. There was no significant difference in the liver rhodanese activity of the three species. Although the rhodanese activity was highest in the liver of *S. galileus*, the distribution of MST among the species showed similar activity in the gut of the tilapia species while the activity of the enzyme in the gill was not significantly different in the *O. niloticus* and *T. zilli.* In the liver, MST in *O. niloticus* was significantly different from *T. zilli* and *S. galileus.* The presence of these sulphurtranferases (rhodanese and 3-MST) in the tissues of these tilapia species is an indication of high cyanide detoxifying mechanism, a protective and possible physiological mechanism for the survival of these organisms in their environment.

In Nigeria, rapid urbanization as well as, high crop yield, severe pest problems, weeds, rodents, locusts and grain eating birds have increased reliance on the use of pesticides, herbicides and fertilizers. Most of which are a cheap sources of cyanides and are widely used. Also co-disposal of industrial, domestic and medical waste in open dumps has contributed immensely to pollution menace in the region. The public awareness on health hazards and risks associated with pesticides and herbicides are also relatively low in Africa, hence the need for continuous monitoring of pollutants such as cyanides in fresh and marine waters and also biota such as fish.

REFERENCES

Agboola FK, Okonji RE (2004). Presence of rhodanese in the cytosolic fraction of the Fruit Bat (*Eidolon helvum*) liver. J. Biochem. Mol. Biol. 37(3): 275-281.

Ali A, AL-Qarawi HM, Mousa BH (2001). Tissue and intracellular distribution of rhodanese and mercaptopyruvate sulphurtranferase in ruminants and birds. Vet. Res. 32: 63-70.

Aminlari M, Gilanpour H (1991). Comparative studies on the distribution of rhodanese in different tissues of domestic animals. Comp Biochem Physiol B 99: 673–677.

Aminlari M, Shahbazi M (1994). Rhodanese (thiosulfate: cyanide sulfur transferase) distribution in the digestive tract of chicken. Poult. Sci. 73: 1465-1469.

Bordo D, Bork P (2002). The rhodanese/Cdc25 phosphatase superfamily. Sequence-structure-function relations. Euro. Mol. Biol. EMBO Rep. 3: 741-746.

Bradford KM (1976). A rapid and sensitive method for the quantitation of micrograme quantities of protein utilizing the principle of protein-dye binding. Anals Biochem. 72: 248-254.

Egekeze JO, Oehme FW (1980). Cyanides and their toxicity: a literature review. Vet. Q. 2:104-114.

Eisler R (1991). Cyanide hazard to fish, wildlife , and invertebrate. A synoptic review. US fish wildlife Service. Biol Rep., 85:1-23.

Jarabak R, Westley J (1974). Human Liver Rhodanese: Nonlinear Kinetic Behaviour. Double Displacement Mechanism. Biochem. 13(16): 3233-3236.

Jarabak R, Westley J (1978). Human Liver Rhodanese: Nonlinear Kinetic Behaviour Double Displacement Mechanism. Biochem. 13(16): 3233-3236.

Keeler RF, Vankampen KR, James LF (1978). Effects of poisonous Plants on Livestock. Academic Press, New York, NY.

Kenneth RO (1991). Vasculature of the fish gills: Anatomical correlates of physiological functions. J. Elec. Microsco. Technol. 19(4): 389-405.

Leduc G (1984). Cyanides in water: toxicological significance. L. J. Weber, ed. Aqu. Toxi. Vol. 2. Raven Press, New York. pp. 153- 224.

Marrs TC, Ballantyne B (1987). Clinical and experimental toxicology of cyanides: an overview. In B. Ballantyne and T. C. Marrs, eds. Clinical and experimental toxicology of cyanides. John Wright, Bristol, England. pp. 473-495.

Nagahara N, Nishino T (1996). Role of Amino acid residues in the active site of rat liver mercaptopyruvate sulphurtransferases. J. Biol. Chem. 271(44): 27395-27401.

Nagahara N, Okazaki T, Nishino T (1995). Cytosolic mecaptopyruvate sulphurtransferase is evolutionarily related to mithochondrial Rhodanese. Striking similarity in active site, amino acid sequence and the increase in mecaptopyruvate sulphurtransferase activity of rhodanese by site directed mutagenesis. J. Biol. Chem. 270: 16230-16235.

Nandi DL, Horowitz PM, Westley J (2000). Rhodanese as a thioredoxin oxidase. Int. J. Biochem. Cell Biol. 32(4): 465-473.

Taniguichi T, Kimura T (1974). Role of 3-mercaptopyruvate sulphurtransferase in the formation of the iron chromophore of adrenal ferridoxin. Biochem. Biophys. Acta. 364: 284-295.

Vennesland B, Castric PA, Conn EE, Solomonson LP, Volini M, Westley J (1982). Cyanide metabolism. Fed. Proc. 41: 2639–2648.

Westley J (1981). Cyanide and sulfane sulphur. In: Cyanide in Biology. Vennesland B, Conn EE, Knowles CJ, Westley J, Wissing F eds., Academic Press, New York. p. 61-76.

Wiemeyer SN, Hill AF, Carpenter JW, Krynitsky AJ (1986). Acute Oral toxicity of sodium cyanide. J. Bacteriol. 107(1): 375-376.

Effect of administration of acetylsalicylic acid on phosphatase enzymes in liver of metabisulphite treated rats

J. E. Olajide[1], M. A. Akanji[2], J. Omale[1*] and N. O. Mohammad[2]

[1]Department of Biochemistry Kogi State University Anyigba, Nigeria.
[2]Department of Biochemistry, University of Ilorin, Nigeria.

Membrane stabilization is an established attribute of most anti-inflammatory drugs. In this work the membrane stabilization capability of acetylsalicylic acid (ASA) was investigated in the liver of metabisulphite insulted experimental rats. Previous reports have shown that sodium metabisulphite administered to rats resulted in labilization of plasma membrane of some rat tissues while acetylsalicylic acid, an anti-inflammatory drug, like certain others, has been reported as membrane stabilizer. The organ of interest (liver) was collected into a 0.25 M sucrose and the homogenate was prepared for enzyme analyses. The activities of the phosphatases were measured using standard methods. Initial administration of metabisulphite alone showed an immediate significant decrease with ($P<0.05$) in alkaline phosphatase activities. Loss in activities was recorded throughout the experimental period but became insignificant after day 3. The activities of the phosphatases in acetylsalicylic acid treated rats were lower when compared with the control group but not as obvious as with the metabisulphite treated rats. The combination treatment (metabisulphite and acetyl salicyclic acid) gave a trend in activities that was in between the administration of individual compounds. This study therefore has shown that acetylsalicyclic acid prevented disruption of liver cellular membrane caused by the administration of sodium metabisulphite.

Key words: Acetylsalicylic acid, metabisulphite, liver, phosphatases, administration.

INTRODUCTION

Acetylsalicylic acid (ASA) is the chemical name for a member of the salicylate drugs commonly known as Aspirin. It is produced by the substitution of the phenolic group of salicylic acid with acetic anhydride (Roger et al., 1981). Acetylsalicylic acid is the prototype for the non-steroidal anti-inflammatory drugs (NSAIDS) which inhibits the production of thromboxane (Wu, 2000). Ibuprofen and related drugs are other members of NSAIDS. Most NSAIDS have been described as potent membrane stabilizers with aspirin being shown (Ngaha and Akanji, 1982) to stabilize kidney lysosomal membrane after its labilization by chloroquine. Though discovered more than 100 years ago, little is known about how aspirin actually works to control inflammation. What is known is that it acts in two pathways, one involving prostaglandins (Abramson and Weissmann, 1989) and the other a more recent discovery, involving nuclear factor-K (Kappa) B (NF-KB) (Holger and Angus, 2001). Acetylsalicylic acid inhibits the activation of NF-KB molecules that activate chemicals to trigger inflammation (Holger and Angus, 2001). However, it was shown in recent works that these anti-inflammatory drugs act by inhibiting prostanoid synthesis through acetylation of fatty acid cyclooxygenase thus irreversibly blocking cyclooxygenase system (Friedrich et al., 2001; Durand et al., 2002a). This effect depends on the type of cells involved due to resynthesis (Dutrand et al., 2002b). The dependant of blockade by acetylsalicylic acid on cell type is the rationale for the recommended dose of (50 - 1500) mg/day with long inter dose intervals (Hla and Bailey, 1989; Patrono et al., 1998; Abouelenin et al., 2002).

Membrane of cells is vulnerable to be attacked by both endogenously produced toxic substances and those con-

*Corresponding author. E-mail: jamesomale123@yahoo.com.

sumed exogenously. Sodium metabisulphite is a food preservative that produces sulphur dioxide (Wedzicha, 1984). Its interaction with dietary components (Bhagat and Lockett, 1964) resulting in toxic products have rescinded its generally regarded as safe (GRAS) status through some toxicological findings (Taylor et al., 1986).

Akanji et al. (1993; 1996) and Akanji and Yakubu (2000) have shown the potentiality of metabisulphite in tissue damage in rats. Metabisulphites generate sulphite and sulphites are known to have inhibitory action on some enzymes such as the lactate dehydrogenases of the heart and malate dehydrogenases (Akanji and Yakubu, 2000). Bisulphites (0.5 mM) have earlier been shown to cause induced oxidation in corn-oil emulsified in 1.5% polysorbate solution (Kaplan et al., 1975) just as un-saturated membrane lipids incubated with a large excess of bisulphite was reported to have different chromatographic pattern indicative of addition of bisulphite across double bonds and hence alteration of membrane structure (Akogyeram and Southerland, 1980). With its high oxygen content metabisulphite generates oxygen radical $^{\cdot}O_2$; a good nucleophile that could react readily with electrophilic sites on biological molecules such as in the membrane of cells (Halliwell, 1974; Akanji et al., 1993; 1996). The potentiality of acetylsalicylic acid in preventing the cellular damage caused by metabisulphite is investigated in this study using changes in pattern of activity of alkaline and acid phosphatases both marker enzymes for specific regions of the cell as index of its effect.

MATERIALS AND METHODS

Experimental animals

Male albino rats (Wister Strain) weighing between 150 – 200 g were acclimatized for four weeks and fed with rat cubes made from animal feed obtained from ECWA feed mills Ltd, Jos, Plateau State, Nigeria, and water ad labium. The experiment was performed according to current guidelines for the care of laboratory animals and the ethical guidelines for the investigation of experimental pain in conscious animals (Zimmerman, 1983). The design/protocol for the study is as approved by the Animal Experiments Ethics Review Board of the Department of Biochemistry, Kogi State University, Anyigba, Nigeria

Acetylsalicylic acid used for the work is a product of Tega Laboratories Chelsea London, while sodium metabisulphite used is a product of May and Baker Ltd, Dagenham England.

Design

The experiment consisted of thirty three rats divided randomly into three groups of 10 rats each, numbered group 1, 2, and 3. The fourth group consisted of 3 rats as the control group. Each group of animals was kept in separate metabolic cages and fed with rat cubes and water *ad libitum*. Each set was replicated trice. The rats in group 1 were administered daily with solution of sodium metabisulphite (10 mg/kg body weight) while rats in group 2 were administered daily with solution of acetylsalicylic acid (10 mg/kg body weight). Group 3 rats were administered daily with solution of the two chemical compounds concurrently while rats in the control Groups were administered with distilled water alone. The route of administration was intraperitoneal.

Statistical analysis

Results were analyzed using Analysis of Variance (ANOVA). A value of $P < 0.05$ was considered as statistically significant.

Drug and metabisulphite administration

Solution of 2 mg/ml Sodium metabisulphite and 2 mg/ml acetylsalicylic acid were prepared in distilled water. The solution of sodium metabisulphite was administered daily to rats in group 1 while the solution acetylsalicylic acid was administered daily to rats in group 2. The rats in group 3 were administered solution of sodium metabisulphite and acetylsalicylic acid concurrently. All administeration lasted for fifteen days. Day 1 represents rats that were given one daily dose of appropriate chemical compound or the combination of the compound and were sacrifice 24 h after the administration while other days (3, 5, 10 and 15) represent rats that were given those numbers of daily doses and sacrifice thereafter (Akanji and Nlumanze, 1987). The rats in the control group were administered with only distilled water daily for 15 days and sacrifice 24 h after the 15th dose.

Preparation of homogenate

The rats were anaesthetized in a glass jar containing cotton wool soaked in chloroform until they go unconscious. They were immediately removed for dissection to avoid damage to the organ of interest. The liver was removed immediately. One gram of the liver was weighed, cut into pieces and homogenized in an ice-cooled 0.25 M sucrose solution (1:5 w/v) using a pre-cooled enamel mortar and pestle. Triton X-100 was added to the homogenate to a final concentration of 1% (Ngaha et al., 1979; Yakubu et al., 2001). The homogenate was frozen over night to allow unbroken cells to lyse before being used for enzyme assay (Akanji and Ngaha, 1989; Adesokan and Akanji, 2003).

Tissue dilution and enzyme measurement

The stock homogenate of the tissue (1:5 w/v; tissue: 0.25 M sucrose) was diluted using 0.25 M as diluents. The dilution factors of the homogenate were 30 and 600 for the assays for alkaline phosphatase and acid phosphotase. Protein determination was carried out with the method of Gromal et al. (1949) and the phosphatase activities were measured as described by Wright et al. (1972). The methods depends on the determination of the amount of phosphate ester that is being hydrolyse within a given period of time in an alkaline pH and acidic pH for alkaline phosphatase and acid phosphatase respectively. In this work, the hydrolysis of para-nitrophenylphosphate (colourless) into para-nitrophenol was followed spectrophotometrically at 400 nm. Para-nitrophenol gives yellow colour in alkaline medium. The intensity of which can be measured maximally at 400 nm. All measurements were carried out using Agilent 8453 UV-visible spectrophotometer. The experiment was perform in triplicate.

RESULTS

The results of administration of sodium metabisulphite (10 mg/kg body wt), acetylsalicylic acid (10 mg/kg) and combination treatment on the activity of alkaline phospha-

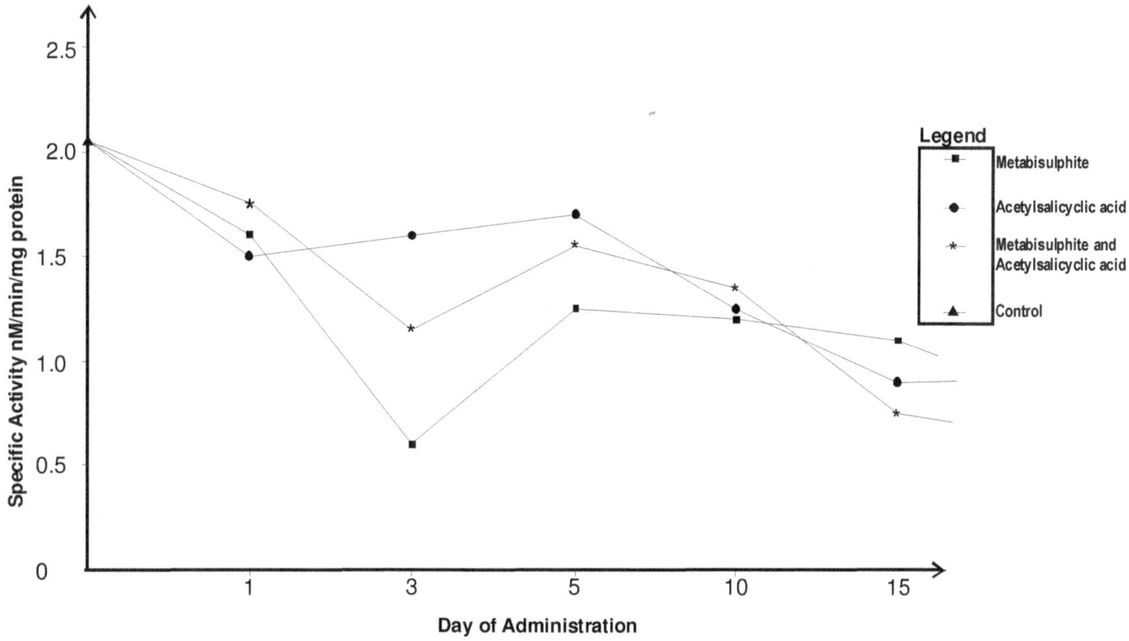

Figure 1. Effect of daily administration of sodium metabisulphate (10 mg/kg b.wt), acetyl salicyclic acid (10 mg/kg b.wt), and combination of the two compounds on the activities of acid phosphatase of rat liver.

tase in the liver of rats are shown in Figure 1. Upon initial administration of sodium metabisulphite alone, there was an immediate decrease ($P<0.05$) in the activity of alkaline phosphatase. The decrease in values of enzyme activity was observed throughout the experimental period when compared with the control value. After the 3rd day, a 70.70% decrease in alkaline phosphatase activity was obtained when compared with the control value. The activities of the enzyme under the influence of administration of acetylsalicylic acid was reduced but was insignificant ($P> 0.05$) as was observed with the administration of metabisulphite.

Administration of both sodium metabisulphite and acetylsalicylic acid gave a trend of activity that was in between the administration of the individual chemical compound. There was a significant: ($P< 0.05$) recovery from loss of enzyme activity under the combination treatment when the results were compared with when metabisulphite alone was administered.

The effect of the individual chemical compound and the combination treatments on the activity of acid phosphatase in the liver of the rats is shown in Figure 2. The three treatments produced same pattern of enzyme activity but with the combination treatment being between the two single treatments. The initial loss of enzyme activity following the first dose of metabisulphite was insignificant ($P>0.05$) with the activity stabilized by the fifth day. On the whole, all experimental values of acid phosphatase activity obtained from the various test administrations showed insignificant ($P>0.05$) decreases when compared with the control value. Summarily, the activities of acid phosphatase in the liver of rats under treatments were not significantly different ($P>0.05$). Results were analyzed using Analysis of Variance (ANOVA). A value of $P < 0.05$ was considered as statistically significant.

DISCUSSION

This work has revealed the membrane stabilization ability of acetylsalicylic acid. This finding confirms further the earlier works of Ignarro (1971), Ngaha and Akanji (1982); Olajide et al., (2005) on the membrane stabilizing role of most non-steroidal anti-inflammatory agents, the class of drugs to which acetylsalicylic acid belongs. The metabolism of xenobiotics is achieved largely in the liver because the drug metabolizing enzyme system is located therein. In this work the potentiality of sodium metabisulphite to disrupt cellular membrane as reported in some earlier works (Akanji et al., 1993; Akanji et al., 1996; Akanji and Yakubu, 2000) was affirmed. In this study, administration of sodium metabisulphite resulted in a significant decrease ($P< 0.05$) in alkaline phosphatase activities in the liver (Figure 1). But the activity of acid phosphatase was not affected to any appreciable extent in the tissue (Figure 2).

Adesokan et al. (2006) showed that the activity pattern of rat liver and kidney enzymes change following administration of xenobiotics. The values of activity of acid phosphatase in the tissue when metabisulphite was administered showed no significant difference from the control value ($P>0.05$). This may imply that the integrity of the lysosomes where acid phosphatase is located was maintained despite the chemical insult.

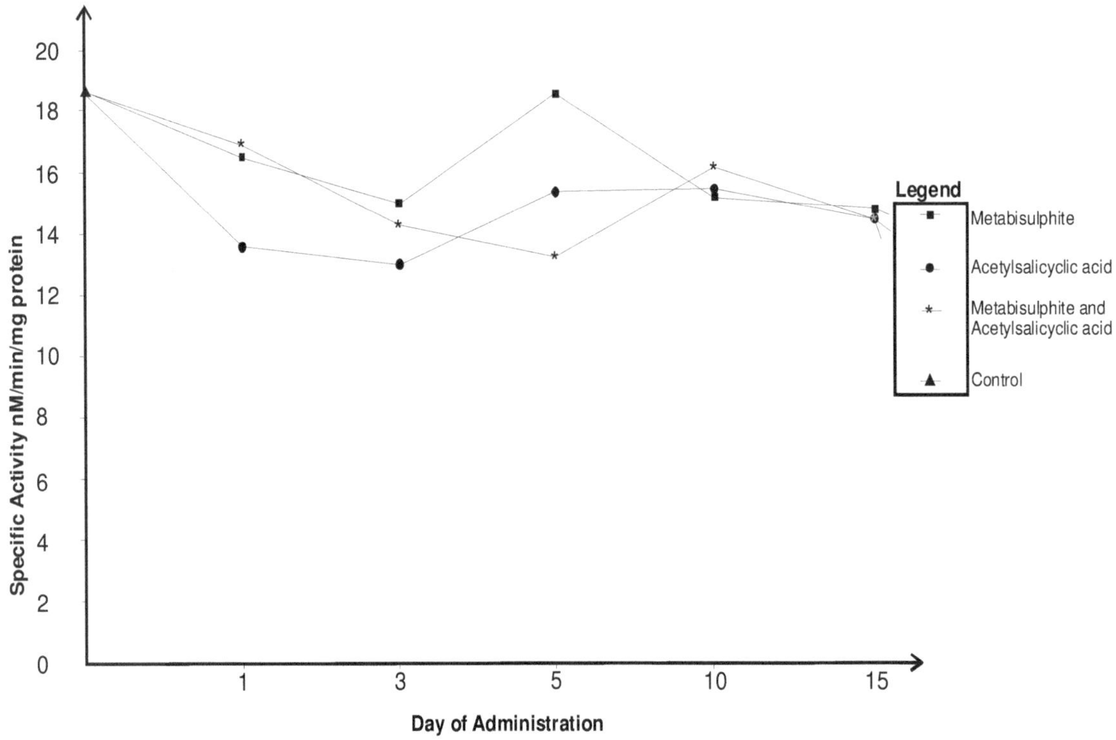

Figure 2. Effect of daily administration of Sodium metabisulphate (10 mg/kg b. wt), acetylsalicyclic acid (10 mg/kg b. wt), and combination of the two compounds on the activities of acid phosphatase of rat liver.

When acetylsalicylic acid alone was administered to the rats, it was observed that the activities of the phosphatases in the liver tissue were not appreciably affected when compared with the control (Figures 1 and 2). This finding supports the work of Ngaha and Akanji (1982) on the stabilizing role of acetylsalicylic acid on rat kidney lysosomal membrane after its labilization by chloroquine.

Despite that the liver tissues contain the enzymes involved in drug metabolism, its cellular membrane damaged by the administration of sodium metabisulphite was repaired by acetylsalicylic acid like drugs by preventing the disruption of kidney cellular membrane (Olajide et al., 2005). From the result, the loss of alkaline phosphatase activities in the liver of rats was prevented following the concurrent administration of sodium metabisulphite and acetylsalicylic acid to the experimental animals (Figure 1). Therefore in this study, acetylsalicylic acid further demonstrated its stabilizing role by preventing the disruption of liver cellular membrane caused by the administration of sodium metabisulphite. This role could be due to the ability of molecules of acetylsalicylic acid to lodge itself in gaps created by sodium metabisulphite between the ordered membrane molecules on the membrane wall

ACKNOWLEDGEMENT

The authors wish to appreciate the technical support of Messrs. Kolade (Chief Technologist Biochemistry Department University of Ilorin) and Friday Emmanuel of Kogi State University Anyigba. The high level of competence exhibited during the typing of the manuscript by Mr. Stephen Jenekwu is highly appreciated. The work would not have been successful without the contribution of Professor Oloyede O. B of the Biochemistry Department University of Ilorin, Ilorin Nigeria.

REFERENCES

Abou-elenin K, Xydakis A, Hamndy O, Horton S (2002). The effect of aspirin and various iontophoresis solution vehicles on skin microvascular reactivity. Microvasc. Res. 63: 91-95.

Abraham SB, Weissmann G (1989). The mechanism of action of nonsteroidal anti inflammatory drugs. Arthritis Rheum. 32: 1-9.

Adesokan AA, Akanji M.A (2003). Effect of administration of aqueous extract of Enantia chloronta on the activities of some enzymes in the small intestine of rats. Nigerian. J. Biochem. Mol. Bio. 18:103-105.

Adesokan AA, Akanji MA, Balogun EA, Aderibigbe A (2006). Effect of oral administration of Aloe Barbadensis-Miller juice on selected biochemical parameters of rat liver and kidney. In : Recent Progress in Medicinal Plants, Volume 13: In Govil et al. (eds) Search for Natural Drugs, Studium Press LLC, Houston, Texas USA. Chapter 3: pp 73-80.

Akanji MA, Nlumanze SE (1987). "Alkaline phosphatase activities following repeated suramin administration in some rats tissues in relation to their functions". Pharmacol. Toxicol. 61: 182-183.

Akanji MA, Ngaha EO (1989). "Effect of repeated administration of berenil on urinary excretion with corresponding tissue pattern in rats". Pharmacol. Toxicol. 64: 272-275.

Akanji MA, Olagoke OA, Oloyede OB (1993). Effect of chronic consumption of metabisulphite on the integrity of the rat kidney cellular system. Toxicology 81: 173-179.

Akanji MA, Olagoke OA, Ipinmoye IT, Oloyede OB (1996). Effect of chronic consumption of metabisulphite on the activities of some enzymes in selected parts of gastro-intestinal tract of rats. Nig .J. Biochem. Mol. Bio. 11: 11-14.

Akanji MA, Yakubu MT (2000). α - Tocopherol protects against metabisulphite induced tissue damage in rats. Nig.J. Biochem. Mol. Bio. 15: 179-183.

Akogyeram C, Southerland WM (1980). "The interaction of bisulphite with membrane lipids". Fed. Proc. Am. Soc. Exp. Biol. Abstract 39: 1836.

Bhaggat B, Lockett M.F (1964). "The effects of sulphites in solid diets on the growth of rats ". Fed. Cosmets Toxicol. 2: 1-13.

Durand S, Fromy B, Koital A, Abraham P, Saumet JL (2002 a). Oral single high-dose aspirin results in a long-lived inhibition of anodal current – induced vasodilation. British .J. Pharmacol. 137: 384-390.

Durand S, Fromy B, Koital A, Abraham P, Saumet JL (2002 b). Vasodilation in response to repeated anodal current application in the human skin relies on aspirin sensitive mechanism. J. Physiol. 540: 261-269.

Friedrich CL, Dominic NM, Ralf Dechend (2001). Aspirin inhibits NF-KB and protects from angiotensin II – induced organ damage. The FASEB .J. Express Article 10. 1096/fj.00-0843 fje.

Gromal AG, Bardwill C. J, David MM (1949) "Determination of serum protein by means of biuret reaction". J. Biol. Chem. 177: 571-766.

Halliwel B (1974). "Super-oxide dismutase, catalase and glutathione peroxidase-solutions to problem of living with oxygen". New Phytol. 73: 1075-1086.

Hla TT, Bailey JM (1989). Differential recovery of prostacyanin synthesis in cultural vascular endothelial versus smooth muscle cells after inactivation of cyclooxygenase with aspirin. Essen. Fatty Acids, 36:175-184.

Holger H, Angus T (2001). Aspirin targets key cells that trigger organ rejection and other immune responses. Science Daily (June 25, 2001). Pittsburgh.

Ignarro LS (1971). "Effects of anti-inflammatory drugs on the stability of rat liver lysosomes in-vitro". Biochem. Pharmacol. 20: 2847-2853.

Kaplan D, Mejilton C, Lutchel D (1975). "Bisulphite induced lipid oxidation". Arch. Environ. Health 30:507-509.

Ngaha EO, Fry M, Plummer DT (1979). "The effect of cephaloridine on the stability of rat kidney lysosomes". Chem. Biol. Interacts. 24: 199-208.

Ngaha EO, Akanji MA (1982). "Effect of chloroquine in the stability of rat kidney lysosomes *in vivo* and *vitro* comp". Biochem. Physiol. 73c: 109-113.

Olajide JE, Akanji MA, Sanni M, Omale J (2005). Acetylsalicylic acid and cellular damage in kidney of metabisulphite treated rats. Anl. Res. Int. 2(3): 388-392.

Patrono C, Collar B, Dalen JE, Roth G (1998). Platelet activating Drugs; The relationship among dose, effectiveness and side effects. Chest. 114: 470s-488s.

Roger HJ, Spectre RG, Trounce JR (1981). "Salicylates. In a textbook of clinical pharmacology". Acad. Press. New York, Pp 281-286.

Taylor SL, Higley NA, Bush RK (1986). Sulfites in food: uses, analytical methods, residue, fate exposure assessment, metabolism, toxicity and hypersensitivity". Adv. Food Res. 30: 1-8.

Wedzicha BL (1984). "In: chemistry of sulphur dioxides in foods". Elsevier Applied Sci. Publisher Borking Essex. Pp 214-215.

Wright PJ, Leathwood FD, Plummer DT (1972). "Enzymes in rat urine: Acid Phosphatase". Enzymologia 42: 317-327.

Wu KK (2000). Aspirin and Salicylate: An old remedy with a new twist. Circulation 102: 2022-2023.

Yakubu MT, Akanji MA, Saliu IO (2001). Protective effect of ascorbic acid on some selected tissues of rantidine-treated rats. Nig. J. Biochem. Mol. Bio. 16: 177-182.

Zimmerman M (1983): Ethical guidelines for investigation of experimental pain in conscious animal. Pain 16: 109- 110

Purification and biochemical properties of lipase from raphia palm fruit mesocarp

T. Okunwaye , J.U. Obibuzor and E.A. Okogbenin

Biochemistry Division, Nigerian Institute for Oil Palm Research, P.M.B. 1030, Benin City, Edo State, Nigeria.

Lipases hydrolyze lipids to yield fatty acids and glycerol which are used as raw materials in industries. The lipase from raphia mesocarp was extracted, purified and characterized. Assay was carried out using *p*- nitrophenyl palmitate as substrate. The lipase was subjected to 80% ammonium sulphate precipitation, purified on ion exchange and gel filtration chromatography. Native and subunit molecular weights were determined on a gel filtration column and 10% sodium dodecyl sulphate polyacrylamide gel electrophoresis respectively. Kinetic parameters (K_m and V_{max}), effect of temperature, pH, sensitivity to metal ions and substrate specificities of the lipase enzyme were studied. Specific activity of 17.40µmol/min/mg and purification fold of 2.68% were obtained.Native and subunitmolecular weightwas 35 kDa. The K_m and V_{max} values were0.01mM and 20.5µmol/min/ml respectively. The lipase enzyme was inhibited by all metallic salts used and enhanced by $CaCl_2$. Optimal pH was 7.0 and temperature for maximal activity was 40°C. The lipase hydrolyzed coconut oil at about twice the rate of palm kernel oil and palm oil, and at a higher rate than olive oil and Raphia oil. The study reveals that raphia mesocarp isa possible source of lipase which can be of industrial use.

Key words: Lipase, raphia palm fruit, mesocarp

INTRODUCTION

Lipases are a group of enzymes that catalyses the hydrolysis (cleavage) and synthesis of various forms of lipids (Svendsen, 2000). They belong to a subclass of the esterases and have been reported to be active at the oil water interface (Ibrahim, 1996). Oilseed lipases have great potential for commercial exploitation as industrial enzymes. The demand of lipolytic enzymes has been increased due to its potentiall use in the various manufacturing processes of industrial goods such as food and pharmaceuticals (Boland et al., 1991; Gandhi, 1997), in the production of biodiesel and industrial detergent (Freire and Castilho, 2008) as well as in the analysis of triacylglycerols (Forgia et al., 1995), raw materials in food, cosmetics and pharmaceutical industries (Hermansyah et al., 2006). Fat splitting, with lipase as a catalyst, is advantageous compared to conventional process due to low energy consumption, low cost, high product quality, ease of purification and safer process.

The yield of plant lipases is low; sometimes it could be cheaper than other sources and does not require fermentation.

Lipase activity has been reported in several oil plant extracts which include castor bean, palm seeds, oleifera seed, sunflower seed, peanuts (Abigor et al., 2002; Henderson and Osborne, 1991; Khan et al., 1991) and palm fruit mesocarp (Abigor et al., 1985).

Raphia is an underutilized and under exploited plant that produces fruits, with improved agronomical practices; tons of the fruits can be produced for cheap source of lipase which can be used industrially. In this study we report the isolation, purification and characterization of lipase from raphia palm mesocarp to hydrolyze vegetable oils for production of fatty acid and glycerol.

MATERIALS AND METHODS

Raphia palm fruits were obtained from Mile 18 village, Edo State, Nigeria. Palm oil, coconut oil, palm kernel oil used for the hydrolysis reaction were obtained from Mega stone farms, Oba town, Ogun State, Nigeria. Raphia oil was extracted by soxhlet using n-Hexane as solvent and olive oil was bought from a supermarket in Benin, Edo State, Nigeria. All chemicals, reagents, resins, standard protein and solvents used in the experiment were of analytical grades and were obtained from Sigma chemical Company, St Louis, Mo., USA, Pharmacia Fine Chemical, Uppsala, Swedenand BDH Chemicals Limited, Poole, England.

Enzyme preparations

The Raphia fruit was allowed to ripe, the exocarp was peeled off and the mesocarp was scraped with a cleaned knife and homogenized for 5 min in 0.15M Tris buffer adjusted to pH7.5 with KOH. The homogenate was filtered through one layer of cheese cloth and centrifuged for 30min at 10,000*g* using High Speed Refrigerated Centrifuge. It yielded a fat layer, a supernatant liquid and a pellet. The supernatant was used for the assay of lipase activity and protein concentration determination. The remaining supernatantwasfractionallyprecipitatedusingammoniumsulphate.

Assay of enzyme activity

Lipase activity was determined by the colorimetric method according to Lotrakul and Dharmsthiti (1997). This method was based on the cleavage of p-nitrophenyl palmitate (p-NPP) at pH 8.0. The reaction contained 180µl of solution A (0.062g of *p*-NPP in 10ml of 2-propanol, sonicated for 2 min before use), 1620µl of solution B(0.4% triton X-100 and 0.1% gum Arabic in 50 mM Tris-Hcl, pH 8.0) and 200µl of properly diluted enzyme sample. The product was detected at 410 nm wavelength after incubation for 15min at 37°C. Under this condition, one unit of lipase activity (U) was defined as the amount of enzyme required to release 1µmol of p- nitrophenol (p-NP) per minute under standard assay condition.

Protein determination

Protein concentrations were determined according to Lowry's method (1951) using BSA (1mg/ml) as the standard protein, absorbance was read at 660 nm and a standard graph was plotted. The amount of protein present in the sample was estimated from the standard graph.

Enzyme purification

The crude extract from the centrifugation step was fractionally precipitated to 80% ammonium sulphate saturation (516 g/L) according to the method of Doonan (1996). After standing for 1 h in an ice bath, the suspension was centrifuged at 10,000*g* for 15 min at 4°C, the supernatant discarded, and the precipitate collected and was dialyzed against Tris buffer for 4 h to remove traces of ammonium sulphate.

Ion exchange chromatography was performed using DEAE Sephadex A-50 as the resin. The resin was packed into a 1.5 x 40 cm column. 5ml of the dialyzed extract was carefully layered on the column bed surface that has been previously equilibrated with about 3 L of Tris buffer, pH 7.5, containing 1mM EDTA and was allowed to drain into the bed. The column was washed with Tris buffer to remove unbound protein, followed by elution with stepwise gradient of 1M NaCl in Tris buffer. Elution was performed at a flow rate of 30ml/h and fraction of 5ml each was collected. Protein was monitored spectrophotometrically at 280nm. The fractions were also assayed for lipase activity. The fractions having lipase activity were pooled stored in the refrigerator for further purification.

The enzyme fraction was further purified on gel filtration chromatography using Sephadex G-200 as the resin. 5ml of the DEAE-Sephadex sample was layered on the column (1.5 × 40cm). Elution was performed at a flow rate of 10ml/h and fractions of 5ml each were collected. Protein was monitored spectrophotometrically at 280nm. The fractions were also assayed for lipase activity. The fractions having lipase activity were pooled.

Determination of native molecular weight

The native molecular weight of lipase wasdetermined by gel filtration on a column of Sephadex G-100 (1.5 x 100 cm) using the followingproteinmarkers:BlueDextran(2,000Da),GammaGlobulin(150,000Da), Creatine phosphokinase (81,000Da), Bovine serum albumin(68,000Da),Ovalbumin(45,000Da),AlphaChymotryp-sinogen(25,700Da).Void volume (V_o) was determined using blue dextran. Apparent molecular weight was determined by semi logarithmic plot of Ve/Vo against molecular weight of the marker proteins.

Determination of subunit molecular weight

The subunit molecular weight of purified lipase extract was analyzed with sodium dodecyl sulphate polyacrylamide gel electro-phoreses (SDS-PAGE) according to the method of Laemmli (1970) in 12% slab gel. About 0.1ml of purified protein was loaded onto the gel. The gel was stained with 0.1% Coomassie brilliant blue R-250.

Determination of kinetic parameters

The Michelis-Menten constant (K_m) and the maximum velocity (V_{max}) were determined by measuring enzyme activity of the purified lipase at various concentrations of emulsified substrates. The kinetic parameters (K_m and V_{max}) were calculated by plotting 1/V against 1/[S] in Lineweaver Burk plot (1934).

Effect of temperature on enzyme activity

A thermostated, water-jacketed reaction chamber was employed in determining the temperature dependence of lipase activity. The

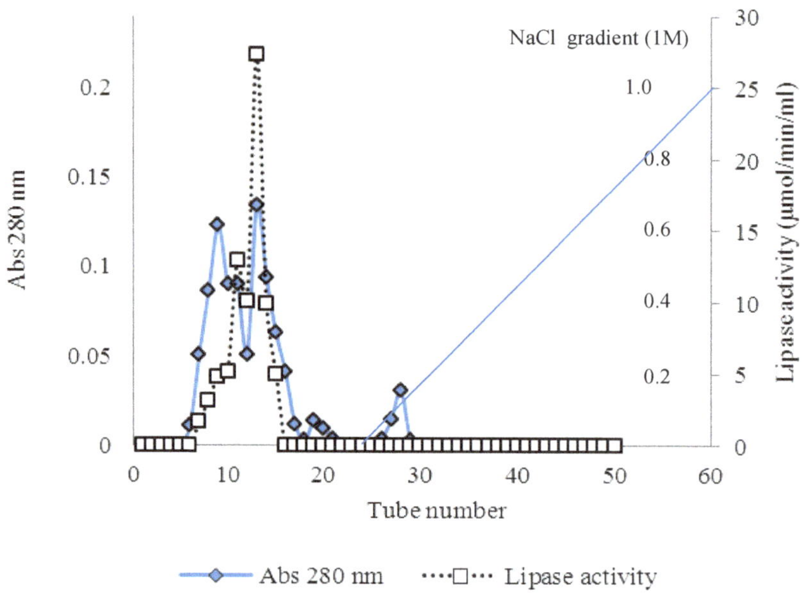

Figure 1. Elution profile of Lipase from raphia palm fruits on DEAE Sephadex A-50.

optimum temperature was determined by measuring the lipase activity at different temperatures of 30, 40, 50, 60, and 70°C after 1 h incubation at pH 7.5 at standard conditions. The values of lipase activities were plotted against temperature.

Thermostability of lipase

Thermal stability was determined by incubating the enzyme at different temperatures ranging from 30 to 70°C for various length of time (30, 60, 90, 120, and 150 minutes) at pH 7.5. Lipase activity was measured and expressed as a percentage of the lipase activity.

Effect of pH on enzyme activity

The effect of pH on the activity of purified enzyme was studied by measuring the enzyme activity at various pH range of 3.5 to 10.0 for 1 h at room temperature. The lipase activity was assayed under standard conditions.

Enzyme inhibition

The following salts at the concentration of 5– 20mM were tested for their possible inhibitory effects on the activity of the purified lipase enzyme: zinc chloride, manganese chloride, calcium chloride, potassium chloride and EDTA at room temperature for 1h and the lipase activities were determined.

Effects of substrates (hydrolysis reactions)

Palm oil, raphia oil, oliveoil, palm kernel oil and coconut oil were used as lipase substrates. The degree of lipolysis was determined by the assay method described previously above.

RESULTS

The elution profile of the crude extract from ion – exchange chromatography column of DEAE Sephadex A-50 is presented in Figure 1. The enzyme was not bound to the DEAE-Sephadex A-50 as it came out in the flow through fractions.No lipase activity was recovered in the fractions obtained by gradient elution. The crude protein was adsorbed on the DEAE- Sephadex A-50 column and this was eluted with a stepwise NaCl gradient(Figure 1), single peak was obtained. The active fractions from the peak was pooled together and used for further purification gelfiltrationchromatography. Single peak was also obtained, active fractions were pooled and used to assay the enzyme. A summary of the purification procedures is presented inTable 1. The partially purified enzyme from the mesocarp of raphia had a specific activity of 17.40 µmol/min/mg, a purification fold of 12.68 and a percentage yield of 5.79.

Five millilitre (5ml) of the dialyzed extract was carefully layered on the DEAE Sephadex A-50 column bed (1.5 x 40 cm) surface that has been previously equilibrated with about 3 L of Tris buffer, pH 7.5, containing 1mM EDTA and was allowed to drain into the bed. The column was washed with Tris buffer to remove unbound protein, followed by elution with 1M NaCl in Tris buffer. Elution was performed at a flow rate of 30ml/h and fraction of 5ml each was collected. Protein was monitored spectrophotometrically at 280 nm. The fractions were also assayed for lipase activity. The fractions 7 to 21 were pooled for further purification.

Five millilitre (5 ml) of the DEAE-Sephadex sample was

Table 1. Summary of purification of lipase from raphia palm mesocarp.

Step	Volume	LPA	LPC	TLA	TLP	SP	Yield	Fold
Crude extract	100.00	19.30	14.06	1930.00	1406.00	1.37	100.00	1.00
$(NH_4)_2SO_4$ ppt	8.00	23.05	16.17	184.40	129.36	1.42	9.55	1.03
Ion Exchange	5.70	31.15	3.18	177.55	18.12	9.79	9.19	7.13
Gel filtration	3.40	32.90	1.89	111.86	6.42	17.40	5.79	12.68

Vol = Volume (ml); LPA = lipase activity (µmol/min/ml); LPC = lipase protein concentration (mg/ml); TLA = total lipase activity (µmol/min/ml); TLP = total lipase protein (mg/ml); SPA = specific activity (µmol/min/mg). Yield = percentage fold = purification fold.

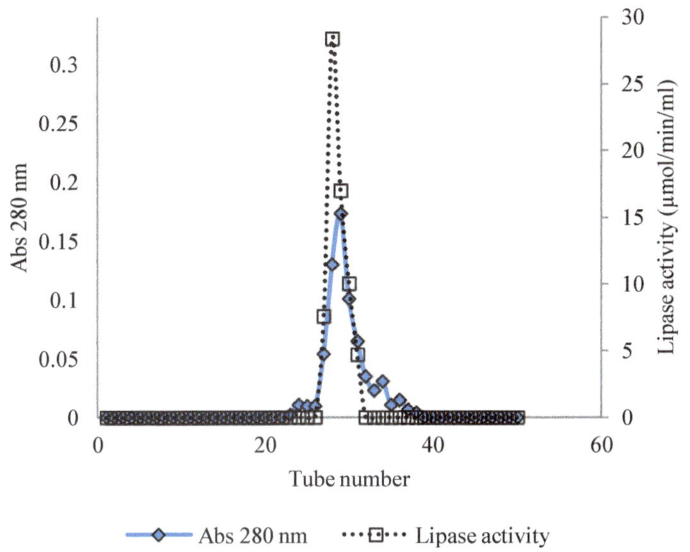

Figure 2. Elution profile of Lipase from raphia palm fruits, (gel filtration on Sephadex G- 200).

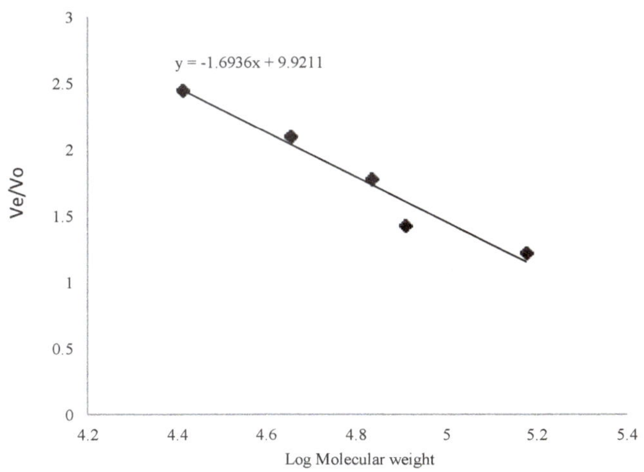

Figure 3. Standard molecular weight curve.

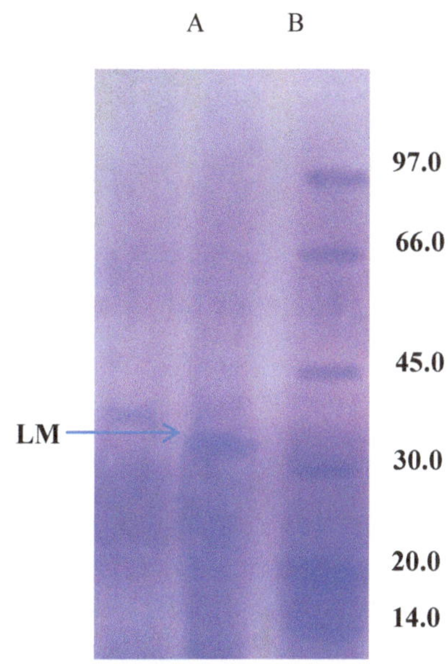

Figure 4. SDS-PAGE of lipase from mesocarp of Raphia palm. LM = lipase from mesocarp of Raphia palm. A= lipase from mesocarp of Raphia. B = standard protein.

layered on the column bed of Sephadex G-200 (1.5 x 40 cm). Elution was performed at a flow rate of 10ml/h and fractions of 5 ml each were collected. Protein was monitored spectrophotometrically at 280nm (Figure 2). The fractions were also assayed for lipase activity. The fractions having lipase activity were pooled and used for analysis.

Molecular weights determination

The calibration curve on Sephadex G-100 (1.5 x 100 cm) for the determination of native molecular weight is shown in Figure 3. The native molecular weight of lipase from raphia mesocarp was 35 kDa.

The picture and calibration curve for the SDS-polyacrylamide gel electrophoresis on 12% slab gels of lipase from Raphia mesocarp is shown in Figure 4. The molecular weight of lipase was estimated to be 35 kDa.

The Lineweaver Burk plot indicating the (K_m and V_{max}) of lipase from raphia mesocarp is shown in Figure 5.

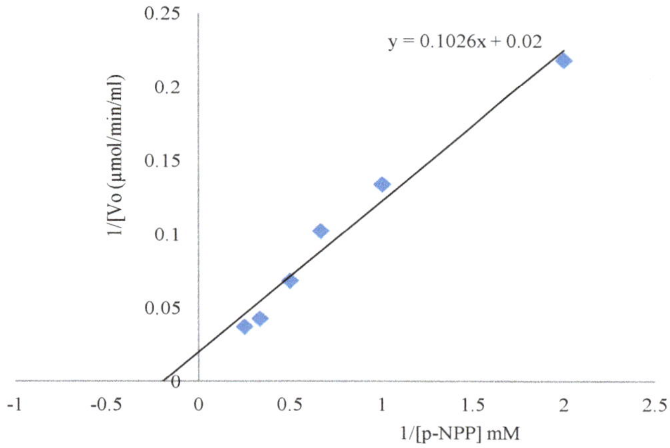

Figure 5. Lineweaver Burk plot of lipase activity (mesocarp of raphia palm.) K_m = 0.01 mM, V_{max} = 20.5 μmol /min/ml.

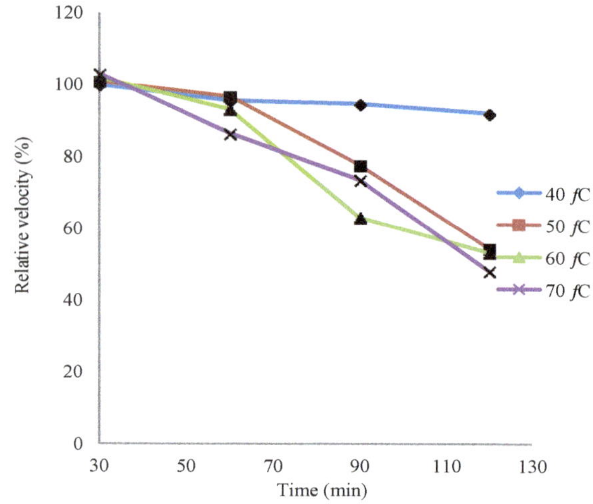

Figure 7. Thermostability of purified lipase from Raphia palm mesocarp. The activity at 30, 40, 50, 60, and 70°C was expressed as a percentage activity of the enzyme incubated at 30°C which was the control.

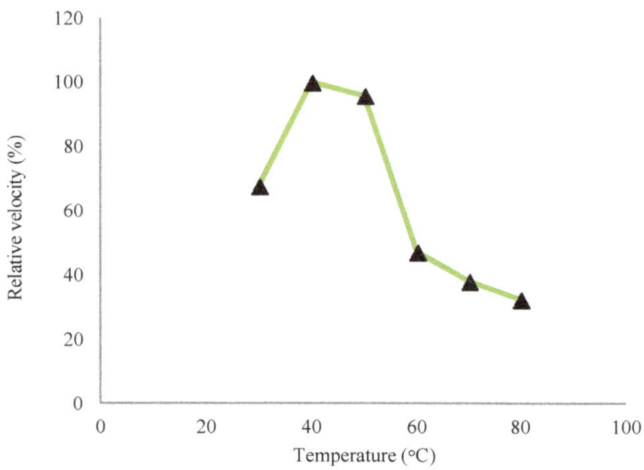

Figure 6. Effect of temperature on purified lipase from Raphia palm mesocarp.

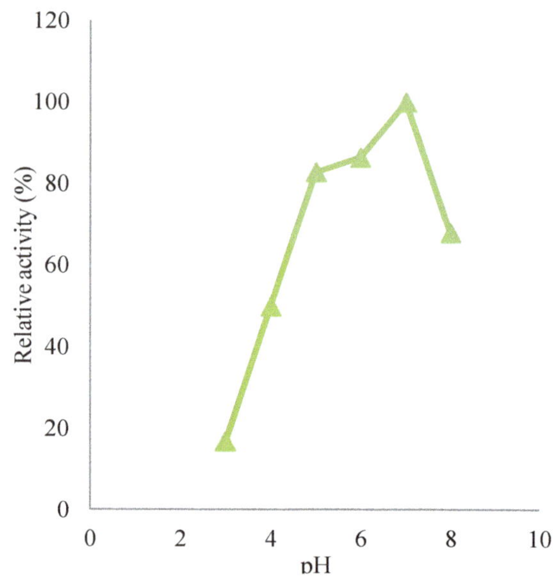

Figure 8. Effect of pH on lipase from raphia palm mesocarp using citrate, pH 3.0-5.0, phosphate, pH 5.0 -7.0 and Tris buffer, pH 7.0-9.0 at 0.05 M.

The results on effect of temperature, thermostability, effect of pH on lipase from Raphia mesocarp are shown in Figures 6, 7 and 8 respectively. Figure 9 shows the hydrolytic activity of the lipase on some vegetable oils and the effect of salt of metal ions on lipase activity are seen in Figure 10.

DISCUSSION

The precipitation of the enzyme was carried out using ammonium sulphate since it was highly soluble in water, cheap and had no deleterious effect on structure of protein.

The specific activity obtained from the purification of lipase from Raphia mesocarp after further purification using Sephadex G 200 and DEAE Sephadex A-50 resins was 17.40, yield of 5.79% and purification fold of 12.68. Lipasefrom*Propionibacteriumacidpropionici*onSephadex G-50 has a 33-fold increase in specific activity (Sarada and Joseph, 1992).

The purified lipase was monomeric and had a subunit and native molecular weight of 35 kDa obtained on gel filtration which is similar to the molecular weight of *Aspergillus niger* lipase 35 kDa on Sephadex G-100 gel filtration chromatography and SDS-PAGE (Sugihara et al., 1988), and also monomeric lipase of *Pseudomonasfragi*,and*Pseudomonasfluorescens*withmolecularweights of 33 kDa.

In many cases, lipases appear to obey Michaelis–

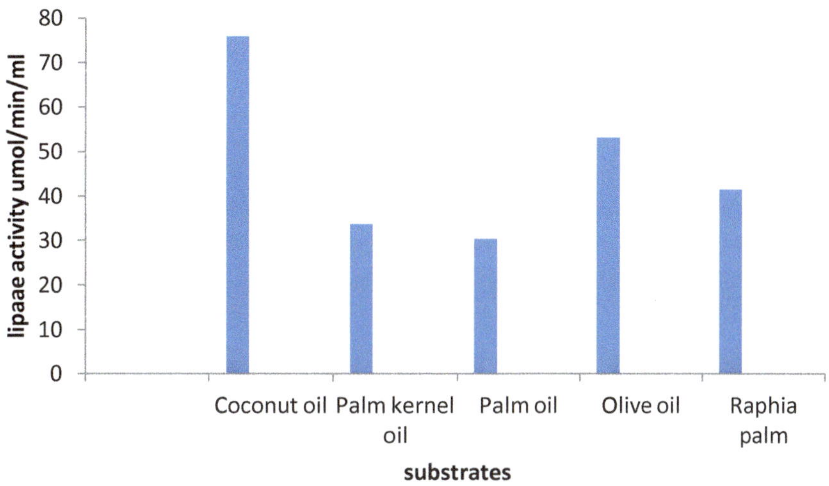

Figure 9. Hydrolytic activity of raphia fruit mesocarp lipase on some vegetable oils.

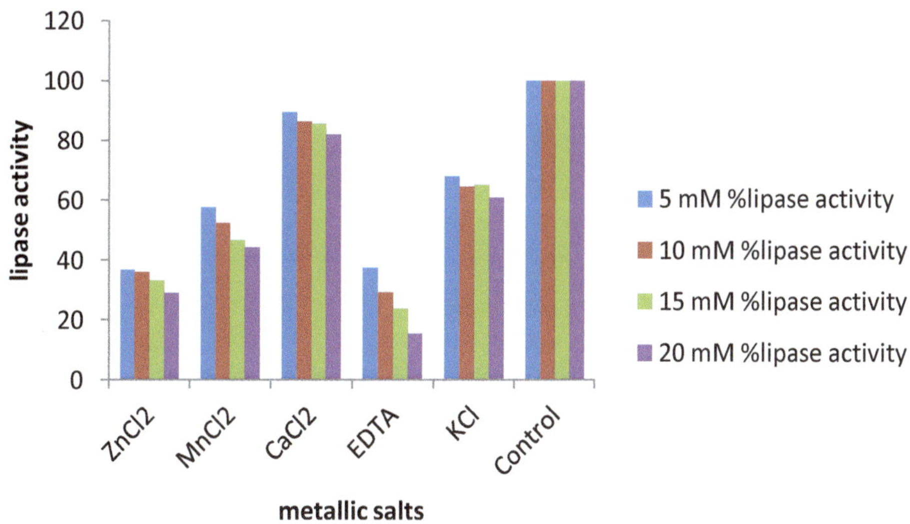

Figure 10. Effect of salts of metal ions on lipase activity from Raphia mesocarp.

Menten kinetics (Guit et al.,1991). Michaelis–Menten kinetics are characterized by two parameters, K_m and V_{max}. The latter is the maximum rate of reaction and K_m is a measure of the affinity of an enzyme for a particular substrate. The kinetic parameters (K_m and V_{max}) of lipase obtained by Lineweaver Burk plot when *p*-nitrophenyl palmitate was used as substrate were 0.01mM and 20.5µmol/min/mg for the mesocarp of raphia.For a *Pseudomonascepacia* lipase, Pancreac'h and Baratti (1996) reported V_{max} value of30 mmol/min when pNPP was used as substrate. The K_m reported in this study is an indication that the enzyme from the germinated palm seeds has high affinity for the substrate and could catalyse the hydrolysis reaction for production of glycerol and fatty acid very fast.

One of the factors that affect the activity of enzymes is temperature and is often used to characterize enzyme. In this study, the optimum temperature of lipase from Raphia mesocarp was 40°C. Activity was rapidly lost at temperatures greater than 50°C. This is similar to the report of Abigoret al.(2003) on lipase from germinating seeds of *Jatropha curcas L.* Iftikhar et al. (2010c) reported that the optimum temperature for lipase from *Rhizopus oligosporus* was 40°C. The relative activity of the partial purified cocoa lipase has an optimum temperature range of 30- 40°C (Ratna, 2008).

The thermal stability experiment showed that lipase from this plant source was stable at 40°C for 60 min, retaining about 95% of its activity. The lipase enzyme was observed to be sensitive to thermal activation; this could be attributed to thermal distortion of its structure.

According to Sharon et al. (1998), the changes in pH affect the protein structure and the enzyme activity. The maximum activity of lipase from raphia mesocarp was observed to be at pH 7 which dropped onwards, this is similar to results of lipase from germinating *Jatropha curcas L.* seed whose pH optima was near neutrality (Abigor et al., 2003), *Pentaclethra* lipase which was also near neutrality (Enujiugha et al., 2004) and partial purified cocoa lipase which had an optimum pH of 7.0 (Ratna, 2008).

Metal ions have been reported to stimulate lipase-catalyzed hydrolysis of oil by removing the fatty acids from the oil-water interface and allowing lipase to act freely on oil molecules (Ohnishi et al., 1994). From this study of the effects of metallic salts on lipase activity from raphia mesocarp, all the metal salts used inhibited the lipase activity at differing levels of inhibition. The highest level of inhibition was observed with Zn^{2+} while the least inhibition was observed with Mn^{2+}. Other salts used in this experiment inhibited the activity of the lipase enzyme. Monnet et al. (2012) reported that calcium ion was found to enhance lipase activity, while Mn^{2+}, Na^{+}, K^{+}, Fe^{3+}, Fe^{2+} and Zn^{2+} strongly inhibited the same. Sharon et al. (1998) reported a lipase of *P. aeruginosa* KKA-5 that retained its activity in presence of Ca^{2+} and Mg^{2+} but was slightly inhibited by Mn^{2+}, Cd^{2+}, and Cu^{2+}. Rahman et al. (2005) stated that metal ions will bind to the enzyme and change the enzyme's conformation to achieve better stability and hence greater activity. EDTA was also observed to inhibit the activity of the lipase and this is in agreement with the report of Monnet et al. (2012) that lipase from dormant ripe and unripe *Terminalia catappaLinn* were strongly inhibited by EDTA, thus suggesting that the lipases are metalloenzymes. According to Enujiugha et al.(2004), the inhibition could be attributed to its chelating process of the system and thereby disrupting the formation of the enzyme substrate complex. This invariably affects the formation of the end product (Enujiugha et al., 2004).

The result of the lipolysis activity shows that the enzyme hydrolyses coconut oil at a higher rate than other oils used. Palm oil and palm kernel oil were hydrolysed at comparable rates; also olive oil and raphia oil were hydrolysed at comparable rates. The result shows that the lipase hydrolyses medium chain triacylglycerol at a faster rate than long chain triacylglycerol. This is in contrast to the findings of Abigor et al. (2003), which reported that lipase from germinating *J. curcas* L. seeds hydrolyzed long-chain TAG than medium chain triacylglycerol.

In conclusion, apart from the fact that this work established the presence of lipase from Raphia mesocarp, it also revealed that this enzyme from these sources had some physico-chemical properties that are similar to what have been obtained from other plant and microbial sources. Some of these properties could make the enzyme to find extensive applications not only in chemical, pharmaceutical, food, or leather industries but also in biodegradation of some materials.

Conflict of interest

Authors declare no conflict of interest.

REFERENCES

Abigor DR, Opute FI, Opoku AR, Osagie AU (1985). Partial purification and some properties of the lipase present in oil palm (Elaeis gunenis Jacq.) mesocarp. J. Sci. Food Agric. 36:599-605.http://dx.doi.org/10.1002/jsfa.2740360711

Abigor RD (2002). Production of biodiesel fuel from vegetable oils using lipases from Pseudomonas cepacia and Jatropha curcas L. Ph.D. Thesis, University of Benin, Benin city, Nigeria. p.234.

Abigor RD, Uadia, PO, Foglia, TA, Haas, MJ, Scott K and Savary, BJ (2003). Partial purification and properties of lipase from germinating seeds of Jatropha cucas. J. Am. Oil Chem. Soc. 79(11):1123-1128. http://dx.doi.org/10.1007/s11746-002-0614-3

Boland W, Frobl C, Lorenz M (1991). Estrolytic and lipolytic enzymes in organic synthesis. Synthesis 12:1049-1072. http://dx.doi.org/10.1055/s-1991-26680

Doonan S (1996). Protein purification protocols. Method Mol. Biol. 59:135-140. http://dx.doi.org/10.1385/0896033368

Enujiugha VN, Thani FA, Sanni JM and Abigor RD (2004). Lipase activity in dormant seeds of African oil bean (Pentaclethra macrophylla Benth). Food Chem. 88(3):405-408. http://dx.doi.org/10.1016/j.foodchem.2004.01.052

Forgia TA, Conkerton EJ, Philip E (1995). Regio selective analysis of triacylglycerols by lipase enzymes with attractive applications for stepwise hydrolysis of triolein using Candida rugosa lipase in biphasic oil-water system. Angew. Chem. Int. Ed. 37:1608-1633.

Freire GDM, Castilho FL (2008). Lipases em Biocatalise in: Bon et al.(org). Enyimas em biotechnologia: Producao Applicacao e Mercado, Rio d Janeiro, Interciencia. 1:369-385.

Gandhi NN (1997). Applications of lipase. J. Am Oil Chem. Soc. 74:621-634. http://dx.doi.org/10.1007/s11746-997-0194-x

Guit RP, Kloosterman M, Menderma GW, Mayer M, Meijer EM (1991). Lipase kinetics: Hydrolysis of triacetin by lipase from Candida cylindracea in a hollow-fiber membrane reactor. Biotechnol. Bioeng. 38:727-732. http://dx.doi.org/10.1002/bit.260380706

Henderson J, Osborne DJ (1991). Lipase activity in ripening and mature fruit of the oil palm. Stability in vivo and in vitro. Phytochemistry 30:1073-1078. http://dx.doi.org/10.1016/S0031-9422(00)95175-6

Hermansyah H, Kubo M, Shibasaki-Kitakawa N, Yonemoto T (2006). Mathematical model Humicola lanuginosa lipase on its thermal stability. J. Biochem. Biophys. Acta 1547:329-338.

Ibrahim CO (1996). Lipolytic enzyme technology for the palm oil industry proceedings of 1996 PORIM International Oil Congress-Chemistry, Technology, Soap and Detergent Industry Conference, 23-28 Sept., 1996 at hotel Isana, Kuala Lumpur, Malaysia. pp.133-149.

Iftikhar T, Niaz M, ZiaMA, Haq IU (2010c). Production of extracellular lipases by Rhizopus oligosporus in a stirred fermentor. Braz. J. Microbiol. 41(4):1124-1132. http://dx.doi.org/10.1590/S1517-83822010000400034

Khan MY, Dahot MV and Noomrio M H (1991). Investigation of lipase activity from Cajanus cajan L seed pack. J. Sci. Ind. Res. 34:384-386.

Laemmli UK (1970). Cleavage of structural proteins during the assembly of the head of bacteriophage T4. Nature 227:680-685. http://dx.doi.org/10.1038/227680a0

Lineweaver H, Burk D (1934). "The determination of enzyme dissociation constants". J. Am. Chem. Soc. 56(3):658-666. http://dx.doi.org/10.1021/ja01318a036

Lotrakul P, Dharmsthiti S (1997). Lipase production by Aeromonassobria LP004 in a medium containing whey and soybean meal. World. J. Mic. Biotechnol.13:163-166. http://dx.doi.org/10.1023/A:1018581512540
Lotschert W, Beese G (1983). Collins Guide to Tropical Plants. London: Collins. ISBN 978-0-00-219112-8. 83-1496.
Lowry O, Rosebrough NJ, Farr AL (1951). Folin- phenol reagent. J. Biol. Chem. 193:265-275.
Monnet TY, Pamphile KBK, Soro RY, Due EA, KouameLP (2012). Biochemical characterization of lipase activities in dormant seeds of T. catappa. J. Appl. Biosci. 24:337-3382.
Ohnishi K, Yoshida Y, Toita J and Sekiguchi J (1994). Purification and characterization of a novel lipolytic enzyme from Aspergillus oryzae. J. Ferm. Bioeng. 78(6):413-419. http://dx.doi.org/10.1016/0922-338X(94)90039-6
Pancreac'h G and Baratti JC (1996). Hydrolysis of p-nitrophenolpamitate in n-heptane by Pseudomonas cepacia lipase: a simple test for the determination of lipase activity in organic media. Enzyme Microb. Technol. 18:412-422.
Rahman R, Baharum SN, Basri M, Salleh AB(2005). High-yield purification of an organic solvent-tolerant lipase from Pseudomonas sp. strain S5. Anal. Biochem. 341(2):267-274. http://dx.doi.org/10.1016/j.ab.2005.03.006
Ratna AS (2008). Partial characterization of llipase from cocoa beans (Theobroma cacao. L.) of clone PBC 159. Ind. J. Chem. 8(3):448-453.

Alpha-amylase inhibitor of amadumbe (*Colocasia esculenta*): Isolation, purification and selectivity toward α- amylases from various sources

R. McEwan[1]*, R. P. Madivha[1], T. Djarova[1], O. A. Oyedeji[2] and A. R. Opoku[1]

[1]Department of Biochemistry and Microbiology, University of Zululand, P/B X1001, KwaDlangezwa 3886, South Africa.
[2]Department of Chemistry, University of Zululand, P/B X1001 KwaDlangezwa 3886, South Africa.

Two proteins (A-1 and B-2) with α-amylase inhibitor activity were extracted and partially purified from *Colocasia esculenta* tubers through 80% ammonium sulphate fractionation, ion-exchange chromatography on DEAE-Sephacel and gel-chromatography on Sephadex G-100. The molecular weight of A-1 and B-2 were estimated to be about 17000 and 19000 daltons, respectively. The inhibitors inactivated α-amylases of animal origin, but had no effect on fungal amylase. Inhibitor A-1 also exhibited activity towards plant amylases, while inhibitor B-2 has no activity on plant amylases. Inhibitor A-1 was the most active against human salivary amylase at pH 6. Inhibitor A-1 was completely destroyed at temperatures above 50°C; while inhibitor B-2 was stable up to 70°C.

Key words: α-amylase inhibitor, amadumbe, *Colocasia esculenta*, diabetes, obesity.

INTRODUCTION

Colocasia esculenta is a tuber commonly known by several common names, such as Amadumbe (South Africa), Taro (Pacific islands) or Cocoyam (West Africa). It is a traditional food crop widely grown in the subtropical parts of South Africa. The tuber could be consumed as a good source of starch and the locally developed cultivars are commonly used as staple food in the KwaZulu-Natal (KZN) province of South Africa. Nutritional evaluation of the tubers showed storage of biologically active substances that have anti-nutritional properties (that is, they reduce the availability of nutrients to animals and humans). Amongst these substances, α-amylase inhibitors were present (McEwan, 2008).

Protein inhibitors of α-amylase occur widely in plants. The inhibitors are believed to make plants less palatable, even lethal to insects, thus contributing some selective advantage to the plants (Sasikiran et al., 2002). Amylase inhibitors are known as starch blockers because they prevent dietary starches from being digested and absorbed by the body. This could be useful for treating obesity and diabetes mellitus – a metabolic disorder characterized by chronic hyperglycaemia resulting from defects in insulin secretion (Ali et al., 2006). Plants play an important role in the treatment of diabetes, particularly in developing countries where most people have limited resources and do not have access to modern treatment. Because of the possible importance of these inhibitors in plant physiology and animal and human nutrition, extensive research has been conducted on their properties and biological effects (Garcia-Olmedo et al., 1987; Silano, 1987). In this study, we report the isolation and characterization of this specific inhibitor in *C. esculenta*.

*Corresponding author. E-mail: ronalda.mcewan@nmmu.ac.za

MATERIALS AND METHODS

Amadumbe (*C. esculenta*) tubers were obtained from the local market at Esikhawini, KZN, South Africa. The amylases (human salivary, type IX-A; porcine pancreatic, type I-A; sweet potato, barley, *Bacillus* species, type II-A; *Aspergillus oryzae*) and all other reagents used were obtained from Sigma Chemical Co., USA.

Protein extraction

Tubers with no physical signs of infection were washed, peeled and

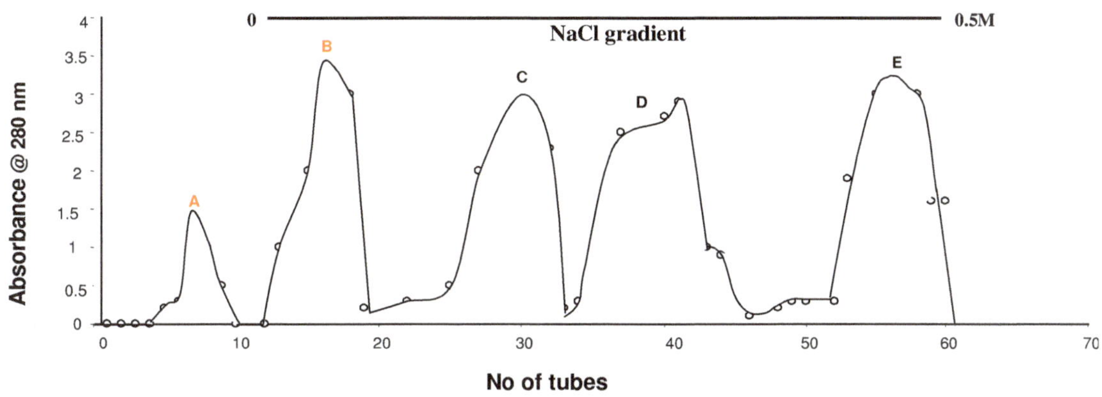

Figure 1. Ion exchange chromatography.

cut into small (2 × 3 cm) pieces and dried at 40°C for 24 h. The dried material was milled (68 mesh) and the flour were defatted with hexane and air-dried. The defatted flour were added (1:5 w/v) to distilled water (containing 1% polyvinylpolypyrrolidone), stirred for 2 h, and filtered. The residue was re-extracted and the combined filtrate was centrifuged at 12 000 × *g* for 20 min. The supernatant (designated as the crude extract) was subjected to 80% $(NH_4)_2SO_4$ saturation and left overnight at 4°C. Protein pellets obtained after centrifugation (12 000 × *g,* 20 min) were re -dissolved in minimum volume of phosphate buffer (0.02 M, pH 6.9, containing 0.3 M NaCl), dialyzed extensively (48 h, with 6 h change of buffer) against the buffer (designated the ammonium sulfate extract) and analyzed for amylase inhibitor (AI) activity.

Ion-exchange chromatography

The ammonium sulfate extract was further purified through ion-exchange chromatography (6 × 1.1 cm DEAE-Sephacel equilibrated with 0.02 M phosphate buffer, the column was eluted with a linear NaCl gradient of 0.2 - 0.5 M at the flow rate of 20 ml/h; 5 ml fractions were collected). The absorbance of the effluent was monitored at 280 nm. Individual peaks were pooled and analyzed for protein and AI activity.

Two peaks (A and B, Figure 1) with AI activity were then separately chromatographed on a Sephadex G-100 column (35 × 1.1 cm, equilibrated with the phosphate buffer and eluted with same buffer at a flow rate of 15 ml/h; 5 ml fractions were collected and the absorbance at 280 nm was determined). Pooled fractions were analyzed for protein and AI activity. Fractions (A-1 and B-2, Figure 2) with AI activities were collected, dialyzed extensively (48 h, with 6 h change of 0.02M phosphate buffer), freeze-dried and dissolved in deionized water.

Enzyme inhibitor assay

Amylase and amylase inhibitor activity assays were based on Bernfeld's method for amylase assay (Bernfeld, 1955). Amylase inhibitor extracts were mixed with amylase and incubated for 30 min at 37°C. The reaction was started by adding extract-enzyme mixture to test tubes containing buffered starch solution (2 mg starch in 20 mM phosphate buffer of pH 6.9 containing 0.4 mM NaCl) and was incubated for 20 min. This reaction was terminated by adding 3,5-dinitrosalicyclic acid (DNS) reagent to the assay mixture. The assay tubes were kept in a boiling water bath for 5 min, cooled under tap water and the colour formed by maltose oxidation was measured at 530 nm. Controls without inhibitor were run simultaneously. One amylase unit is defined as the amount of enzyme that will liberate 1 µmol of maltose from starch under the assay conditions (pH 6.9, 37°C, 5 min). Inhibitory activity is expressed as the percentage of inhibited enzyme activity out of the total enzyme activity used in the assay.

Gel chromatography on Sephadex G-100

Molecular weight was determined by gel chromatography on Sephadex G-100, under the conditions described by Sigma. Cytochrome C (12.4 kDa), carbonic anhydrase (29 kDa), alcohol dehydrogenase (150 kDa), and β-amylase (200 kDa) were used as molecular weight markers. Proteins were determined with the Sigma kit.

pH and temperature stability of inhibitors

The inhibitors (70 µg) were separately mixed with 20 µmol of phosphate buffer (pH 6) and incubated at the various temperatures (20 - 100°C) for 30 min. Aliquots were withdrawn and assayed for residual inhibitory activity against human salivary amylase as already described.

Inhibitor A1 (70 µg) was incubated with 20 µmol of each of HCl (pH 1, 2), citrate buffer (pH 3, 4), acetate buffer (pH 5), phosphate buffer (pH 6, 7, 8), glycine-NaOH buffer (pH 9, 10) in a volume of 0.2 ml at 25°C for 30 min. Aliquots were assayed for inhibitory activity against human salivary amylase as described. pH stability of inhibitor B2 was not determined.

RESULTS

Two proteins (A and B-1) with α-amylase inhibitory activities were purified from Amadumbe tubers by ion-exchange chromatography on a DEAE-Sephacel column. The elution pattern of the ion-exchange chromatography is shown in Figure 1. Only two (A and B) of the five protein peaks showed α-amylase inhibitory activity.

The inhibitor peaks (Figure 1) from the DEAE-Sephacel chromatography were subsequently separated by gel filtration. Two (A1 and B2) of these proteins had inhibitory activity against α-amylase (Figure 2).

AI-A and AI-B were purified (3.89- and 4.42-fold, respectively from the crude extract with a yield of approximately

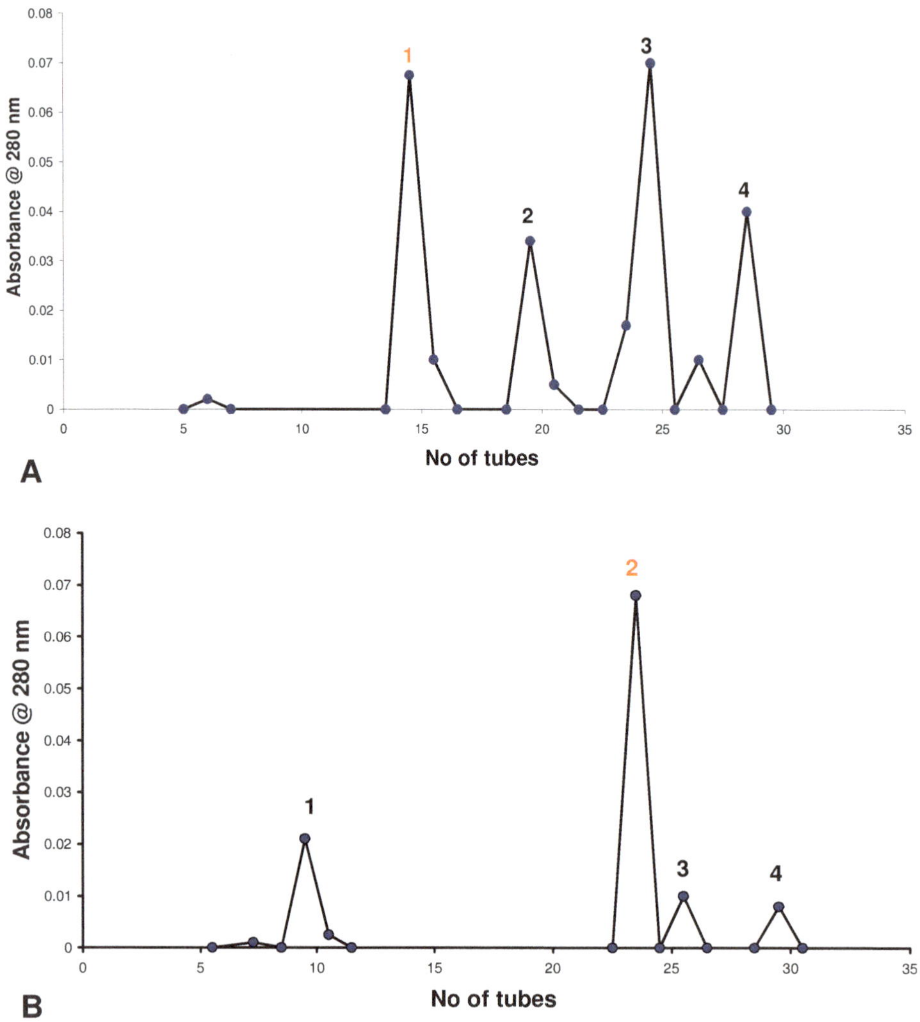

Figure 2. Gel chromatography for sample A and B.

from the crude extract, with a yield of approximately 18 and 15%, by the procedures shown in Table 1. The proteins had specific activity of 11.29 and 12.83 U/mg, respectively.

The purified AI-A1 and AI-B2 fractions showed a molecular weight of approximately 17 and 19 kDa, respectively, as estimated by Sephadex G-100 gel filtration (Figure 3) and the use of standard proteins (cytochrome C, carbonic anhydrase, alcohol dehydrogenase, and β-amylase).

The results obtained from the studies on the effect of temperature, pH, and the inhibitory activity against various amylases are presented in Table 2. The two AI-A1 and B2 showed no inhibition to the fungal amylase and AI-B2 showed no inhibition to plant amylases from barley and sweet potato. However, both inhibitors did show activity against amylase from human saliva and porcine pancreas. The two proteins displayed temperature optima for activity at 40°C and a complete loss of activity at 80°C. The purified proteins showed optimum activity at pH 4.0 and 6.0 with a rapid decline as the pH increase above 7.5.

DISCUSSION

While proteinaceous inhibitors of α-amylase from cereals (Roy and Gupta, 2000; Heidari et al., 2005; Muralikrishna and Nirmala, 2005) and legumes (Giri and Kachole, 1998; Melo et al., 1999) have been well characterized, little is known of these inhibitors in starchy tubers. The present work describes the purification and characterization of the two α-amylase inhibitors present in an ammonium sulphate extract of *C. esculenta.* The little available information on α-amylase inhibitors present in *C. antiquorum* (Sharma and Pattabiraman, 1980), taro (Seltzer and Strumeyer, 1990) and sweet potato (Rekha et al., 1999) seemed to indicate that most tubers had two proteins that showed inhibiting activity. The molecular weight of these proteins ranged between 11 and 25 kDa. Temperature is

Table 1. Purification of amylase inhibitors from Amadumbe.

Fraction	Total protein (mg)	Inhibitor (units)	Specific activity (units/mg)	Yield (%)	Purification
Crude extract	20.9	62	2.9	100	
Ammonium sulphate	18.5	56	3.02	88.5	1.04
DEAE Sephacel					
Peak A	7.2	54	7.5	33.44	2.59
Peak B	5.9	50.5	8.56	28.22	2.93
Sephadex G-100					
Peak A1	3.7	41.8	11.29	17.70	3.89
Peak B1	5.6	39.8	12.83	14.83	4.42

The inhibitor units were calculated using human salivary amylase.

Table 2. Kinetic studies of *C. esculenta* α-amylase inhibitors.

α-Amylase inhibitor activity								
Action of inhibitors on different amylases			Action of inhibitors on different temperature			Action of inhibitors using different pH with human salivary amylase		
Source of amylases	% Inhibition		Temperature (°C)	% Inhibition		pH	% Inhibition	
	A1	B2		A1	B2		A1	
Human saliva	62	56	20	15	35.7	1	0	nd
Barley	56	0	30	0	17.8	2	0	nd
Sweet potato	4.7	0	40	13	28.5	3	0	nd
Bacillus species	10.2	23	70	0	2.07	4	17.2	nd
Aspergillus species	0	0	80	0	0	5	14.2	nd
Porcine pancreas	28.5	48.5	100	0	0	6	33.8	nd
						7	7.3	nd
						8	1.8	nd
						9	0	nd
						10	0	nd

nd – not determined.

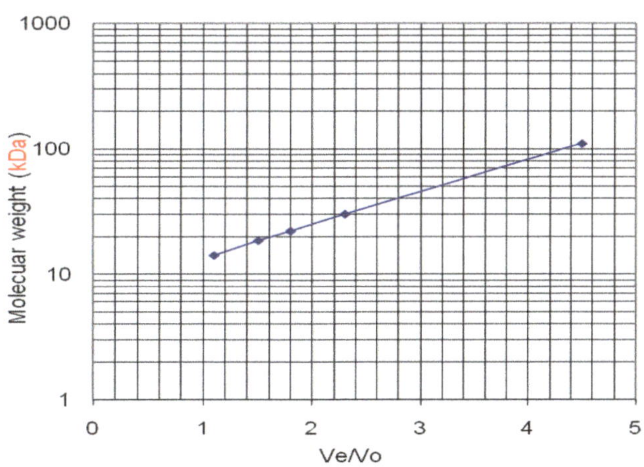

Figure 3. Standard (Mw of standards in Daltons) log molecular weight graph using gel filtration. Standards cytochrome C (12 400) carbonic anhydrase (29 000) alcohol dehydrogenase (150 000) β-amylase (200 000).

one of the most important parameters that affect the rate of enzyme hydrolysis. The optimum temperature displayed for both the *C. esculenta* inhibitors are similar to those reported for the navy-bean amylase inhibitor (Hoover and Sosulski, 1984) and for the α-amylase inhibitor from *Pachyrhizus erosus* tuber (Noman et al., 2006). We can also conclude that alpha-amylase inhibitors are fairly heat-stable (Prathibha et al., 1995). The residual activity in processed Amadumbe could be attributed to this property (McEwan, 2008).

α-Amylase inhibitors show strict target enzyme specificity and recognize only one out of several closely related isoenzymes or enzymes from different species (Weselake et al., 1983; Franco et al., 2000). In most cases the mechanism of inhibition occurs through the direct blockage of the active centre at several sub sites of the enzyme (Payan, 2004). Literature (Sharma and Pattabiraman, 1980, 1982; Ida et al., 1994) indicates that many AI present in tubers are active against mammalian amylases, but exhibit no activity on plant amylases. Alpha-amylase inhibitors present in Amadumbe also had no effect on the *Aspergillus* amylase. Amadumbe grow in moist, humid conditions and fungal infections are prevalent. It is possible that fungi have become resistant to the action of these α-amylase inhibitors. Sharma and Pattabiraman (1982) have reported similar results for *Dioscorea alata*.

Bifunctional properties have been demonstrated by a number of inhibitors and have therefore, received a particular attention as the appealing candidates for pest-control (Maskos et al., 1996). α-Amylase inhibitors found in wheat (Franco et al., 2000), barley (Richardson, 1991) and Indian finger millet (Campos and Richardson, 1983) have been shown to efficiently inhibit α-amylases from different insect sources. It is possible that the two AI present in Amadumbe may complement each other, with defense against a wider spectrum of intruders.

However, as reported, increases in urbanization and the associated changes in life-style, incidences of obesity and diabetes are on the increase. Inhibitory activity of plant amylases against mammalian amylases could cause a marked decrease in the availability of digested starch (Ali et al., 2006). This could suggest a potential in the prevention and treatment of diabetes and nutritional problems, which result in obesity. Based on the results of this study, the two α-amylase inhibitors of *C. esculenta* may have similar potential in the prevention and therapy of obesity and diabetes. Additional studies are needed to further investigate this possibility.

REFERENCES

Ali H, Houghton PJ, Soumyanath A (2006). α-Amylase inhibitory activity of some Malaysian plants used to treat diabetes; with particular reference to *Phyllanthus amarus*. J. Ethnopharmacol., 107: 449-455.

Bernfeld P (1955). Amylases α and β. In: Colowick, S.P. and Kaplan, N.O. (eds), Methods in Enzymology, Academic Press, New York, (1): 149-158.

Campos FAP, Richardson M (1983). The complete amino acid sequence of bifunctional α-amylase/trypsin inhibitor from seeds of ragi (Indian finger millet, *Eleusine coracana* Gaertneri). FEBS Lett., 152: 300-304.

Franco OL, Rigden DL, Melo FR, Bloch C, Silva CP, Grossi-de-Sá MF (2000). Activity of wheat α-amylase inhibitors towards bruchid α-amylases and structural explanation of observed specificities. Eur. J. Biochem., 267: 2166-2173.

Garcia-Olmedo F, Salcedo G, Sanchez-Monge R, Gomez L, Royo J, Carbonero P (1987). Plant proteinaceous inhibitors of proteinases and alpha-amylases. In: Miflin, B. (Ed.), Oxford University Press, Oxford, Oxford Surveys Plant Mole. Cell Biol., 4: 275-334.

Giri AP, Kachole MS (1998). Amylase inhibitors of pigeonpea (*Cajanus cajan*) seeds. Phytochemistry. 47: 197-202.

Heidari R, Zareae S, Heidarizadeh M (2005). Extraction, purification, and inhibitory effect of alpha-amylase inhibitor from wheat (*Triticum aestivum* Var. *Zarrin*). Pakistan J. Nutr., 4: 101-105.

Hoover R, Sosulski F (1984). Characteristics and concentrations of α-amylase inhibitor in *Phaseolus vulgaris* biotypes. Starch/Starke. 36: 246-250.

Ida EI, Finardi-Filho F, Lajolo FM (1994). Purification and partial characterization of two proteinaceous α-amylase inhibitors from triticale. J. Food Biochem., 18: 83-102.

Maskos K, Huber-Wunderlich M, Gloskshuber R (1996). RBI, a one-domain α-amylase/trypsin inhibitor with completely independent binding sites. FEBS Lett., 397: 11-16.

McEwan R (2008). Anti-Nutritional content of *Colocasia esculenta* (Amadumbe) a traditional crop food in Kwazulu-Natal, pp 59-74. PhD thesis, University of Zululand, Empangeni, South Africa.

Melo FR, Sales MP, Pereira LS, Bloch C, Franco OL, Ary MB (1999). Alfa-Amylase inhibitors from cowpea seeds. Protein and Peptide Lett., 6: 385-390.

Muralikrishna G, Nirmala M (2005). Cereal α-amylases – an overview. Carbohydr. Polym., 60: 163-173.

Noman ASM, Hoque MA, Sen PK, Karim MR (2006). Purification and some properties of α-amylase from post-harvest *Pachyrhizus erosus* L. tuber. Food Chem., 99: 444-449.

Payan F (2004). Structural basis for the inhibition of mammalian and insect α-amylases by protein inhibitors. Biochem et Biophys., 1696: 171-180.

Prathibha S, Bala N, Leelama S (1995). Enzyme inhibitors in tuber crops and their thermal stability. Plant Foods Hum. Nutr., 48: 247-257.

Rekha MR, Padmaja G, Easwari Amma CS, Sheela MN (1999). Genotype differences in the α-amylase inhibitor activity in sweet potato and yam tubers. J. Root Crops, 25: 95-101.

Richardson M (1991). Seed storage proteins: the enzyme inhibitors In: Rogers, L.J. (Ed). Methods in Plant Biochemistry, vol. 5, Academic Press, New York, pp 259-305.

Roy I, Gupta MN (2000). Purification of a 'double-headed' inhibitor of alpha-amylase / proteinase K from wheat germ by expanded bed chromatography. Bioseparation, 9: 239-45.

Sasikiran K, Rekha MR, Padmaja G (2002). Proteinase and alpha-amylase inhibitors of sweet potato: Changes during growth phase, sprouting, and wound induced alterations. Bot. Bull. Acad. Sinica, 43: 291-298.

Seltzer RD, Strumeyer DH (1990). Purification and characterization of Esculentamin, A proteinaceous alpha-amylase inhibitor from the taro root, *Colocasia esculenta*. J. Food Biochem., 14: 199-217.

Sharma KK, Pattabiraman TN (1980). Natural plant enzyme inhibitors. Isolation and characterization of two α-amylase inhibitors from *Colocasia antiquarum* tubers. J. Sci. Food Agric., 31: 981-991.

Sharma KK, Pattabiraman TN (1982). Natural plant enzyme inhibitors. Purification and properties of an amylase inhibitor from yam (*Dioscorea alata*). J. Sci. Food Agric., 33: 255-262.

Silano V (1987). Alpha-amylase inhibitors. In: Kruger JE, Lineback D, Stauffer CE (Eds.), Enzymes and their roles in Cereal Technology, American Association of Cereal Chemists, St. Paul, MN., pp 141-199.

Weselake RJ, MacGregor AW, Hill RD (1983). An endogenous α-amylase inhibitor in barley kernel. Plant Physiol., 72: 809-812.

Lecithin: Cholesterol acyltransferase, lipoprotein lipase and lipoproteins in adult Nigerians with sickle cell disease

M. A. Emokpae[1, 2*], P. O. Uadia[3] and H. B. Osadolor[2]

[1]Department of Chemical Pathology, Aminu Kano Teaching Hospital, Kano, Kano State, Nigeria.
[2]Department of Medical Laboratory Science, School of Basic Medical Sciences, University of Benin, Benin City, Edo State, Nigeria.
[3]Department of Biochemistry, University of Benin, Benin City, Edo State, Nigeria.

Lecithin: cholesterol acyltransferase (LCAT) is a glycoprotein enzyme involves in the esterification of plasma cholesterol to form cholesteryl ester and is important in the maturation of high density lipoprotein. Lipid metabolism in sickle cell anaemia patients may be affected. The objective of this study was to determine the activities of serum lecithin: cholesterol acyltransferase, lipoprotein lipase, lipids and lipoprotein levels in sickle cell anaemia patients in a steady state. The study population consisted of 144 confirmed sickle cell disease patients; 68 males aged 22.2 ± 3.8 years and 76 females aged 21.0 ± 3.0 years. Fifty age matched males; 25 sickle cell trait (HbAS), 25 normal haemoglobin (HbAA) and fifty females; 25 HbAS and 25 HbAA were used as controls. Statistically significant decrease in LCAT and LPL activities ($p < 0.001$) were observed in sickle cell anaemia patients compared with HbAS and HbAA controls in both sexes. Decrease levels of cholesterol ($p < 0.001$) and HDL cholesterol ($p < 0.001$) were also observed in sickle cell anaemia patients. The activities of lecithin: cholesterol acyltransferase and lipoprotein lipase were lower in subjects with sickle cell anaemia than sickle cell trait and normal haemoglobin. This may contribute to the changes observed in lipid metabolism in sickle cell anaemia. The exact cause is not known but appears to be multifactorial.

Key words: Sickle cell disease, Lecithin: cholesterol acyltransferase, lipoprotein lipase, Lipoproteins.

INTRODUCTION

Lecithin: cholesterol acyltransferase (LCAT) is a 63KDa glycoprotein enzyme which is synthesized in the liver and secreted in the plasma (Francone and Fielding, 1991; Wang et al., 1997). LCAT plays an important role in reverse cholesterol transport. It binds to high density lipoprotein (HDL-cholesterol) to catalyze the transfer of a fatty acid residue from the sn-2 position of lecithin to cholesterol to form cholesterol ester and lysolecithin (Saku et al., 1999).The enzymatic activity of LCAT depends on the presence of apolipoprotein A-1 which serves as cofactor (Vaziri and Parks, 2001). It thus faciletates free cholesterol uptake by HDL particles from the peripheral tissues and maturation of lipoprotein poor HDL_3 to cholesterol ester rich HDL_2 particles (Glomset, 1968). A lack of LCAT activity leads to increase in free cholesterol and phospholipids and decrease in esterified cholesterol. This phenomenon results in abnormalities of all kinds of lipoprotein particles (Nichols et al., 1975; King et al., 1980) which are observed in some organs such as kidney, cornea and erythrocytes in patients with LCAT deficiency (Gjone, 1988; Godin et al., 1978; Hosoda et al., 1984). These changes clinically correspond to renal insufficiency, cornea opacity and haemolytic anaemia respectively (Saku et al., 1999). Lipoprotein lipase (LPL) is an enzyme involved in the regulation of fatty acid influx, responsible for the hydrolysis of triglyceride rich chylomicrons, very low density lipoprotein (VLDL) and intermediate density lipoprotein (IDL), liberating free fatty acid and glycerol for uptake in the in the target tissue. LPL is functional at the vascular surface of the capillary endothelium anchored by proteoglycan chains of heparin sulphate (Vaziri and Parks, 2001). Sickle cell anaemia

*Corresponding author. E-mail: biodunemokpae@yahoo.com.

Table 1. Lipid, lipoproteins, LCAT and LPL in male sickle cell disease subject s compared with HbAS and HbAA controls.

	HbSS males	HbAS Males	P-value	HbAA males	P-value
No of subjects	68	25		25	
Age of years	22.2 ± 3.8	28.7 ± 7	-	28.8 ± 7	-
Triglyceride (mmol/L)	1.10 ± 0.4	1.19 ± 0.18	N/S	1.4 ± 0.12	P < 0.001
Total cholesterol (mmol/L)	3.06 ± 0.5	4.05 ± 0.06	P < 0.001	4.3 ± 0.12	P < 0.001
HDL cholesterol (mmol/L)	0.72 ± 0.17	1.18 ± 0.03	P < 0.001	1.2 ± 0.06	P < 0.001
LDL cholesterol (mmol/L)	1.92 ± 0.54	2.15 ± 0.14	NS	2.52 ± 0.16	P < 0.001
VLDL cholesterol (mmol/L)	0.48 ± 0.06	0.42 ± 0.07	N/S	0.41 ± 0.08	NS
LPL (µmol/glycerol liberated/hr/L)	4.12 ± 1.2	5.12 ± 0.4	P < 0.001	5.56 ± 0.23	P < 0.001
LCAT(µmol/cholesterol liberated/hr/L)	66.8 ± 2.8	69.2 ± 3.0	P < 0.001	70.2 ± 2.96	P < 0.001

(SCA) is a chronic haemolytic disorder and manifests by a wide variety of clinical, biochemical and haemotological features. The chronic haemolytic anaemia often leads to limited oxygen availability to peripheral tissues. This lack of oxygen has the potential to modify events in the intracellular metabolism (Buchwoski et al., 2007) and metabolism of lipids may be altered in individuals with SCA (Rahimi et al., 2006).

Sickle cell disease (SCD) is the most common and most important haemoglobin variant accounting for over 60% of the world's major haemoglobinopathies. It is estimated that 2 - 3 millions Nigerians are affected (Olatunji, 2002). Plasma lipids have been studied in Nigerians (Oforofuo and Adedeji, 1994; Akinyaju and Akinyaju, 1976) and adult African Americans (Shores et al., 2003) with SCD and none has studied the activities of lecithin: cholesterol acyltransferase (LCAT) and lipoprotein lipase (LPL) in conjunction with lipoproteins in this group of individuals. The objective of this study is to determine the concentration of plasma lipids and lipoproteins, the activities of LCAT and LPL in adult sickle cell disease patients in Kano, northern Nigeria.

MATERIALS AND METHODS

The study was conducted in Aminu Kano teaching hospital, Kano from January 2006 to December, 2008.

The study was approved by the ethical committee of the hospital and consent was also obtained from the patients. The study population was made up of 144 confirmed sickle cell disease subjects. They consisted of 68 males aged 22.2 ± 3.8 years and 76 females aged 21.0 ± 3.0 years. They were SCD patients on steady state attending sickle cell disease clinics of the hospital. The controls subjects consisted of 25 HbAS males aged 28.7 ± 7.0, 25 HbAA males aged 28.8 ± 7 years, 25 HbAS females aged 29.0 ± 7 years and 25 HbAA females aged 26.8 ± 8 years.

Fasting blood specimens were obtained from the patients and were allowed to clot at room temperature for 30 min. The specimens were centrifuged at 3000 rpm for 10 min to obtain sera. Serum triglyceride and total cholesterol were determined using enzymes catalyzed colorimetric methods by Randox laboratories, UK. While friedwald (Friedwald et al., 1972) formula was used to calculate LDL cholesterol levels. HDL- cholesterol was determined in the supernatant after precipitation with magnesium chloride-phosphotungstic acid solution. LCAT was measured using the Anasolv LCAT assay obtained from DAICCHI chemicals, japan (Saku et al., 1999; Korn, 1955), in which proteoliposome was used as substrate. The substrate (0.2 ml) was incubated with 0.02 ml plasma at 37°C for 60 min. Aliquot was used at 30 min interval to measure free cholesterol. Isopropanol was added as an arresting agent. Then 0.01 ml of the incubating mixture was added to 1.0 ml cholesterol oxidase reagent and was incubated for 10minutes at 37°C. The absorbance was read at 540 nm against reagent blank. Recrystallized cholesterol solution (500 µmol/L) in isopropanol was used as the calibrator. The enzyme activity was expressed as the function of free cholesterol liberated during the incubation period. LPL was determined by incubating the serum in glyceryl trioleate substrate, the glycerol liberated was measured in an aliquot of the incubating mixture (Korn, 1955; Anurag and Anuradha, 2002). The substrate was incubated with 0.2 ml of serum at 37°C for 60 min and 0.1 ml 0.1 NH_2SO_4 was added directly in a centrifuge tube. The content of the tube was allowed to equilibrate at room temperature and 0.1 ml of 0.5 M sodium periodate was added. After 5 min, 0.1 ml of 0.5 M sodium arsenite was added to reduce the excess periodate. After additional 10 min, 9 ml chromotrophic acid reagent was added and the reaction tube place in covered boiling water for 30 min. The cool tube was adjusted to a volume of 10 ml with distilled water and the absorbance read at 570 nm. Glycerol standards were used as calibrators. The results were expressed as micromole glycerol liberated per hour per litre.

Student's t-test for unpaired means was used for statistical comparison of the means. Results were expressed as mean ± SD and value was considered significantly different at $p < 0.05$.

RESULTS

Results are as indicated in Tables 1 and 2. In the male SCD patients, there was a decrease in the mean triglyceride level compared with male HbAS (NS) and male HbAA ($p < 0.001$). Statistically significant decrease in mean cholesterol level was observed in male SCD patients compared with male HbAS ($p < 0.001$), and HbAA ($p < 0.001$). The mean HDL-cholesterol level was also significantly decreased in male SCD patients compared with HbAS ($p < 0.001$) and HbAA ($p < 0.001$). The mean LDL cholesterol level in male SCD patients was also decrease compared with male HbAS (NS) and HbAA ($p < 0.001$). The mean LPL activities in male SCD patients were significant lower than those of the male HbAS ($p < 0.001$) and HbAA ($p < 0.001$). In the same vein, the mean activity of LCAT in SCD patients was

Table 2. Lipid, lipoproteins LCAT and LPL in female SCD patients compared with HbAS and HbAA controls.

	HbSS females	HbAS females	P-value	HbAA females	P-value
No of subjects	76	25		25	
Age (years)	21 ± 3.0	29.0 ± 7	-	26.8 ± 8	-
Triglycerides (mmol/L)	1.05 ± 0.56	1.3 ± 0.18	P < 0.001	1.42 ± 0.12	P < 0.001
Total cholesterol (mmol/L)	3.43 ± 0.4	4.1 ± 0.2	P < 0.001	4.4 ± 0.14	P < 0.001
HDL cholesterol (mmol/L)	0.78 ± 0.15	1.26 ± 0.05	P < 0.001	1.31 ± 0.03	P < 0.001
LDL cholesterol (mmol/L)	2.0 ± 0.39	2.62 ± 0.04	P < 0.001	2.68 ± 0.04	P < 0.001
VLDL cholesterol (mmol/L)	0.41 ± 0.08	0.62 ± 0.08	NS	0.42 ± 0.07	NS
LPL (μmol/glycerol liberated/hr/L)	3.39 ± 0.21	5.22 ± 0.5	P < 0.001	5.36 ± 0.3	P < 0.001
LCAT (μmol/cholesterol liberated/hr/L)	59 ± 1.3	69 ± 3.2	P < 0.001	71.2 ± 2.2	P < 0.001

significantly lower than those the male HbAS ($p < 0.001$) and HbAA ($p < 0.001$) controls (Table 1).

In the female SCD patients, there was statistical significant decrease in the mean triglyceride compared with the female HbAS ($P < 0.001$) and HbAA ($P < 0.001$). The mean total cholesterol level was significantly lower than the female HbAS ($P < 0.001$) and HbAA ($P < 0.001$). The mean LDL cholesterol was lower than the female HbAS ($P < 0.001$) and HbAA ($P < 0.001$).

Similarly, the mean LPL activity in the female SCD patients was significantly lower than HbAS ($P < 0.001$) and HbAA ($P < 0.001$).

The LCAT activity in SCD patient was also significantly lower than HbAS ($P < 0.001$) and HbAA ($P < 0.001$) controls (Table 2). Comparing Tables 1 and 2, it was observed that mean triglyceride level was significantly lower in female SCA than Hb AS and HbAA whereas in the male subjects, the concentration was only significantly lower in HbAA and not in HbAS.

DISCUSSION

The finding of low serum cholesterol and other lipoprotein levels in SCD patients is consistent with other reports both in Nigeria (Oforofuo and Adedeji, 1994; Akinyaju and Akinyaju, 1976) and elsewhere (Rahimi et al., 2006; Zailare et al., 2003; Elmehdawi, 2008). It is important to mention that the levels in Nigerian adult sickle cell disease patients is much more lower compared with the lipid levels reported in both African American16 and Saudi Arabian20 patients with SCD. There is an indication of progressive decrease in plasma cholesterol level wherever HbS gene is inherited as sickle cell trait (HbAS) or as SCD (HbSS) irrespective of sex. This is an indication that the low cholesterol and other lipoproteins levels are intrinsic to the disease. The exact etiology of hypocholesterolaemia in SCA is not known but several studies postulated different mechanisms (Oforofuo and Adedeji, 1994; Shalev et al., 2007; Lakhani et al., 2001).

Hypocholesterolaemia tends to occur in patients with chronic anaemia, increased erythropoietic activity and it has been suggested that this may be due to increase cholesterol requirements by the proliferative erythroid cells (Elmehdawi, 2008).

The chronic anaemia may leads to an increased plasma volume and having "dilution effect" on plasma consistuents 21. There may be down regulation in the pathway of cholesterol biosynthesis in the hepatic cells of SCD subjects (Shalev et al., 2007).

A Low LCAT and LPL activity was observed in SCD subjects compared with HbAS and HbAA subjects in this study. Little is known about LCAT metabolism and regulation in health and disease (Lakhani et al., 2001). LCAT appears to be exclusively secreted by the liver and few perturbations such as inflammatory mediators affect its biosynthesis (Saku et al., 1999). It was observed that SCA patients in steady state may undergo constant inflammatory process which may occur as a result of subclinical vaso-occlusion which may in turn leads to inflammatory response (Bourantas et al., 1998). The biosynthesis of this enzyme may be down regulated in sickle cell disease due to subclinical inflammatory episodes. Because LCAT plays an important and central role in plasma lipoprotein metabolism, the effects of a significant decrease in LCAT activity in SCD patients as observed in this study may be seen in a wide range of lipid and lipoprotein changes.

These altered lipolytic enzymes and lipoproteins reflect changes in composition and concentration of different lipoproteins and can also lead to alterations in lipid composition of membrane such as in red blood cells. The low LCAT activity in SCA patients may affect reverse cholesterol transport, HDL maturation and oxidation of atherogenic lipids in SCA patients.

Conclusion

Lipid metabolism in SCD appears to be different from that in sickle cell trait and normal haemoglobin in adult Nigerian SCD patients. The exact cause is not known but appears to be multifactorial.

REFERENCES

Akinyaju PA, Akinyaju CO (1976). Plasma and red cell lipids in sickle cell disease. Annals Clin. Lab. Sci. 6(6): 521-524.

Anurag P, Anuradha CV (2002). Metiformin improves lipid metabolism and attenuates lipid peroxidation in high fructose fed rats. Diab. obesity metabol. 4(1): 36-42.

Bourantas KL, Dalekos GN, Kakis A, Chaidos A, Tsiara S and Mavridis A (1998). Acute phase proteins and interleukins in steady state sickle cell disease. Eur. J. Haematol. 61(1): 49-54.

Buchwoski MS, Swift LL, Akohoue SA, Shaukar SM, Flakoll PJ, Abumrad N (2007). Defects in postabsorptive plasma homeostasis of fatty acids in sickle cell disease. J. Parenter. Enteral. Nutr. 31(4): 263 – 268.

Elmehdawi RR (2008). Hypolipidemia: A word of caution. Libyan J. Med. p. 1-7.

Francone OL, Fielding CJ (1991). Structure function relationships in human lecithin:cholesterol acyltransferase. Site-directed mutagenesis at serine residues 181 and 216. Biochem. 30: 10074-10077.

Friedwald WT, Levy RL, Fredickson DS (1972). Estimation of concentration of low density lipoprotein cholesterol in plasma without use of preparative ultracentrifuge. Clin. Chem. 10: 499-502.

Gillet MPT, Obineche EN, Lakhani MS, Abdulle AM, Amirlak I, Rukhaimi MA, Suleiman MN (2001). Levels of cholesterol ester and lipids in the plasma of patients with end stage renal failure. Ann. Saudi Med. 21(5-6): 283-286.

Gjone E (1988). Familial lecithin:cholesterol acyltransferase(LCAT) deficiency: an updated review. Ophthalmic Paediatric Genet. 9: 167-169.

Glomset JA (1968). The plasma lecithin:cholesterol acyltransferase reaction. J. lipid Res. 9: 155-167.

Glomset JA, Mitchell CD, King WC, Applegate KR, Forte T, Norum KR, Gjone E (1980). *In vitro* effects of lecithin:cholesterol acyltransferase on apolipoprotein distribution in familial lecithin:cholesterol acyltransferase deficiency. Ann. NY Acad. Sci. 348: 224-243.

Godin DV, Gray GR, Frohlich J (1978). Erythrocte membrane alterations in lecithin:cholesterol acyltransferase deficiency. Scand J. Clin. Lab. Invest. 150(suppl): 162-167.

Jimi N, uesugi N, Saku K, Itabe H, Zhang B, Arakawa K, Takebayashi S (1999). Possible induction of renal dysfunction in patients with lecithin: cholesterol acyltransferase deficiency by oxidized phosphotidylcholine in Glomeruli. AHAJ Arterioscler Thromb vascular Biol. 19: 794-801.

Korn ED (1955). Clearing factor: a heparine activated lipoprotein lipase. Isolation and characterization of enzyme from normal rat heart. J. Biol. Chem. 215: 1-26.

Murayama N, Asano Y, Hosoda S, Maesawa M, Saito M, Takaku F, Sugihara T, Miyashima K, Yawata Y (1984). Decreased sodium influx and abnormal red cell membrane lipids in a patient with familial plasma lecithin:cholesterol acyltransferase deficiency. Am. J. Hematol.16: 129-137.

Norum KR, Glomset JA, Nichols AV, Forte T, Alberts J, King WC, Mitchell CD (1975) . Plasma lipoproteins in familial lecithin:cholesterol acyltransferase deficiency: effects of incubation with lecithin:cholesterol acyltransferase *in vitro*. Scand J. Clin. Lab. Invest. 35(142 suppl): 31-35.

Oforofuo IAO, Adedeji MO (1994). Effect of sickle cell gene expression in plasma cholesterol in a Nigerian population. Clin. Biochem. 27: 505-508.

Olatunji PO (2002). Sickle cell disease in developing countries: Magnitude and Challenges. Postgrad. Doc. Afr. 25(3): 61-64.

Rahimi Z, Merat A, Haghshenass M, Madani H, Rezeal M, Nagel RL (2006). Plasma lipids in Iranians with sickle cell disease: Hypocholesterolemia in sickle cell anaemia and increase of HDL-cholesterol in sickle cell trait. Clin. Chimica Acta 365(1-2): 217 – 220.

Shalev H, kapelushnik J, Moser A, Knobler H, Tamary H (2007). Hypocholesterolemia in chronic anaemias with increased erythropoietic activity. Am. J. Hematol. 829(3): 199-202.

Shores J, Peterson J, Vander jagt D, Glew RH (2003). Reduced cholesterol levels in African American Adults with sickle cell disease. J. Nat. Med. Ass. 95(11): 813-817.

Vaziri ND, Parks JS (2001). Down regulation of hepatic lecithin:cholesterol acyltransferase gene expression in chronic renal failure. Kidney int. 59: 2192-2196.

Wang J, Gebre AK, Anderson RA, Parks JS (1997). Cloning and *in vitro* expression of rat lecithin:cholesterol acyltransferase. Biochem. Biophs. Acta 1346: 207-211.

Zailare MZ, Marzouki ZM, Khoja SM (2003). Plasma and red blood cell membrane lipid concentration of sickle patients Saudi Med. J. 24(4): 376-379.

14

Purification and biochemical characterization of a xylanase purified from a crude enzyme extract for the determination of active site residues

Sacha Tabosa-Vaz, Ajit Kumar, Kugen Permaul and Suren Singh*

Department of Biotechnology and Food Technology, Durban University of Technology, Box 1339, Durban 4001, South Africa.

This paper describes the purification of a thermostable xylanase (Xyn) from a commercially available xylanase stock (Buzyme 2511 from Buckman Laboratories International, South Africa). The purified enzyme showed a Mr. of 66 kDa, and presented a dimeric form with two similar molecular weight chains of 33 kDa. The optimum pH and temperature were found to be 6.0 and 50°C, respectively. The enzyme was active over a broad pH (5 to 10) and temperature (50 to 90°C) range. The enzyme showed 90% residual activity at 70°C for more than 35 min. The substrate progress curve showed that the enzyme substrate reaction attains V_{max} within 30 min. The K_m value determined by Lineweaver-Burk Plot was 1 mg/ml by using birchwood xylan as a substrate. The inhibition of enzyme activity by chemical modifiers like *o*-phthaldehyde, trinitrobenzenesulphonic acid, diethyl pyrocarbonate and woodword's reagent K suggest the importance of Cys, His, Lys and Glu residues for the catalytic activity of the enzyme.

Key words: Xylanase, buzyme 2511, enzyme thermostability, steady state kinetics.

INTRODUCTION

The endo-1, 4-β-xylanase (EC 3.2.1.8) is the key enzyme (Wong et al., 1988) that depolymerizes xylan, which is the main component of hemicellulose and consists of 1, 4-glycosidically linked β-D-xylose. Xylanases are used in paper manufacturing, waste treatment, deinking, fuel and chemical production (Biely, 1985). The major interest of thermostable xylanases in pulp bleaching is that it helps in reducing the kappa number and increases the brightness of the pulp. Hence, the requirement of chlorine as a bleaching agent could be substantially reduced and this could lead to the development of environmentally friendly technologies (Coughlan and Hazelwood, 1993). There are some reports of xylanases that are active and stable at both alkaline pH and elevated temperatures (Nakamura et al., 1993; George et al., 2001a). Most of the commercial xylanases showing activity in the alkaline range are only active at 50°C (Viikari et al., 1994).

Investigations involving chemical modification of an enzyme can potentially yield insights into structure - function relationships. In comparison to the extensive studies on the biotechnological applications of xylanases, inadequate information is available concerning the mechanism of action of xylanases. The mechanism of action of xylanases is similar to the double displacement mechanism of lysozyme (Vernon and Banks, 1963). The role of carboxyl groups (Chauthaiwale and Rao, 1994; Ko et al., 1992) in catalysis and aromatic residues such as tryptophan (Nath and Rao, 1998) and tyrosine (Bray and Clarke, 1995) in substrate binding of xylanases are well documented. Crystal structure of family 10 xylanases has shown the presence of a His residue in the active site (White et al., 1994; Harris et al., 1996). Charnock et al. (1997) suggested the possible role of a Lys residue in positioning the substrate in the active site of a xylanase from *Pseudomonas fluorescence*.

The role of a Trp residue in the active site of the xylanase from the extremophilic *Bacillus* sp. was

*Correspondence author. Email: singhs@dut.ac.za.

illustrated by Nath and Rao (1998). George et al. (2001b) reported the involvement of a Lys residue in the active site of a xylanase from Thermomonospora sp. Chemical modifiers such as OPTA (*o*-phthaldehyde), TNBS (trinitrobenzenesulphonic acid), DEP (diethyl pyrocarbonate) and WRK (woodword's reagent K) have been used to study the mechanism of enzymatic catalysis and determining the active site residues of many enzymes (George et al., 2001a; George et al., 2001b; Nath and Rao, 2001; Rawat and Rao, 1997; Anish and Rao,2007). In this paper, we describe the purification and biochemical properties of a thermostable xylanase (Xyn) and also identify active site amino acid residues by using chemical modifiers and fluorescent labeling. Buzyme 2511 is an impure commercial enzyme and needed to be characterized and purified to homogeneity which could help in molecular cloning of this enzyme. Further the data will help in understanding the mechanism of action of substrate hydrolysis by this xylanase.

MATERIALS AND METHODS

Materials

Commercially available xylanase stock (Buzyme 2511) was purchased from Buckman Laboratories International, South Africa. Ammonium sulphate, DNS (dinitrosalicylic acid), BSA (bovine serum albumin), sephadexG-100, CBB R-250 (coomassie brilliant blue R-250), β-mercaptoethanol, OPTA (*o*-phthalaldehyde), L-cysteine, TNBS (2, 4, 6-trinitrobenzene sulfoic acid), DEP (diethyl pyrocarbonate, WRK (Woodward's reagent K, N-ethyl-5-phenylisoxazolium-3-sulfonic) were purchased from Sigma-Aldrich, USA. Centricon$^+$20 30kDa cut off columns were from Millipore Corporation USA, Birchwood xylan was from Roth chemicals, Germany and precast 12% SDS-PAGE gels were from Bio-Rad Laboratories. All other chemicals were of analytical grade.

Enzyme assays

Xyn activity assays were carried out by incubating the enzyme with 0.5 ml of xylan (1%) in a final volume of 1 ml and incubating the mixture at 50°C, pH 7 for 30 min. The reducing sugar released was determined by the DNS method (Baily et al., 1992). One unit of Xyn activity was defined as the amount of enzyme that released 1 µmol of xylose equivalent per minute from xylan, under assay conditions. Protein concentration was determined according to the method of Bradford (1976).

Production and purification of Xyn

A xylanase from commercially available xylanase stock (Buzyme 2511) was purified to homogeneity. All purification steps were carried out at 4°C unless otherwise stated. The broth was subjected to fractional ammonium sulphate precipitation (40 to 60%). The precipitate was dissolved in a minimum amount of 50 mM sodium phosphate buffer, pH 7 and dialyzed against 1 L of the same buffer with several changes for 24 h. The dialyzed fraction was applied to a Sephadex G-100 column (150×2 cm), previously equilibrated with 50 mM sodium phosphate buffer at pH 7. Elution was carried out by using the same buffer at a flow rate of 1 ml/min a nd 4-ml fractions were collected during a period of 640 min.

The fractions having maximum specific activity were pooled and concentrated. The purity of the enzyme was checked by SDS-PAGE (Laemmli, 1970) followed by staining with CBB R-250. To further purify and concentrate the samples, active fractions were run through Centricon$^+$20, 30 kDa cut off columns. The supernatant was then assayed to check for activity to ensure the presence of active protein. Centricon$^+$20, 30 kDa cut off columns were used according to the information provided in the manufacturers manual.

Properties of purified Xyn

To determine the optimum temperature and pH of the enzyme, the activity of Xyn (1 µM) was determined at different temperature (20, 37, 50, 60, 70, 80 and 90°C) and pH (4 to 10) values. To determine the temperature stability, the enzyme (1 µM) was incubated at different temperatures (50, 70, 80 and 90°C) for 35 min at pH 6. The aliquots were withdrawn at different time intervals and the residual enzymatic activity was then measured. To determine its pH stability, the enzyme (1 µM) was incubated in buffer of desired pH (5 to 10) for 2 h at 50°C. Aliquots were withdrawn at different time intervals and the residual enzymatic activity was measured at 50°C and pH 7 for 30 min using xylan as a substrate (0.5 ml of 1%) in a final reaction mixture volume of 1 ml.

Determination of molecular weight of Xyn

The molecular weight of the enzyme was determined by gel filtration chromatography using Sephadex G-100 previously equilibrated with 50 mM phosphate buffer pH 7. The column was calibrated using the following marker proteins: rabbit muscle phosphorylase B (97,000), bovine serum albumin (66,000 Da), ovalbumin (44,000 Da), carbonic anhydrase (30,000 Da), soybean trypsin inhibitor (20,100) and lysozyme (14,400 Da) obtained from Amersham, USA. The void volume (V_o) of the column was determined by the elution volume (V_e) of blue dextran. The presences of subunits in the protein were determined using a 12% SDS-PAGE.

Steady-state kinetic studies using birchwood xylan

Substrate hydrolysis was measured by incubating the enzyme (1 µM) with substrate (1 mg/ml). At different time intervals, 100 µl aliquots were withdrawn and the presence of product was measured as described above. To obtain Lineweaver-Burk plot at different pH values, Xyn (1 µM) was incubated at pH values of 3 to 10 and assayed with increasing concentrations of birchwood xylan (0.5 to 10 mg/ml) at 50°C for 30 min. The reciprocals of substrate hydrolysis (1/v) at each pH value were plotted against the reciprocals of the substrate concentrations. The corresponding V_{max} and K_m values were obtained for each pH value. A graph of log V_{max}/K_m against pH was plotted to obtain the pKa values. The approximate pKa values of ionizing groups were determined by fitting straight lines to plots of the log V_{max}/K_m against pH (Kulkarni et al., 2008).

Modification of Xyn with OPTA

Modification was carried out by incubating 1 µM of Xyn in 50 mM sodium phosphate buffer, pH 7 with varying concentrations of OPTA (0, 100 , 200 and 400 mM), at 25±1 °C. Fresh OPTA solution was prepared in methanol for each experiment. At different time intervals, aliquots were withdrawn from the reaction mixture and the residual activity was measured on termination of the reaction by

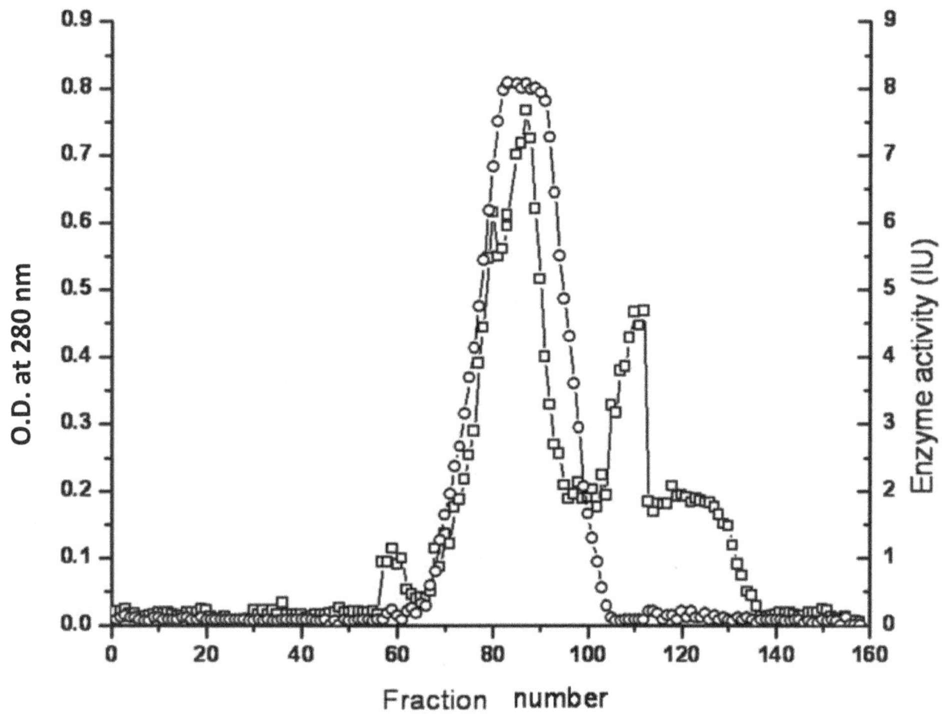

Figure 1. The Sephadex G-100 gel filtration column profile showing total protein at 280 nm (□) and xylan hydrolyzing activity (○) in the collected fractions.

adding 5 µl of 10 mM Cys. Methanol had no effect on the activity of the enzyme and was always present in concentrations of less than 2 % (v/v). The formation of Xyn-isoindole derivative was followed spectrofluorometrically by monitoring the increase in fluorescence at 410 nm with the excitation wavelength fixed at 338 nm. The modified enzyme was passed through a Biogel P-2 column to separate the isoindole derivative from the free OPTA and Cys in the reaction mixture. The isoindole content of the modified enzyme was calculated spectrophotometrically at 338 nm using a molar absorption coefficient of 7.66 $mM^{-1}cm^{-1}$ (Simons Jr. and Johnson, 1978).

Modification of Xyn with TNBS

Xyn (2 µM) was incubated with varying concentrations of TNBS (0, 100, 200 and 300 mM) in the presence of 0.25 ml 4% sodium bicarbonate at 37°C in a reaction volume of 0.5 ml (George et al., 2001b). Aliquots were withdrawn at suitable time intervals and the reaction was terminated by adjusting the pH to 4.5. The degree of inactivation in presence of 1% xylan was also determined. Control tubes with only enzyme, TNBS (0, 100, 200 and 300 mM) and TNBS/substrate were incubated under identical conditions.

Modification with DEP

Xyn (2 µM) in 2 ml of 50 mM sodium phosphate buffer at pH 7 was incubated with varying concentrations of DEP (0, 100, 200 and 300 mM) at 25±1°C. Freshly prepared DEP in absolute ethanol was used. Samples were removed periodically at different time intervals and the reaction was arrested by the addition of 50 ml of 10 mM imidazole buffer, pH 7.5. The residual activity of the diluted enzyme derivative was determined and expressed as a percentage of the control. The DEP concentration was determined spectrophotometrically at 230 nm. The amount of N-carbethoxyimidazole formed was calculated by using a molar absorption coefficient of 3000$M^{-1}cm^{-1}$ (Melchior and Fahrney, 1970). Decarbethoxylation of the DEP modified enzyme was carried out according to Miles, 1977.

The DEP modified enzyme samples were incubated with 50 mM hydroxylamine, pH 7, at 25±1°C for 4 h and the enzyme activity was determined as described above.

Modification with WRK

Xyn (2 µM) was incubated with increasing concentrations of WRK (0, 10, 20, 30, 40 and 50 mM) in the presence of phosphate buffer (50 mM) pH 7 at 37°C for 0 to 60 min. The aliquots were withdrawn at specific time intervals and the residual activity of the diluted enzyme derivative was determined. The Xyn-WRK adduct was separated from the free enzyme by gel filtration on Sephadex G-25.

RESULTS

Purification of Xyn

The Xyn from Buzyme 2511 was purified to homogeneity by a 15.18 fold purification factor and a specific activity of 3267 IU/mg of total protein. Sephadex G-100 gel filtration column fractions 55 to 130 showed the presence of protein while fractions 70 to 99 showed Xyn hydrolyzing activity (Figure 1). The purity of the enzyme was

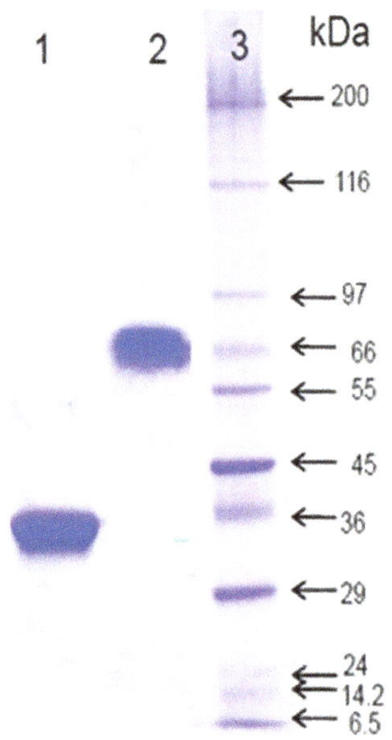

Figure 2. SDS - PAGE of Xyn. Lane (1), purified Xyn by sephadex G-100 column treated with DTT for 3 min in boiling water bath, (2) enzyme without DTT treatment and (3) molecular weight markers.

analyzed by SDS-PAGE. The difference in molecular weight of the purified Xyn as determined by by SDS–PAGE (33 kDa) (Figure 2) and gel filtration (66 kDa) (Figure 3) suggests the presence of two subunits in this protein. In the presence of DTT and boiling conditions, the disulphide bonds present in the enzyme are broken down and SDS-PAGE shows a band of 33kDa proving the presence of two equal subunits (Figure 2).

Characterization of Xyn

The temperature optimum of the purified Xyn was 50°C (Figure 4a) and retained 52% relative activity at 37°C and 10% relative activity at 90°C. The enzyme was thermostable, retaining 95% activity at 50°C and 90% activity at 70°C for more than 35 min (Figure 4b). The Xyn was active in the range of pH 5 to 9, with optimum activity at pH 6.0 (Figure 5a). It was stable in a broad range of 5 to 10 with more than 60% residual activity after incubation at pH 5 for 2 h and about 20% residual activity after incubation at pH 10 for 2 h (Figure 5b).

Study state kinetics and pH dependence of the enzyme activity

The substrate progress curve analysis showed that V_{max} for the enzyme substrate reaction is attained within 30 min at pH 7 and at 50°C (Figure 6). The Lineweaver-Burk plot at pH 7 showed a K_m value of 1mg/ml and a V_{max} of 33 µM/min^{-1} (Figure 7a). The plot of initial velocity and substrate concentration was hyperbolic at each pH investigated and all Lineweaver-Burk plots were linear. The plots indicate the dependence of Xyn activity on the ionization of at least four groups. The ionizable group essential for the function of Xyn activity was determined through the effects of pH on V_{max}/K_m values and the plots of V_{max}/K_m against pH are shown in Figure 7b. The log V_{max}/K_m versus pH plot shows pK values of 4.1, 6.0, 8.3 and 10.4 for the enzyme substrate complex. The values obtained from the acidic limb of the curves are consistent with the participation of a carboxylate group of Glu and Asp residue in the Xyn. However, the basic limb of the rate profile may reflect the ionization of a Lys residue since Lys groups in proteins have a pK value of 10.54 (Raetz and Auld, 1972).

The pK values of 6.0 and 8.3 represents ionization of His and Cys residues. The slopes of the acidic and basic limb of the curves are less than unity, indicating that the simple model of the pH dependence of enzyme action does not describe adequately the Xyn enzyme system. The divergence from unit slopes may arise from electrostatic perturbation of ionization constants, multiple intermediates or conformational effects. This may indicate the interaction of ionizable groups on the enzyme suggesting a complex mechanistic pathway for the Xyn system (Wakim et al., 1969). The nature of the essential residues for the function of Xyn activity was resolved through specific chemical modification of the enzyme.

Modification of Xyn by OPTA: Role of Cys and Lys residue in the active site

Fluorescence excitation and emission spectroscopic data showed that an isoindole derivative was formed following the reaction between Xyn and OPTA. The modification of Xyn by OPTA resulted in concomitant increase in fluorescence at 410 nm (excitation wavelength, 338 nm), which is characteristic for the formation of an isoindole derivative (Figure 8a). Therefore, this indicates that the SH and NH_2 groups of Xyn involved in the reaction with OPTA are situated at the active site of Xyn. OPTA forms a fluorescent isoindole derivative by cross-linking the proximal thiol and amino groups of the protein. Upon complete inactivation, an average of one isoindole derivative per molecule of enzyme was found, based on the increased absorbance at 410 nm. As shown in Figure 8b the increase in the amount of isoindole derivative formed correlated well with the decrease in enzyme activity, suggesting that OPTA causes modification of the Xyn by the formation of a single isoindole derivative. OPTA has been used as a fluorescent probe and a modifier of Cys and Lys residues (Levi et al., 1963;

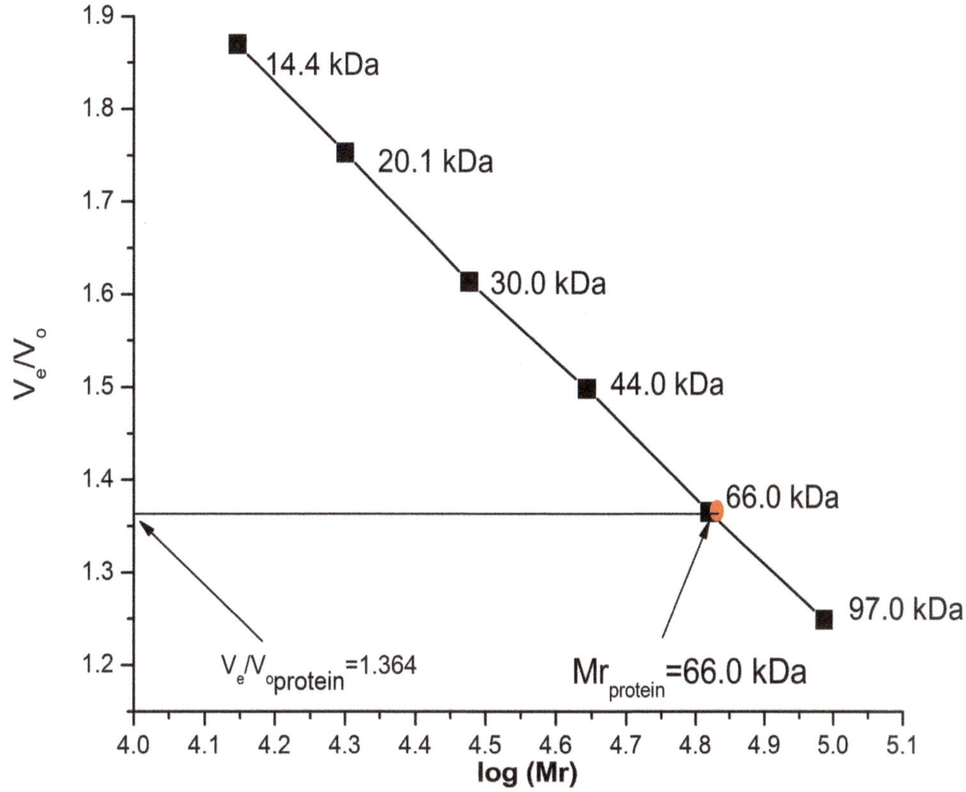

Figure 3. Gel filtration chromatography by Sephadex G 100: rabbit muscle phosphorylase B (Mr 97,000); bovine serum albumin (Mr 66,000); ovalbumin (Mr 44,000); carbonic anhydrase (Mr 30,000); soybean trypsin inhibitor (Mr 20,100); lysozyme (Mr 14,400) Red circle indicate the V_e/V_o and Mr of purified enzyme.

Simons and Johnson, 1978; Palczewski et al., 1983). The incubation of Xyn with increasing concentrations of OPTA (0 to 400 mM) resulted in a time dependent decrease in enzyme activity.

The natural logarithm of percent activity remaining was plotted against time to obtain linear pseudo first order plots (Figure 8b). When the pseudo first order rate constants obtained at each concentration were re-plotted against OPTA concentration, a linear relationship was observed. Analysis of the order of inactivation with respect to OPTA concentration using methods described by Levy et al. (1963) yielded a slope of 1.09, indicating that one molecule of OPTA binds to one molecule of enzyme at the active site (Figure 8b). Modification of Lys by TNBS and Cys residue by DTNB of the enzyme abolished the ability of the enzyme to form an isoindole derivative with OPTA supporting the above assumption.

Modification of Lys residue by TNBS: Role of Lys residue in the active site

Incubation of the Xyn (2 µM) with different concentrations of TNBS resulted in a time and concentration dependent loss of enzyme activity as shown in Figure 9. The reaction followed pseudo first order kinetics. The pseudo first order rate constants (*k*) were linearly related to the concentrations of the reagent, suggesting that no reversible complex was formed during the inactivation process. Furthermore a reaction order of 2.2 with respect to the modifier was determined from the slope of the double logarithmic plots (Figure 9), indicating that 2 mol of TNBS inactivated 1 mol of enzyme. Hence the TNBS induced inactivation of Xyn is a result of direct chemical modification of two essential Lys residues.

There was no change in the k_{cat} values of native and modified enzyme, while the K_m values were 1 and 6 mg/ml respectively. The comparable k_{cat} values suggested that Lys residues were not present in the catalytic site, while the significant increase in the K_m indicated the presence of Lys at the substrate-binding site. The modification of the Lys residue present in the substrate binding site results in a decrease in the efficiency of binding of the substrate to the substrate binding site and results in an increase of K_m. The complete protection of the Xyn against inactivation by TNBS and OPTA, by the substrate, confirms the presence of Lys at the substrate-binding site.

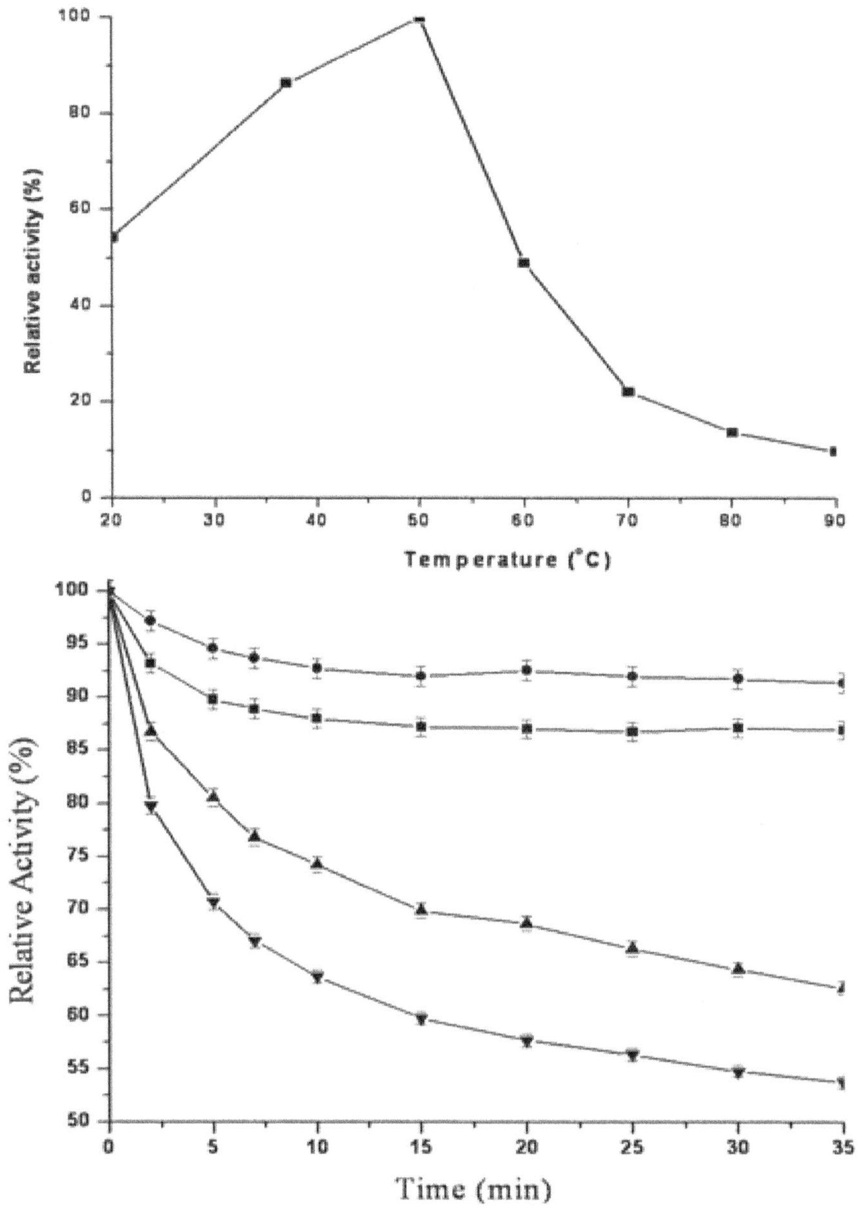

Figure 4. (A) Optimum temperature of Xyn. (B) Stability of Xyn at 50°C (●), 70°C (■), 80°C (▲) and 90°C (▼) for 35 min at pH 6.

Modification of Xyn with DEP: Role of His in the active site

Reaction of the Xyn with DEP resulted in progressive loss in activity. The inactivation was dependent on the concentration of DEP. The pseudo first order rate constants for different concentrations of DEP (0, 100, 200 and 300 mM) were calculated and the slope of the plots of log K against log (DEP) produced a value of 1.15 (Figure 10). These results showed that the loss in activity was due to the modification of a single His residue. DEP is specific for His at neutral pH, but it also reacts to a lesser extent with tyrosine and Lys (Miles, 1977). The modification of tyrosine residues by DEP was excluded, as there was no decrease in the absorbance of the modified protein at 278 nm. Moreover, there was no loss in activity on treating the purified enzyme with N-acetylimidazole, which is a tyrosine specific reagent. The recovery of total activity in the presence of hydroxylamine further proved the involvement of DEP in His modification and negated the modification of Lys by DEP.

The kinetic parameters for the Michaelis-Menten constant (K_m) and turnover number (k_{cat}) were obtained from the Lineweaver-Burk plots for native and DEP

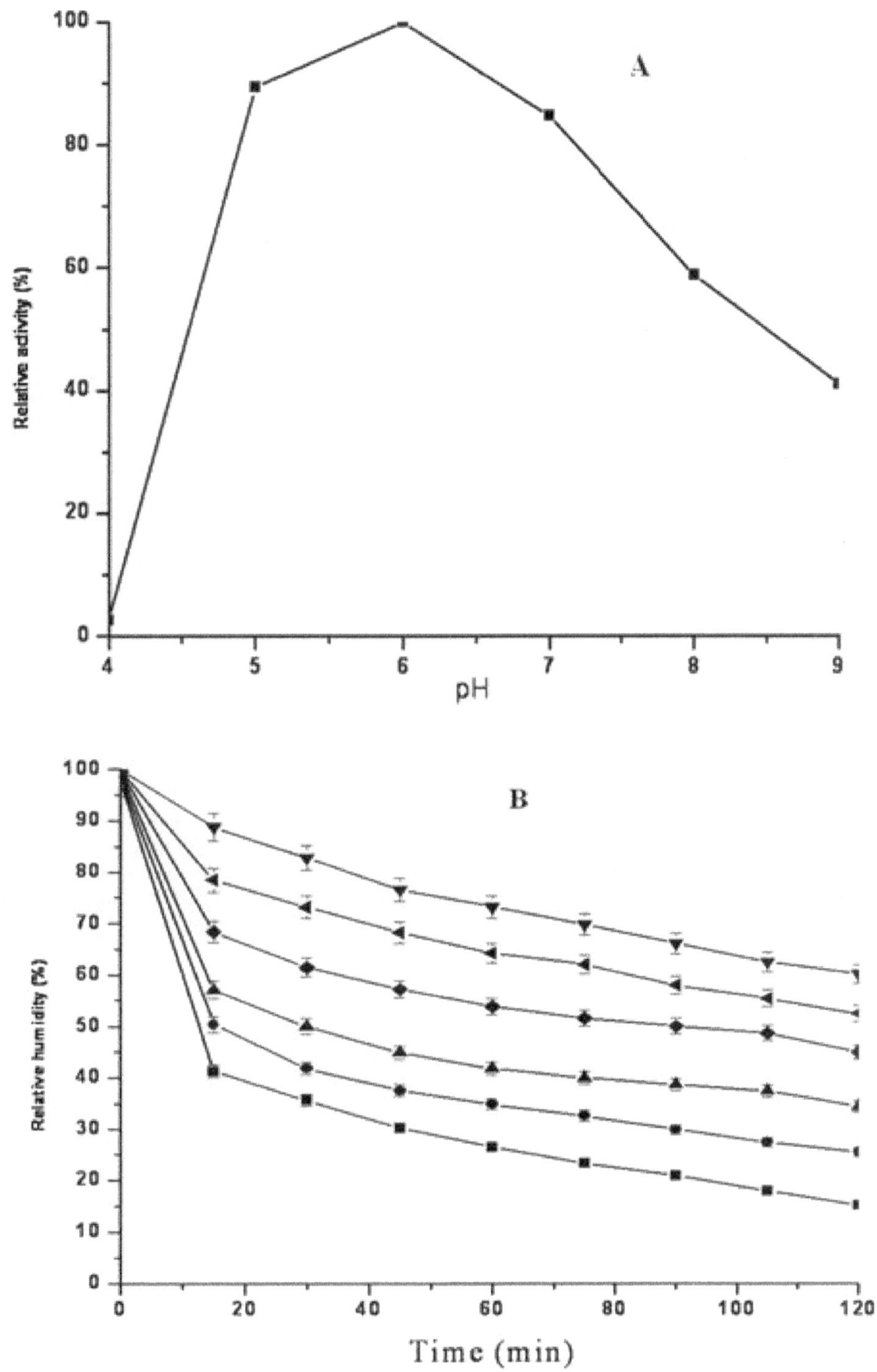

Figure 5. (A) Optimum pH of Xyn. (B) Stability of Xyn at pH 5 (◄), pH 6 (▼), pH 7 (♦), pH 8 (▲), pH 9 (●) and pH 10 (■) for 2 h and 10.4 for the active ionizable groups at the active site of the enzyme corresponding to Glu, His, Cys and Lys residues, respectively.

modified Xyn. There was no change in the K_m values of modified and native enzyme, while the k_{cat} values were 6530 and 23420 min^{-1} respectively. The similar K_m values and the significant decrease in the k_{cat} suggested that His was not present in the substrate binding site and was present at or near the catalytic site of the Xyn. Xylan did not provide any protection to the Xyn against inactivation by DEP, which may be because xylan binds only at the

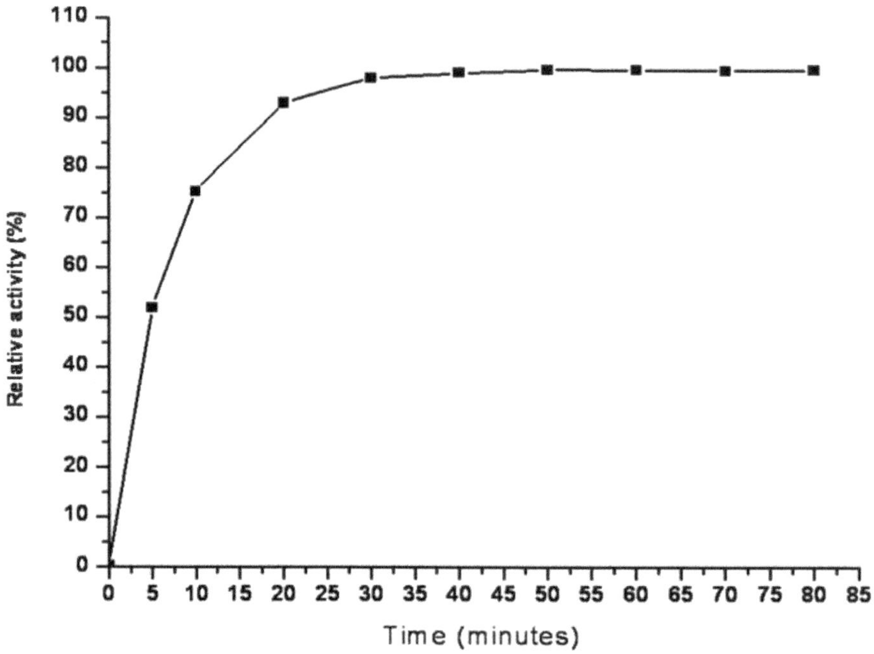

Figure 6. % relative activity of purified Xyn on substrate over the time.

substrate binding site, therefore making the catalytic site susceptible to attack by DEP. The His residue in the active site may be involved in a network of hydrogen bonds, which are responsible for maintaining the ionization state of the two catalytic residues responsible for the hydrolysis of the β-1,4 glycosidic bonds.

Inhibition of Xyn with WRK

WRK has been generally used as a chemical modifier of enzyme carboxyl groups. Incubation of the Xyn (2 μM) with different concentrations of WRK (0 to 50 mM) resulted in a time and concentration dependent loss of enzyme activity as shown in Figure 11. The reaction followed pseudo first order kinetics. The pseudo first order rate constants (k) were linearly related to the concentrations of the reagent, suggesting that no reversible complex was formed during the inactivation process. Furthermore a reaction order of 1.18 with respect to the modifier was determined from the slope of the double logarithmic plots (Figure 11), indicating that 1 mol of WRK inactivated 1 mol of enzyme. Hence, the WRK induced inactivation of Xyn is a result of direct chemical modification of essential Asp or Glu residue.

DISCUSSION

In this paper, the purification of Xyn from a commercially available xylanase stock, Buzyme 2511 was described. A plethora of xylanases have been reported which contain carboxyl and tryptophan residues in the active site and there are also a few reports of Lys and His. To the best of our knowledge there are few reports of the presence of Lys in the active site of a xylanase from *P. fluorescens* and Thermomonospora sp. (Harris et al., 1996; Charnock et al., 1997; Nath and Rao, 1998). Our studies suggest the presence of essential Lys, His and Asp or Glu residues in the active site of the studied Xyn by fluorescent chemo-affinity labeling and group specific modification. The xylanase purified in this study was more thermostable when compared to most of the other reported thermostable xylanases from wild bacterial strains. It was more stable than the xylanase isolated from Thermotoga sp. strain FjSS-B.1, which had a half life of 8 min at 100°C (Simpson et al., 1991). However, a recombinant xylanase whose gene has been isolated from *Thermotoga neopalitina* and expressed in *E. coli* had a half-life of 30 min at 100°C (Velikodvorskaya et al., 1997).

The xylanase isolated from *Thermus thermophilus* was also highly thermostable with 61% residual activity after incubation at 90°C for 48 h (Lyon et al., 2000). Many authors have reported several xylanases from bacterial and fungal sources which can be used in paper and pulp industry. *Bacillus* SSP-34 was reported to produce a xylanase with activity of 506 IU/ml (Subramaniyan, 2000). A *Bacillus* sp. strain NCL 87-6-10 produced 93 U/ml of xylanase in a zeolite induced medium (Balakrishnan et al., 2000). Streptomyces sp. QG-11-3 was also found to produce a xylanase with an activity of 96 U/ml (Beg et al.,

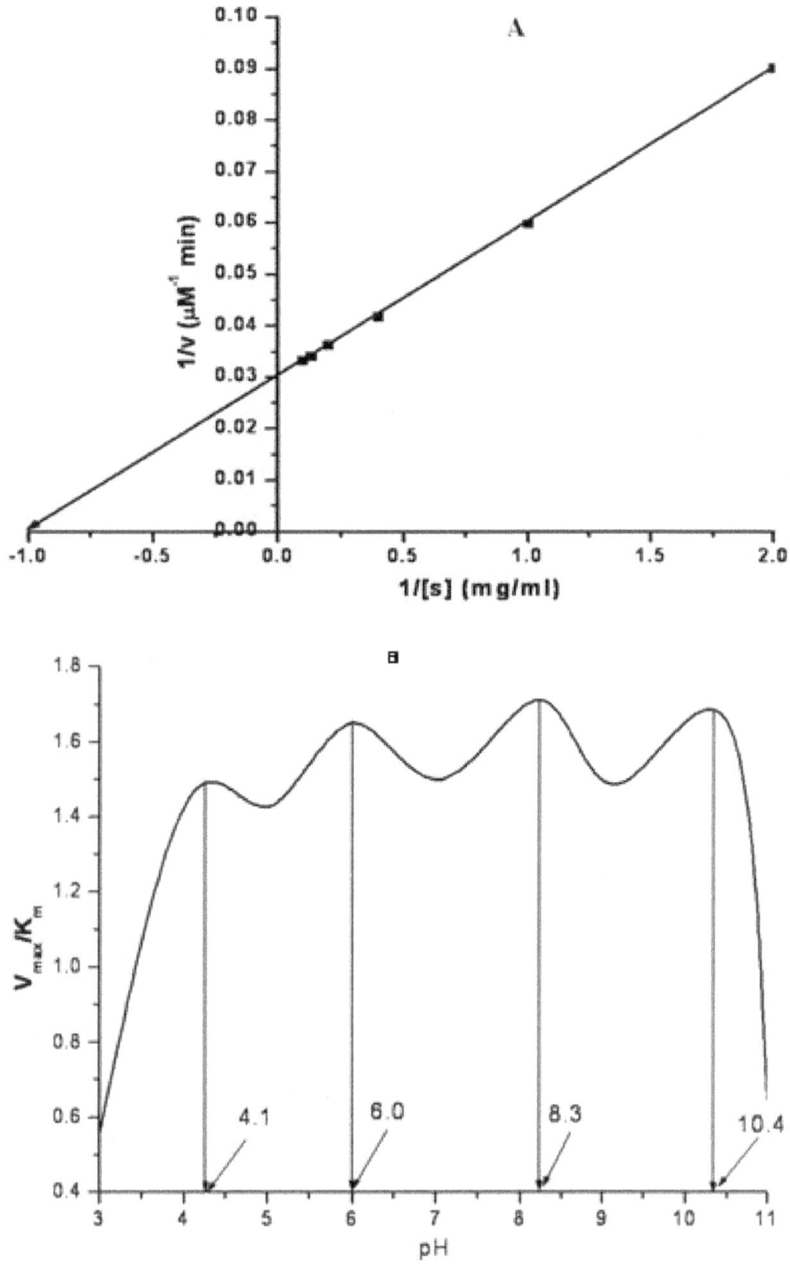

Figure 7. (A) Lineweaver-Burk plot at pH 7 to determine the K_m and V_{max} value of Xyn- (B) Identification of p*K*a values of ionizable side chains of active site amino acids. The secondary plot of the log V_{max}/K_m versus the pH provided p*K*a values of 4.1, 6.0, 8.3.

2000). *Rhodothermus marinus* was found to be producing thermostable xylanases of approximately 1.8 to 4.03 IU/ml (Dahlberg et al., 1993; Hreggvidsson et al., 1996). *Caldocellum saccharolyticum* possesses xylanases with optimum activities at pH values 5.5 to 6.0 and at a temperature of 70°C (Luthi et al., 1990). Xylanases from *Dictyoglomus* sp. showed optimum activities at pH 5.5 and 90°C, however, merits the significant pH stability at pH values 5.5 to 9.0. Gomes et al. (1992) reported xylanase activity (188.1 U/ml-optimum pH 5.2) from *Trichoderma viride*. Similar to *T. viride*, *T. reesei* was also known to produce higher xylanase activity – approximately 960 IU/ml (Bailey et al., 1993). Singh et al. (2000) reported a xylanase activity of 59,600 nkat/ml (approximately 3576 U/ml) from a *Thermomyces lanuginosus* strain. Column chromatographic techniques, mainly ion exchange and size exclusion are the generally utilized schemes for xylanase purification (Wong and Saddler,

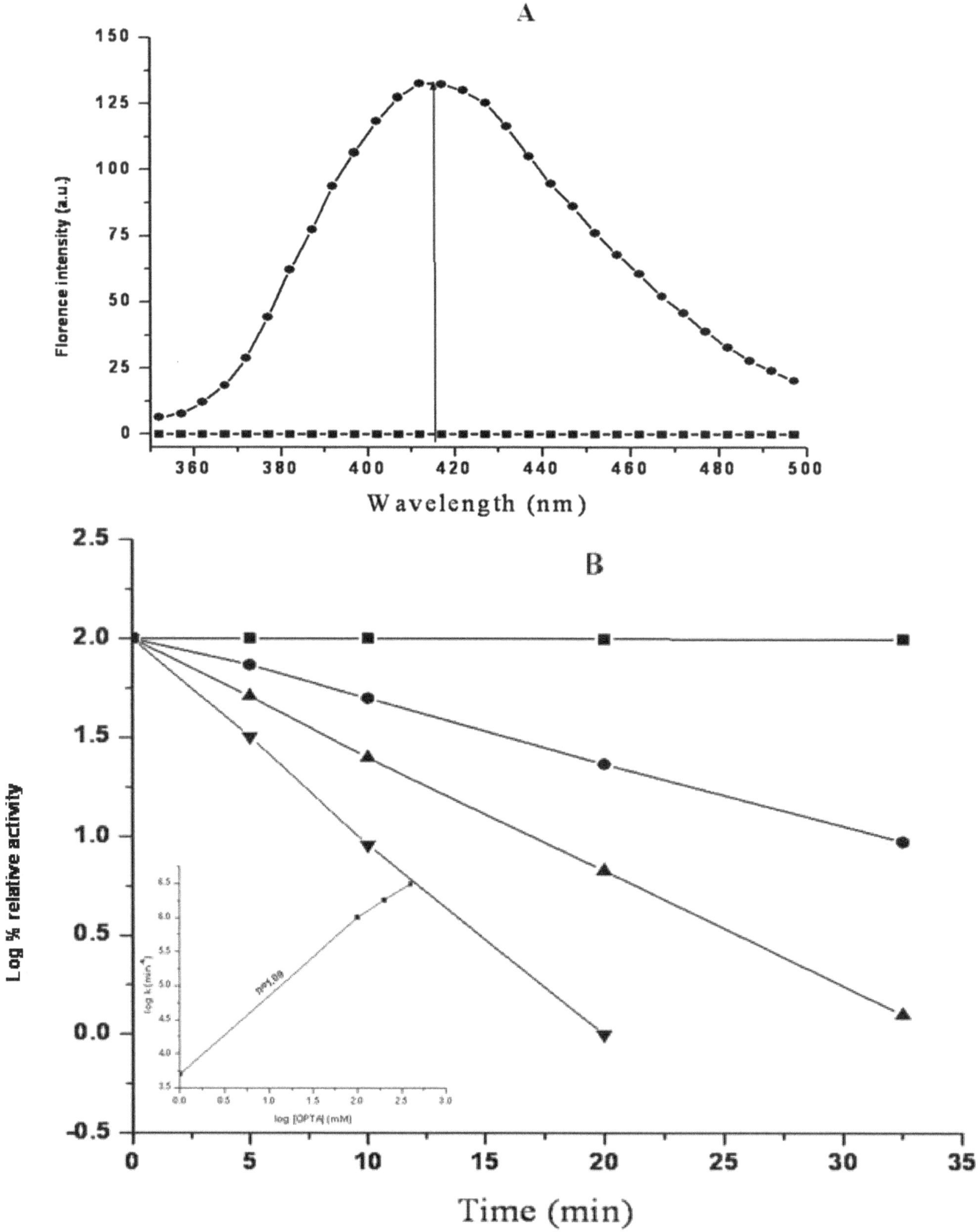

Figure 8. (A) Modification of Xyn with OPTA (fluorescence measured at 415 nm. excitation wavelength= 338 nm) for native Xyn (■) and for Xyn incubated with 400 mM OPTA (•) for 32 min. (B) Xyn activity after incubation with 0 (■),100 mM (•), 200 mM (▲) and 400 mM (▼) OPTA . Inset: Double logarithmic plots of pseudo first order rate constants as a function of OPTA concentration.

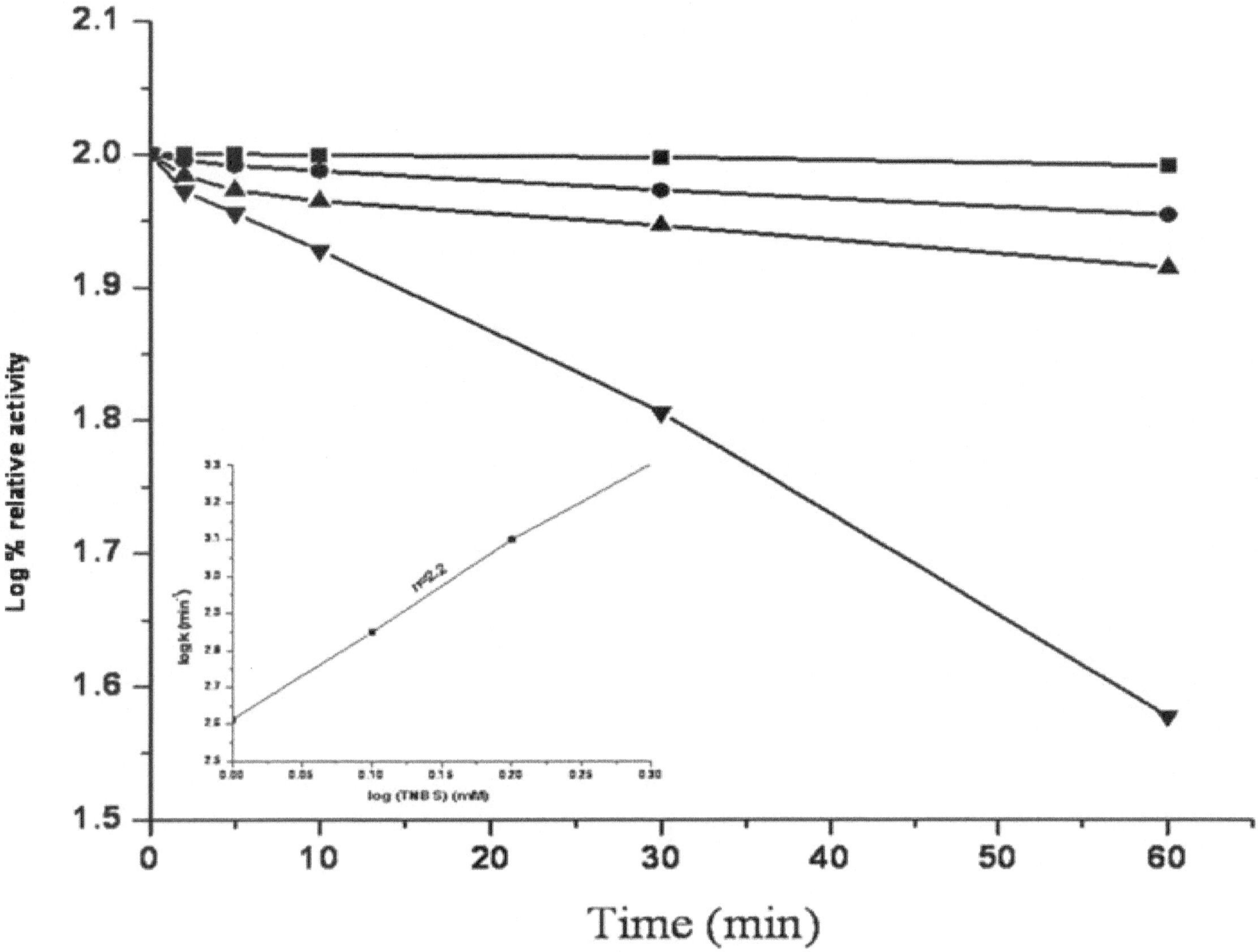

Figure 9. Modification of Xyn by TNBS: Xyn activity after incubation with 0 (■), 100 mM (●), 200 mM (▲) and 300 mM (▼) TNBS. Inset: Double logarithmic plots of pseudo first order rate constants as a function of TNBS concentration.

1992).

There are several reports regarding the purification of xylanases to electrophoretic homogeneity, however, the yield and purification fold varies in different cases. In all the cases the culture supernatants are initially concentrated using precipitation or ultrafiltration techniques. A moderately thermostable xylanase was purified from *Bacillus* sp. Strain SPS-0 using ion exchange, gel and affinity chromatographies (Bataillon et al., 2000). Thermostable xylanases from thermophilic organisms like Dictyoglomus and Thermotoga spp which grow at a temperature higher than 80°C could be easily purified by the inclusion of one additional step of heating (Bergquist et al., 2001). Takahashi et al. (2000) purified a low molecular weight xylanase (23 kDa) from *Bacillus* sp. strain TAR-1 using CM Toyopearl 650 M column. This xylanase with optimum activity at 70°C had broad pH profile. Kimura et al. (2000) purified *Penicillium* sp. (Beg et al., 2000) xylanase with molecular weight 25 kDa which was induced by xylan and repressed by glucose. A dimeric xylanase also reported from *Thermotoga thermarum* with a Mr of 266 kDa with subunits of 105 kDa. Thorough information about the xylanase can be obtained from a review by Subramaniyan and Prema (2002). The three dimensional structure of family 10 xylanases have revealed several highly conserved residues that are on the surface of the active site. Charnock et al. (1997), have suggested that Lys-47 plays an important role in positioning the substrate into the active site of xylanase A from *Pseudomonas fluorescens*.

The structural analysis of the active site of family 10 xylanases from *Cellulomonas fimi*, *P. fluorescens* and *Penicillium simplicissimum* have shown the presence of a hydrogen bonding network involving His residues in the

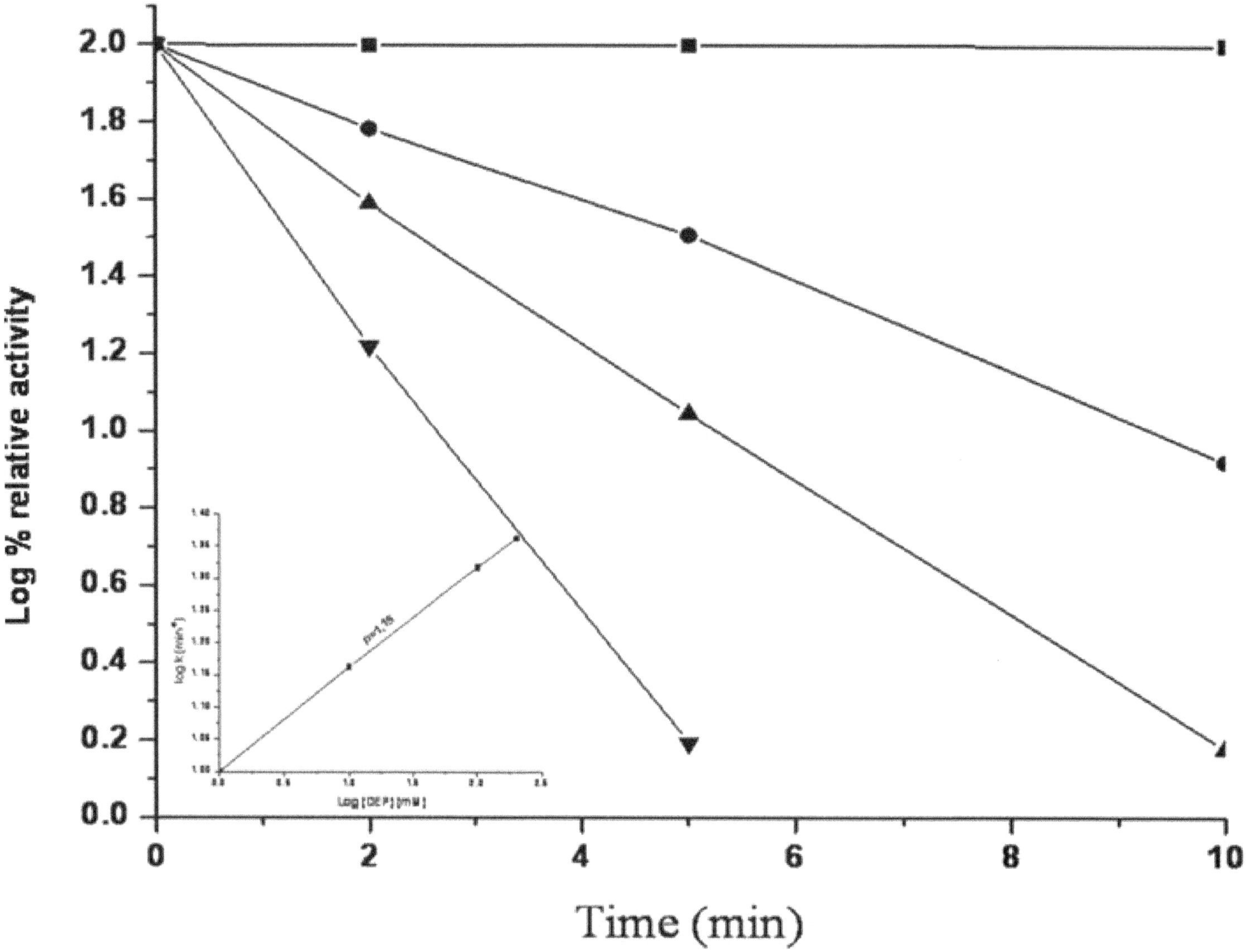

Figure 10. Modification of Xyn by DEP: Xyn activity after incubation with 0 (■), 100 mM (●), 200 mM (▲) and 300 mM (▼) DEP. Inset: Double logarithmic plots of pseudo first order rate constants as a function of DEP concentration.

active site (White et al., 1994; Harris et al., 1996; Schmidt et al., 1998). White et al. (1994) reported that the trios of residues Glu233-His205-Asp235 are highly conserved within family 10. Roberge et al. (1997), have undertaken the site directed mutagenesis for xylanase A derived from *Streptomyces lividans*. They have proved that His81 and His207 are very important in the hydrogen bonding network of the active site. Mutation of these residues modifies the interaction necessary to maintain the ionization state of the two catalytic Glu residues of xylanaseA. His residue present in the active site of the Xyn may play a role similar to that played by the conserved His residues in other family 10 xylanases. The insight into the active site of the enzyme will provide the information which will be helpful in enhancing its catalytic activity.

In conclusion, the enzyme showed its thermostable nature by exhibiting 90% residual activity at 70°C for more than 35 min. Cys, His, Lys and Glu residues were found to be important for the catalytic activity of the enzyme. After characterization, this enzyme can be used effectively in the pulp and paper industry. The N- and C-terminal sequencing of the enzyme are worth investigating and accordingly primers can be designed for the purpose of cloning and hence, higher production of the enzyme can be achieved.

ACKNOWLEDGEMENTS

The National Research Foundation (NRF) and Durban University of Technology are gratefully acknowledged for

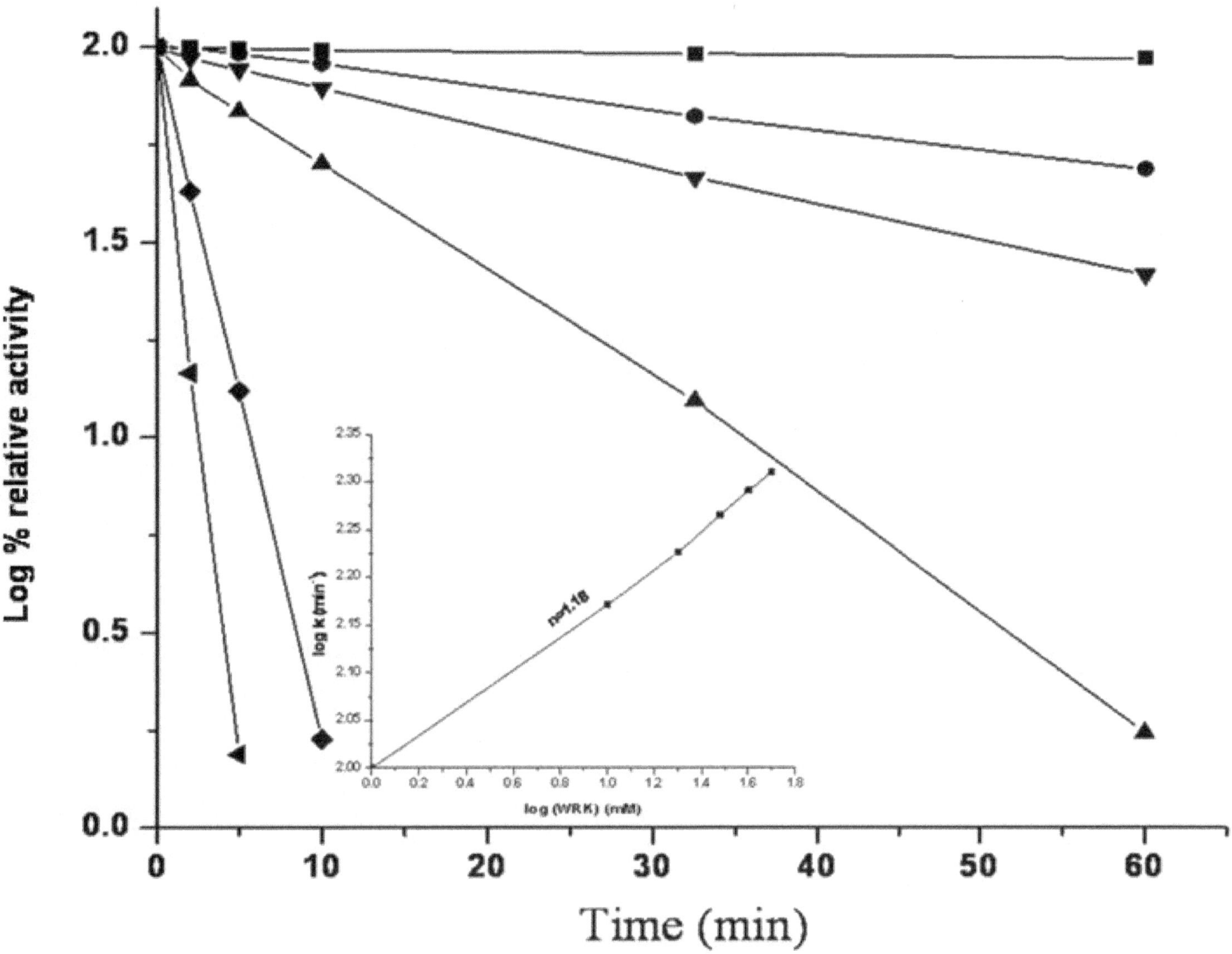

Figure 11. Modification of Xyn by WRK: Xyn activity after incubation with 0 (■), 10 mM (●), 20 mM (▼), 30 mM (▲), 40 mM (♦) and 50 mM (◄) WRK. Inset: Double logarithmic plots of pseudo first order rate constants as a function of WRK concentration.

funding this project. We also thank Buchman laboratories international, South Africa for giving permission to publish the work on purification of Buzyme 2511.

REFERENCES

Anish R, Rao M (2007). Biochemical characterization of a novel β-1-4 glucan 4-glucanohydrolase from *Thermomonospora sp.* having a single active site for lichenan and xylan Biochimie, 89: 1489-1497.

Bailey MJ, Buchert J, Viikari L (1993). Effect of pH on production of xylanase by *Trichoderma reesei* on xylan and cellulose-based media. Appl. Microbiol. Biotechnol., 40: 224.

Baily MJ, Biely P, Poutanen K (1992). Inter-laboratory testing of methods for assay of xylanase activity. J. Biotechnol., 23: 257-270.

Balakrishnan H, Srinivasan MC, Rele MV, Chaundhari K, Chandwadkar AJ (2000). Effect of synthetic zeolites on xylanase production from an alkalophilic *Bacillus* sp. Curr. Sci., 79: 95.

Bataillon M, Nunes-Cardinali AP, Castillon N, Duchiron F (2000). Purification and characterization of a moderately thermostable xylanase from *Bacillus* sp. strain SPS-0. Enzyme Microb. Technol., 26: 187.

Beg QK, Bhushan B, Kapoor M, Hoondal GS (2000). Production and characterization of thermostable xylanase and pectinase from *Streptomyces* sp. QG-11-3. J. Ind. Microbiol. Biotechnol., 24: 396.

Bergquist P, Gibbs MD, Morris DD, Thompson DR, Uhl AM, Daniel RM (2001). Hyperthermophilic xylanases. Methods Enzymol. 330: 301.

Biely P (1985). Microbial xylanolytic systems. Trends Biotechnol., 3: 286-290.

Bradford MM (1976). A rapid and sensitive method for the quantitation of microgram quantities of protein utilizing the principle of protein-dye binding. Anal. Biochem., 72: 248-54.

Bray MR, Clarke AJ (1995). Identification of an essential tyrosyl residue in the binding site of *Schizophyllum commune* xylanaseA. Biochem., 34: 2006-2014.

Charnock SJ, Lakey JH, Virden R, Hughes N, Sinnott ML, Hazlewood GP, Pikersgill R, Gilbert HJ (1997). Key residues in subsite F play a critical role in the activity of *Pseudomonas fluorescens* subspecies cellulosa XylanaseA against xylooligosaccharides but not against highly polymeric substrates such as xylan. J. Biol. Chem., 272: 2942-2951.

Chauthaiwale J, Rao M (1994). Chemical modification of xylanase from alkalothermophilic *Bacillus* sp.: Evidence for essential carboxyl group. Biochim. Biophys. Acta, 1204: 164-168.

Coughlan MP, Hazelwood GP (1993). β-1,4-D-Xylan degrading enzyme systems: biochemistry, molecular biology and applications. Biotechnol. Appl. Biochem., 17: 259-289.

Dahlberg L, Holst O, Kristjansson JK (1993). Thermostable xylanolytic enzymes from *Rhodothermus marinus* grown on xylan. Appl. Microbiol. Biotechnol., 40: 63.

George SP, Ahmad A, Rao M (2001a). A novel thermostable xylanase from *Thermomonospora* sp.: influence of additives on thermostability. Bioresour. Technol., 78: 221-224.

George SP, Ahmad A, Rao M (2001b). Involvement of a lysine residue in the active site of a thermostable Xylanase from *Thermomonospora sp.* Biochem. Biophys. Res. Commun., 282: 48-54.

Gomes I, Gomes J, Steiner W, Esterbauer H (1992). Production of cellulase and xylanase by a wild strain of *Trichoderma viride*. Appl. Microbiol. Biotechnol., 36: 701.

Harris GW, Jenkins JA, Connerton I, Pickersgill RW (1996). Refinedcrystal structure of the catalytic domain of xyalanaseA from *Pseudomonas fluorescens* at 1.8 Å resolution. Acta Crystallogr., D52: 393-401.

Hreggvidsson GO, Kaiste E, Holst O, Eggertsson G, Palsdottir A, Kristjansson JK (1996). An extremely thermostable cellulase from the thermophilic eubacterium *Rhodothermus marinus.* Appl. Environ. Microbiol., 62: 3047.

Kimura T, Ito J, Kawano A, Makino T, Kondo H, Karita S, Sakka K, Ohmiya K (2000). Purification, charcaterization and molecular cloning of acidophilic endo-1,4-α-D-xylanase from *Penicillium* sp. 40. Biosci. Biotechnol. Biochem., 64: 1230.

Ko EP, Akatsuka H, Moriyama H, Shinmyo A, Hata Y, Katsube Y, Urabe I, Okada H (1992). Site directed mutagenesis at aspartate and glutamate residues of xylanase from *Bacillus pumilus.* Biochem. J., 288: 117-121.

Kulkarni A, Gaikwad S, Rao M (2008). pH induced structural alterations in an aspartic protease from *Vigna radiate* indicating an alkali induced molten globule state Int. J. Biol. Macromol., 43: 373-376.

Laemmli UK (1970). Cleavage of structural proteins during the assembly of the head of bacteriophage T4. Nature, 227: 680-685.

Levy HM, Leber PD, Ryan EM (1963). Inactivation of myosin by 2,4-dinitrophenol and protection by adenosine triphosphatase and other phosphate compounds. J. Biol. Chem., 238: 3654-3659.

Luthi E, Jasmat NB, Bergquist PL (1990). Xylanase from the extremely thermophilic bacterium *Caldocellum saccharolyticum*: over expression of the gene in *Escherichia coli* and characterisation of the gene product. Appl. Environ. Microbiol., 36: 2677.

Lyon PF, Beffa T, Blanc M, Auling G, Aragno M (2000). Isolation and characterization of highly thermophilic xylanolytic *Thermus thermophilus.* Can. J. Microbiol., 46: 1029-1035.

Melchior WB, Fahrney D (1970). Ethoxyformylation of proteins. Reaction of ethoxyformic anhydride with α-chymotrypsin, pepsin, and pancreatic ribonuclease at pH 4. Biochem. 9: 251-258.

Miles EW (1977). Modification of histidyl residues in proteins by diethylpyrocarbonate. Meth. Enzymol., 47: 431-442.

Nakamura S, Wakabayashi K, Nakai R, Ano R, Horikoshi K (1993). Purification and some properties of an alkaline xylanase from alkaliphilic *Bacillus* sp. strain 41 M-1. Appl. Environ. Microbiol., 59: 2311-2316.

Nath D, Rao M (1998). Structural and Functional Role of Tryptophan in Xylanase from an Extremophilic *Bacillus*: Assessment of the Active Site. Biochem. Biophys. Res. Commun., 249: 207-212.

Nath D, Rao M (2001). Acid-induced partly folded conformation resembling a molten globule state of Xylanase from an alkalothermophilic *Bacillus* sp. Biochem. Biophys. Res. Commun., 288: 1218-1222

Palczewski K, Hargrave PA, Kochman M (1983). *o*-Phthalaldehyde, a fluorescence probe of aldolase active site. Eur. J. Biochem., 137: 429-435.

Raetz CRH, Auld DS (1972). Schiff bases of pyridoxal phosphate with active center lysines of ribonuclease A. Biochem., 11: 2229-2236.

Rawat U, Rao M (1997). Site and significance of cysteine residues in xylose reductase from *Neurospora crassa* as deduced by fluorescence studies. Biochem. and Biophys. Res. Commun., 239: 789-793.

Roberge M, Shareck F, Morosoli R, Kluepfel D, Dupont C (1997). Characterization of two important histidine residues in the active site of xylanase A from *Streptomyces lividans*, a family 10 glycanase. Biochem., 36: 7769-7775.

Schmidt A, Schlacher A, Steiner W, Schwab H, Kratky C (1998). Structure of the xylanase from *Penicillium simplicissimum.* Protein Sci., 7: 2081-2088.

Simons SS Jr., Johnson DF (1978). Reaction of o-phthalaldehyde and thiols with primary amines: Fluorescence properties of 1-alkyl(and aryl)thio-2-alkylisoindoles. Anal. Biochem., 90: 705-725.

Simpson HD, Haufler UR, Daniel RM (1991). An extremely thermostable xylanase from the thermophilic eubacterium *Thermotoga.* Biochem. J., 277: 413-417.

Singh S, Pillay B, Dilsook V, Prior BA (2000). Production and properties of hemicellulases by a *Thermomyces lanuginosus* strain. J. Appl. Microbiol., 88: 975.

Subramaniyan S (2000). Studies on the Production of Bacterial Xylanases. Ph.D. Thesis, Cochin University of Science and Technology, Kerala, India.

Subramaniyan S, Prema P (2002). Biotechnology of Microbial Xylanases: Enzymology, Molecular Biology and Application. Crit. Rev. Biotechnol., 22: 33-46.

Takahashi H, Nakai R, Nakamura S (2000). Purification and partial characterization of a basic xylanase produced by thermoalkaliphilic *Bacillus* sp. strain TAR-1. Biosci. Biotechnol. Biochem., 64: 887.

Velikodvorskaya TV, Volkov IY, Vasilevko VT, Zverlov VV, Piruzian ES (1997). Purification and some properties of *Thermotoga neapolitana* thermostable xylanase B expressed in *E. coli* cells. Biochem., (Moscow) 62: 66-70.

Vernon CA, Banks B (1963). The enzymic hydrolysis of glycosides. Biochem. J., 86: 7.

Viikari L, Kantelinen A, Sundquist J, Linko M (1994). Xylanases in bleaching: From an idea to the industry. FEMS Microbiol. Rev. 13: 335-350.

Wakim J, Robinson M, Thoma JA (1969). The active site of porcine pancreatic α-amylase: Factor contributing to catalysis. Carbohydr. Res. 10: 487-503.

White A, Withers SG, Gilkes NR, Rose DR (1994). Crystal structure of the catalytic domain of the β-1,4-glycanase Cex from *Cellulomonas fimi.* Biochem. 33: 12546-12552.

Wong KKY, Saddler JN (1992). Trichoderma xylanases, their properties and application. Prog. Biotechnol. 7: 171.

Wong KKY, Tan LUL, Saddler JN (1988). Multiplicity of β-1,4-xylanase in microorganisms: function and applications. Microbiol. Rev. 52: 305-317.

Malate dehydrogenase from blood stream *Trypanosoma vivax* exists in isoenzyme forms

A. U. Wurochekke[1]*, A. J. Nok[2], H. M. Inuwa[2] and C. E. Gimba[3]

[1]Biochemistry Department, Federal University of Technology, Yola, Adamawa State, Nigeria.
[2]Biochemistry Department, Ahmadu Bello University, Zaria, Kaduna State, Nigeria.
[3]Chemistry Department, Ahmadu Bello University, Zaria, Kaduna State, Nigeria.

Malate dehydrogenase was isolated and partially purified from blood stream *Trypanosoma vivax*. The enzyme exists in isoenzyme forms (MDH_1 and MDH_2). The relative molecular weights of MDH_1 and MDH_2 determined by Sodium dodecyl sulphate polyacrylamide gel electrophoresis were 61 and 63 kdal. respectively. MDH_2 lost activity immediately after purification. It has purification fold of 15 and percentage recovery of 31%. Similarly, MDH_1 was purified 21 fold from the crude with 54% recovery. This isoenzyme was considerably stable. There was no MDH activity when the live parasites were incubated with the assay medium suggesting that the enzyme was not secreted by parasites. However, addition of detergent led to a surge in MDH activity suggesting that the enzyme is membrane bound. Malate dehydrogenase from the infective form of *T. vivax* has a stable and non – stable isoforms.

Key words: Malate dehydrogenase, isoenzyme, parasites, *Trypanosoma vivax.*

INTRODUCTION

Typanosoma vivax is one of the protozoan parasites that retard economic growth and agricultural production in Africa. The parasite affects wide range of domestic animals in the region. Available chemotherapeutic agents, which are less efficient, (Gutteridge and Coom, 1977) are targeted at cellular components enzymes or metabolic pathways of the parasite. Parasite mitochondria have therefore, been one of the targets for chemotherapy. Gutteridge and Coom (1977), reported that the enzyme pyruvate kinase of the trypanosome is likely the target of inhibition by the arsenicals. Trypanosome alternative oxidase, which is the terminal oxidase of glycerol-3-phosphate oxidase system does not exist in the host and has been attracting attention as target for trypanocides (Kita et al., 2001). Nok (2002) also reported the specific inhibition of the mitochondrial glycerol-3-phosphate dependent electron transport of *Trypanosoma congolense* by azanthraquinone.

The blood stream forms of the parasite possess non-functional mitochondrion and survive exclusively by substrate-level phosphorylation with glucose readily available in the environment as t he energy source. In contrast, the procyclics living in the tse-tse flies where amino acids are the predominant substrates possess well developed mitochondria with citric acid cycle enzymes and a respiratory chain.

Malate dehydrogenase is an oxido-reductase enzyme that is vital in carbohydrate metabolism. It catalyses the oxidation of L-malate to oxaloacetate in the TCA cycle of cells. Information about this vital enzyme from the blood stream *T. vivax* was not documented. This work was designed to isolate and partially purify malate dehydrogenase from blood stream *T. vivax* and also to determine whether the enzyme is membrane bound or secreted by the parasite.

MATERIALS AND METHODS

Experimental animals

Four adult goats purchased from Zaria market were screened for possible infection by parasites. They were acclimatized for two weeks in the department of Veterinary Parasitology, A.B.U. Zaria. The animals were well fed and water was provided *ad libitum.*

The parasite

T. vivax (Agai strain) was kindly provided by the National Institute of Trypanosomiasis (NITR), Jos, Nigeria.

*Corresponding author. E-mail: wchekke@yahoo.co.uk.

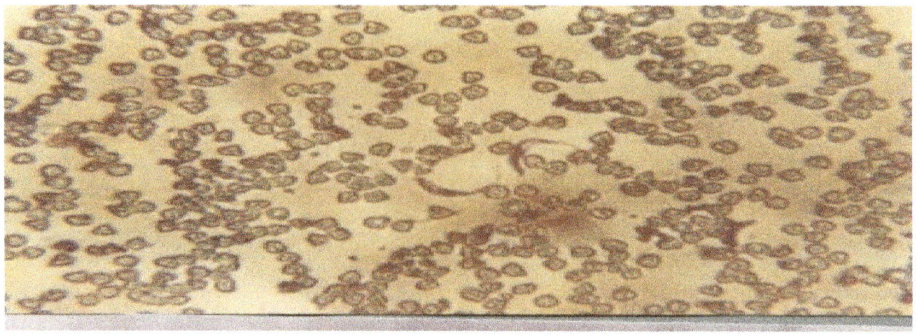

Figure 1. Photograph of *Trypanosoma vivax* in the infected blood.

Reagents

Substrates (oxaloacetic acid, L – Malate), coenzymes (NADH, NAD+), α - ketogluterate, Succinic acid, pyruvic acid, isosuccinic acid, DEAE – cellulose – 52, TEMED and bromophenol blue were products of Sigma chemical company, U.S.A. Acrylamide, N, N, methyl bis acrylamide (Bis), ammonium persulphate, coomaise blue, sodium dodecyl sulphate (SDS), mercaptoetyhanol, glycine, sodium dihydrogen phosphate, disodium hydrogen phosphate and heparin were obtained from BDH chemical, pool, England. All other chemical used were of analytical grade.

Inoculation of goats with the *Trypanosome*

A suspension of *T. vivax* of approximately 10^3 parasites per field was intraperitoneally injected into the apparently healthy goats. The level of parasiteamia was monitored everyday for two weeks.

Parasiteamia determination

Parasiteamia was determined by the wet blood film technique. A smear of the infected blood was made on a slide and covered with a cover slip. The blood film was then examined under the microscope (mg × 40). Motile parasites were counted per field by the method of Herbert and Lumsden (1976).

Isolation of the *Trypanosomes*

Trypanosomes were isolated from the infected goats by the method of Lanham and Godfrey (1970). At peak parasiteamia, blood was drawn through the jugular vein using heparanised needles. The blood was loaded unto previously packed column (1.5 x 30 cm) with DEAE- cellulose and the parasites were eluted using phosphate buffered saline.

Preparation of the crude enzyme

The reconstituted pure parasites were divided into two portions. One of the portions was frozen and thawed three times to solubilize the *Trypanosome*. To the other portion, 1% triton X – 100 was added by intermittent shaking for about 30 min. The two portions were assayed for malate dehydrogenase activity and protein content.

Assay for malate dehydrogenase

Malate dehydrogenase activity was assayed spectrophotometrically in the direction of oxaloacetate (OAA) reduction, by following the disappearance of NADH at 340 nm. The assay medium in the cuvette contained 0.134 mM NADH, 0.2 mM OAA in 25 mM phosphate buffer (pH 7.2) in a total volume of 500 µl.

Protein determination

Protein concentrations were determined by the biuret method, using bovine serum albumin BSA as standard.

Ion-exchange chromatography

Charged slurry of DEAE – 52 cellulose was packed into a chromatographic column (1.5 × 30 cm) and equilibrated with 25 mM phosphate buffer pH 7.2. 5 ml of the crude enzyme was dispensed into the column. The enzyme was eluted by linear phosphate gradient of 0.02 – 1 M. Fifty fractions of 5 ml each were collected and assayed for MDH activity and protein content.

SDS polyacrylamide gel electrophoresis

Sodium dodecyl sulphate polyacrylamide gel electrophoresis of the active eluents from the ion-exchange chromatography was performed as described by Webster and Pringle (1972). Prepared test samples and reconstituted protein standards were separately dispensed on to the gels (12% running gel and 6% stacking gel) and electrophoresis was performed at a current of 5 mA/gel tube for about 1.5 h.

At the end of electrophoresis, the molecular weights of the partially purified enzymes were extrapolated from a calibration curve of $\log_{10}$ of the molecular weights of the standard proteins against the RF values.

MDH activity from the live parasite

Harvested live parasites were incubated with the assay medium and MDH activity determined in the direction of oxaloacetate reduction as described before. Various volumes containing different concentration of the life parasites were used. Drops of triton – x – 100 were added to the harvested parasites in an increasing order as the activity of the enzyme was determined at intervals.

RESULTS

Figure 1 shows the photograph of blood stream *T. vivax* before separation from the blood. After separation, the

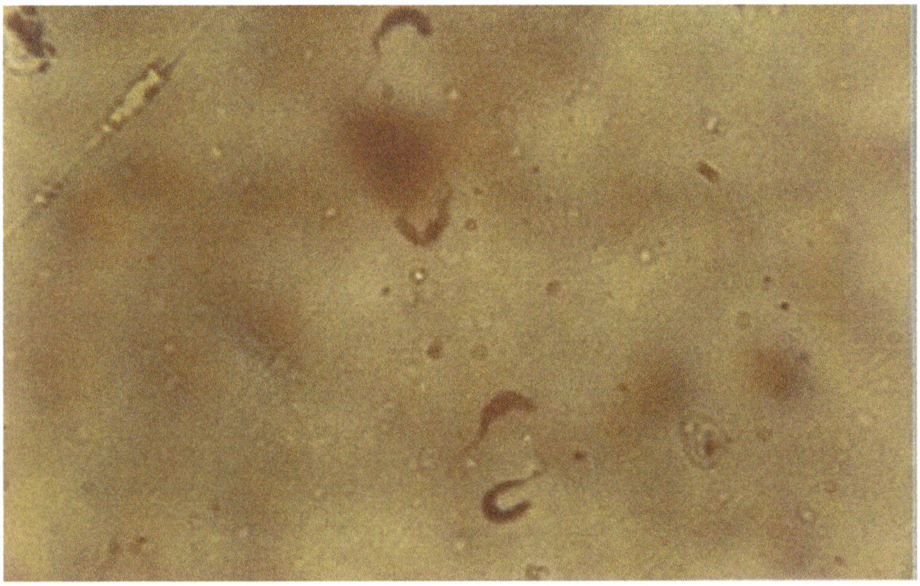

Figure 2. Photograph of *Trypanosoma vivax* after separation from the blood.

Table 1. Purification table.

Step	Volume (cm^3)	Protein (mg/ml)	Total protein (mg)	Activity/ min ml	Total activity (units)	Specific activity (units/mg)	Purification fold	% yield
Crude	12	2.0	24.0	0.0051	0.0612	0.00255	1	100
DEAE cellulose chromatographyPeak1(MDH_1)	5	0.1	0.5	0.0069	0.0345	0.069	27.1	56.4
DEAE – Cellulose chromatographyPeak2 (MDH_2)	5	0.1	0.5	0.0016	0.0080	0.016	6.3	13

parasites were motile and retained their morphology as shown in Figure 2.MDH was released into the solution after three cycles of freezing and thawing.

Total and specific activity of the enzyme is presented in Table 1. Elution of the enzyme by linear phosphate gradient on DEAE – cellulose resulted into two distinct peaks Figure 3. The active peak gave purification fold of 27.1 and percentage recovery of 56.4% (Table 1).

When the two distinct peaks were separately subjected to electrophoresis single bands were observed corresponding to the two bands out of the several bands observed from the crude enzyme.

The molecular weights of the partially purified MDH1 and MDH_2 were determined by the sodium dedocyl sulphate polyacrylamine gel electrophoresis. The logarithms of the molecular weight of the standard protein were plotted against the relative mobility expressed as percentage of bromophenol blue (Figure 4). The relative molecular weight of the MDH_1 and MDH_2 were found to be 61 and 63 kDal respectively.

Figure 5 show the activity of MDH when various concentrations of the life parasites were incubated with the substrates NADH and OAA in the reaction medium. There was no activity observed initially, however, when different volumes of triton X-100 were added, the activity of MDH was observed and it increases with increase in parasite concentration.

DISCUSSION

Various isoenzymes of malate dehydrogenase have been

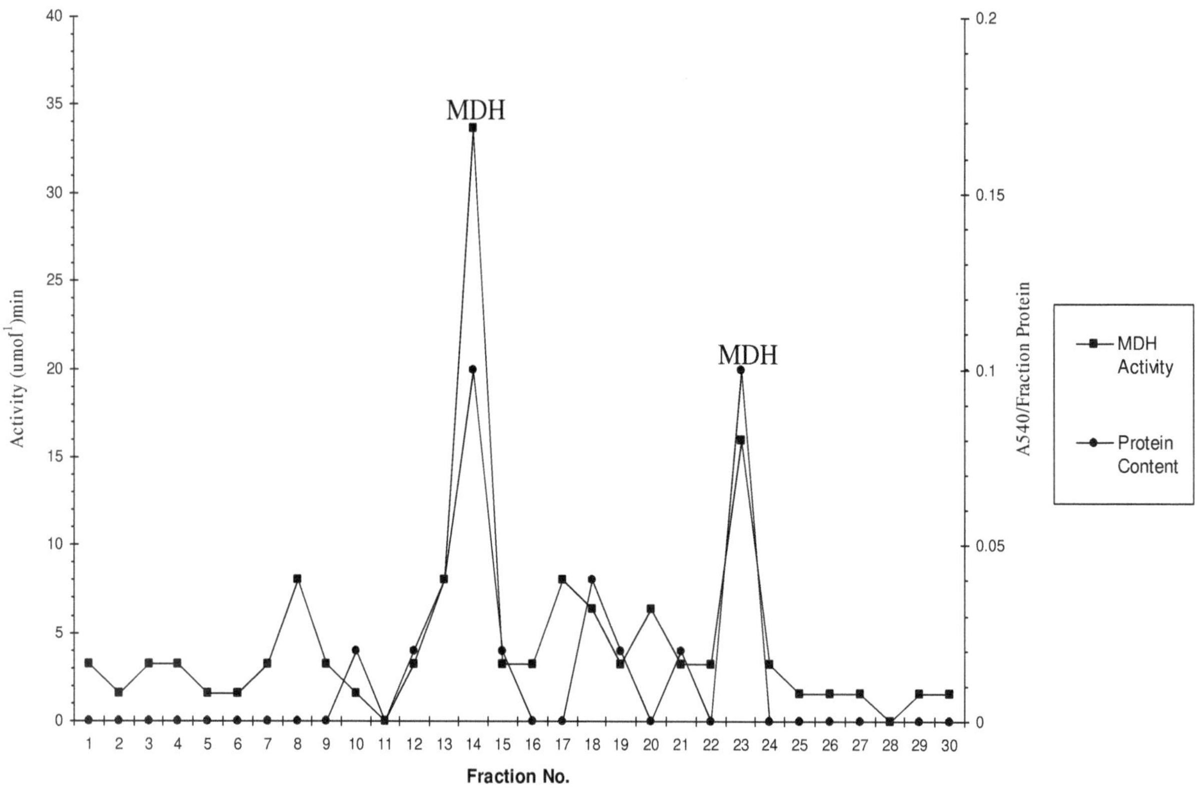

Figure 3. Elution profile of MDH by linear phosphate gradient from DEAE 52 Cellulose.

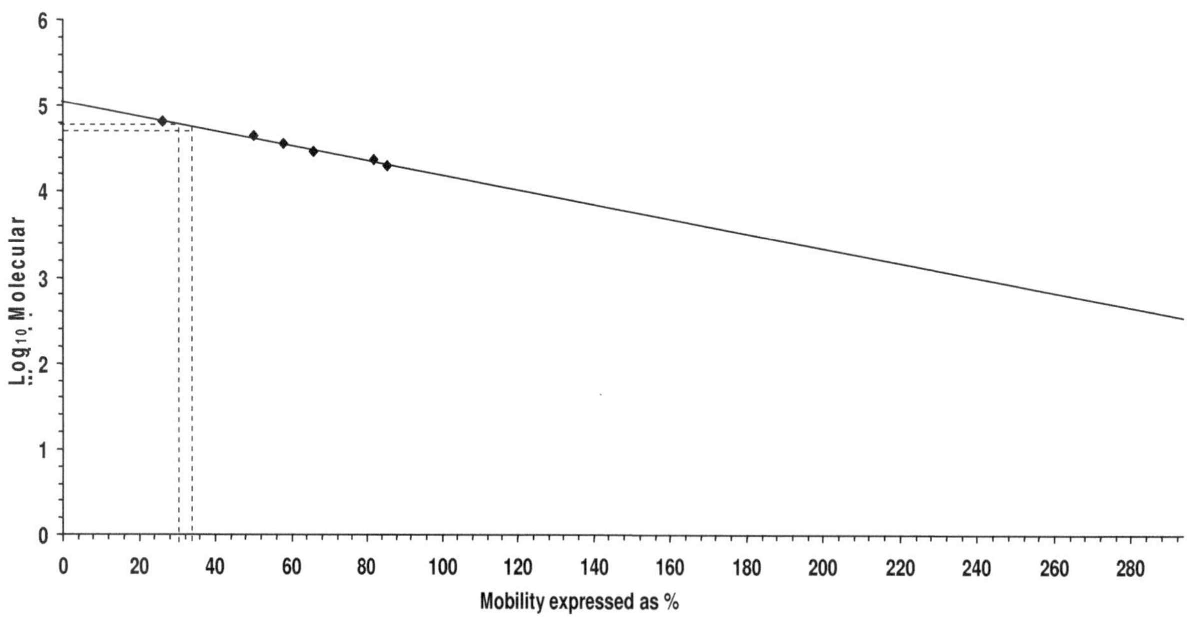

Figure 4. SDS – PAGE Calibration curve for determination of molecular weight of MDH.

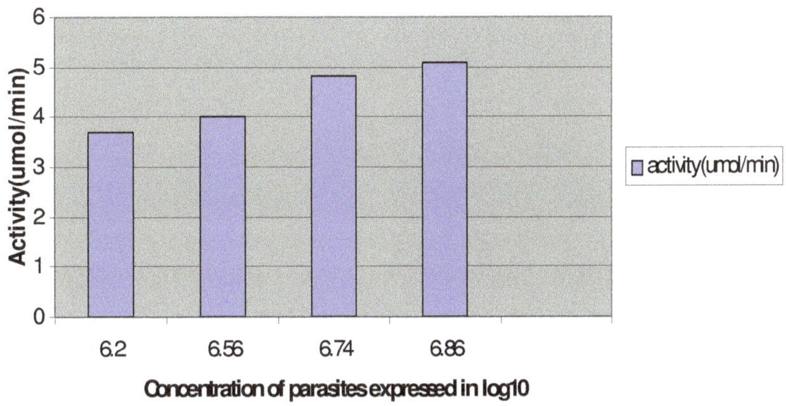

Figure 5. MDH Activity from the live parasites.

purified from different sources (Genda et al 2006; Bokato et al., 1995; Yelena et al., 2008; Lang-Unnasch, 1992). Studies have shown low degree of primary structural similarity among isoenzymes of malate dehydrogenases, however, positions for nucleotide binding, catalysis and sub unit interface were found to be conserved (Van Hellemond et al., 2005; Embley et al., 2003). This may account for the identical properties observed in isoenzymes of malate dehydrogenases. The differences observed between MDH from organelles and those from the cytosol is that a common ancestral MDH - gene may have been duplicated before inversion of premodial eukaryotes by bacteria to produce mitochondria according to the probable endosyonbiotic origin of these organelles (Hrdy et al., 2004).

MDH_1 was stable while MDH_2 lost its activity immediately and could not be characterized. It might be that MDH_2 is the mitochondrial enzyme that lost its activity due to thermal inactivation or post translational modification. MDH_2 could have been glycosylated after translation because of the environmental conditions. This could account for the higher molecular weight of the MDH_2 observed. Under blood steam condition substrate availability could be limited and the mitochondria are not fully functioning. It is also important to note that the parasite life cycle alternates between the vector and the host and it must adjust to the different environmental conditions. Therefore, it might be possible during the developmental stage of the parasite in the vector, where the mitochondria is fully functioning and substrates for both catabolism and anabolism are available, MDH_2 may not be glycosylated, may be active and its molecular weight may not be high as observed.

When MDH activity was assayed on the live parasites, no activity was observed indicating that the parasites do not secrete the enzyme. However, addition of detergent led to a surge in MDH activity. Increased activity was also observed as the concentration of the parasite was increased in the reaction medium. Both observations strongly suggest that MDH from *T. vivax* is membrane linked. Variant surface glycoprotein which protects the blood stream parasite from attack by the host immune system (Vickerman, 1985) and procyclic acidic repetitive protein are anchored to the plasma membrane by glycosyl phosphatidyl inositol (GPI) (Ferguson et al., 1985). In the present work it is obvious that MDH from the blood stream *T. vivax* is membrane bound but whether the linkage is via GPI remains to be elucidated.

REFERENCES

Bokato G, Kochan Z, Swierczynski J (1995). Purification and Properties of cytosolic and mitochondrial malic enzyme isolated from human brain. Int. J. Biochem. Cell Biol., 27(1)47-54.

Embley TM, Vander Giezen M, Horner DS, Dyal PS, Foster P (2003). Mitochondria and hydrogenosomes are two forms of the same foundermental organelle. Philos. Trans. R Soc. Lond. B Biol Sci., 358: 191-201.

Ferguson MA, Low MG, Cross GA (1985). Glycosyl-sn-1,2-Dimyristyl phosphatidylinisitol is covalently linked to *trypanosoma brucei* variant surface glycoprotein. J. Biol. Chem., 260: 145-147.

Genda T, Watabe S, Ozaki H (2006) Purification and characterization from Corynebacterium glutamicum. Biosci. Biotechnol. Biochem., 70; 1102-11089

Gutteridge WE, Coomb GH (1977). Catabolism and the Generation of energy. In Biology of parasitic protozoa. Macmillan Press Ltd., London, pp. 32-58.

Herbert WJ, Lumsden WAR (1976). *Trypanosoma brucei* A rapid "Matching" method of estimating lost parasiteamia. Experimental parasitol.

Hrdy I, Hirt RP, Dolezal P, Bardonova L, Foster PG, Tachezy J, Embley TM (2004). Trichomonas hydrogenosomes contain the NADH dehydrogenase module of mitochondrial complex 1 Nature 432: 618-622.

Kita K, Miyadera H, Saruta F, Miyoshi H (2001). Parasite mitochondria as target for chemotherapy. J. Health Sci., 47(3): 219-239.

Lang-Unnasch N (1992). Purification and properties of plasmodium Falciparum malate dehydrogenase. Mol. Biochem. Parasitol., 50(1): 17-25.

Lanham SH, Godfrey DG (1970). Isolation of Salivarian trypanosomes From man and other mammals using DEAE cellulose. Expl. Pasasit., 28: 521-534.

Nok JA (2002). Azaan thraquinone inhibits respiration and *in vitro* growth of long slender bloodstream forms of *Trypanosoma Congolense*. Cell biochem. Funct., 20: 205-212.

Van Hellomond JJ, Opperdoes FR, Tielens, AG (2005). The extra ordinary mitochondrion and unusual citric acid cycle in trypanosoma brucei. Biochem. Soc. Trans., 33: 967-971.

Vickerman K (1985). Developmental cycles and Biology of pathogenic trypanosomes. Br. Med. Bull., 41: 105.

Webster K, Pingle TR, Osborn M (1972). Measurement of molecular weights by electrophoresis on SDS. Acrylarmide gel. In: methods in enzymol, 26, (Fleischer, S. and Parker, L, eds). Academic press New York, pp 3-18.

Production of cuticle - degrading proteases by *Beauveria bassiana* and their induction in different media

Priyanka Dhar and Gurvinder Kaur*

Department of Biotechnology, Indian Institute of Technology Guwahati, Guwahati, Assam, 781039, India.

The concern for the development of hyphomycete fungi as suitable biocontrol agents of insect pests leads to the isolation of various insect pathogenic fungi. Amongst them, one of the most studied entomopathogenic fungus is *Beauveria bassiana*. The conidia of mitosporic fungi adhere to the host cuticle and germinate to produce an infectious propagule, and produce a sequential release of extra cellular enzymes to breach the insect cuticle. Protease is one of the most important and earliest enzymes involved in the host invasion. Extracellular protease production by seventeen *B. bassiana* isolates was investigated in the present study. High protease activity was observed during four to six days of culture incubation. Induction mechanism of subtillisin type Pr1 and trypsin type Pr2 activity were investigated utilizing different media. Minimal medium supplemented with casein (1%) showed high protease production and minimal medium supplemented with colloidal chitin (2%) was also able to induce Pr1 activity. The pH, ammonia and oxalic acid production in *in vitro* conditions was also investigated and the alteration in pH for protease production was not significant irrespective of the medium used. The protease activity gel was also studied and a common 66 kDa protease was observed in all the seven isolates studied.

Key words: Ammonia, *Beauveria bassiana*, oxalic acid, subtillisin type Pr1, trypsin type Pr2, protease, pH.

INTRODUCTION

The entomopathogenic fungi are natural soil borne insect pathogens and are currently being reviewed as an effecttive alternative to chemical pesticides. Amongst them, *Beauveria bassiana*-based mycoinsecticides have been developed and registered worldwide for control of agricultural pests (Hajek et al., 2001; St. Leger et al., 1992). These fungi are unique in their pathogenesis, where the infection of the host initiates by the attachment of the spores on the hydrophobic cuticle and subsequent development of the infectious structures called appresoria, which releases an array of cuticle solubilizing enzymes (Inglis et al., 2001). During the fungal penetration through the host cuticle, hydrolytic enzymes such as proteases, chitinases and lipases are produced and secreted and are important for the initiation of the infection process (St. Leger et al., 1986, 1996). The secretion of proteolytic enzymes is believed to be an important pathogenic factor for fungal attachment on cuticle (St.Leger et al., 1987b). The best understood model of fungal entomopathogencity determinant is based on the subtilisin like endoprotease designated Pr1 (St.Leger et al., 1995). Several studies indicated that Pr1 is essential for cuticle penetration that results in pathogenicity (Shah et al., 2005; Wang et al., 2002). It is an important virulence determinant and is induced by insect cuticle, derepressed under starvation and repressed in the presence of excess nutrients although the typsin - like protease Pr2 is the first to appear during *in vitro* growth on the cuticle (Butt et al., 1998; Wang et al., 2002; Gillespie et al., 1998). The role of Pr2 in insect parasitism is not yet elucidated completely, although St. Leger et al. (1994) reported that Pr2 as well the other cuticle - degrading proteases may complement each other in the splitting of peptide bonds in the insect cuticle. Results obtained by Gillespie et al. (1998) are consistent with a role

*Corresponding author. E-mail: gurvinder@iitg.ernet.in.

of Pr2 in the induction or activation of Pr1, since Pr2 was detected before Pr1 in a culture containing ground locust cuticle. St. Leger et al. (1988) observed the regulation mechanism of Pr1 and Pr2 in *M. anisopliae* and reported the catabolic repression of the enzymes. Bidochka and Khachatourians (1988) reported a high protease production from *Beauveria bassiana* in medium containing 1% gelatin as sole source of carbon and nitrogen. Induction of protease by protein is also reported in the fungus *Neurospora crassa* (Drucker, 1972). The caseinolytic activity of *M. anisopliae* in the culture filtrates is mainly due to the presence of two proteases, Pr1 and Pr2 both capable of hydrolyzing casein (St. Leger et al., 1987a). Two chymotrypsin like proteinases were produced when *B. bassiana* was grown on different carbon and nitrogen sources (Chrzanowska et al., 2001).

The relationship between gene regulation and pH sensibility by the organism is a well established fact. Regulation of gene expression by pH is observed in many fungi (Caddick et al., 1986; St Leger et al., 1998). Mutations in the pH regulated gene effecting the expression was also observed in *Y. lipolytica* (Otero and Gaillardin; 1996). In *M. anisopliae* regulation of gene expression by pH include the cuticle - degrading subtilisin proteases and other putative virulence factors (St. Leger et al., 1999). Insect cuticle infected with *M. anisopliae* demonstrated an increase in pH compared to unaffected cuticles (St. Leger et al., 1998). This suggests that *M. anisopliae* produces one or more basic compounds. Subtilisins like proteases from *M. anisopliae* are only produced under alkaline conditions (St. Leger et al., 1996). The production of cuticle degrading proteases is often subjected to the alteration of the pH by the production of ammonia in the media (St. Leger et al., 1999). Ammonia and oxalic acid production has also been studied by many workers but concluded that it does not affect the virulence (Bidochka and Khachatourians, 1993). Ammonia production by *M. anisopliae* is tightly regulated by amino acids which as protein degradation products could serve as signals for the presence of proteinaceous nutrients in the environment (St. Leger et al., 1999). Given that ammonium is linked with nitrogen catabolite repression there are presumably additional regulatory mechanisms that allow protease production in the presence of ammonia (Arst and Cove, 1969). Various protease isozymes were observed for *M. anisopliae* (St. Leger et al., 1998). A 35 kDa serine protease was produced by *B. bassiana* grown in a liquid medium containing gelatin as sole carbon nitrogen source (Bidochka and Khachatourians, 1987).

In this study the variation in the proteolytic activity of seventeen *B. bassiana* isolates was investigated for a period of ten days to study the effect of incubation time on the protease production. Subtilisin type Pr1 and trypsin like Pr2 activity of the high protease producing isolates was studied in media containing different carbon and nitrogen source to elucidate the induction of these two enzymes. At the same time studies on pH, ammonia and oxalic acid release into the media and their effect on protease production was also studied.

MATERIALS AND METHODS

Fungal isolates and culture conditions

A total of sixteen isolates were obtained from ARSEF (USDA-ARS Plant Protection Unit, Ithaca, NY) and one isolate was from India (Table 1). The isolates were routinely sub cultured on SDA (Sabouraud Dextrose Agar) slants at 28°C and maintained at 4°C. Seven day old SDA slants were used for the preparation of conidial suspension (1 x 10^6 conidia/ml) to inoculate SDY broth (4% Dextrose, 1% Peptone, and 1% Yeast extract) and incubated at 28°C and 180 rpm for 3 days. The harvested mycelium was washed twice with sterilized distilled water and inoculated into induction medium containing casein (1%) (Sigma) in basal salts medium (0.1% KH_2PO_4, 0.05% MgSO4, 0.05% NaCl) at 20% v/v based on the final volume (50 ml) of the culture for the determination of proteolytic activity. The pH of the culture media was adjusted to 8.0 and the cultures were incubated at 28°C and 180 rpm for a period of 10 days. The mycelia were harvested by centrifugation and the supernatant was assayed for proteolytic activity on alternate days of growth till tenth day of culture.

For the determination of subtilisin (Pr1) activity and trypsin (Pr2) activity three different media; minimal media, minimal media supplemented with casein (1%) and minimal media supplemented with colloidal chitin (2%) were used and the pH was maintained at 7.0. For the enzyme production, conidia at a concentration of 1 x 10^6 conidia/ml were inoculated into each media (three replicates were maintained) and incubated as shake cultures at 180 rpm, 28°C for 72 h. The mycelia were harvested by centrifugation and the supernatants assayed for enzyme activity (Pr1 and Pr2).

Enzyme assays and protein determination

Proteolytic activity

The protease assay was done by the method of Kunitz (1947). Casein substrate was prepared for enzyme assay by dissolving 2 gm of casein (Sigma) in 90 ml of 0.01 M Tris HCl (pH 8.0) containing 10 mM $CaCl_2$ (pH 8.0) and the volume made up to 100 ml with H_2O. A 400 µl of casein substrate was added to 200 µl of the culture supernatant in 0.01M Tris HCl pH 8, 10 mM $CaCl_2$. The reaction mixture was incubated at 35°C for 10 min and the reaction terminated by the addition of 1.2 M TCA. The reaction mixture was centrifuged at 4000 rpm for 5 min and the supernatant was taken and the absorbance at 280 nm (A_{280}) was observed against water as blank. One unit of protease activity was defined as the amount of enzyme that produced 1 mM of Tyrosine per minute under the above conditions. Tyrosine was taken as standard for all the enzyme assay calculations.

Subtilsin (Pr1) and Trypsin (Pr2) activity

The subtilisin (Pr1) activity and typsin (Pr2) activity was assayed by the method of St. Leger et al. (1987a). The Pr1 activity was assayed using specific synthetic substrate, N-succinyl-Ala-Ala-Pro-Phe p-nitroanilide and Pr2 activity was assayed using specific synthetic substrate, Benzoyl-phenylalanine-valinearginine-*p*-nitroanilide (BAPNA). Each substrate (50 µl) was mixed with 0.85 ml of Tris-HCl buffer (pH 8.5) and 100 µl of culture supernatant. The reaction mixture was incubated for 1h at 28°C and the reaction was terminated using 0.25 ml of 30% acetic acid. The reaction mixture was left to stand for 15 min on ice prior to spectrophotometry. Absorbance at 410 nm (A_{410}) was observed and concentration of para-nitro aniline was determined. One unit of protease activity was defined as the amount of enzyme that produced 1 µM of para-nitro aniline per minute under the above conditions. Protein content was measured by the method of Bradford (1976) using Bovine Serum Albumin as a standard for proteolytic, Pr1 and Pr2 activity.

Table 1. Source of *B. bassiana* isolates.

Isolates	Code No./ Accession No. ARSEF	Host insect	Geographical location
UB1	1788	*H. virescens*	Spain
UB2	2041	*C. medinalis*	Philippines
UB3	5278	*B. tabaci*	U.S.A
UB4	2417	*H. puer*	India
UB5	2597	*S. litura*	India
UB6	6646	*C. septumpunctata*	India
UB7	4027	*C. septempunetata.*	Denmark
UB8	1166	*H. armigera.*	Spain
UB9	2033	*Coccinella* sp.	U.S.A
UB10	2034	*Coccinella* sp.	U.S.A
UB11	4018	*C. septempunetata*	Denmark
UB12	1886	*C. infuscatellus*	India
UB13	2412	*X. jamaicensis*	India
UB14	8250	*B. fulvicornis*	India
UB15	6650	*S. litura*	India
UB16	2660	*Adult Coleoptera*	India
AB1	Local	Unknown	India

Quantification of ammonia

Ammonia concentration was measured by the colorimetric method of Chaney and Marbach (1962). An uninoculated broth was used as a negative control. For this study two solutions were prepared, solution I (Phenol 10 gm/l, Sodium nitroprusside 0.05 gm/l) and solution II (NaOH 5 gm/l, NaCl 0.42 gm/l) and added in equal volumes to 1 ml of culture supernatant. Absorbance at 625 nm (A_{625}) was determined after incubating the solution at room temperature for 30 min.

Quantification of oxalic acid

Oxalic acid concentration was determined by the method of Yan et al. (2004). To the culture supernatant (1ml), 1 M H_2SO_4, 400 µl of 0.03 M $K_2Cr_2O_7$ and 8 ml of ddH_2O was added. After a gentle mix, 400 µl of 1×10^4 M Victoria blue B was added and the total volume of the solution were made to 10 ml by ddH_2O. The solution was incubated for 9 min at 60°C. Mixture was cooled to quench the reaction with tap water for 2 min and absorbance at 610 nm (A_{610}) of the solution was determined.

Protease activity gel

Molecular weight of the native enzyme was observed using gelatin acrylamide gel electrophoresis. SDS-PAGE was performed on 12.5% gels containing 0.3% gelatin as a copolymerized substrate under non-reducing conditions. After electrophoresis, the gels were shaken gently at room temperature for 1 h in 2.5% Triton X -100 solution to remove SDS and denaturize the enzymes. Gels are then incubated in 0.01 M Tris-HCl (pH 8) – 10 mM $CaCl_2$, buffer at 37°C for 5 - 8 h. The gels were fixed and stained with Coomassie brilliant blue R - 250 for 3 h and then detained for 1 h. clear proteolytic zones in the gel were visible against dark blue background.

Statistical analysis

Statistical analyses were performed by SPSS software. Test of significance were carried out using Tukey's test.

RESULTS

Proteolytic activity

B. bassiana isolates incubated for ten days showed variation among different isolates studied and the enzymatic activity was maximum from day four to day eight (Table 2). On the second day of incubation, isolates UB8, UB10 and UB 14 showed a activity of 0.58, 0.60 and 0.52 U/ml respectively. On the fourth day, isolates UB 3, UB 9, UB13 produced maximum enzymatic activity of 0.76, 0.90 and 0.77 U/ml respectively. Isolates UB3 (0.75 U/ml), UB 8 (0.74 U/ml), UB 11 (0.76 U/ml) on sixth day and isolates UB11 and UB16 showed an activity of 0.89 and 0.84 U/ml respectively on 8th day and with decreased protease activity on 10th day. Isolates UB3, UB9 and UB13 showed high enzymatic activity on all days of study. High specific activity was observed on 10th day in almost all the isolates with highest specific activity by isolates UB 3 (10.41 mU/mg) and UB 9 (14.58 mU/mg) respectively. High enzyme producing strains were screened based on the enzyme activity on 4th day and further analysis of Pr1 and Pr2 activity was determined along with pH, ammonia and oxalic acid release.

Table 2. Enzymatic activity and specific activity of *B. bassiana* isolates

Isolates	Enzyme and specific activity as on									
	Day 2		Day 4		Day 6		Day 8		Day 10	
	Enzymatic activity (U/ml)	Sp. activity (mU/mg)	Enzymatic activity (U/ml)	Sp. activity (mU/mg)	Enzymatic activity (U/ml)	Sp. activity (mU/mg)	Enzymatic activity (U/ml)	Sp. activity (mU/mg)	Enzymatic activity (mU/ml)	Sp. activity (mU/mg)
UB1	0.4247^{c}	1.26^{f}	0.5748^{d}	1.39^{g}	0.5718^{c}	2.37^{f}	0.5710^{d}	5.19^{c}	0.5705^{b}	0.87^{f}
UB2	0.4422^{c}	1.61^{f}	0.4367^{e}	2.46^{f}	0.3967^{de}	5.54^{c}	0.4010^{e}	7.65^{a}	0.4137^{d}	1.39^{e}
UB3	0.3538^{d}	2.01^{e}	0.7633^{b}	3.45^{e}	0.7466^{a}	3.37^{e}	0.6985^{c}	7.61^{a}	0.6088^{b}	10.41^{a}
UB4	0.0156^{h}	7.08^{a}	0.1109^{h}	4.32^{d}	0.1243^{g}	4.79^{d}	0.1279^{h}	4.38^{d}	0.1313^{f}	1.13^{f}
UB5	0.1757^{f}	1.57^{f}	0.3497^{f}	9.43^{a}	0.3467^{e}	1.59^{g}	0.3421^{f}	2.49^{f}	0.3401^{e}	3.71^{d}
UB6	0.1141^{g}	0.518^{g}	0.2081^{g}	7.43^{b}	0.2548^{f}	9.78^{a}	0.2842^{g}	1.65^{g}	0.3022^{e}	3.27^{d}
UB7	0.1423^{g}	1.17^{f}	0.1476^{h}	1.15^{g}	0.2355^{f}	1.75^{g}	0.3211^{f}	3.23^{e}	0.3371^{e}	3.39^{d}
UB8	0.5889^{a}	3.63^{c}	0.5679^{d}	2.87^{f}	0.7347^{a}	6.19^{b}	0.7582^{b}	5.30^{c}	0.6084^{b}	6.01^{b}
UB9	0.4336^{c}	3.25^{c}	0.9021^{a}	3.09^{e}	0.6216^{b}	3.89^{e}	0.5864^{d}	6.69^{b}	0.5211^{c}	14.58^{a}
UB10	0.6059^{a}	4.71^{b}	0.6536^{c}	6.74^{c}	0.4105^{d}	5.60^{c}	0.4526^{e}	6.24^{b}	0.4269^{d}	6.10^{b}
UB11	0.4204^{c}	3.43^{c}	0.4925^{e}	3.02^{e}	0.7583^{a}	5.03^{c}	0.8946^{a}	5.07^{c}	0.8277^{a}	5.05^{c}
UB12	0.1575^{g}	1.01^{f}	0.3723^{f}	1.42^{g}	0.4224^{d}	2.01^{f}	0.4014^{e}	4.31^{d}	0.3515^{e}	3.99^{d}
UB13	0.0524^{h}	0.546^{g}	0.7747^{b}	2.49^{f}	0.6706^{b}	2.96^{f}	0.6660^{c}	4.05^{d}	0.6454^{b}	9.22^{a}
UB14	0.5216^{b}	3.35^{c}	0.5531^{d}	3.53^{e}	0.5179^{c}	4.51^{d}	0.4745^{e}	4.26^{d}	0.4429^{d}	4.57^{c}
UB15	0.1755^{f}	1.92^{e}	0.2845^{g}	3.71^{e}	0.2059^{f}	2.50^{f}	0.1831^{h}	2.12^{f}	0.1819^{f}	2.06^{e}
UB16	0.2672^{e}	2.86^{d}	0.5222^{d}	3.23^{e}	0.5982^{c}	4.47^{d}	0.8405^{a}	5.43^{c}	0.8309^{a}	4.95^{c}
AB1	0.0360^{h}	0.070^{h}	0.0817^{i}	0.15^{h}	0.0804^{h}	0.16^{h}	0.1024^{h}	0.46^{h}	0.2031^{f}	1.71^{e}

Values followed by same lower case alphabets in the column are statistically equivalent according to Tukey's test.

pH, ammonia, oxalic acid, Pr1 and Pr2 activity in minimal medium

High Pr1 activity was observed for isolate UB13 (4.44 U/ml) followed by isolate UB 3 (3.88 U/ml) (Table 3). High specific activity of Pr1 enzyme was observed for isolate UB 5 (3.89 U/mg) followed by isolate UB 3 (1.26 U/mg). Almost all the isolates showed same level of Pr2 activity except isolate UB 1 (2.03 U/ml). The pH of the medium increased from the initial (7.0) for all the isolates except UB 1 and a high Pr1 activity was observed at pH 7.02 and 7.05 respectively. The ammonium concentration was high for isolate UB 3 followed by UB 2. Oxalic acid concentration was least (26.0 µg/ml) for the isolate UB 13 that showed high Pr1 activity (4.44 U/ml).

pH, ammonia, oxalic acid, Pr1 and Pr2 activity in minimal medium supplemented with Casein (1%)

High Pr1 enzyme activity was observed for isolate UB 1 and UB 2 (4.77 U/ml) followed by isolate UB 9 (4.04 U/ml) (Table 4). Specific activity was observed to be extremely low compared to enzyme activity for both Pr1 and Pr2 enzymes. High Pr2 activity was observed for isolate UB 5 (0.75 U/mg) and the rest of the isolates showed similar enzymatic activity. The pH of the medium showed slight change in case of isolates UB 1 (7.17), UB 2 (7.08) and a pH of 7.06 for the isolate UB 9. The ammonia concentration was the least for the isolate UB 1 (0.015) which showed high Pr1 activity. The other two isolates which showed

Table 3. pH, ammonia, oxalic acid concentration, Pr1 and Pr2 activity of *B. bassiana* isolates in minimal medium.

Isolates	pH	Ammonia concentration (µg/ml)	Oxalic acid concentration (µg/ml)	Pr1		Pr2	
				Enzyme activity (U/ml)	Sp. activity (U/mg)	Enzyme activity (U/ml)	Sp. activity (U/mg)
UB1	6.99	0.022	33.5	0.76[e]	0.17[e]	2.03[a]	0.45[b]
UB2	7.01	0.065	43.0	1.49[c]	0.38[d]	0.15[b]	0.04[c]
UB3	7.05	0.068	60.5	3.88[b]	1.26[b]	0.16[b]	0.05[c]
UB5	7.02	0.054	85.5	0.35[f]	3.89[a]	0.16[b]	1.78[a]
UB9	7.01	0.059	75.5	1.24[d]	0.67[c]	0.17[b]	0.09[c]
UB10	7.03	0.055	94.5	0.28[f]	0.12[e]	0.17[b]	0.07[c]
UB13	7.02	0.052	26.0	4.44[a]	0.33[d]	0.16[b]	0.01[c]

Values followed by same lower case alphabets in the column are statistically equivalent according to Tukey's test.

Table 4. pH, ammonia, oxalic acid concentration, Pr1 and Pr2activity of *B. bassiana* isolates in MM supplemented with casein (1%).

Isolates	pH	Ammonia concentration (µg/ml)	Oxalic acid concentration (µg/ml)	Pr1		Pr2	
				Enzyme activity (U/ml)	Sp. activity (U/mg)	Enzyme activity (U/ml)	Sp. activity (U/mg)
UB1	7.17	0.015	-	4.77[a]	0.05[a]	0.14[b]	0.001[a]
UB2	7.08	0.027	-	4.77[a]	0.05[a]	0.09[b]	0.009[a]
UB3	7.06	0.037	-	2.56[d]	0.04[a]	0.13[b]	0.002[a]
UB5	7.05	0.033	-	3.77[c]	0.05[a]	0.75[a]	0.009[a]
UB9	7.06	0.026	-	4.04[b]	0.06[a]	0.18[b]	0.003[a]
UB10	7.02	0.029	-	1.87[e]	0.03[a]	0.11[b]	0.002[a]
UB13	7.11	0.035	-	0.07[f]	0.001[b]	0.14[b]	0.009[a]

high Pr1 activity also showed less ammonia concentration when compared to other isolates. Interestingly, oxalic acid production was not observed in any of the isolates in this medium.

pH, ammonia, oxalic acid, Pr1 and Pr2 activity in minimal medium supplemented with Colloidal chitin (2%)

Pr1 activity was moderately high for all the *B. bassiana* isolates. Isolate UB 13 showed high Pr1 activity (4.36 U/ml) and UB 5 showed high specific activity for Pr1 (0.39 U/mg) (Table 5). High Pr2 activity was observed for the isolate UB 3 (0.18 U/ml). Pr2 specific activities were indistinguishable among the isolates. The pH of the medium showed slight changes, and the isolates showing high Pr1 activity also showed relatively high concentration of ammonia. Interestingly, even in this medium oxalic acid production was not observed.

Protease activity gel

Isolate UB 1 and UB 5 both showed same kind of zymograms with two bands corresponding to 97 and 66 kDa of native protease enzymes (Figure 1). A 97.4 kDa protease was observed in the zymogram for the isolates UB 2, UB 3 and UB 9. A 66 kDa native protease was observed in common in all the isolates. A 55 kDa protease was observed in isolate UB 3 which was absent in all other isolates studied.

DISCUSSION

Protein is one of the most primary constituent in insect cuticle after chitin. Protease is one of the most important and earliest enzymes involved in the invasion followed by chitinase after the eventual exposure of chitin in the host cuticle after the proteolytic degradation of cuticular proteins (Smith et al., 1981; St.Leger et al., 1986). Extracellular proteases were even found in insect haemolymph (Shimizuy et al., 1993). An immense variation in the protease activity was observed among the seventeen *B. bassiana* isolates. Variation in the time of maximum protease production was also observed with different isolates. The variation in protease production might be attributed to the geographical origin of the isolates as three among four high protease producing isolates viz. UB3, UB9, UB10 were from same geographical origin (USA), although variation cannot be directly correlated with the insect host. Most of the isolates showed

Table 5. pH, ammonia, oxalic acid concentration, Pr1 and Pr2 activity of *B. bassiana* isolates in MM supplemented with colloidal chitin (2%).

Isolates	pH	Ammonia concentration (µg/ml)	Oxalic acid concentration (µg/ml)	Pr1		Pr2	
				Enzyme activity (U/ml)	Sp. activity (U/mg)	Enzyme activity (U/ml)	Sp. activity (U/mg)
UB1	7.17	0.015	-	3.53^{c}	0.24^{b}	0.11^{bc}	0.01^{a}
UB2	7.08	0.027	-	3.59^{c}	0.34^{a}	0.09^{bc}	0.01^{a}
UB3	7.06	0.037	-	3.02^{d}	0.26^{b}	0.18^{a}	0.02^{a}
UB5	7.05	0.033	-	3.79^{b}	0.39^{a}	0.09^{bc}	0.01^{a}
UB9	7.06	0.026	-	2.66^{e}	0.21^{b}	0.10^{bc}	0.01^{a}
UB10	7.02	0.029	-	1.53^{f}	0.15^{c}	0.05^{c}	0.01^{a}
UB13	7.11	0.035	-	4.36^{a}	0.26^{b}	0.09b^{c}	0.01^{a}

Values followed by same lower case alphabets in the column are statistically equivalent according to Tukey's test.

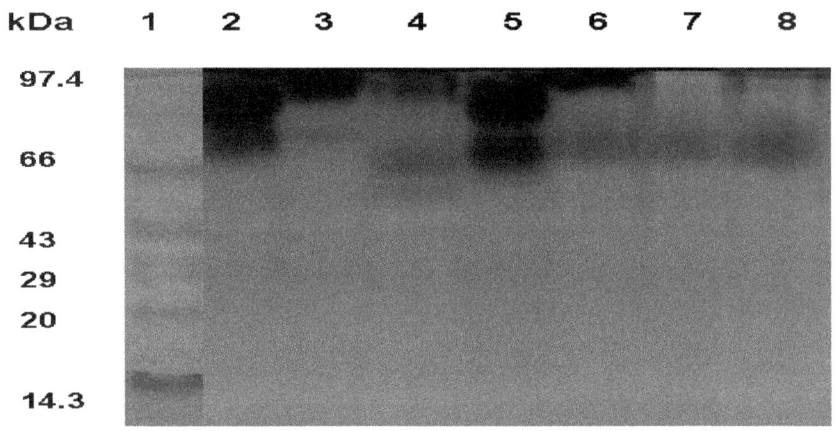

Figure 1a. Protease activity in gelatin polyacrylamide gel electrophoresis. Lane 1 corresponds to MW marker after complete destaining of the gel. Lane 2,3,4,5,6,7,8 corresponds to isolates UB 1, UB 2, UB 3, UB 5, UB 9, UB 10 and UB 13, respectively.

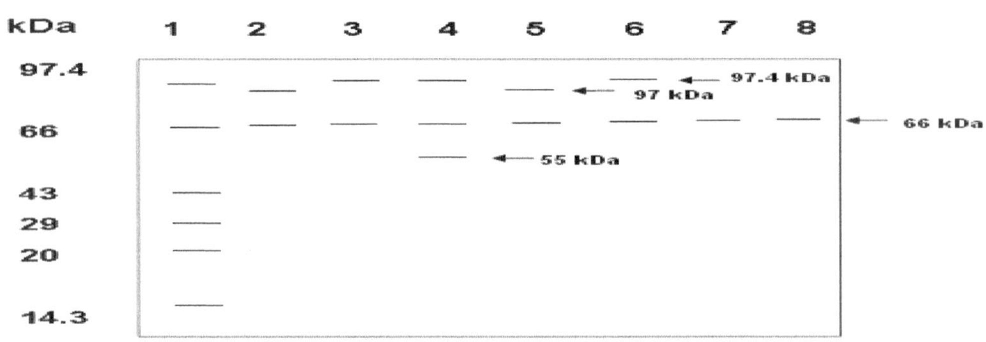

Figure 1b. Schematic representation of Figure 1a with the corresponding MW for each isozyme.

high protease production during four to six days of culture which is comparable with the observations by Ito et al. (2007), demonstrating high protease activity on fifth day of culture in *B. bassiana*. The time for maximum protease production does not necessarily depend upon the media constituents and Kucera (1971) observed high amounts of protease release 3 days post inoculation in the media with different nitrogen sources. Proteolytic activity of *B. bassiana* isolates decreased with increase in culture age probably due to nutrient limitation or autolysis of the culture. Caseinolytic and elastolytic activity of an *M. anisopliae* isolate was observed to be high during four to six days of culture and then a steep decrease up to sixteenth day of culture probably due to increasing degree of autolysis of the aging culture (Braga et al., 1999).

The caseinolytic activity can be attributed to the presence of Pr1 and Pr2 activity in the culture filtrate (St. Leger, 1987a). The Pr1 and Pr2 activity in three media with different carbon and nitrogen sources was investigated for the selected seven isolates. The use of minimal medium for protease production could possibly emulate the enzyme activity in the initial stages of host infection when readily metabolizable compounds are extremely low. Almost all the isolates showed moderate to high Pr1 activity and low levels of Pr2 activity even in the minimal medium indicating that the enzyme production is constitutive. The Pr1 and Pr2 activity was observed to increase rapidly during carbon and nitrogen depression for entomopathogenic fungus *M. anisopliae* (St. Leger, 1988), although Bidochka and Khachatourians (1988) reported the lack of protease production from one *B .bassiana* isolate in the absence of exogenous protein. Constitutive production of protease was also found in *Aspergillus* sp. (Cohen and Drucker, 1977). In the present study, the Pr1 and Pr2 activity was observed to be high in the medium containing casein (1%) as a sole source of carbon and nitrogen. The reason perhaps is the presence of protein in the minimal medium acting as an inducer for both the enzymes. Similar results of protease induction by protein were observed by Drucker (1972) in *Neurospora crassa*. The Pr2 production could be enhanced by the addition of proteins like gelatin, BSA or casein (Paterson et al., 1993). The most notable observation was that the isolates that showed maximum Pr1 production in minimal medium viz., UB 3 (3.88 U/ml) and UB 13 (4.44 U/ml) showed decreased Pr1 activity in minimal medium supplemented with casein (1%), UB 3 (2.56 U/ml) and UB 13 (0.07 U/ml), which was found to be contrary in case of other five isolates that showed low Pr1 activity in minimal medium and high in minimal medium supplemented with casein (1%). These results indicate the presence of catabolite repression of Pr1 activity produced by different isolates. St. Leger et al. (1988) reported that modifying the concentrations of carbon and nitrogen independently could repress protease production even in the absence of other repressor molecule. All the isolates showed high Pr1 activity in minimal medium supplemented with colloidal chitin (2%).

The exploitation of colloidal chitin as sole source of carbon and nitrogen appeared to induce the Pr1 activity which is indeed comparable with the results of St. Leger et al. (1988) who observed that protease activity could be induced by using polymeric substances like cellulose and insect cuticle. Chitin, the main constituent of insect cuticle enhances the Pr1 activity either by inducing the protease or possibly that colloidal chitin was unable to produce catabolite repression. However the role of colloidal chitin to induce or repress Pr2 enzyme was unclear as very low levels of Pr2 activity was detected in this medium containing colloidal chitin (2%).

Genes that are involved in virulence and pathogenesis are often regulated by environmental signals like pH (De Bernardis et al., 1998). An increase in pH was observed on insect cuticle after 60 h growth of fungus which is consistent with the pH regulation of the protease production during growth by *M. anisopliae* in liquid culture containing cuticle. The pH of the culture media rises due to the release of ammonical by - products of cuticle protein degradation (St. Leger et al., 1998; St.Leger et.al., 1989). Protease production in *M. anisopliae* was also correlated with increased pH of culture medium caused by the production of ammonia; however oxalic acid producing mutants showed decreased activity (St Leger et al., 1999). However, in this investigation no such correlation was observed in relation to pH of the culture medium and subtilisin type protease production (Pr1). The pH of the unbuffered media changed diminutively from the initial pH 7 in the three different media. High ammonia production in minimal medium supplemented with casein (1%) could be correlated with the high Pr1 production, although high Pr1 production in minimal medium supplemented with colloidal chitin (2%) cannot be associated with low ammonia concentration. Dias et al. (2008) reported moderate production of Pr1 from *B. bassiana* in unbuffered minimal medium at pH 7 after 72 h of culture. High concentration of oxalic acid should not be directly correlated with the low protease production in nutrient deprived minimal medium. However oxalic acid was not detected in other two media suggesting a favorable condition for high Pr1 secretion by ammonia production alone. The native proteases from the seven *B. bassiana* isolates differed in their molecular weights.

Almost all the isolates showed a 66 kDa protease. St. Leger et al. (1996) observed two trypsin like proteases in *M. anisopliae* with 30 kDa and 27 kDa molecular weight. A 103 kDa and 12 kDa metalloprotease was reported from *M. anisopliae* conidia (St. Leger et al., 1998).

The results represented in this study would increase the knowledge about the degree of protease production in various *B. bassiana* isolates and the incubation time as well as the Pr1 and Pr2 production by *B. bassiana* isolates could lead to new course of understanding of the role of proteases in virulence of the entomopathogenic fungi.

ACKNOWLEDGEMENTS

We acknowledge the financial support provided by Department of Science and Technology (Project No.SR/FT/L-144) and Ministry of Human Resource and Development (Project No. F. 26-11/2004 TS V). We also thank Dr. Richard A. Humber (USDA-ARS) for providing the fungal cultures.

REFERENCES

Arst HN, Cove DJ (1969). Methyl ammonium Resistance in *Aspergillus nidulans*. J. Bacteriol. 98:1284-1293.

Butt TM, St. Leger RJ, Leal SCM, Kerry BR (1998). Variation in the subtilisins of fungal pathogens of insects and nematodes. In: Couteaudier BP, Clarkson J (eds.) Molecular variability of fungal pathogens. CAB International, Wallingford pp. 149-169.

Bidochka MJ, Khachatourians GG (1987). Purification and Properties of an Extracellular Protease Produced by the Entomopathogenic Fungus *Beauveria bassiana*. Appl. Environ. Microbiol. 53:1679-1684.

Bidochka MJ, Khachatourians GG (1988). Regulation of extracellular protease in the entomopathogenic fungus *Beauveria bassiana*. Exp. Mycol. 13: 161-168.

Bidochka MJ, Khachatourians GG (1993). Oxalic acid hyper production in *Beauveria bassiana* mutants is related to an utilizable carbon source but not to virulence. J. Invertebr. Pathol. 62: 53-57.

Bradford MM (1976). A rapid and sensitive method for quantification of microgram quantities of protein utilizing the principle of protein dye binding. Anal. Biochem. 72: 248-254.

Braga GUL, Destefano RHR, Messias CL (1999). Protease production during growth and autolysis of submerged *Metarhizium anisopliae* cultures. Rev. Microbiol. 30: 107-113.

Chrzanowska J, Banas J, Kolaczkowska M (2001). Purification and Characterization of *Beauveria bassiana* Proteinases. Acta Biotechnol. 21: 73-81.

Caddick MX, Brownlee AG, Arst HN (1986). Regulation of gene expression by pH of the growth medium in *Aspergillus nidulans*. Mol. Gen. Genet. 203: 346-353.

Chaney AL, Marbach EP (1962). Modified reagents for the determination of urea and ammonia. Clin. Chem. 8:130-132.

Cohen BL, Drucker M (1977). Regulation of exocellular protease in *Neurospora crassa:* Induction and repression under conditions of nitrogen starvation. Arch. Biochem. Biophys. 182:601-613.

Drucker H (1972). Regulation of exocellular proteases in *Neurospora crassa:* Induction and repression of enzyme synthesis. J. Bacteriol. 110:1041-1049.

De Bernardis F, Muhlschlegel FA, Cassone A, Fonzi WA (1998). The pH of the host niche controls gene expression in and virulence of *Candida albicans*. Infect. Immun. 66:3317-3325.

Dias BA, Neves PMOJ, Furlaneto-Maia L, Furlaneto MC (2008). Cuticle degrading proteases produced by the entomopathogenic fungus *Beauveria bassiana* in the presence of coffee berry borer cuticle. Brazilian J. Microbiol. 39: 301-306.

Gillespie JP, Bateman R, Charnley AK (1998). Role of cuticle-degrading proteases in the virulence of *Metarhizium* sp. for the desert locust, *Schistocerca gregaria*. J. Invertebr. Pathol. 71:128-137.

Hajek AE, Wright SP, Vandenberg JD (2001). Control of arthropods using pathogenic fungi. In: Pointing SB, Hyde KD (eds.). Bioexploitation of filamentous fungi. Fungal, diversity research series. 6:309-347.

Inglis DG, Goettel MS, Butt TM, Strasser H (2001). Use of Hyphomyecete fungi for managing insect pests. In: Butt TM, Jackson CW, Magan N (eds.). Fungi as Biocontrol Agents-Progress Problems and Potential. Wallingford, UK: CAB International pp. 23-69.

Ito ET, Varea-Pereira G, Miyagui DT, Pinotti MHP, Neves PMOJ (2007). Production of Extracellular Protease by a Brazilian Strain of *Beauveria bassiana* Reactivated on Coffee Berry Borer, *Hypothenemus hampei*. Braz. Arch. Biol. Technol. 50:217-223.

Kunitz M (1947). Crystalline soyabean trypsin inhibitor. II. General properties. J. Gen. Physiol. 30:291-310.

Kucera M (1971). Toxins of the entomophagous fungus *Beauveria bassiana*. II. Effect of nitrogen sources on formation of the toxic protease in submerged culture. J. Invertebr. Pathol. 17:211-215.

Otero RC, Gaillardin C (1996). Dominant mutations affecting expression of pH-regulated genes in *Yarrowia lipolytica*. Mol. Gen. Genet. 252:311- 319.

Paterson IC, Charnley AK, Cooper RM, Clarkson JM (1993). Regulation of production of a trypsin-like protease by the insect pathogenic fungus *Metarhizium anisopliae*. FEMS Microbiol. Lett. 109:323-328.

St. Leger RJ, Frank DC, Roberts DW, Staples RC (1992). Molecular cloning and regulatory analysis of the cuticle-degrading protease structural gene from the entomopathogenic fungus *Metarhizium anisopliae*. Eur. J. Biochem. 204:991-1001.

St.Leger RJ, Cooper RM, Charnley AK (1986). Cuticle – degrading enzymes of entomopathogenic fungi: Synthesis in culture on cuticle. J. Invertebr. Pathol. 48:85-95.

St.Leger RJ, Joshi L, Bidochka MJ, Rizzo NW (1996). Biochemical Characterization and Ultrastructural Localization of Two Extracellular Trypsins Produced by *Metarhizium anisopliae* in Infected Insect Cuticles. Appl. Environ. Microbiol. 62:1257-1264.

St. Leger RJ, Cooper RM, Charnley AK (1987b). Production of cuticle-degrading enzymes by the Entomopathogen *Metarhizium anisopliae* during infection of cuticles from *Colliphora vomitoria* and *Manduca sexta*. J. Gen. Microbiol. 133:1371-1382.

St. Leger RJ, Joshi L, Bidochka MJ, Roberts DW (1995). Protein synthesis in *Metarhizium anisopliae* growing on host cuticle. Mycol. Res. 99:1034-1040.

Shah FA, Wang CS, Butt TM (2005). Nutrition influences growth and virulence of the insect-pathogenic fungus *Metarhizium anisopliae.* FEMS Microbiol. Lett. 251:259–266.

St Leger RJ, Bidochka MJ, Roberts DW (1994). Isoforms of the cuticle degrading Pr1 protease and production of a metalloproteinase by *Metarhizium anisopliae.* Arch. Biochem. Biophys. 313:1-7.

St. Leger RJ, Charnley AK, Cooper RM (1987a). Characterization of cuticle-degrading proteases produced by the entomopathogen *Metarhizium anisopliae*. Arch. Biochem. Biophys. 253:221-232.

St. Leger RJ, Durrands PK, Cooper RM, Charnley AK (1988). Regulation of production of proteolytic enzymes by the entomopathogenic fungus *Metarhizium anisopliae.* Arch Microbiol. 150:413-416.

St Leger RJ, Joshi L, Bidochka MJ, Roberts DW (1998). Ambient pH is a major determinant in the expression of cuticle degrading enzymes and hydrophobin by *Metarhizium anisopliae.* Appl. Environ. Microbiol. 64: 709-713.

St.Leger RJ, Nelson JO, Screen SE (1999). The entomopathogenic fungus *Metarhizium anisopliae* alters ambient pH, allowing extracellular protease production and activity. Microbiology. 145:2691-2699.

St. Leger RJ, Butt TM, Staples RC, Roberts DW (1989). Production in vitro of appressoria by the entomopathogenic fungus Metarhizium anisopliae. Exp. Mycol. 13:274-288.

Smith RJ, Pekrul S, Grula EA (1981). Requirement for sequential enzymatic activities for penetration of the integument of the corn earworm (*Heliothis zea*). J. Invertebr. Pathol. 38:335-344.

Shimizuy S, Tsuchitan I, Matusumoto T (1993). Production of an extracellular protease by *Beauveria bassiana* in the haemolymph of the silkworm, *Bombyx mori.* Lett. Appl. Microbiol. 16:291-294.

Wang C, Typas MA, Butt TM (2002). Detection and characterization of pr1 virulent gene deficiencies in the insect pathogenic fungus *Metarhizium anisopliae.* FEMS Microbiol. Lett. 213:251-255.

Yan ZY, Xing GM, Li ZX (2004). Quantitative Determination of Oxalic Acid Using Victoria Blue B Based on a Catalytic Kinetic Spectrophotometric Method. Microchim. Acta. 144:199-205.

Investigating structure-activity relationships within *Enterococcus faecalis* 2-deoxyribose aldolase

Nathan Wymer*, **Jeremy Steflik and Maria Brown**

Pfizer Global Research and Development, Groton, Connecticut (CT), USA.

The 2-deoxyribose aldolase (DERA) is an enzyme used for the synthesis and decomposition of carbohydrates. The DERA has been shown in previous studies to be utilized in an alternative metabolic pathway during stress. This article uses the *Enterococcus faecalis* DERA as a model system to study the structure-function relationship of amino acids towards DERA activity in hopes of garnering a better understanding of catalytic activity and possible mutations outside the binding pocket that might affect this activity. The active site was probed using saturation mutagenesis and identified two positions that increased activity (His187, Ser206). The outer shell amino acids were selected using bioinformatics. Mutagenesis of these outer shell amino acids combined with statistical analysis identified whether mutation of amino acids were advantageous or deleterious towards overall DERA activity without the need for high-throughput screening. These results will provide a basis to further study the intricate structure-activity relationships of amino acids within this DERA.

Key words: *Enterococcus faecalis*, 2-deoxyribose aldolase, structure-function relationships.

INTRODUCTION

The 2-deoxyribose aldolase (DERA) is a ubiquitous enzyme used in the synthesis and decomposition of carbohydrates, particularly nucleic acids (Wang et al., 1990; Heine et al., 2004; Liu, et al., 2004; Tao and Xu, 2009). The DERA catalyzes the reversible cleavage of 2-deoxyribose into glyceraldehyde-3-phosphate (G3P) and acetaldehyde, before it gets to metabolism. Both prokaryotic and eukaryotic cells can use nucleic acid degradation during metabolic stress to slow the progression of cell damage (Sgarrella et al., 1997). As such, the DERA may be a target to combat pathogenic bacteria metabolism. The *Enterococcus faecalis* DERA was used as a model system to better understand the DERA enzyme and to understand the protein structure-function relationships that drive DERA catalytic activity.

E. faecalis was chosen as a model system, because it is a common Gram-positive bacterium found in the gastrointestinal tracts of humans and other mammals. *Enterococcus* infections are well known in patients with prolonged hospital stays and undergoing multiple courses of antibiotics (Huycke et al., 1998; Murray, 2000). Like other pathogenic bacteria, *Enterococcus* bacteria have become resistant to a variety of antibiotics, including vancomycin (Huycke et al., 1998; Murray, 2000).

This study intends to understand one aspect of the metabolism of pathogenic bacteria in order to combat this increasing threat.

To better understand the structure-function relationships within the DERA, this study used bioinformatics and structure-guided selection methodology to identify amino acid positions that might negatively affect DERA catalytic activities. With a better understanding of the protein structure-function relationships of this DERA, this study can increase our knowledge of the precise catalytic and binding functions of amino acids within the active site as well as possible mutations that could provide resistance in the future.

*Corresponding author. E-mail: nathan.wymer@pfizer.com.

Abbreviations: DERA, 2-Deoxyribose aldolase; **G3P,** glyceraldehyde-3-phosphate; **TB,** terrific broth.

MATERIALS AND METHODS

Selection of mutations

Mutations outside the active site were selected using methods

Table 1. List of the mutations within DERA investigated during this study. The list contains both mutations selected using bioinformatics and structure-guided techniques as well as their overall impact on catalytic activity.

Bioinformatics selected mutations				Active site mutations			
Negative	Neutral		Positive	Negative		Neutral	Positive
I11L	I11V	Y75F	I24L	T10A	S154C	H187A	H187N
V21I	E15D	I81V	S35A	T10R	S154Q	H187R	H187C
F34Y	Q22E	I94V	Y104L	T10C	S154G	H187E	H187Q
F34A	I24V	S99A	S158A	T10Q	S154I	H187L	H187G
N86D	I25V	Q100E	R171K	T10G	S154L	H187K	H187M
I92L	D26E	Q100K	V186I	T10L	S154K	H187S	H187F
I92V	K29R	Q100G	I196V	T10P	S154M	S206G	H187T
Y102F	I38V	Q101D		T10S	S154P	S206M	H187W
Y104F	W42Y	Q101E		T10W	S154T	S206T	H187V
R106K	F45L	Q107E		H187L	S154W		S206A
D177E	S47A	Q110E		H187P	S154V		S206N
E191D	Q49A	G111A		S154A	S206R		S206C
A205T	Q49K	K119R		S206K	S206K		S206H
T207S	Q49E	V137I		S206F	S206F		S206I
T207A	D52G	K138R		S206W	S206W		S206V
V209I	D52E	I167V		S206Y	S206Y		
	T53S	R168K		S154D			
	I61V	M178L		S154C			
	E71A			S154Q			

previously described (Jonsson et al., 1993; Fox et al., 2003; Gustafsson et al., 2003). This selection process uses protein sequence alignments and bioinformatics to identify positions that are not conserved in a consensus sequence thus suggesting mutations that might not negatively impact activity. Sixty potential mutations were identified (Table 1). These mutations were then combined in a statistically significant variant set to provide relevant information into a machine-learning algorithm (Jonsson et al., 1993; Fox et al., 2003).

For the selection of positions within the active site, amino acids were identified using protein sequence alignments and relevant DERA crystal structures (Heine et al., 2004; Sakuraba et al., 2007). Four sites were hypothesized to hydrogen-bond to a G3P molecule (Table 1). These four sites were then subjected to saturation mutagenesis.

Gene synthesis and manipulation

The gene for the *E. faecalis* DERA enzyme (NCBI: NP_813976) was codon optimized for expression in *Escherichia coli* and then synthesized. For the statistical-based mutation screening, each of the DERA variants was synthesized *de novo* (DNA2.0, Menlo Park, CA). For the saturation mutagenesis of the active site amino acids, the mutations were prepared using the Quick-Change protocol (Stratagene, Santa Clara, CA). A degenerate NNK codon was used within the quick-change primers (IDT, Coralville, IA) to simultaneously prepare mutations at a specific site. Ninety six colonies were picked for each mutant site. These colonies were added to 1 ml TB broth (MoBio, Carlsbad, CA) containing antibiotic in a 2 ml microplate (VWR, Radnor, PA) and incubated overnight with shaking (37°C, 210RPM). The cells were then pelleted (20 min, 3000 xg's), and the medium was decanted. The plasmid DNA was extracted using miniprep technologies (Qiagen, Hilden, Germany) and sent for DNA sequencing. Usually, the full complement of saturation mutants was not completed from this initial 96 member library. Those missing mutants were prepared individually using new quick-change primers with appropriate codon changes.

DERA expression and activity

The DERA gene was cloned into a pSTRC18 plasmid. The pSTRC18 plasmid is a proprietary expression plasmid containing a modified inducible *trc* promoter and aaDA (streptomycin) antibiotic resistance gene. The DERA enzymes were expressed in BL21 (DE3) Gold cells (Stratagene, Santa Clara, CA). Single colonies were picked into 2 ml LB medium containing streptomycin (50 mg/L) and incubated overnight with shaking (37°C, 210RPM). After overnight incubation, 1 ml of seed culture was added into 50 ml autoinduction TB medium (Invitrogen, Carlsbad, CA) containing streptomycin (50 mg/L) in a 250 ml baffled-flask. The culture was then incubated overnight with shaking (30°C, 210RPM). The DERA enzyme expression was monitored with 4-20% Tris-Glycine SDS-PAGE (Invitrogen, Carlsbad, CA) and staining with coomassie. DERA activity was measured using methods similar to previously described (Wang et al., 1990; Heine et al., 2004; Liu et al., 2004; Sakuraba et al., 2007; Tao and Xu, 2009).

Statistical analysis of mutations

The activity contributions of the statistically prepared variant set were analyzed using machine learning by methods previously described (Jonsson et al., 1993; Fox et al., 2003).

Molecular structure model of DERA

A molecular structure model for the DERA was created using the *Thermotoga maritima* DERA crystal structure (1O0Y) as a template. (Sakuraba et al., 2007) The model was minimized using the in-house software package MoVIT (Figure 1). The DERA model shows a classical $\alpha\beta$-barrel fold with the catalytic amino acids in the center. As can be deduced by other DERA structural studies, K152 forms the imine intermediate with the acetaldehyde, D89 is the catalytic base, and K181 assists in modulating the pKa of the K152 side-chain amine (Heine et al., 2004; Sakuraba et al., 2007).

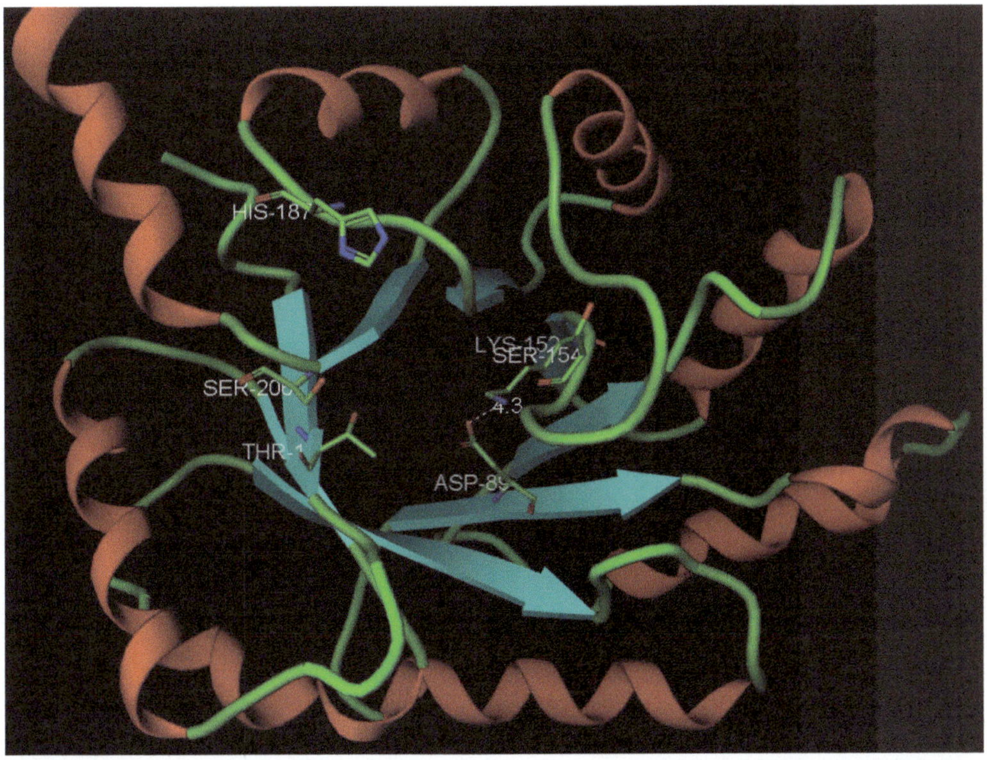

Figure 1. Molecular model of *E. faecalis* DERA. The structure has a classic αβ-barrel fold. The catalytic (D89, K152, K181) and binding pocket (T10, S154, H187, S206) amino acids are highlighted.

RESULTS AND DISCUSSION

Statistical analysis of the DERA activity assay shows that 16 of the 60 mutations chosen by bioinformatics had a clear negative influence on product formation (Table 1). Interestingly, several of these deleterious mutations occur at the same position (F34, I92, T207). These findings suggest how critical these positions can be to catalytic activity despite the original bioinformatics data suggesting these amino acid positions could accept mutation. Explaining the reasoning for their deleterious effects can be difficult, at best. The molecular modeling structure shows the F34 position helps form the rear cap of the active site. Modifying this amino acid could disrupt the catalytic residue K181 thus affecting the pKa of the imine-forming amino acid K152. Mutations at I92 and T207 are clearly detrimental, but like most of the deleterious mutations, they occur outside the active site. The exact reasons for these negative affects most likely occur from the disruption of the hydrogen bonding network of the active site, but more study is needed to determine the exact relationships.

The amino acids positions within the active site chosen for saturation mutagenesis are part of the hydrogen-bonding network for the G3P substrate (Heine et al., 2004; Sakuraba et al., 2007).

These residues help to stabilize the phosphate group as well as hydrogen bond to the C2-hydroxyl. All mutations at T10 and S154 gave mutant DERAs with lower catalytic activities than the WT. The H187 and S206 mutations also gave wider variety of changes in activities. Interestingly, the mutation H187R did not improve activity. Alignments of DERA protein sequences as well as crystal structures usually show lysine or arginine amino acids at this position, presumably to form an ionic interaction with the phosphate group of G3P. Simply adding a positive charge into the active site is not enough, thus the steric consequences of adding the larger arginine group within the active site must overcome the added benefits of this new ionic interaction.

Understanding which mutations increase activity can also be of importance. These mutations could provide a mechanism for resistance to inhibition while maintaining overall DERA activity. We are also currently investigating how these mutations individually or in combination could increase activity.

Conclusions

This study looked at ways to disrupt the catalytic activity of the *E. faecalis* DERA enzyme. New structure-function relationships can be identified within the DERA by combining bioinformatics and structure-guided approaches to search for amino acids that have both positive and negative impacts on activity. These findings suggest that the most efficient method to disrupt catalytic activity is to disrupt the hydrogen-bonding network of the G3P substrate. Thus, the hydrogen-bonding network of the DERAs might be investigated further since mutation would most likely lead to significantly decreased DERA activity. This initial study helps to better understand the DERA as part of the overall carbohydrate metabolism of pathogenic bacteria and could lead to a new generation of antibiotics in the future.

ACKNOWLEDGEMENTS

We acknowledge Van Martin (Pfizer) for scientific discussion and leadership, David Bauer (Pfizer) for DERA expression support, Bradley Poland (Pfizer) for molecular modeling support, and DNA 2.0 with bioinformatics and DNA synthesis.

REFERENCES

Fox R, Roy A, Govindarajan S, Minshull J, Gustafsson C, Jones JT, Emig R (2003). Optimizing the search algorithm for protein engineering by directed evolution. Protein Eng., 16(8): 589-597.

Gustafsson C, Govindarajan S, Minshull J (2003). Putting engineering back into protein engineering: bioinformatic approaches to catalyst design. Curr. Opin. Biotechnol., 14(4): 366-370.

Heine A, Luz JG, Wong CH, Wilson IA (2004). Analysis of the class I aldolase binding site architecture based on the crystal structure of 2-deoxyribose-5-phosphate aldolase at 0.99A resolution. J. Mol. Biol., 343(4): 1019-1034.

Huycke MM, Sahm DF, Gilmore MS (1998). Multiple-drug resistant enterococci: the nature of the problem and an agenda for the future. Emerg. Infect. Dis., 4(2): 239-249.

Jonsson J, Norberg T, Carlsson L, Gustafsson C, Wold S (1993). Quantitative sequence-activity models (QSAM)--tools for sequence design. Nucleic Acids Res., 21(3): 733-739.

Liu J, Hsu CC, Wong CH (2004). Sequential Aldol Condensation Catalyzed by DERA Mutant Ser238Asp and a Formal Total Synthesis of Atorvastatin. Tetrahedron Lett., 45: 2439-2441.

Murray BE (2000). Vancomycin-resistant enterococcal infections. N. Engl. J. Med., 342(10): 710-721.

Sakuraba H, Yoneda K, Yoshihara K, Satoh K, Kawakami R, Uto Y, Tsuge H, Takahashi K, Hori H, Ohshima T (2007). Sequential aldol condensation catalyzed by hyperthermophilic 2-deoxy-d-ribose-5-phosphate aldolase. Appl. Environ. Microbiol., 73(22): 7427-7434.

Sgarrella F, Poddie FPA, Meloni MA, Sciola L, Pippia P, Tozzi MG (1997). Channelling of deoxyribose moiety of exogenous DNA into carbohydrate metabolism: role of deoxyriboaldolase. Comp. Biochem. Physiol., 117B, 251-257.

Tao J, Xu JH (2009). Biocatalysis in development of green pharmaceutical processes. Curr. Opin. Chem. Biol., 13(1): 43-50.

Wang YF, Wong CH, Carlos FB III (1990). Deoxyribose-5-phosphate Aldolase as a Synthetic Catalyst. J. Am. Chem. Soc., 112: 2013-2014.

18

Effects of 4-nonylphenol on metabolic enzymes, some ions and biochemical blood parameters of the African catfish *Clarias gariepinus* (Burchell, 1822)

Alaa El-Din H. Sayed[1]*, Imam A. A. Mekkawy[1,2] and Usama M. Mahmoud[1]

[1]Department of Zoology, Faculty of Science, Assiut University, Egypt.
[2]Department of Biology, Faculty of Science, Taif University, Taif, Saudi Arabia.

In present work, some biochemical characteristics of the catfish, *Clarias gariepinus* were studied under the effect of different sublethal doses of 4-nonylphenol (0, 0.05, 0.08 and 0.1 mgl^{-1}). Liver enzymes ALT and AST increased insignificantly at $P<0.05$ whereas the ALP decreased insignificantly. The activities of G6PDH showed significant increase with the increased sublethel doses of 4-nonylphenol while the activities of LDH decreased insignificantly with increase of such doses. The concentrations of serum glucose and total cholesterol significantly increased after exposure to 4-nonylphenol, although Hyperglycemia is evident. The kidney function parameters such as total serum protein and uric acid increased insignificantly after exposure to 4-nonylphenol in comparison with the control fish whereas creatinine exhibited significant increase ($P<0.05$). Some serum ions decreased significantly (HCO_3^- and Na^+) or insignificantly (Cu^{+2}, Cl^- and Ca^{+2}) and others increased significantly (Fe^{+2}) or insignificantly (K^+) under 4-nonylphenol stress on *C. gariepinus*. Increased serum anion gap was associated with increased hyperglycemia and insignificant hypocalcemia. Such increased gap referred to nonylphenol-induced metabolic acidosis. The adverse impact of the sublethal doses of 4-nonylphenol on the molecular structure of the protein was evident by electrophoresis. In conclusion, in addition to its tissue-specific estrogenic effects, 4-nonylphenol has non-estrogenic adverse effects on liver and kidney functions through activation of other metabolism-related genes distinct from estrogen-responsive genes.

Key words: Biochemistry, liver, kidney, anion gap, 4-Nonylphenol.

INTRODUCTION

Pollutants especially chemical ones are one of the major problems in different countries all over the world. A large number of chemicals are found to contaminate aquatic ecosystems and their animals including fish and amphibians during their adult life and sensitive stages of development (Radhaiah et al., 1987). Chemical toxicity is concentration-dependent and related to chemical persistent nature and accumulation in the aquatic ecosystem components.

Nonylphenol ethoxylate (NPE) is one of the most dangerous chemicals that are recorded in aquatic environments (Clark et al., 1992; Tsuda et al., 2000; Rivero et al., 2008). Such chemical is widely used in the production and formulation of many commercially sold products (e.g. industrial and commercial detergent, polymer resin, cosmetic products). Bacterial degradation of nonylphenol ethoxylates produces more toxic nonylphenol (NP) (Hano et al., 2009). Such compound is also estrogenic both *in vitro* and *vivo* assays (Folmar et al., 2002). Nonylphenol is not a single chemical compound but is used to refer to a family of compounds all of which have a central aromatic or benzene ring and a nine carbon chain (Cox, 1996). The stability and persistence of 4-nonylphenol (NP) in the aquatic environment is evident (Uguz et al., 2003; Sone et al., 2004; Rivero et al., 2008) and a wide variety of animals including fish, mollusks and crustaceans are affected by

*Corresponding author. E-mail: alaa_h254@yahoo.com.

its toxicity especially due to its estrogenic-like behavior (Flouriot et al., 1995; Cox, 1996). So many countries including European Union have banned the use of NPEs in detergents and other commercial products (Environment Canada, 2002; European Union, 2002).

The adverse effects of nonylphenols and its ethoxylates increase due to their bioaccumulation in fish (Ahel et al., 1993; Cox, 1996). The bioconcentration factor (BCF) of nonylphenol in fish varies from 3 to 1300 (Ekelund et al., 1990; Ahel et al., 1993). Relatively low concentrations of nonylphenol and nonylphenol ethoxylates can cause death to fish (Staples et al., 2004; Vazquez-Duhalt et al., 2005). The environmental protection agency (EPA) stated that the freshwater life should not be affected if the one-hour average concentration of nonylphenol does not exceed 28 ppm on an average more than once every three years (Vincent and Sneddon, 2009). Lethal concentration of 0.032 ml/l nonylphenol was previously determined for *Orychroeis niloticuls* (Rivero et al., 2008) and 96 h LC_{50} for *Clarias gariepinus* exposed to NP was 3.48 mg l^{-1} (Satyanarayanan et al., 2011).

Physical and chemical blood changes are another indicator of fish pollution (Wilson and Taylor, 1993). So, blood characteristics and hematological parameters are important in diagnosing the functional status of organs of exposed animal to toxicants (Joshi et al., 2002; Adedeji et al., 2009). The complex of unspecified biochemical indicators of blood reveals the general effect of pollutants on fish and makes it possible to forecast the consequence of long-term exposure to chemical pollutants (Adedeji et al., 2009). Moreover, evaluation of blood biochemistry was considered as a useful tool for the diagnosis of diseases and assessing the physiological status of fish (Stoskopf, 1993).

Many studies have investigated changes in many physiological and biochemical blood indices induced by environmental conditions and presence of contaminants (Kori-Siakpere et al.,, 2006; Maheswaran et al., 2008; Ololade and Oginni, 2010).The biochemical parameters in fish are valid for physiopathological evaluation and sensitive for detecting potential adverse effects and relatively early events of pollutant damage (Almeida et al., 2002; Matos et al., 2007; Osman et al., 2010).

The liver plays an important role in several vital functions of basic metabloishm and it is also the major organ of accumulation, biotransformation and degradation of contaminants (Matos et al., 2007). The evaluation of biochemicals in fish liver has become an important tool for monitoring environmental exposure of fish to such chemicals in experimental studies (Matos et al., 2007). Several investigations demonstrated that changes in the levels of antioxidants enzyme activities can be used as possible biomarkers in different aquatic organisms (Regoli et al., 2004; Matos et al., 2007). Such biomarkers could be used to identify possible environmental contamination before the health of aquatic organisms is seriously affected (Barnhoorn and van Vuren, 2004) and to develop water quality indices (Pickering and Pottinger, 1995). Changes in the concentrations and activities of metabolic key factors such as LDH and G6PDH enzymes (Long et al., 2003; Osman et al., 2007b; Mekkawy et al., 2010a) were used to identify heavy metal contamination in fishes. G6PDH has long been recognized as an antioxidant enzyme (Pandolfi et al., 1995; Salvemini et al., 1999) and as a biomarker of pollution induced carcinogenesis in fish (Winzer et al., 2001). The cytoplasmic enzyme LDH is widely used as marker of organ or tissue lesions in toxicology and in clinical chemistry (Das et al., 2004). In most cases of tissue damage, whether due to disease or toxic compound, the activity of LDH was reported to be significantly affected (Singh and Sharma, 1998).

Several changes in plasma ion concentrations were measured for different fish species under different types of stresses to indicate osmoregulatory disruptions (Grosell, 2006; Fernandes et al., 2007; Fast et al., 2009). Carrera et al. (2007) studied the effect of 4-nonylphenol on osmoregulation referring to its action and its ecological significance.

The African catfish *C. gariepinus* is distributed throughout Africa (Nguyen and Janssen, 2002). Such a species is of commercial importance due to its high growth rate, high stocking-density capacities, high consumer acceptability and high resistance to poor water quality and oxygen depletion (Adewolu et al., 2008; Karami et al., 2010). Some records have shown that Clarias fishery contributes about 17% of the annual fish production from all fisheries sectors (Ololade and Oginni, 2010). Moreover, the African catfish has been used in fundamental researches and considered as an excellent model for toxicological studies (Osman et al., 2007a; Mahmoud et al., 2009), since it has a well documented biology (Osman et al., 2007b; Mahmoud et al., 2009). So, the present work aimed at evaluating the adverse impacts of different sublethal doses of 4-nonylphenol on some biochemical characteristics of the African catfish, *C. gariepinus*.

MATERIALS AND METHODS

Specimen collection

Specimens of adult Catfish *C. gariepinus* were collected from the River Nile at Assiut and then were transported to Fish Biology Laboratory of Zoology Department, Faculty of Science, Assiut University. The fish [males (500 to 600) and females (1000 to 1200 g] were fed on a commercial pellet diet (3% of body weight per day) and kept together in 100 L rectangular tanks containing tap water (conductivity 2000 ls cm^{-1}; pH 7.5; oxygen saturation 88 to 95%; temperature 27 to 28°C; photoperiod 12:12 light: *C. gariepinus* dark). After two weeks of acclimatization, fishes were used for experimental setup.

Experimental setup

The adapted adult fish were classified into four groups (10 fish each): control, 0.05 mgl^{-1} 4-nonylphenol (4NP), 0.08 and 0.1 mgl^{-1} -

4NP. The time of exposure was 15 days with a daily change of 4NP. These sublethal doses of 4NP were considered according to Bennie (1999). The experimental conditions were the same as those previously mentioned with daily change of water and 4NP. 4-Nonylphenol was purchased from Sigma-Aldrich (Schnelldorf, Germany) with purity of 99.3%.

Blood sampling and biochemical parameters

After 15 days, four fish from each group were randomly selected and anesthetized to bring down the stress due to handling using tricaine methanesulfonate (0.4 g l^{-1}). Blood was collected by cardiac puncture using sterilized syringes (2 ml). The serum was then removed by subjecting the tubes to centrifugation at 3000 rpm for 5 min and then stored at –80°C until further analyses of the following blood parameters.

Creatinine (Cr), Uric acid, Aspartic Amino Transferase (AST), Alanine Amino Transferase (ALT), Alkaline Phosphatase (ALP), glucose, cholesterol and total protein were determined by kits of SGMitalia Company. Enzyme activities of Lactate Dehydrogenase (LDH) and Glucose-6-Phosphate Dehydrogenase (G6PDH) were determined in serum with spectrophotometer (Micro lab 200 vital scientific Dieren–The Netherlands) at a wavelength of 340 nm and at 37°C using kits, Stanbio LDH (UV- Rate) procedure No. 0940 USA for the quantitative determination of Lactate Dehydrogenase (Kachmar and Moss, 1976) and RANDOX Laboratories Ltd., PD410, United Kingdom BT294QY for the quantitative determination of Glucose-6-Phosphate Dehydrogenase, (Kornberg, 1955). The concentrations of Hco_3^-, Na^+, K^+, Cl^-, Cu^{+2}, Fe^{+2} and Ca^{+2} were measured in serum using an atomic absorption spectrophotometer (GBC Model 300).

Protein analysis by polyacrylamide gel electrophoresis (SDS-PAGE)

Serum of each treatment in addition to control were suspended in 1.0 ml lysing buffer (100 ml 0.625 M (0.706 g/100 ml) Tris-HCl (pH 6.8) - 2% SDS (w/v) - 10% glycerol (v/v) - 5% 2- mercaptoethanol (v/v)). Heated at 100°C for 5 min, centrifuged at 10,000 rpm for 30 min and 50 µl of each extracted protein treatment was used for protein analysis using SDS-PAGE according to Laemmli (1970) in the first dimension. The low molecular weight standards (Pierce, USA) were run concurrently and the protein molecular mass was determined using Gel-Pro Analyzer package (Media Cybernetica, 1993 to 1997)

Statistical analysis

The basic statistics, means, standard errors and ranges of the measured parameters were estimated. The patterns of variation due to 4-nonylphenol doses were studied by one way analysis of variance using SPSS package (SPSS, 1998) at the 0.05 significance level. Levene's test of equality of error variance of the dependent variables was applied, with rejection of the null hypothesis for raw, log-transformed and SQRT-transformed data. So, the homogeneity of variance was assumed for raw data. The Tukey-HSD test was considered for multiple comparisons.

Ethical statement

All experiments were carried out in accordance to the Egyptian laws and University guidelines for the care of experimental animals. All procedures of the current experiment have been approved by the committee of the Faculty of Science of Assiut University, Egypt.

RESULTS

Biochemical blood parameters

Changes in the biochemical parameters in experimental fish are shown in Table 1. Liver enzymes ALT and AST increased insignificantly at $P<0.05$ whereas the ALP decreased insignificantly. The activities of G6PDH showed significant increase with the increased sublethel doses of nonylphenol while the activities of LDH decreased insignificantly with the increase of such doses. The concentrations of serum glucose and total cholesterol significantly increased after exposure to 4-nonylphenol, although Hyperglycemia is evident.

The kidney function parameters such as total serum protein and uric acid increased insignificantly after exposure to 4-nonylphenol in comparison with the control fish whereas creatinine exhibited significant increase ($P<0.05$). Under the effect of 4-nonylphenol, HCo_3^- and Na^+ decreased significantly whereas Cl^-, Cu^{+2} and Ca^{+2} decreased insignificantly. K^+ increased insignificantly whereas Fe^{+2} exhibited significant increase. So, nonylphenol affects the transport of chloride and calcium ions and hence osmoregulation in fish gills. The serum anion gap (Table 1) increased with increase 4-nonylphenol doses referring to metabolic acidosis and renal tubule dysfunction. But such increase in anion gap did not implicated in fish mortality of *C. gariepinus*. The increase in anion gap, creatinine and glucose may suggest a shock response under stress of higher doses of 4-nonyphenol.

Electrophoresis of serum protein

Under 4-nonylphenol stress, the protein fraction patterns (Figures 1 and 2 and Tables 2 and 3) show variability in the number, molecular weight and percentage of protein fractions of serum of *C. gariepinus* in comparison with the control. Few sex-related variations are evident especially under stress. Two protein fractions (r3 and r4) were recorded in the control male and female and disappeared under stress. Nonylphenol stress lead to the appearance of other protein fraction which not recorded in the control fish (r7, r8, r12 and r16). Protein fraction, r6 is the only fraction which persists in the control and treated fishes with highest percentage (32.75 to 47.75%) of the total protein in each individual. Based on presence/absence criterion, similarity between treated and control fish can be expressed in clusters (Figure 2). There are two main clusters: one includes the control sex and the other includes the treated fishes. The later main clusters are divided into three subclusters. Such variability in protein fractions reflects the variation in gene activity (activation-inactivation mechanism) and protein synthesis under nonylphenol stress. Protein recorded fractions suggest that the albumin and globulins fractions were affected by this pollutant.

Table 1. Effect of different doses of 4-nonylphenol on the biochemical parameters (mean ± SD, range) of the African catfish *Clarias gariepinus*.

Biochemical parameter	Groups			
	Control mean±SD (Max-Min)	0.05 mg^{-1} 4-nonylphenol mean±SD (Max-Min)	0.08 mg^{-1} 4-nonylphenol mean±SD (Max-Min)	0.1 mg^{-1} 4-nonylphenol mean±SD (Max-Min)
AST (IU l^{-1})	34.41±2.16 (32.12-36.81)^{a}	35.15±1.78 (33.18-36.91)^{a}	35.68±1.58 (33.71-37.11)^{a}	36.97±0.71 (36.1-37.8)^{a}
ALT(IU l^{-1})	16.97±1.14 (15.8-18.1) ^{a}	17.65±0.8 (16.4-18.5)^{a}	18.27±1.09 (17.2-19.8) ^{a}	18.87±0.77 (17.8-19.6)^{a}
ALP (IU l^{-1})	47±3.65 (43-51) a	45.75±2.98 (42-49) ^{a}	43.75±2.75 (41-47) ^{a}	41.25±2.87 (38-45)^{a}
LDH (IU l^{-1})	66.78±3.37 (62.14-70.05) ^{a}	64.9±3.18 (62.16-68.12)^{a}	63.36±2.15 (61.12-66.01)^{a}	61.13±2.57 (58.11-64.11)^{a}
G6PDH (g/dl)	65.49±1.11 (64.12-66.78) ^{a}	66.64±1.26 (65.14-68.11)^{a}	67.88±0.95 (67.12-69.12)^{ab}	69.9±1.7 (68.15-72.18)^{b}
Glucose (Mg/dl)	72.9± 3.2 (68.6-75.8)^{a}	74.92±4.34 (68.6-78.4)^{ab}	79.4±0.9 (78.2-80.4)^{bc}	81.7± 0.78(80.8-82.6)^{c}
Total protein (Mg/dl)	4.1±0.316 (3.8-4.5) ^{a}	4.37±0.4 (3.9-4.8) ^{a}	4.6±0.57 (3.9-5.2) ^{a}	4.97±0.38 (4.6-5.5)^{a}
Total cholesterol (Mg/dl)	213±1.81 (211-215) ^{a}	217.75±2.87 (214-221)^{ab}	225.5±6.66 (220-235)^{bc}	232.25± 8.18 (221-240)^{c}
Creatinine (Mg/dl)	0.367±0.022 (0.34-0.39) ^{a}	0.3925±0.012 (0.38-0.41)^{ab}	0.457±0.0 4 (0.41-0.51)^{b}	0.467±0.058 (0.4-0.54)^{b}
Uric acid (Mg/dl)	22.67±0.74 (21.8-23.6) ^{a}	22.9±0.245 (22.6-23.1)^{a}	23.5±0.74 (22.8-24.5)^{a}	25.27±0.61 (24.6-25.8)^{a}
HCO^{-} (mEq l^{-1})	19.75±1.5 (18-21)^{a}	18.75±0.957 (18-20)^{ab}	16.25±1.7 (14-18)^{bc}	13.75±1.7 (12-16)^{c}
Na^{+} (mEq l^{-1})	126.25±3.09 (122-129)^{a}	124±2.16 (121-126) ^{ab}	120.5±1.29 (119-122)^{bc}	117.25±3.59(112-120)^{c}
K^{+} (mEq l^{-1})	4.10±0.18 (3.9-4.3) ^{a}	4.37±0.368 (4.02-4.89) ^{a}	4.53±0.31 (4.15-4.91)^{a}	4.78±0.41 (4.18-5.11)^{a}
Cl^{-} (mEq l^{-1})	95.75±3.3 (92-99) ^{a}	93.25±3.59 (90-98)^{a}	91.75±1.7 (90-94)^{a}	90.25±1.25 (89-92)^{a}
Cu^{+2} (µg/ml)	108.89±8.85 (103.14-122.1)^{a}	106.36±9.31 (100.12-120.1)^{a}	105.47±9.06 (100.21-119.01)^{a}	99.105±1.12 (98.11-100.12)^{a}
Anion gap (mEq l^{-1})**	10.8±0.12 (7.8-12.8)a	12.0±0.18 (10-14)^{ab}	12.5±0.18 (12-13)^{bc}	13.2±0.12 (13-15)^{c}
Fe^{+2} (µg/ml)	15.65±0..59 (14.8-16.2) ^{a}	16.42±0.28 (16.1-16.8)^{ab}	16.77±0.784 (15.9-17.8)^{ab}	17.35±0.51 (16.9-17.8)^{b}
Ca^{+2} (µg/ml)	52.63±1.93 (50.12-54.16) ^{a}	52.86±2.75 (50.11-56.12) ^{a}	50.36±1.17 (49.21-52.01)^{a}	48.84±1.45 (47.14-50.11)^{c}

* Different letters indicates there is a significant difference at ($p \leq 0.05$; ** Anion gap= $Na^{+}+K^{+}-Cl^{-} - HCO^{-}$.

DISCUSSION

Nonylphenol (NP) is a more toxic persistent degradation product of nonylphenol ethoxylates (Jigang et al., 2010). It is suspected to be endocrine-disrupting chemicals. Nonylphenol was found to have tissue-specific estrogenic adverse effects and non-estrogenic effects on different organ functions of aquatic animals, especially fishes (Madigou et al., 2001; Marin et al., 2008). In the present work, different non-estrogenic adverse effects of nonylphenol were recorded on the biochemical characteristic of the African Catfish, *C. gariepinus* including liver function enzymes, osmoregulation and other metabolic parameters; such effects were dose-dependent. Mekkawy et al. (2011) reported dose-dependent 4-nonylphenol-induced changes in some blood parameters of such species including apoptosis, erythrocyte morphological alterations, micronucleus formation and blood counts. The present authors also reported nonylphenol hormonal effects on the thyroid and reproductive hormones and reproduction tissues.

Different studies referred to the common effects of variable types of stress including xenobiotics on the liver function and metabolic enzymes (Mekkawy and Lashien, 2003; Öner et al., 2008; Adedeji et al., 2009, Mekkawy et al., 2010b; Osman et al., 2010). In different fish species including *C. gariepinus*, ALT and AST enzymes activity were found to increase in response to pesticides (Adedeji et al., 2009), heavy metals (Öner et al., 2008; Mekkawy and Lashien, 2003; Mekkawy et al., 2010b) and nonylphenols (Bhattacharya et al., 2005; Bhattacharya et al., 2008); such response was dose-dependent. In present work, the sublethal concentrations of 4NP caused insignificant increase of these enzymes

Figure 1. Protein fractions identified in serum of male and female catfish, *C. gariepinus* under different doses of 4-nonylphenol in comparison with control. M; unstained protein marker (200 kda), 1; male control, 2; female control, 3; male treated with 0.05 mgl^{-1} of 4-nonylphenol for 15 days, 4; female treated with 0.05 mgl^{-1} of 4-nonylphenol for 15 days, 5; male treated with 0.08 mgl^{-1} of 4-nonylphenol for 15 days, 6; female treated with 0.08 mgl^{-1} of 4-nonylphenol for 15 days, 7; male treated with 0.1 mgl^{-1} of 4-nonylphenol for 15 days, 8; female treated with 0.1 mgl^{-1} of 4-nonylphenol for 15 days.

activity. Satyanarayanan et al. (2011) working on the same species stated that ALT and AST increased with increase of NP doses till 0.5 and 0.75 mgl^{-1} then decreased with 1 mgl^{-1}. Increase in the levels of ALT and AST has been shown to reflect liver damage (Bhattacharya et al., 2005; Bhattacharya et al., 2008) but to a limited action. The 4-nonylphenol-induced ALP enzyme exhibited insignificant decrease. The alkaline phosphatases are a group of enzymes mainly associated with the liver and bone, but are also found in cells of intestine and kidney. Accordingly, 4-nonyphenol can affect indirectly to some extent different ALP-related metabolic processes. The increase in serum ALT and AST and decrease in ALP under some heavy metal stress on *Oreochromis niloticus* were recorded by Öner et al. (2008). The pesticide Diazinone induced a signify-cant decrease in blood alkaline phosphatase and acid phosphatase of *C. gariepinus* whereas, ALT and AST were comparable in the control and treated fishes (Adedeji et al., 2009). In other fish species these enzymes increased with stresses (Bhattacharya et al., 2005, 2008).

The blood glucose, LDH and G6PDH have been used as bioindicators of stress in the catfish *C. gariepinus* (Osman et al., 2007b, 2010; Sayed et al., 2007; Adedeji et al., 2009; Mekkawy et al., 2010a; Satyanarayanan et al., 2011) and other different fish species (Mekkawy and Lashien, 2003; Mekkawy et al., 2010b). In present study, the blood glucose level increased significantly ($P<0.05$) in fish exposed to 4-nonylphenol as compared to the control fish. Similar increased level of glucose was recorded in blood of fishes exposed to ultraviolet radiation (Sayed et al., 2007; Mekkawy et al., 2010a), heavy metals (Mekkawy and Lashien, 2003; Osman et al., 2007b; Mekkawy et al., 2010b) and other pollutants (Poléo and Hyherød, 2003; Rosety-Rodriguez et al., 2005; Adedeji et al., 2009). The chemical pollutants modulate the metabolism of carbohydrates, causing hyperglycemia by stimulating the glycogenolysis in fish (Levesque et al., 2002; Osman et al., 2010).

The literatures in concern with the adverse effects of 4-nonylphenol or its allies on the G6PDH and LDH enzymes activity of fishes are rare (Satyanarayanan et al., 2011) but there are many studies about the effects of

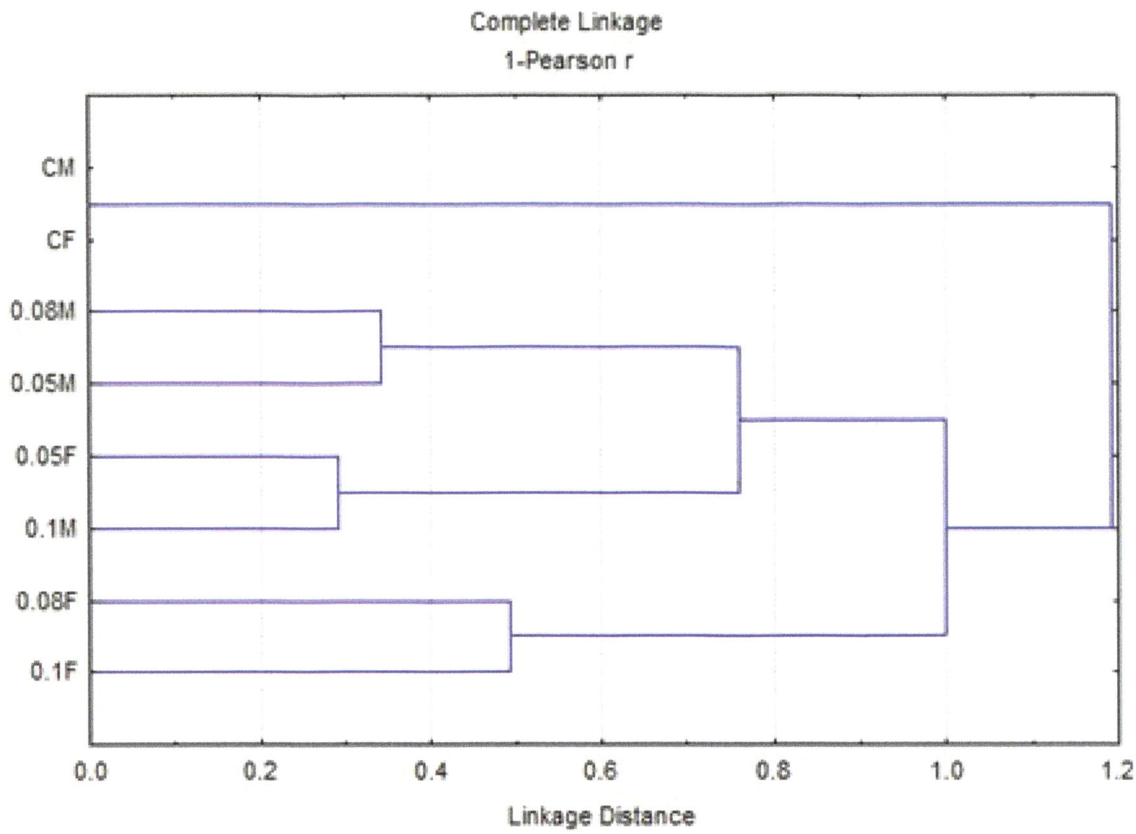

Figure 2. Clustering of electrophoretic protein patterns on the basis of band presence-absence criterion to illustrate the relationship between the control and 4-nonyphenol treated sexes of *C. gariepinus*.

other chemical pollutants on these enzymes (Mekkawy et al., 2010a). The stress-induced LDH and G6PDH exhibited variable patterns toward increase, decrease or no trend (Pandey et al., 2003; Rosety-Rodriquez et al., 2005; Adedeji et al., 2009; Mekkawy et al., 2010a; Satyanarayanan et al., 2011). In present work, LDH exhibited insignificant decrease with increase of 4NP doses while G6PDH level increased significantly. Satyanarayanan et al., (2011) reported that LDH increased with increase of NP to higher value (159.09±2.84) at 0.5 mg l^{-1} NP then decreased to 52.18± 0.95 at 1 mgl^{-1} NP. Leopold et al., (2003) reported that the increase in G6PDH activity represents protection against elevated levels of reactive oxygen species in cells exposed to an oxidative stress through the increased NADH production. LDH is generally associated with cellular metabolic activity and so, its inhibition may be due to ion imbalance or plasma membrane damage (Sastry and Gupta, 1980) and may also due to formation of enzyme inhibitor complex (Singh and Sharma, 1998; Rajanna et al., 1999).

Increase in the blood cholesterol concentration is used as an indicator of liver dysfunction because homeostasis of lipids is one of the principle liver functions (Kaplan et al., 1988). Many studies are in concern with such parameters to estimate the nature and degree of stress in fishes (Poléo and Hyherød, 2003; Mekkawy and Lashien, 2003; Rosety-Rodriguez et al., 2005; Osman et al., 2007b; Sayed et al., 2007; Mekkawy et al., 2010a). In present work, the blood cholesterol level was significantly ($P<0.05$) increased in *C. gariepinus* exposed to 4-nonylphenol as compared with control fish. The increase of cholesterol as a response to pollution could be due to fact that excess energy reserves are required by organisms to mediate the effects of stress (Lee et al., 1983).

Total protein, creatinine and uric acid are biomarkers for muscle and purine metabolism, liver damage and kidney acid (Goss and Wood, 1988; Sayed et al., 2007; Hadi et al., 2009; Mekkawy et al., 2010a). The protein make up of an organism is of important diagnostic significance (Shalaby et al., 2006; Hadi et al., 2009) because of protein's involvement in enzymes, hormones and antibodies as well as osmotic pressure balance and maintaining acid-base balance. Total protein level may increase, decrease or exhibit no significant trend (Sayed et al., 2007; Mekkawy et al., 2010a, b). In present study, 4-nonylphenol-induced total protein level significantly increased with dose increase referring to limited effect on protein-related metabolic processes. The rise in the creatinine of *C. gariepinus* referred to 4NP-induced effects on muscle metabolism whereas the insignificant

Table 2. Protein fractions (%) identified in serum of *C. gariepinus* under the effect of different doses of 4-nonylphenol in comparison with the control.

Lanes	Dose							
	Control		0.08 mgl^{-1} 4-nonylphenol		0.05 mg^{-1} 4-nonylphenol		0.1 mg^{-1} 4-nonylphenol	
	Male	Female	Male	Female	Male	Female	Male	Female
Protein fractions	%							
r1								
r2								
r3	6.57	6.32						
r4	6.20	7.40						
r5								
r6	35.22	32.75	46.89	47.75	39.39	44.22	44.38	42.14
r7					7.34			
r8			12.24	16.18	8.92	5.05		
r9	10.56	12.29		9.85		13.10	18.73	16.91
r10	5.07	9.32	8.01		9.21	6.43	6.91	
r11	7.08	4.39	4.81	6.27	7.65			7.36
r12				8.84				
r13	8.05	9.32	11.84		9.66	10.18	9.45	12.83
r14	6.17	3.21						
r15	15.07	14.99	16.20	11.11	17.83	21.02		20.76
r16							20.52	
r17								

Table 3. Molecular weight (kda) identification of protein fractions in serum of *C. gariepinus* under the effect of different doses of 4-nonylphenol in comparison with the control.

Lanes	Dose								
	Marker	Control		0.08 mg^{-1} 4-nonylphenol		0.05 mg^{-1} 4-nonylphenol		0.1 mg^{-1} 4-nonylphenol	
		Male	Female	Male	Female	Male	Female	Male	Female
Protein fractions	Molecular weight (kda)								
r1	200								
r2	150								
r3	120	118.98	119.59						
r4	100	105.1	102.65						
r5	85								
r6		76.449	75.607	77.29	78.131	77.29	76.168	75.607	76.168
r7	70					70.701			
r8	60			57.101	61.667	57.826	59.565		

Table 3. Cont.

r9		55.652	56.377		52.899		55.362	54.348	55.362
r10	50	48.974	50.435	51.159		51.159	50.725	47.692	
r11		38.974	37.692	38.205	42.051	40			38.205
r12	30				27.586				
r13	25	24.667	26.379	25.517		24.667	25.862	23.867	24.333
r14	20	19.638	20	20.333					
r15	15	14.348	15.362	14.493	14.855	15.217	14.348		13.768
r16								13.116	
r17	10								

increase in uric acid may be due the limited effect of 4NP on purine metabolism; renal tubules may also be damaged. Hadi et al. (2009) reported that the increase of creatinine level might be induced by glomerular insufficiency, increased muscle tissue catabolism or the impairment of carbohydrates metabolism.

Ions are very important for any organism because the involved in most biological processes such as respiration, muscle contraction, absorption, nerve impulses transmission, osmoregulation, acid-base balance and execration in fish. Different studies considered the adverse effects of different types of stress on these biological processes in fishes (Pǎrt et al., 1985; Burkhardt-Holm et al., 2000; Hughes et al., 2000; McCormick et al., 2005). In present work, some ions decreased significantly (HCO_3^- and Na^+) or insignificantly (Cu^{+2}, Cl^- and Ca^{+2}) and others increased significantly (Fe^{+2}) or insignificantly (K^+) under 4-nonylphenol stress on *C. gariepinus*. Pǎrt et al. (1985) reported increased ion permeability and sodium efflux of gill epithelial cells of rainbow trout due to the toxicity of nonylphenol-ethoxylate. Accordingly, McCormick et al. (2005) indicated that estrogenic compounds may cause a general shift toward increased capacity for ion uptake (Na^+, Cl^- and Ca^{+2}). Calcium seems to play an important response to direct as well as indirect effects of nonylphenol (Burkhardt-Holm et al., 2000). Such Nonylphenol could cause cell death through inhibition of the calcium pump in the endoplasmic reticulum (Hughes et al., 2000). Histopathological alterations in the liver of rainbow trout as necrosis, cell number decrease and vacuolation were mediated by inhibition of calcium pump after exposure to NP (Uguz et al., 2003).

The serum anion gap in fishes was considered by many authors to describe the degree of stress on the kidney and accumulation of organic acids (Converse et al., 1994). In present work, the increased serum anion gap in *C. gariepinus* referred to metabolic acidosis and renal tubule dysfunction. But the increase in anion gap did not implicated in fish mortality of that species exposed to 4-nonyphenol. The increase in anion gap, creatinine and glucose may suggest a shock response under stress of higher doses of 4-nonyphenol. Such increased anion gap was associated with insignificant hypocalcemia.

Iron is necessary for the formation of some proteins, hemoglobins, myoglobins and cytochromes (Arain et al., 2006). Also, it is necessary for oxygen transport, cellular respiration and peroxide deactivation. In present work, increased Fe^+ was recorded under 4-nonylphenol stress in comparison with the control fish.

Serum protein of fishes is representing by variable fractions including prealbumin, albumin and globulins fractions (Hussein and Mekkawy, 2001). Electrophoresis (or SDS-PAGE) of serum protein and identification of their fractions by molecular weight, number and concentration are valid in postulating stress-induced variations (Hanson and Tabita, 2003; Mekkawy et al., 2010a). Such variations reflect the activation and inactivation of genes under pollutant stress (Mekkawy and Lashin, 2003; Mekkawy et al., 2010a). In present work, the 4-nonylphenol doses reflect variability in band patterns, fraction molecular weight and fraction appearance and disappearance. Only one protein fraction persist with the increased doses of 4-nonylphenol. Moreover, sexual dimorphism was evident in the control and treated fish.

Interestingly, In spite of the tissue-specific estrogenic effects that have reported for 4-nonylphenol, it showed non-estrogenic adverse effects on liver and kidney functions of the studied species, *C. gariepinus*.

REFERENCES

Adedeji OB, Adeyemo OK, Agbede SA (2009). Effects of diazinon on blood parameters in the African catfish (*Clarias gariepinus*). Afr. J. Biotechnol., 8(16): 3940-3946.

Adewolu MA, Adeniji CA, Adejobi AB (2008). Feed utilization, growth and survival of *Clarias gariepinus* (Burchell, 1822) fingerlings cultured under different photoperiods. Aquaculture, 283: 64-67.

Ahel M, McEvoy J, Giger W (1993). Bioaccumulation of the lipophilic metabolites of nonionic surfactants in freshwater organisms. Environ. Pollut., 79: 243-248.

Almeida JA, Diniz YS, Marques SFG, Faine IA, Ribas BO, Burneiko RC, Novelli EIB (2002). The use of the oxidative stress responses as biomarkers in Nile tilapia *(Oreochromis niloticus)* exposed to in vivo cadmium contamination. Environ. Int., 27: 673-679.

Arain SI, Kazi TG, Bhanger MI, Memon AN (2006). Evaluation of the status of iron in different honey samples. J. Chem. Soc. Pak., 28(5): 448-450.

Barnhoorn IEJ, van Vuren JHJ (2004). The use of different enzymes in feral freshwater fish as a tool for the assessment of water pollution in South Africa. Ecotoxicol. Environ. Safe, 59: 180-185.

Bennie DT (1999). Review of the environmental occurrence of alkylphenols and alkylphenol ethoxylates. Water Qual. Res. J. Can., 34(1): 79-122.

Bhattacharya H, Lun L, Gomez RGD (2005). Biochemical effects to toxicity of CCl4 on rosy barbs (*Puntius conchonius*). Our Nat., 3: 10-25.

Bhattacharya H, Xiao Q, Lun L (2008). Toxicity studies of nonylphenol on rosy barb (*Puntius conchonious*): A biochemical and histopathological evaluation. Tissue Cell, 40: 243-249.

Burkhardt-Holm P, Wahli T, Mejer W (2000). Nonylphenol affects the granulation pattern of epidermal mucous cells in rainbow trout, *Onchorynchus mykiss*. Ecotoxicol. Environ. Safe, 46: 34-40.

Canada E (2002) Canadian Environmental Quality Guidelines for Nonylphenol and its Ethoxylates (Water, Sediment, and Soil). In Scientific Supporting Document. Ecosystem Health: Sciencebased Solutions. National Guidelines and Standards Office, Environmental Quality Branch, Environment Canada, Ottowa.

Carrera EP, Garcia-Lṗez A, Martin del Rio Mdel P, Martinez-Rodriguez G, Solé M, Mancera JM (2007). Effects of 17 beta-estradiol and 4-nonylphenol on osmoregulation and hepatic enzymes in gilthead sea bream (*Sparus auratus*). Comp. Biochem. Physiolog. C Toxicol. Pharmacol., 145(2): 210-217.

Clark LB, Rosen RT, Hartman TG, Louis JB, Suffet IH, Lippincott RL (1992). Determination of alkylphenol ethoxylates and their acetic acids derivatives in drinking water by particle beam liquid chromatograph/electroscopy. Int. J. Environ. Anal. Chem., 47(3): 167-180

Converse LJ, Fernandes PJ, MacWilliams PS, Bossart GD (1994). Hematology, Serum Chemistry, and Morphometric Reference Values for Antillean Manatees (*Trichechus manatus manatus).* J. Zoo Wildlife Med., 25(3): 423-431.

Cox C (1996). Nonylphenol and related chemicals. Journal of pesticide reform/ Spring, 16(1) (corrected 4/2003).

Das PC, Ayyappan, S, Das BK, Jena JK (2004). Nitrite toxicity in Indian major carps: sublethal effect on selected enzymes in fingerlings of *Catla catla*, *Labeo rohita* and *Cirrhinus mrigala*. Comp. Biochem. Physiol. C, 138: 3-10.

Ekelund R, Bergman A, Granmo A, Berggren M (1990). Bioaccumulation of 4-nonylphenol in marine animals - A reevaluation. Environ. Pollut., 64: 107-120.

Fast J, Aiken AC, Allan J, Alexander L, Campos T, Canagaratna MR, Chapman E, DeCarlo PF, de Foy B, Gaffney J, de Gouw J, Doran JC, Emmons L, Hodzic A, Herndon SC, Huey G, Jayne JT, Jimenez JL, Kleinman L, Kuster W, Marley N, Russell L, Ochoa C, Onasch TB, Pekour M, Song C, Ulbrich IM, Warneke C, Welsh-Bon D, Wiedinmyer C, Worsnop DR, Yu X-Y, Zaveri R (2009). Evaluating simulated primary anthropogenic and biomass burning organic aerosols during MILAGRO: implications for assessing treatments of secondary organic aerosols. Atmos. Chem. Phys., 9: 6191-6215.

Fernandes C, Fontaınhas-Fernandes A, Monteiro SM, Salgado MA (2007). Changes in plasma electrolytes and Gill Histopathology in Wild Liza saliens from the Esmoriz-Paramos Coastal Lagoon, Portugal. Bull. Environ. Contam. Toxicol., 79: 301-305.

Flouriot G, Pakdel F, Ducouret B, Valotaire Y (1995). Influence of xenobiotics on rainbow trout liver estrogen receptor and vitellogenin gene expression. J. Mol. Endocrinol., 15(2): 143-151

Folmar LC, Hemmer MJ, Denslow ND, Kroll K, Chen J, Cheek A, Richman H, Meredith H, Grau EG (2002). A comparison of the estrogenic potencies of estradiol, ethinylestradiol, diethylstilbestrol, nonylphenol and methoxychlor in vivo and in vitro. Aqua. Toxicol., 27: 361-372.

Goss G, Wood C (1988). The effects of acid and acid/aluminum exposure on circulating plasma cortisol levels and other blood parameters in the rainbow trout *Salmo gaivdneri*. J. Fish Biol., 32: 63-76.

Grosell M (2006). Intestinal anion exchange in marine fish osmoregulation. J. Exp. Biol., 209: 2813-2827.

Hadi A, Shokr A, Alwan S (2009). Effects of aluminum on the biochemical parameters of fresh waterfish *Tilapia zillii*. J. Sci. Appl., 3: 33-41.

Hano T, Oshima Y, Kinoshita M, Tanaka M, Wakamatsu Y, Ozato K, Nassef M, Shimasaki Y, Honjo T (2009). In ovo nanoinjection of nonylphenol affects embryonic development of atransgenic see-through medaka *Oryzias latipes* olvas-GFP/STII-YI strain. Chemosphere, 77: 1594-1599.

Hanson TE, Tabita FR (2003). Insights into the stress response and sulfur metabolism revealed by proteome analysis of a Chlorobium tepidum mutant lacking the Rubisco-like protein. Photosynth. Res., 78: 231-248.

Hughes PJ, McLellan H, Lowes DA, Khan SZ, Bilmen JG, Tovey SC, Godfrey RE, Michell RH, Kirk CJ, Michelangeli F (2000). Estrogenic alkylphenols induce cell death by inhibiting testis endoplasmic reticulum Ca^{2+} pumps. Biochem. Biophys. Res. Commun., 277: 68-574.

Hussein SY, Mekkawy IAA (2001). The effects of leadexposure and lead-clay interaction on the growth performance, biochemical and physiological characteristics and histopathology of Tilapia zillii. Bull. Fac. Sci. Assiut Univ., 30(1-E): 65-97.

Jigang X, Cuijuan N, Xuejiao P (2010). Effects of chronic exposure to nonylphenol on locomotor activity and social behavior in zebrafish (*Danio rerio*). J. Environ. Sci., 22(9): 1435-1440.

Joshi PK, Bose M, Harish D (2002). Changes in haematological parameters in a siluroid catfish *Clarias batrachus* (Linn) exposed to mercuric choloride. Pollut. Resour., 21(2): 129-131.

Kachmar JF, Moss DW (1976). In Foundamentals of Clinical Chemistry, 2nd ed. NW Tietz, Editor. WB Saunders, Philadelphia, p. 652.

Kaplan A, Ozabo L, Ophem K, Febiger L (1988). Clinical Chemistry: Interpretation and Techniques, third ed., Lea and Febiger, Philadelphia.

Karami A, Christianus A, Ishak Z, Courtenay SC, Sayed MA, Noor Azlina M, Noorshinah H (2010). Effect of triploidization on juvenile African catfish (*Clarias gariepinus*). Aquacult. Int., 18: 51-858.

Kori-Siakpere O, Ake JEG, Avworo UM (2006). Sub-lethal effects of some selected haematological parameters of Heteroclarias (A hybrid of *Heterobranchus bidorsalis* and *Clarias gariepinus*). Int. J. Zool. Res., 2: 77-83.

Kornberg A (1955). Lactic dehydrogenase of muscle. In: J.M. Lowestein, Editor, Methods in Enzymology, Academic Press, New York, pp. 441-442.

Laemmli UK (1970). Cleavage of structural proteins during assembly of head of bacteriophage-T4. Nature, 227: 680-685.

Lee RN, Gerking SD, Jezierska B (1983). Electrolyte balance and energy mobilization in acid stressed rainbow trout *Salmo gairdneri* and their relation to reproductive stress. Environ. Biol. Fish., 8: 115-123.

Leopold JA, Zhang YY, Scribner AW, Stanton RC, Loscalzo J (2003). Glucose-6-phosphate dehydrogenase overexpression decreases endothelial cell oxidant stress and increases bioavailable nitric oxide. Arterioscler. Thromb. Vasc. Biol., 23: 411-417.

Levesque HM, Moon TW, Campbell PGC, Hontela A (2002). Seasonal variation in carbohydrate and lipid metabolism of yellow perch (*Perca flavescens*) chronically exposed to metals in the field. Aquat. Toxicol., 60: 257-267.

Long SM, Ryder KJ, Holdway DA (2003). The use of respiratory enzymes as biomarkers of petroleum hydrocarbon exposure in *Mytilus edulis planulatus*. Ecotoxicol. Environ. Safe, 55: 261-270.

Madigou T, Le Goff P, Salbert G, Cravedi JP, Segner H, Pakdel F, Valotaire Y (2001). Effects of nonylphenol on estrogen receptor conformation, transcriptional activity and sexual reversion in rainbow trout (*Oncorhynchus mykiss*). Aqua. Toxicol., 53: 173-186.

Maheswaran R, Devapanl A, Muralidharan S, Velmurugan B, Ignaeimuthu S (2008). Haematological studies of fresh water fish, *Clarias batradrus* (L) exposed to mercuric chloride. IJIB 2(1): 49-54.

Mahmoud UM, Mekkawy IAA, Sayed AH (2009). Ultraviolet radiation-A (366 nm) induced morphological and histological malformations during embryogenesis of *Clarias gariepinus* (Burchell, 1822). J. Photochem. Photobio. B: Biol., 95(2): 117-128.

Marin MG, Rigato S, Ricciardi F, Matozzo V (2008). Lethal and estrogenic effects of 4-nonylphenol in the cockle *Cerastoderma glaucum*, Marine Pollut. Bull., 57: 552-558.

Matos P, Fontaعnhas-Fernandes A, Peixoto F, Carrola J, Rocha E (2007). Biochemical and histological hepatic changes of Nile tilapia *Oreochromis niloticus* exposed to carbaryl. Pestic Biochem. Phys., 89: 73-80.

McCormick SD, O'Dea MF, Moeckel AM, Lerner DT, Bjornsson BT (2005). Endocrine disruption of parr-smolt transformation and seawater tolerance of Atlantic salmon by 4-nonylphenol and 17β-estradiol. Gen. Comp. E, 142: 280-288.

Media Cybernetica (1993-1997). L.P.; 8484 Georgia Avenue, Silver Spring, Maryland 20910.

Mekkawy IAA, Lashein FE (2003). The effect of lead and cadmium on LDH and G-6-PDH isozyme patterns exhibited during the early embryonic development of the teleost fish, *Ctenopharyngodon idellus* with emphasis on the corresponding morphological variations. The Big Fish Bang, the proceeding of the 26th Annual Larval Fish Conference (LFC2002), 22-26 July, 2002, Bergen, Norway, edited by: Howard I. Browman and Anne Berit Skiftesvik.

Mekkawy IAA, Mahmoud UM, Osman AG, Sayed AH (2010a). Effects of ultraviolet A on the activity of two metabolic enzymes, DNA damage and lipid peroxidation during early developmental stages of the African catfish, *Clarias gariepinus* (Burchell, 1822). Fish physiol. Biochem., 36: 605-626.

Mekkawy IAA, Mahmoud UM, Sayed AH (2011). Effects of 4-nonylphenol on blood cells of the African catfish *Clarias gariepinus* (Burchell, 1822). Tissue Cell, doi:10.1016/j.tice.2011.03.006.

Mekkawy IAA, Mahmoud UM, Wassif ET, Naguib M (2010b). Effects of cadmium on some haematological and biochemical characteristics of *Oreochromis niloticus* (Linnaeus, 1758) dietary supplemented with tomato paste and vitamin E. Fish physiol. Biochem., 37: 71-84.

Nguyen LTH, Janssen CR (2002). Embryo-larval toxicity tests with the African catfish (*Clarias gariepinus*): comparative sensitivity of endpoints. Arch. Environ. Con. Toxic., 42: 256-262.

Ololade IA, Oginni O (2010). Toxic stress and hematological effects of nickel on African catfish, *Clarias gariepinus*, fingerlings. J. Environ. Chem. Ecotoxic, 2(2): 014: 019.

ضner M, Atli G, Canli M (2008). Changes in serum biochemical parameters of freash water fish Oreochromis niloticus following prolonged metal (Ag, Cd, Cr, Cu, Zn) exposures. Environ. Toxicol. Chem., 27(2): 360-336.

Osman AG, Mekkawy IAA, Verreth J, Kirschbaum F (2007a). Effects of lead nitrate on the activity of metabolic enzymes during early developmental stages of the African catfish, *Clarias gariepinus* (Burchell, 1822). Fish Physiol. Biochem., 33: 1-13.

Osman AG, Wuertz S, Mekkawy IAA, Exner H, Kirschbaum F (2007b). Lead Induced Malformations in Embryos of the African Catfish *Clarias gariepinus* (Burchell, 1822). Environ. Toxicol., 22(4): 375-389.

Osman GM, Koutb M, Sayed AH (2010). Use of hematological parameters to assess the efficiency of quince (*Cydonia oblonga* Miller) leaf extract in alleviation of the effect of ultraviolet -A radiation on African catfish *Clarias gariepinus* (Burchell, 1822). J. Photoch. Photobio. B, 99: 1-8.

Pandey S, Parvez S, Sayed I, Haque R, Bin-Hafeez B, Raisuddin S (2003). Biomarkers of oxidative stress: a comparative study of river Yamuna fish *Wallago attu* (Bl. & Schn.). Sci. Total Environ., 309: 105-115.

Pandolfi PP, Sonati F, Rivi R, Mason P, Grosveld F, Luzzatto L (1995). Targeted disruption of the housekeeping gene encoding glucose-6-phosphate-dehydrogenase (G6pd)-G6pd is dispensable for pentose synthesis but essential for defense against oxidative stress. Embo. J., 14: 5209-5215.

Părt P, Svanberg O, Bergstrom E (1985). The influence of surfactants on gill physiology and cadmium uptake in perfused rainbow trout gills. Ecotoxicol. Environ. Safe, 9: 135-144.

Pickering A, Pottinger T (1995). Biochemical effects of stress. Elsevier Science, New York.

Poléo A, Hytterød S (2003). The effect of aluminium in Atlantic salmon (*Salmo salar)* with special emphasis on alkaline water. J. Inorg. Biochem., 97: 89-96.

Radhaiah V, Girija M, Rao KJ (1987). Changes in selected biochemical parameters in the kidney and blood of the fish, *Tilapia mossambica* (Peters), exposed to heptachlor. Bull. Environ. Contam. Toxicol., 39: 1006-1011.

Rajanna B, Chetty CS, McBride V, Rajanna S (1999). Effects of lead on K+-para-nitrophenyl phosphatase activity and protection by thiol reagents. Biochem. Int., 20: 1011-1018.

Regoli F, Frenzilli G, Bochetti R, Annarumma F, Scarcelli V, Fattorini D, Nigro M (2004). Time-course variations of oxyradical metabolism, DNA integrity and lysosomal stability in mussels, *Mytilus galloprovincialis*, during a filed translocation experiment. Aquat. Toxicol., 68: 167-178.

Rivero CLG, Barbosa AC, Ferreira MN, Dorea JG, Grisolia CK (2008). Evaluation of genotoxicity and effects on reproduction of nonylphenol in *Oreochromis niloticus* (Pisces: cichlidae). Ecotoxicology, 17: 732-737.

Rosety-Rodriguez M, Ordonez F, Rosety I, Rosety J, Rosery M (2005). Erythrocyte antioxidant enzymes of gilthead as early-warning bio-indicators of oxidative stress induced by malathion. Haema, 8: 237-240.

Salvemini F, Franze A, Iervolino A, Filosa S, Salzano S, Ursini MV (1999). Enhanced glutathione levels and oxidoresistance mediated by increased glucose-6- phosphate dehydrogenase expression. J. Biol. Chem., 274: 2750-2757.

Sastry KV, Gupta PK (1980). Alterations in the activities of a few dehydrogenases in the digestive system of 2 teleost fishes exposed to lead nitrate. Ecotoxicol. Environ. Safe, 4: 232-239

Satyanarayanan SK, Chokkalingam K, Mathan R, Tamara G (2011). Toxicity studies of nonylphenol and octylphenol: hormonal, hematological and biochemical effects in *Clarias gariepinus*, J. Appl. Toxicol., DOI 10.1002/jat.1629.

Sayed AH, Ibrahim ATh, Mekkawy IAA, Mahmoud UM (2007). Acute effects of Ultraviolet-A radiation on African Catfish *Clarias gariepinus* (Burchell, 1822). J. Photoch. Photobiol. B, 89: 170-174.

Shalaby A, Khattab Y, Abdel-Rahman A (2006). Effects of garlic Allium sativum chloramphinicol on growth performance, physiological parameters and survival of Nile tilapia *Oreochromis niloticus*. J. Venom Anim. Toxins incl. Trop. Dis., 12: 172-201.

Singh RK, Sharma B (1998). Carbofuran-induced biochemical changes in *Clarias batrachus*. Pest Sci., 53: 285-290.

Sone K, Hinago M, Kitayama A, Morokuma J, Ueno N, Watanabe H, Iguchi T (2004). Effects of 17 B-estradiol, nonylphenol, and bisphenol-A on developing *Xenopus laevis* embryos. Gen. Comp. Endocrinol., 138: 228-236.

SPSS (1998) SPSS for Windows. SPSS Inc, Headquarters. Chicago.

Staples C, Mihaich E, Carbone J, Woodbrun K, Klecka G (2004). A weight of evidence analysis of the chronic ecotoxicity of nonylphenol ethoxylates, nonylphenol ether carboxylates, and nonylphenol. Hum. Ecol. Risk Assess., 10: 999-1017.

Stoskopf MK (1993). Clinical pathology. In Stoskopf, M.K. (ed.). Fish Medicine Saunders, Philadelphia, pp. 113-131.

Tsuda T, Takino A, Kojima M, Harada K, Muraki T, Tsuji M (2000). 4-Nonylphenols and 4-terc-octylphenol in water and fish from rivers flowing into lake Biwa. Chemosphere, 41: 757-762

Uguz C, Iscan M, Erguven A, Isgor B, Togan I (2003). The

bioaccumulation of nonyphenol and its adverse effect on the liver of rainbow trout (*Onchorynchus mykiss*). Environ. Res., 92: 262- 270.
Union E (2002). 4-Nonylphenol (branched) and Nonylphenol Risk Assessment Report, Vol. 10. Institute for Health and Consumer Protection European Chemicals Bureau, Italy.
Vazquez-Duhalt R, Marquez-Rocha F, Ponce E, Licea AF, Viana MT (2005). Nonylphenol, and integrated vision of a pollutant. Appl. Ecol. Environ. Res., 4: 1-25.
Vincent MDV, Sneddon J (2009). Nonylphenol: An overview and its determination in oysters and wastewaters and preliminary degradation results from laboratory experiments. Microchem. J., doi:10.1016/j.microc.2009.02.005.
Wilson RW, Taylor EW (1993). The physiological responses of freshwater rainbow trout, *Onchorynchus mykiss*, during acute exposure. J. Comp. Physiol., 163: 38-47.
Winzer K, Van Noorden, CJF, Kohler A (2001). Quantitative cytochemical analysis of glucose-6-phosphate dehydrogenase activity in living isolated hepatocytes of European flounder for rapid analysis of xenobiotic effects. J. Histochem. Cytochem., 49: 1025-1032.

19

Extraction and comparison of fibrolytic enzyme additives from gut of 11 ungulates

Fabian Nde Fon[1], Ignatius Verla Nsahlai[2] and Peter Frank Scogings[1]

[1]Department of Agriculture, University of Zululand, KwaDlangezwa, 3886, South Africa.
[2]Animal and Poultry Science, University of KwaZulu-Natal, Pietermaritzburg, South Africa.

Microbial populations in herbivores gut attack, degrade and ferment structural carbohydrates in forage cell walls, producing volatile fatty acids and microbial proteins for the host. Exocellulases, endocellulases and cellobiases are the major cellulolytic enzymes while hemicellulase (xylanase) exposes cellulose for fermentation. This study aimed to isolate proteins that hydrolyse fibre or assist fibrolysis in any way from 11 herbivores gut microbial ecosystems (cow, sheep, horse, camel, elephant, zebra, llama, wildebeest, giraffe, impala and buffalo), optimize their working conditions and compare fibrolytic activities as potential feed additives. Exocellulase, endocellulase, xylanase and cellobiase in rumen and fecal crude enzyme extracts were assayed and compared. A broad *in vitro* pH range (4.5 to 8.0) of endocellulase activity was observed for all ecosystems. Enzymes from horse, zebra, impala, wildebeest and elephant showed the highest potential for degrading fibre encouraging further investigation as feed additives.

Key words: Crude cellulase extraction, hemicellulase, rumen and faecal inoculum, ungulates.

INTRODUCTION

Cellulolytic microbes possess three types of enzymes which enable partial or complete solubilisation of plant cell walls. These enzymes are exocellulases (EC 3.2.1.91), endocellulases (EC 3.2.1.4) and cellobiases (EC 3.2.1.21). These complex enzyme consortia function synergistically to hydrolyse plant cell walls. Exocellulases remove cellobiose units from the non-reducing ends of the cellulose chain, endocellulases act in a random fashion on the region of low crystallinity on the cellulose chain and β-glucosidase produces glucose units from cellobiose (Lynd et al., 2002). Although glucosidase is not strictly necessary for cell wall solubilisation, its presence enhances cellulose hydrolysis because these three enzymes function synergistically and inhibition of endocellulase and exocellulase by cellobiose is avoided. Solubilisation of plant cell walls does not solely depend on cellulases because of the complexity of its association with other macromolecules (Lynd et al., 2002). Apart from celluloses, hemicelluloses are the second most abundant plant fraction available in nature with xylan being the most abundant of all the hemicelluloses (Petzold et al., 2008). Lignin, pectin and tannins are examples of macromolecules that are often found in close association with cellulose (Mussatto et al., 2008). Therefore for solubilisation of cellulose in roughages to be successful the enzymes hemicellulase, ligninase, tannase and pectinase are required (Fon and Nsahlai, 2012).

Not all the microbes (bacteria, fungi and protozoa) that are found in both the fore- and hindguts have fibrolytic or hemicellulolytic properties but can assist in the fermentation process either directly or indirectly. Fermentation yields volatile fatty acids which can reduce the pH of the rumen fluid to less than 6, thereby halting the activity of most rumen microbes which function optimally at pH 6.2 and above (Russell and Dombrowski 1980). The presence of non-fibrolytic microbes that utilise acetic acid as a source of energy reduces its concentration hence maintaining the optimal rumen pH.

Different studies have shown that adding exogenous enzymes to herbivore diet increases milk production due to an increase in feed digestion (Schingoethe et al., 1999; Kung et al., 2000; Yang et al., 2000). In a similar study on fibre rich forages, Feng et al. (1996), Krause et al. (1998), Yang et al. (1999) and Beauchemin et al. (2000) observed only small increments in fibre digestion. *In vitro* increases in dry matter (DM) digestibility have also been reported by Nakashima et al. (1988), Feng et al. (1996), and Yang et al. (1999).

It is possible that microbial enzymes from herbivorous species consuming different food items in the pasture can vary in plant cell wall hydrolysis (Antonio et al., 2011). The major limitations in investigating the fibrolytic potential of the different microbial ecosystems especially that from the wild includes; animals being expensive to kill for rumen fluid, expensive extraction and purification techniques. Several methods have been applied to isolate proteins from the rumen fluid or faecal samples including ammonium sulfate precipitation, three phase partitioning precipitation, gel filtration and acetone precipitation. Ammonium sulphate is one of the most used methods in precipitating crude enzymes because of its simplicity, availability and cost effectiveness (Henry et al., 1974). The main objective of this study was to identify potential fibrolytic enzyme additives secreted by microbes from gut of 11 ungulates.

MATERIALS AND METHODS

Experimental animals and nutrition

Animals were chosen with no preference to sex or species. *Bostaurus* (fistulated Jersey cows), *Equus caballus* (horse) and *Ovis aries* (fistulated sheep) were from Ukilinga Research farm, University of KwaZulu Natal (UKZN), Pietermaritzburg. *Llama glama* (llama), *Loxodonta africana* (elephant) and *Camelus dromedarius* (camel) were provided by Brian Boswell Circus, Pietermaritzburg. *Equus quagga* (zebra), *Connochaetes taurinus* (wildebeest), *Aepyceros melampus* (impala), *Syncerus caffer* (buffalo) and *Giraffa camelopardalis* (giraffe) were provided by the Tala Game Reserve, Umbumbulu, KwaZulu-Natal (SA).

This study was conducted in winter (April – August). Camel, elephant and llama were kept in an enclosed area and fed with grass hay (*Eragrostis* dominant) on a regular basis. Game animals such as the impala, wildebeest, zebra and buffalo were grazing on a dry land in an open field where *Pennisetum clandestinum* (Kikuyu grass) dominated, with other grasses. The giraffe browsed on a variety of tree leaves (various acacia species). At the Ukulinga research farm, horse, sheep and cow were fed entirely on grass hay (*Eragrostis* dominant).

Faecal and rumen fluid collection

A modified procedure previously described by Smith et al. (1974) was employed when collecting rumen samples from fistulated cows and sheep. Rumen digesta (200 ml) was collected through a fistula, strained through four layers of cheese cloth, and treated immediately with 150 µl of phenylmethylsulfonyl fluoride (0.1 mmol/L PMSF) to inhibit proteases from lysing enzymes of interest (Owolabi et al., 1888; Vercoe et al., 2003). On the other hand, faeces were collected *in situ* from the horse, giraffe, buffalo, impala, wildebeest, zebra, elephant, llama and and camel and transported in an airtight insulated flask maintained at 38°C. All experimental animals sample collection was governed by the UKZN ethical roles (083/10/animal).

Protein exclusion and concentration measurements

Protein isolation was done using a modified procedure described by Henry et al. (1974). Both rumen fluid (100 ml) and faeces were used for protein isolation. In the case of faeces, 20 g was dissolved in 80 ml of a homogenization buffer (50 mmol/L sodium-acetate buffer, 0.02% (m/v) NaN_3, 0.1 mmol/L PMSF and 0.1 mmol/L EDTA at pH 5.0) before topping to 100 ml. Sample solutions in sealed centrifuge tubes were placed on a shaker for 30 min to facilitate bacteria detachment from fibres before centrifuging (7500 x*g*, 30min at 4°C) to sediment particulate matter (Figure 1). The supernatant was centrifuged (30 000 x*g*, 15 min at 4°C) to sediment bacteria cells. The sedimented particles and bacteria cells were dissolved in 10 ml and 5 ml of homogenisation buffer, respectively, sonicated for 3 min at 1 min intervals using a Q125 Utrasonic Processor (to lyse bacteria cells) and centrifuged (30 000 x*g*, 15 min at 4°C). The supernatants of the three different steps were pooled and centrifuged (30 000 x*g*, 15 min at 4°C) to sediment any unlysed cells. Ammonium sulfate (60% (m/v) $(NH_4)_2SO_4$) was dissolved in the sample solution to facilitate protein precipitation before centrifuging (7000 x*g*, 15 min at 4°C). The precipitate was dissolved in 10 ml of storage buffer (20 mmol/L sodium acetate, 0.02 % (m/v) NaN_3, and 0.1 mmol/L EDTA at pH 5.0) before dialyzing (12 h in storage buffer). Dialyzed sample solutions were concentrated using polyethylene glycol 20 000.

Bradford (1976) dye-binding assay was used to determine crude enzyme concentrations. A standard curve for a micro assay was prepared with bovine serum albumin concentrations of 0, 5, 10, 20, 30 and 40 µg/100 µl of reaction buffer (20 mmol/L sodium acetate, 0.02% (m/v)NaN_3 and 0.1 mmol/L EDTA at pH 5.0; Goeser et al., 2009). Bradford reagent (900 µl of 600 mg Coomassie Brilliant Blue G-250 dissolved in 1 L of 2% perchloric acid and filtered through Whatman number 1 filter paper) was pipetted onto the standard solution (100 µl) and absorbance read at 595 nm after 3 min. For the unknown protein sample solutions, 5 µl was diluted into 100 µl of the reaction buffer before adding Bradford reagent (900 µl) and allowed to stand for 3 min. Each assay was performed in triplicates and absorbance was measured at 595 nm. Unknown protein concentrations were determined from the standard curve.

Activities of crude enzyme samples and pH optimization

The method described by Seyis and Aksoz (2005) for the optimisation of xylanase activity from *Trichoderma* 1073 D3 was modified slightly and used in this assay. Enzyme stability within a pH range of 4.0 to 8.0 was determined by pre-incubating the crude enzyme extracts in 500 µl sodium-acetate buffer (400 ug/500 µl 20 mmol/L sodium-acetate buffer) at different pH values for 30 min before adding 500 µl of carboxymethyl cellulose (1% (m/v) CMC in 20 mmol/L sodium acetate, 0.02% (m/v) NaN_3 and 0.1 mmol/L EDTA) and incubating for 2 h at 39°C. The substrate concentrations as well as the crude enzyme concentration were constant throughout the experiment. Enzyme reactions were stopped by heating at 100°C. Incubated samples were centrifuged at maximum speed (12875 xg) on a desktop centrifuge for 5 min and 400 µl of the sample solution was used for reducing sugar analysis.

Quantification of reducing sugars

Dinitrosalicylic (DNS) method (Miller, 1959) was chosen for analysis because of its sensitivity, simplicity and availability. A standard curve

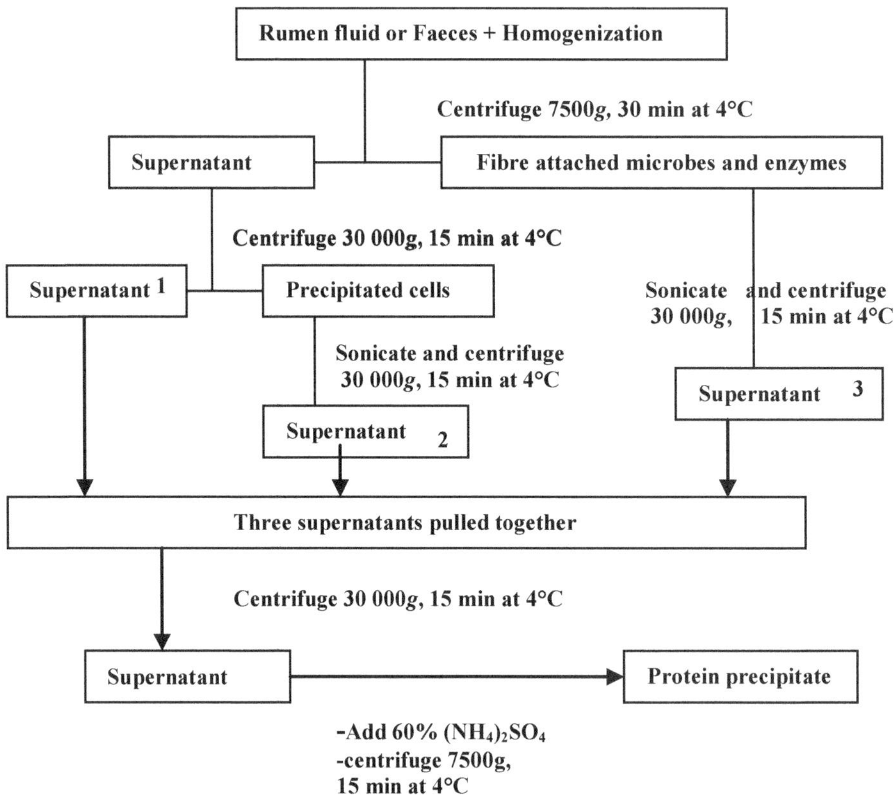

Figure 1. A flow chart for the extraction of fibrolytic enzymes from rumen fluid and faeces.

of micrograms of reducing sugars (glucose or xylose) against absorbencies at 540 nm was plotted as described by Wood and Bhat (1988). DNS reagent, 600 µl (0.001 M sodium metabisulfate, 0.708 M potassium sodium tatrate, 0.25 M sodium hydroxide, and 0.021 M phenol) was pipetted into 400 µl of the sample solution and boiled for 5 min. The reaction mixture was then cooled under running water and the absorbance measured at 540 nm. The absorbance values were translated into concentration of reducing sugars using the standard curve.

Enzyme assays

Exocellulase activity was assayed following a modified version of the method described by Gerrit et al. (1985). Exocellulase activity was measured by pipetting 0.5 ml of 1% (m/v) crystalline cellulose in the reaction buffer into 0.5 ml of crude enzyme solution obtained from different herbivores and incubating for 48 h at 39°C and pH 5.5 (standard assay condition). The enzyme reaction was stopped by boiling at 100°C, and the reaction mixture was centrifuged (6000 *xg*, 5 min at 4°C) and 400 µl sample analysed for reducing sugars using DNS method. Each ecosystem was represented by three samples each of which was analyzed in triplicates. Specific enzyme activity was measured in µg of glucose/µg protein.

For endocellulase activity, the reaction mixture contained 0.5 ml of 0.5% (m/v) carboxymethyl cellulose (CMC) in the reaction buffer at pH 5.5 and 0.5 ml of crude enzyme. The enzyme reaction was stopped as described for exocellulases. Samples were centrifuged at 6000 *xg* at RT for 5 min, and then 400 µl of the reaction mixture was analyzed for reducing sugars using DNS method.

Cellobiase activity was measured in a mixture of 0.5 ml of 0.1% (m/v) p-nitrophenyl β-D-glucopyranoside(*p*NP-G) in the reaction buffer and 0.5 ml of crude enzyme solution. The mixture was incubated in a water bath for 1 h at 39°C. Each enzyme reaction was stopped with 100 µl of sodium carbonate solution (200 mmol/L $NaHCO_3$) and the reaction mixture centrifuged (6000 *xg* for 5 min), following which *p*-nitrophenol (pNP) liberated was analysed by reading the absorbance at 407 nm. The molar absorption coefficient (ε) of *p*NP at 407 nm was taken to be 18300 mol^{-1} cm^{-1} from similar calculations done by Frutos et al. (2002). The number of moles of *p*NP released was assumed to be directly proportional to the amount of glucose.

Xylanase hydrolyses xylan into xylose and oligosaccharides (Chivero et al., 2001). A modified procedure described by Khanna (1993) was used to assay xylanase activity. The assay was carried out by pipetting 0.6 ml of 0.1% (m/v) xylan solution in the reaction buffer (pH 5.0) into 0.4 ml of crude enzyme solution and incubating at 39°C for 1 h. The reaction was stopped with 100 µl of 200 mmol/L $NaHCO_3$, centrifuged (at 6000 xg for 5 min) and analysed for reducing sugars using the DNS method. Each ecosystem was represented by three samples each of which was analysed in triplicate. Specific activities of the above enzymes were defined as µg of xylose/mg crude enzyme.

Statistical analysis

The results from exocellulase, endocellulase, cellobiase and hemicellulase specific activities as well as total protein, purification fold and yields were subjected to analysis of variance (ANOVA) using the general linear model of SAS. The model was: $Y_{ij} = \mu + T_i + e_{ij}$, where Y_{ij} is the individual observation, μ is the overall mean, T_i

Table 1. Purification of crude enzyme extracts from herbivores using ammonium sulphate.

Sample source	Sample fraction	Total protein (mg)	Total activity (nmol glucose)	Specific activity[1]	Purification (fold)	Yield (%)
Cow	-Supernatant	310^{ab}	2928^{f}	9.4^{f}	1	100
	-$(NH_4)_2SO_4$ precipitate	123^{b}	1982^{g}	16.1^{b}	1.706	39.7
Sheep	-Supernatant	313^{ab}	2783^{f}	8.9^{g}	1	100
	-$(NH_4)_2SO_4$ precipitate	138^{b}	1378^{g}	10.0^{f}	1.125	43.9
Horse	-Supernatant	588^{a}	5553^{c}	9.4^{f}	1	100
	-$(NH_4)_2SO_4$ precipitate	233^{b}	8263^{a}	35.0^{a}	3.765	39.6
Camel	-Supernatant	396^{ab}	3300^{e}	8.3^{g}	1	100
	-$(NH_4)_2SO_4$ precipitate	333^{ab}	2960^{f}	8.9^{g}	1.067	84.1
Elephant	-Supernatant	462^{ab}	5390^{c}	11.7^{d}	1	100
	-$(NH_4)_2SO_4$ precipitate	146^{b}	4536^{d}	31.1^{a}	2.667	31.6
Zebra	-Supernatant	876^{a}	7787^{b}	8.9^{g}	1	100
	-$(NH_4)_2SO_4$ precipitate	159^{b}	5231^{c}	33^{a}	3.688	18.2
Llama	-Supernatant	518^{ab}	4317^{d}	8.3^{g}	1	100
	-$(NH_4)_2SO_4$ precipitate	201^{b}	2010^{g}	10.0^{f}	1.2	38.8
Wildebeest	-Supernatant	938^{ab}	8337^{a}	8.9^{g}	1	100
	-$(NH_4)_2SO_4$ precipitate	182^{b}	3855^{e}	21.1^{e}	2.375	19.5
Giraffe	-Supernatant	336^{ab}	2800^{f}	8.3^{g}	1	100
	-$(NH_4)_2SO_4$ precipitate	84^{b}	747^{h}	8.9^{g}	1.067	25
Impala	-Supernatant	451^{ab}	375^{i}	8.3^{b}	1	100
	-$(NH_4)_2SO_4$ precipitate	135^{b}	2025^{g}	15.0^{c}	1.8	29.9
Buffalo	-Supernatant	486^{ab}	4320^{d}	8.9^{g}	1	100
	-$(NH_4)_2SO_4$ precipitate	170^{b}	2742^{f}	16.1^{b}	1.813	35.0
SEM		52.1	290.1	1.51	0.13	6.81
P-value		*	**	**	*	**

Substrate = Carboxymethyl cellulose, *P<0.05, **P < 0.01, values with different superscripts letters in the same column and animal differ at P<0.05 ($^{a, b, c, d, e, f, g}$), [1] = nmol glucose/mg/min.

is the effect of the treatment (enzyme sources, pH or crude enzyme concentrations) and e_{ij} is the random variation.

RESULTS

Crude enzymes (endocellulases) activities

The colour of the protein precipitates consisted of variable shades of brown and green. The horse, wildebeest, zebra, buffalo and impala were characterized by a light reddish brown precipitate while those from the elephant, giraffe, llama and camel were characterised by a whitish-brown precipitate. Precipitates from the cow and sheep were characterized by a greenish colour. The crude enzyme extracts were active to the extent that their enzyme specific activities on carboxymethyl cellulose increased (P<0.05) as protein concentration decreased following purification (Table 1). The activity recovered (purification fold) was highest (P<0.05) in the horse and zebras, intermediate in the wildebeest, impala and buffalo, and lowest in the camel, sheep and giraffe. It was interesting to note that the camel with the highest percentage yield (84.1%) had the lowest recovery (1.06 fold).

pH optimization

Endocellulase had a broad (P<0.05) pH range of activity for all the experimental microbial ecosystems (Figure 2). Despite the broad pH range, some distinct (P<0.05) peaks of endocellulase activities were observed within the pH range 4.5–5.0 for the elephant, at pH 5 for the horse and the zebra, within pH 5.0-6.0 for the sheep, and within pH 5.0–7.0 for the wildebeest. Endocellulase activity for the rest of the animals (cow, llama, giraffe and camel) responded minimally within pH 4.0 to 8.0 range.

Optimization of crude enzyme concentrations for enzyme assay

Generally, product formation increased (P < 0.05) with increasing crude enzyme concentration from the sampled herbivore microbial ecosystems. Glucose liberated from crystalline cellulose, CMC and *p*NP-G following incubation with crude extracts, increased (P < 0.05) non-linearly (Figure 3). Xylose liberated from xylan by xylanase from crude enzyme extracts also increased (P < 0.05) non-

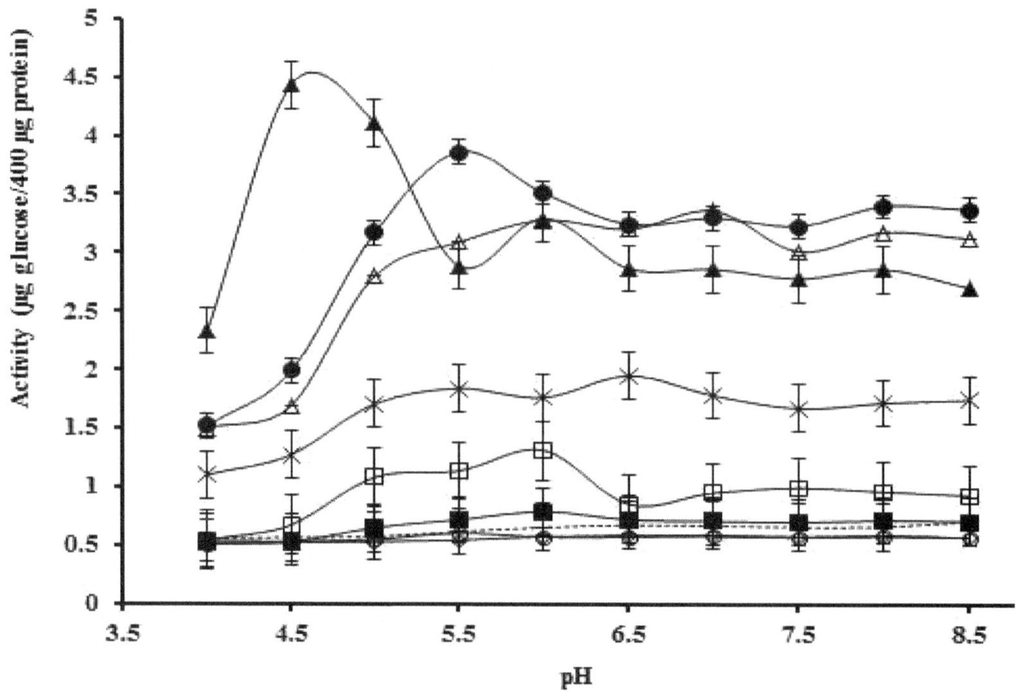

Figure 2. The effect of pH on endocellulase activity. Product formation was monitored for samples from the reaction mixture of cow (■),sheep (□), horse (●), camel (○), elephant (▲), zebra (Δ), llama (◊), wildebeest (·), giraffe (+), impala (--) and buffalo (▪).

linearly (Figure 3D). In Figure 3A, the highest increase was obtained in the zebra, horse and wildebeest while the elephant and impala showed moderate increases. High ($P < 0.05$) levels of product formation were also observed in horse, impala and wildebeest when endocellulase concentrations were varied (Figure 3C). A similar situation was also observed when cellobiase was incubated with *p*NP-G (Figure 3B). Product formation from the reaction mixtures of the horse, zebra, impala, wildebeest and sheep were dominant while the rest were crowded together at low product formation. The result was not very different when hemicellulases were incubated with xylan (Figure 3D). The product formed in the sheep was as high as those observed from the horse and zebra, while the cow, camel and giraffe activities tended to be high. An enzyme concentration of 340 µg/ml for optimal activity was established as suitable for exocellulase and endocellulase.

DISCUSSION

Collecting fresh samples with limited exposure to oxygen maximised the presence of anaerobic cellulolytic and hemicellulolytic microbes as well as their enzymes. Limited exposure to oxygen prevents cell death, which was vital for the release of periplasmic proteins during sonication (Dunn et al., 2000). Ammonium sulfate precipitation reduced the amount of unwanted proteins in the samples. This was confirmed by the increase in specific activity of crude enzyme samples with a decrease in protein concentration (Groleau and Forsberg 1983). The reddish-brown protein precipitate observed in horse, wildebeest, impala and zebra was due to high cellobiase content while the greenish colour in cow and sheep precipitates was probably due to the plant pigment chlorophyll.

Specific activity and enzyme stability were both affected by pH and temperature. The rate of enzyme reactions might be compromised if these factors were neglected. pH optimisation was vital as enzymes were isolated from 11 different ecosystems which might vary greatly. The pH profile showed that activity occurs with a wide range as opposed to the narrow *in vivo* pH range (5.6 to 8) described by many authors (Russell and Dombrowski, 1980). This disparity can be ascribed to the presence of enzymes only in *in vitro* systems unlike *in vivo* where very low or high pH might compromise microbial growth and survival. However, the peaks observed in the horse (pH 5.5) and elephant (pH 5) demonstrated that these ecosystems were harbouring microbes that were active in slightly acidic conditions. Variations can also be attributed to the genetic differences among strains of microbes in various ecosystems in the expression of these enzymes.

Other factors also affecting activity include enzyme crystalline cellulose, CMC and *p*NP-G following incubation x substrate concentrations can be observed in the work

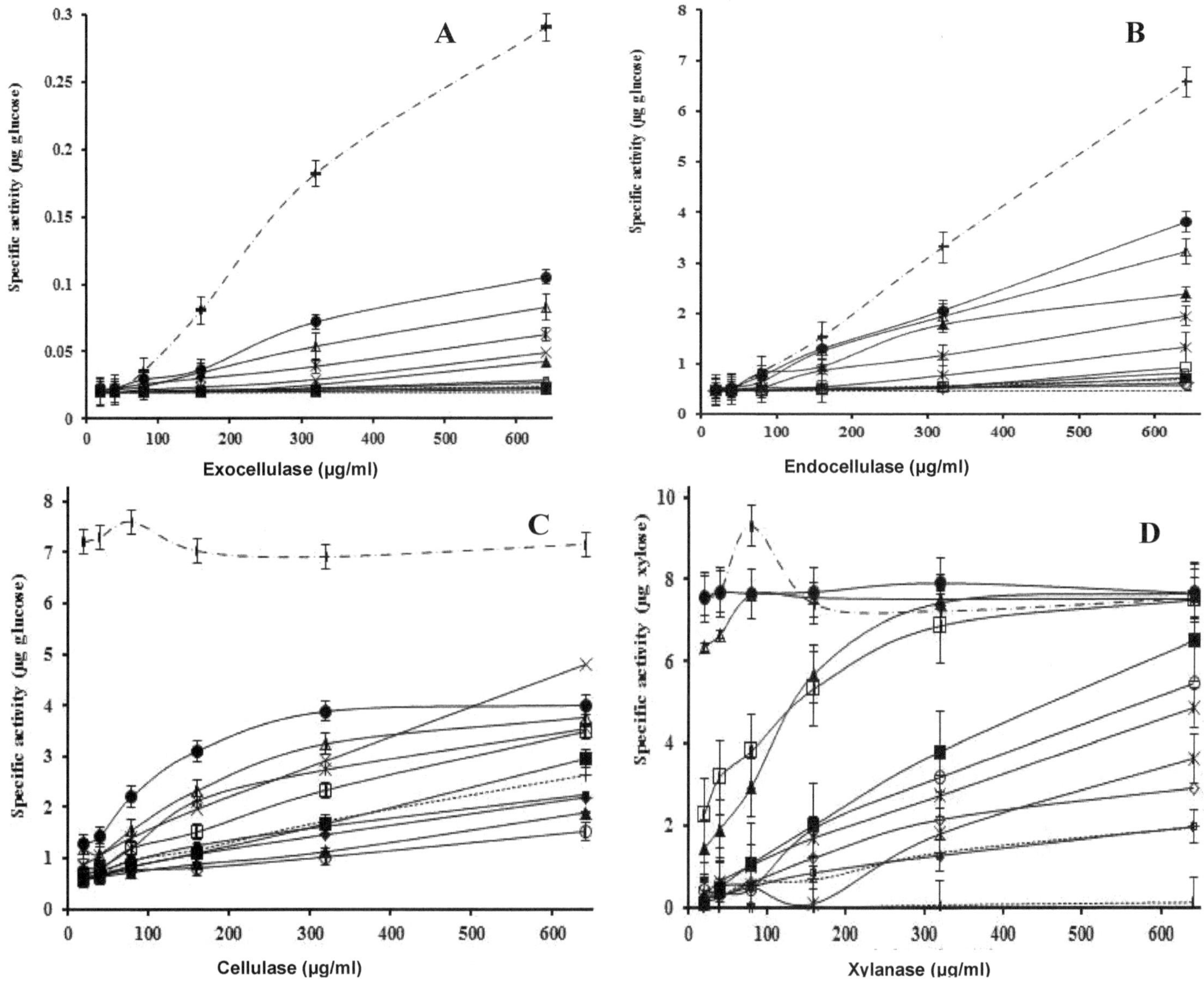

Figure 3. The effect of enzyme concentrations on product formation. Product formation from crystalline cellulose (3A), carboxymethylcellulose (3B), p-nitrophenyl ß-1,4-glucopyranoside (3C) or xylan (3D) were monitored from the reaction mixture of cow (■), sheep (□), horse (●), camel (○), elephant (▲), zebra (Δ), llama (◊), wildebeest (·),giraffe (+), impala (×) and buffalo (▫). The positive controls (-··-) were commercial cellulase from *Aspergillusniger* (3A (exocellulase) and 3B (endocellulase)) cellobiase (3C) and xylanase from *Thermomyces ianuginosus* (3D). The negative control (**...**) was BSA for all experiments. Exocellulase = crystalline cellulase, Endocellulase = carboxymethyl cellulase.

of Henry et al. (1974) but should be optimised for every sample as enzyme concentration and activity vary relatively to its purification method (Jahir et al., 2011). The results arising from incubating different crude enzyme concentrations with the different substrates (crystalline cellulose, CMC, xylan and cellobiose) established that 340 µg was the ideal concentration to apply in enzyme assays. It was also noted that as the crude enzyme concentration increased the rate also increased linearly but linearity was lost at high crude enzyme concentrations. This was expected because at a high crude enzyme concentration, substrate concentration would become rate limiting. This was similar to the results obtained by Groleau and Forsburg (1983) in studies on cellobiase and endocellulase.

In an attempt to optimise extract concentrations for enzyme assays with the different substrates, it was noted that the rate at which protein samples from the different ecosystems hydrolyse specific substrates was not the same. Crude enzyme samples from the horse, zebra and wildebeest were the most active when incubated with all four substrates but for the elephant which was also very active when incubated with CMC. Interestingly, the hindgut fermenters contained the most active protein samples. Although, many studies have stated that there are no major differences among the microbes in both fore- and hindgut fermenters (Smith and Mackie, 2004; (Smith and Mackie, 2004, Alexander, 1993), the results obtained from this study clearly indicate that there might be some factors influencing activity in these chambers. If time spent by digesta in the hingut is a major factor (Mould et al., 2005; Schingoethe, 1993) then it is imperative to postulate that microbes in the hindgut will be more efficient in order to extract sufficient nutrients from the food before elimination as faeces. However, not all the foregut fermenters exhibited low activities as the wildebeest; the impala and cow showed intermediate activities with all four substrates. Variation in these activities might have been influenced by differences in ecological niches or composition of microbial species. For herbivores within the same geographical region, the differences in enzyme activity could be attributed to differential genetic evolution of microbes which might have resulted in variable fibrolytic efficiencies.

Conclusion

An enzyme concentration of 340 µg/ml for optimal activity was established as suitable for exocellulase and endocellulase. Crude enzyme samples from the zebra, horse and wildebeest were the most active when incubated with all four substrates while those from the elephant, impala, cow and giraffe were intermediate. Future studies to confirm high activity among these ecosystems would include determining the affinity of enzyme to substrate, maximum velocity as well as their catalytic efficiencies. These results can assist in establishing cellulases from microbial population of zebra, horse, impala and wildebeest as feed additives.

ACKNOWLEDGEMENTS

We would like to acknowledge National Research Foundation for funding this project and the University of Zululand and University KwaZulu-Natal for facilitating the project.

REFERENCES

Alexander RM (1993). The relative merits of foregut and hindgut fermentation. J. Zool. 231:391-401.

Antonio F, Fernanda D, Vici AC, Amin Reis, Ricardo V, Jorge JA, Terenzi HF, Reis RA, Teixeira de Moraes Polizeli, Lourdes Md (2011). Production of fibrolytic enzymes by Aspergillus japonicus C03 using agro-industrial residues with potential application as additives in animal feed. Bioprocess Biosyst. Eng. 34:347-355.

Beauchemin K, Rode L, Maekawa M, Morgavi D, Kampen R (2000). Evaluation of a nonstarch polysaccharidase feed enzyme in dairy cow diets. J. Dairy Sci. 83:543-553.

Bradford MM (1976). A rapid and sensitive method for the quantitation of microgram quantities of protein utilizing the principle of protein-dye binding. Anal. Biochem. 72:248-254.

Chivero ET, Mutukumira AN, Zvauya R (2001). Partial purification and characterisation of a xylanase enzyme produced by a micro-organism isolated from selected indigenous fruits of Zimbabwe. Food Chem. 72:179-185.

Dunn RV, Reat V, Finney J, Ferrand M, Smith JC, Daniel RM (2000). Enzyme activity and dynamics: xylanase activity in the absence of fast anharmonic dynamics. Biochem. J. 346:355.

Feng P, Hunt CW, Pritchard GT, Julien WE (1996). Effect of enzyme preparations on in situ and in vitro degradation and in vivo digestive characteristics of mature cool-season grass forage in beef steers. J. Anim. Sci. 74:1349-1357.

Fon F, Nsahlai I (2012). Effect of direct-fed microbial consortia on ruminal fermentation of maize stover in sheep. Small Rumin. Res. 111:71-75.

Frutos C, Sonsoles P, Chiquinquirá C, Eduardo C (2002). Reverse micelles in organic solvents: a medium for the biotechnological use of extreme halophilic enzymes at low salt concentration. Archaea 1:105-111

Gerrit B, Marjo F, Frank M, Fons G (1985). The cellulase of Trichoderma Viride. Purification characterization and comparison of all detectable endoglucanases, exoglucanases and β-glucosidases. Eur. J. Biochem. 146:301-308.

Goeser JP, Hoffman PC, Combs DK (2009). Modification of a rumen fluid priming technique for measuring in vitro neutral detergent fiber digestibility. J. Dairy Sci. 92:3842-3848.

Groleau D, Forsberg C (1983). Partial characterization of the extracellular carboxymethylcellulase activity produced by the rumen bacterium Bacteroides succinogenes. Can. J. Microbiol. 29:504-517.

Henry B, Nick V, Desh P, Gordon A (1974). Purification and characterisation of two cellulases from Auxin-treated Pea Epicoltyls. J. Biol. Chem. 10:1012-1018.

Jahir AK, Ram KR, Varun R, Priyanka G (2011). Deciphering Cow Dung for Cellulase Producing Bacteria. Eur. J. Exp. Biol. 1:139-147.

Khanna S (1993). Regulation, purification, and properties of xylanase from Cellulomonas fimi. Enzyme Microb. Technol. 15:990-995.

Krause M, Beauchemin K, Rode L, Farr B, Nørgaard P (1998). Fibrolytic enzyme treatment of barley grain and source of forage in high-grain diets fed to growing cattle. J. Anim. Sci. 76:2912-2920.

Kung L, Treacher R, Nauman G, Smagala A, Endres K, Cohen M (2000). The effect of treating forages with fibrolytic enzymes on its nutritive value and lactation performance of dairy cows. J. Dairy Sci. 83:115-122.

Lynd LR, Weimer PJ, van Zyl WH, Pretorius IS (2002). Microbial Cellulose Utilization: Fundamentals and Biotechnology. Microbiol. Mol. Biol. Rev. 66:506-577.

Miller GL (1959). Use of Dinitrosalicylic Acid Reagent for Determination of Reducing Sugar. Anal. Chem. 31:426-428.

Mould FL, Kliem KE, Morgan R (2005). Alternative methodologies - stretching the in vitro box. Anim. Feed Sci. Technol. 123-124:501-515.

Mussatto SI, Fernandes M, Milagres AMF, Roberto IC (2008). Effect of hemicellulose and lignin on enzymatic hydrolysis of cellulose from brewer's spent grain. Enzyme Microb. Technol. 43:124-129.

Nakashima Y, Ørskov E, Hotten P, Ambo K, Takase Y (1988). Rumen degradation of straw 6. Effect of polysaccharidase enzymes on degradation characteristics of ensiled rice straw. Anim. Prod. 47:421-427.

Owolabi B, Beguin P, Kilburn G, Miller C, Antony J (1988). Expression in Escherichia coli of the cellulomonas fimi structural Gene for Endoglucanase B. Appl. Environ. Microbiol. 54:518-523.

Petzold K, Gunther W, Kotteritzsch M, Heinze T (2008). Synthesis and characterization of methyl xylan. Carbohydr. Polym. 74:327-332.

Russell JB, Dombrowski D (1980). Effect of pH on the efficiency of growth by pure cultures of rumen bacteria in continuous culture. Appl. Environ. Microbiol. 39:604-610.

Schingoethe DJ, Stegeman GA, Treacher RJ (1999). Response of lactating dairy cows to a cellulase and xylanase enzyme mixture

applied to forages at the time of feeding. J. Dairy Sci. 82:996-1003.
Schingoethe J (1993) Nutrient needs of ruminants versus monogastric species, in: C. Church (Ed.), The Ruminant Animal Digestive Physiology and Nutrition, Waveland Press, USA. pp. 448.
Seyis I, Aksoz N (2005). Investigation of factors affecting xylanase activity from Trichoderma harzianum 1073 D3. Braz. Arch. Biol. Technol. 48:187-193.
Smith RC, Moussa NM, Hawkins GE (1974). Utilization of the nucleic acids of Escherichia coli and rumen bacteria by sheep. Br. J. Nutr. 32:529-537.
Vercoe P, Kocherginskaya S, White B (2003). Differential protein phosphorylation-dephosphorylation in response to carbon source in Ruminococcus flavefaciens FD-1. J. Appl. Microbiol. 94:974-980.
Wood TM, Bhat KM (1988). Methods for measuring cellulase activities. Methods Enzymol. 160:87-112.
Yang WZ, Beauchemin KA, Rode LM (1999). Effects of an Enzyme Feed Additive on Extent of Digestion and Milk Production of Lactating Dairy Cows. J. Dairy Sci. 82:391-403.
Yang WZ, Beauchemin KA, Rode LM (2000). A comparison of methods of adding fibrolytic enzymes to lactating cow diets. J. Dairy Sci. 83:2512-2520.

Altered protease activity due to heat shock in the whole organism *Bombyx mori* L.

Rajesh R. K.[1], Manjunatha H. B.[2]* and Aparna H. S.[1,3]

[1]Department of Biochemistry, Karnatak University, Dharwad 580 003, Karnataka, India.
[2]Department of Sericulture, University of Mysore, Mysore 570 006, Karnataka, India.
[3]Department of Biotechnology, University of Mysore, Mysore 570 006, Karnataka, India.

The domesticated insect *Bombyx mori* respond well to environmental fluctuations that govern the success of cocoon production due to its piokilothermic nature. Since, the interaction between heat shock proteins and proteases in the whole organism that complete larval life in natural environment remain obscure, we have employed zymography technique to examine different classes of proteases and its activity under imposed temperature stress (heat shock) condition. The total protein extracted from the whole larvae was analyzed in SDS-polyacrylamide gel copolymerized with gelatin, casein and BSA. Discrete proteolytic bands aptly visualized in the gel contained gelatin as substrate revealed strong activity of five proteases with molecular mass of 116 (P1), 83 (P2), 57 (P3), 35 (P4) and 24 kDa (P5) and the only protease that exhibits activity in all the substrates was P5. Various protease inhibitors like leupeptine, aprotonin, pepstatin, EDTA and PMSF were used for identification and most of the proteases activity eventually was inhibited by PMSF while other inhibitors had no effect. With this we could determine the presence of large group of serine proteases in the whole body of *B. mori*. Further, for the first time, we have employed zymographic technique to evaluate protease activity in the heat shock (40°C) induced larvae. Interestingly, the activity of three proteases (P3 to P5) found inhibited on gelatin substrate while high molecular weight proteases (P1 and P2) were not affected. The protease assay further confirmed decreased protease activity in the heat shock induced larvae compared to control. The differential and altered proteases activities in normal and heat-shocked larvae respectively envisaged its significance in stress physiology and adaptation of whole organism to elevated temperature.

Key words: *Bombyx mori*, heat shock, lepidoptera, serine-protease, zymography.

INTRODUCTION

Silkworm, *Bombyx mori* is the only insect that is truly domesticated and well exploited commercially for the production of silk. In recent years, it is considered as central model insect among lepidopterans (Goldsmith et al., 2004) for comparative proteome and genome studies due to its rich genetic resources. And also used as a potential tool for expression of foreign genes of interest in synthesis of recombinant proteins of pharmaceutical importance (Tomita et al., 2003). Consequently, using advanced proteome tools and techniques protein map has been established from silkglands (Jin et al., 2004; Zhang et al., 2006), midgut (Kajiwara et al., 2005) and whole egg (Manjunatha et al., 2005) of *B. mori*. All the information's are stored in the proteome database KAIKO-2D and accessible through http://kaiko2ddb.dna.affrc.go.jp/. The advent of integrated approach following electrophoretic [single dimensional electrophoresis (1-DE) and two dimensional electrophoresis (2-DE)] techniques, liquid chromatography-mass spectrometry and application of bioinformatics tools could unravel the structure and

*Corresponding author. E-mail: manjunathahb@gmail.com

functional properties of biomolecules. Hence, the concept of whole organism has been followed in analysis of differential expression of heat shock proteins (HSPs) (Vasudha et al., 2006; Kundapur et al., 2009; Aparna et al., 2010) in *B. mori* and finding its evolutionary significance in post-genomic era (Sorensen and Loeschcke, 2007). Hitherto, temperature and pH known to govern the catalytic activity of an enzyme either in cells or tissues but their activity and its correlation during the period of heat shock and/or any stress conditions in the whole organism remain enigmatic.

Briefly, although heat shock response and expression of HSPs in *B. mori* have been investigated over two decades (Lohmann and Riddiford, 1992; Joy and Gopinathan, 1995; Evgen'ev et al., 2005), their biological significance and role in acquired thermotolerance remain unclear (Manjunatha et al., 2010). However, HSPs involvement in physiological and cytoprotective function (Kregel, 2002) thermotolerance in mammalian cells (Landry et al., 1989), *Drosophila* (Landry and Huot, 1995), *Musca domestica* (Tiwari et al., 1997), *Lucilia cuprina* (Tiwari et al., 1995), apoptosis (Arrigo, 1998, 2005), cell growth and differentiation (Mehlen et al., 1997) is obvious. As a consequence, molecular and thermal biologist's attention aids to determine the role of HSPs in the whole organism adaptation to heat including species inhabiting in various environmental conditions (Sorensen and Loeschcke, 2007).

Obviously, the structural and functional properties of proteins, which are significant in cell function and protection, are governed by different classes of proteases and analyzed employing a simple and sensitive technique "Zymography". Normally, many proteins are synthesized as inactive precursors known as zymogens and are subsequently converted to physiologically active forms by selective enzymatic cleavage of peptide bonds called zymogen activation (Neurath and Walsh, 1976). As SDS activates the zymogen into active protease, various proteases like matrix metalo proteases (MMPs) expressed during cancer development were identified and characterized (Tanriverdi et al., 2009; Lokeshwar et al., 1993). Interestingly, insect (lepidoptera) larval mid gut hosts a complex proteolytic environment of various proteases like trypsin, chymotrypsin, elastases, cathepsin-B like proteases, aminopeptidases and carboxypeptidases, which are responsible for protein digestion, killing bacteria and viruses (Nakazawa et al., 2004). In *B. mori*, serine proteases are known to dominate the larval midgut environment and contribute about 95% of the total digestive activity. Interestingly, silkworm egg also posses some proteases that are involved in vitellin and yolk degradation during embryo development (Ikeda et al., 1990; Maki and Yamashita, 2000). Since most of these studies are confined to specific tissue, an interaction between protein and proteases in the whole organism that complete its larval life in natural environmental conditions remain enigmatic. Thus, we have employed zymography technique to examine different classes of proteases present in the whole body of *B. mori* larvae and their altered activity under stress conditions due to heat shock.

MATERIALS AND METHODS

Experimental insect

The domesticated silkworm, *B. mori* eggs procured from Central Sericultural Germplasam Centre, Hosur, Tamil Nadu, India were incubated under optimum environmental conditions (25 ± 1°C and 75 ± 5% relative humidity) until hatching. Hatched larvae were reared on mulberry leaves following the standard procedure until spinning of cocoons (Jolly, 1987).

Sample preparation and electrophoresis

Day 3 of fifth instar silkworm larvae were given heat shock at 40°C for 2 h in BOD incubator with 70 ± 5% humidity followed by 2 h recovery at room temperature. The total protein from the whole body of control and heat shocked larvae was extracted separately as described by Vasudha et al. (2006) without phenylemethanesulphonylfluoride (PMSF) and dithioerythritol/β-mercaptoethanol. The extracted total protein sample was incubated with various protease inhibitors like leupeptin (100 μM), aprotonin (100 μg/ml), pepstatin (10 μM), EDTA (10 mM), PMSF (5 mM) separately for 30 min at 4°C prior to electrophoresis. None of these inhibitors were added to the protein sample extracted from the control and heat shocked larvae for comparative analysis. SDS-PAGE (12%) was carried out at 4°C incorporating different substrates, gelatin (0.1%), BSA (0.1%) and casein (1%) into the running gel, whereas only gelatin was used as substrate for the protein sample derived from heat shocked larvae.

Zymography

The zymography was performed as described by Heussen and Dowdle (1980) with suitable modifications. The electrophoresed gel was rinsed twice in Triton X-100 (2.5%) at room temperature. In-gel protease activity was carried out incubating the gel at 37°C for 18 h in reaction buffer (50 mM Tris- HCl, pH 6.8). The gel was stained with Coomassie brilliant blue R-250 in isopropanol (0.25%) followed by destaining in acetic acid (7%). The protease activity was determined based on lytic bands in blue background and images were captured using Alpha Imager UV Gel Doc System and Alpha EaseFC software for molecular weight determination.

Protease assay

The protease assay was carried out using different substrates as described by Cupp-Enyard (2008) with necessary modifications. The proteases present in the sample extracted from both normal and heat shocked silkworm larvae digest the substrates and the resultant peptides were estimated using tyrosine as standard (Lowry et al., 1951).

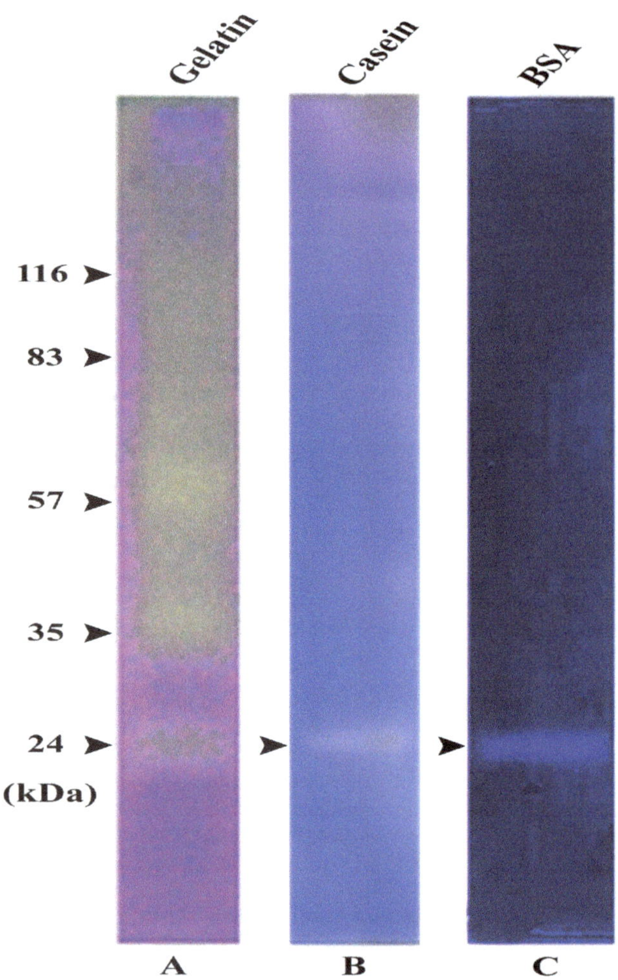

Figure 1. SDS-zymograph of proteases derived from the whole body of *Bombyx mori*. SDS-polyacrylamide gel copolymerized with gelatin (A), casein (B) and BSA (C) as substrates. Proteolytic bands along with their respective predicted molecular weight are denoted with arrows.

RESULTS AND DISCUSSION

Proteases and their activity

Total protein extracted from the whole body of day 3 fifth instar larvae of *B. mori* was analyzed for protease activity by zymography considering the recent concept of whole organism to understand molecular basis underlay in cellular protection and stress physiology of living organisms (Sorensen and Loeschcke, 2007; Ulmasov et al., 1992). This approach facilitated us to detect few proteolytic bands that contrasted well in the blue background of un-degraded protein substrate in Coomassise stained polyacrylamide gel. Figure 1 explicit discrete lytic zone that represents differential proteolytic activity of different proteases on various substrates gelatin, casein and BSA.

Comparatively, five discrete proteolytic bands were noticed in the gel copolymerized with gelatin as substrate (Figure 1A), where as only one lytic zone (band) was observed in the gels supplemented with casein (Figure 1B) and BSA (Figure 1C) as substrates (Table 1). These proteolytic zones were designated as P1-P5 (P stands for protease) for comparative analysis. The molecular weight determined for the proteolytic bands was 116 (P1), 83 (P2), 57 (P3), 35 (P4) and 24 kDa (P5) (Figure 1A). Of the five different proteases, strong activity with large proteolytic band was observed with P3, where as P5 exhibited similar activity in all the substrates used (Figure 1 and Table 1). However, 24 kDa (P5) protease activity was obvious in gelatin, casein and BSA embedded gels indicating broad specificity, while other proteases have narrow specificity with gelatin. Thus, most of the proteases detected in the protein sample of day 3 of fifth instar larvae of *B. mori*, were gelatinizes rather than caseinolyases and BSA degrading proteases.

Classification of proteases using protease inhibitors

Five different types of proteases investigated in the present study (P1 to P5) were identified and their activity was measured in presence of various protease inhibitors in the gel-containing gelatin as substrate. In the present experiment, we have incorporated various protease inhibitors (leupeptin, aprotonin, pepstatin, EDTA and PMSF) into the extraction buffer and analyzed independently. Among different protease inhibitors, PMSF inhibited the activity of all the proteases (P2 to P5) except high molecular weight protease P1 that showed partial inhibition (Figure 2 and Table 2). More interestingly, the P3 maintained its strong activity in presence of other inhibitors except in PMSF, where as P5 exhibited partial inhibition with pepstatin (Figure 2; Lane 4). Since, each well of the gel received same quantum of sample, differences in intensity and size of the lytic zones was proportional to the protease activity per sample (Naggie et al., 1997). This differential inhibition of protease activity suggested that of the five proteolytic bands (P1 to P5) are detected in the present study, three bands P2 to P4 found belong to class serine proteases as evident by their inhibition with PMSF, which is a known inhibitor of serine proteases (Figure 2; Lane 2).

Whereas, 24 kDa protease could be either serine or aspartate protease as its activity was inhibited by PMSF and pepstatin. Interestingly, a serine protease of 37 kDa was reported as zymogen from the larval midgut of *B. mori*, which is then activated after pupation. Its sequence analysis and cDNA determined the p37k as trypsin-type serine protease (Kaji et al., 2009). However, the biological significance of 24 kDa protease in fifth instar silkworm larvae is unknown, hitherto, serine proteases

Table 1. Differential activity of proteases on different substrates in the whole body of *Bombyx mori.*

Proteolytic bands	Mol. Wt. (kDa)	Gelatin	Casein	BSA
P1	116	+	-	-
P2	83	+	-	-
P3	57	+	-	-
P4	35	+	-	-
P5	24	+	+	+

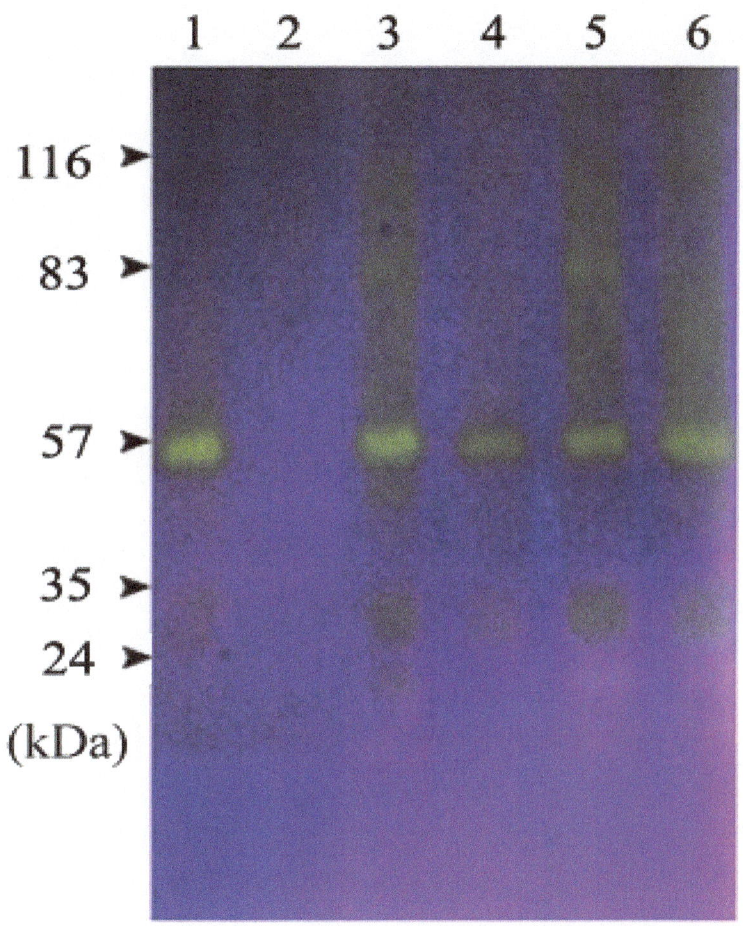

Figure 2. Effect of inhibitors on proteases of *Bombyx mori.* Lane 1 - Protein sample devoid of inhibitor; Lane 2 - Protein sample with PMSF; Lane 3 - Protein sample with EDTA; Lane 4 - Protein sample with Pepstatin; Lane 5 - Protein sample with Aprotonin; Lane 6 - Protein sample with Leupeptine. All these protease inhibitors were added to extraction buffer and loaded into the gel copolymerized with gelatin as substrate.

play pivotal role in blood coagulation systems of both arthropods and mammals. More importantly, the high molecular weight protease (116 kDa – P1) detected in the whole body of *B. mori* was not inhibited by any of the inhibitors used in the present study even with increased concentrations. This clearly indicates that this would be a new protease, which opens ample scope for molecular characterization and to determine its biological significance. Coincidently, five different serine proteases from the haemolymph of the mosquito *Anapheles*

Table 2. Effect of different inhibitors on protease activity in *Bombyx mori.*

Inhibitors (concentration)	Proteolytic bands				
	P1 (116 kDa)	P2 (83 kDa)	P3 (57 kDa)	P4 (35 kDa)	P5 (24 kDa)
PMSF (5 mM)	±	+	+	+	+
EDTA (10 mM)	-	-	-	-	-
Pepstatin (10 μM)	-	-	-	-	±
Aprotonin (100 μg/ml)	-	-	-	-	-
Leupeptin (100 μM)	-	-	-	-	-

('+' Complete inhibition; '-' no inhibition; '±' partial inhibition with residual activity).

gambiae (Gorman and Paskewitz, 2001) and a zymogen of a serine-type protease in *B. mori* (Katsumi et al., 1995) were reported, which has an important role in insect immunity. Hence, the present data suffice an insight for detail investigation to uncover its cellular and biological significance in *B. mori.*

Effect of heat shock on protease activity

Conventionally, zymography technique is routinely followed to identify proteolytic activity in SDS-polyacrylamide gel copolymerized with a substrate and fibrin zymography method identifies enzymes in enzyme reaction buffer containing protease inhibitors (Kim and Choi, 1998). But in the present study, we have adopted a new strategy that evaluate the protease activity as influenced by heat shock wherein the total protein extracted from the heat induced *B. mori* larvae was used without protease inhibitors either in the gel or reaction buffer.

Total proteins of the 3rd day fifth instar heat shock given and control larvae were subjected to zymography analysis devoid of protease inhibitors. Conditions maintained in zymography for heat shocked and control samples were identical to make comparative analysis more precise and accurate in different protein substrates. Interestingly, of the five active proteases (P1 to P5) extracted from the larvae without heat shock, a complete inhibition of P3, P4 and P5 and no inhibition of P1 and of P2 were observed in the protein sample derived from heat shock induced larvae (Figure 3A and Table 3). In casein and BSA embedded gels, the activity of P5 was found normal in control larvae, while a partial inhibition in its activity observed in the protein samples of heat shock induced larvae (Figures 3B, C and Table 3). Surprisingly, high molecular weight proteases P1 (116 kDa) and P2 (83 kDa) were not affected due to heat shock. During this experiment, care was taken not to boil the protein samples extracted from the heat shocked and control silkworm larvae before loading them into the gel for electrophoresis in order to substantiate the effect of heat shock on differential inhibition of protease activity. Correspondingly, the protease assay performed for the protein samples derived from heat shock induced larvae

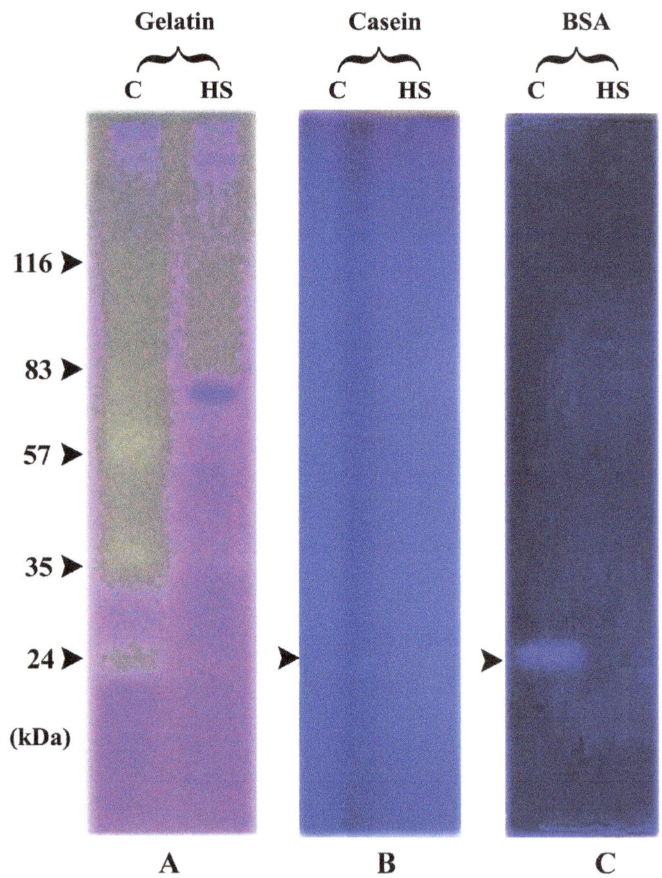

Figure 3. Effect of heat shock on protease activity in fifth instar larvae of *Bombyx mori* in presence of different substrates, gelatin (A), casein (B) and BSA (C). C - Protein sample from non-heat shock larvae; HS – Protein sample of silkworm larvae subjected to heat shock at 40°C for 2 h followed by 2 h recovery. Here, protein sample devoid of protease inhibitor was not boiled prior to electrophoresis.

Table 3. Altered protease activity on different substrates due to heat shock in the larvae of *Bombyx mori*.

Proteolytic bands	Mol. Wt. (kDa)	Gelatin		Casein		BSA	
		C	HS	C	HS	C	HS
P1	116	+	+	-	-	-	-
P2	83	+	+	-	-	-	-
P3	57	+	-	-	-	-	-
P4	35	+	-	-	-	-	-
P5	24	+	-	+	±	+	±

(C- control; HS- heat shock; '+' strong activity; '±' weak activity; '-' no activity).

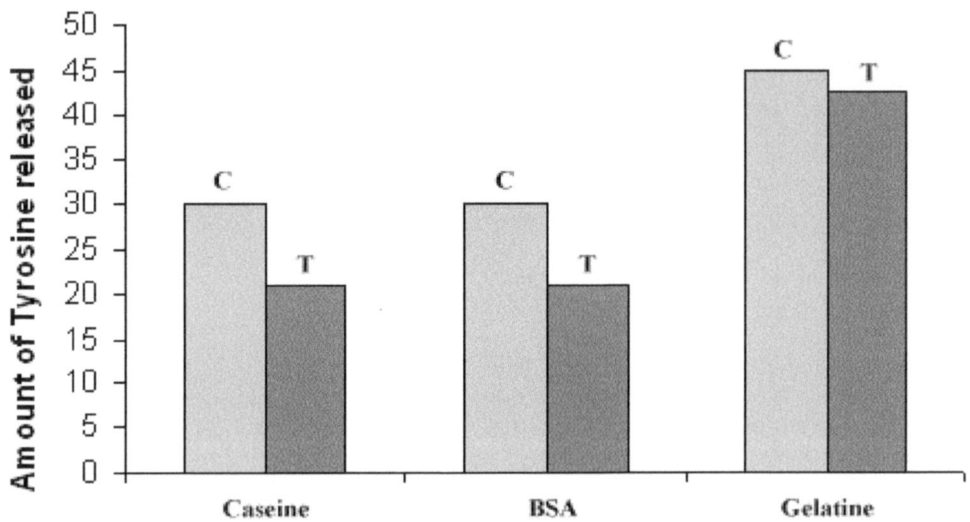

Figure 4. Altered protease activity (µg/ml) in the presence of different substrates (gelatin, casein and BSA) due to heat shock in *Bombyx mori*. Protein sample derived from non-heat shock larvae (C) and larvae subjected to heat shock (T) at 40°C for 2 h followed by 2 h recovery.

revealed decreased protease activity compared to that of control protein sample in all the substrates tested (Figure 4), which is in conformity with that of zymography analysis. Interestingly, the total arginine and lysine-specific protease activities of whole cultures and cell-free supernatant of *Porphyromonas gingival* decreased as the growth temperature increased. Thus, *P. gingivalis* responds to an increased temperature by adopting a less inflammatory and aggressive phenotype while retaining its population (Rimondia et al., 1999). With these inferences, we suppose that a reduction in most active protease P3 and others, no alteration in P1 and P2 activity due to heat shock might have significant role in acclimation to elevated temperature that silkworm *B. mori* experience during rearing in the field conditions. Although, its interaction with the expression of different sets of HSPs in *B. mori* is enigmatic (Aparna et al., 2010; Vasudha et al., 2006), but protein complexes consists of HSPs and other proteins are one of the most common products derived from protein interactions. Evidently, while in *Thermoanaerobacter brockii* cells exposed to temperature increases from 60 to 65°C, three enzymes (malate dehydrogenase, isocitrate dehydrogenase and alcohol dehydrogenase) were protected from aggregation through interaction HSP60 and HSP10, which were active in protein folding assays (Truscott et al., 1994). Further, several proteins including sHSP detected in most temperature-dependent complexes in *T. tengcongensis* could strengthen the interaction of the bacterial proteins to prevent protein aggregation and degradation (Meng et al., 2009). Thus, it is worth noting that formation of protein complexes at high temperature is not only for avoiding aggregation but also for maintaining protein activities and native structures as have been observed in the present investigation.

Conclusion

Application of zymography facilitated detection of four

active serine and one unknown proteases in fifth instar larvae, which is the most active feeding and silk protein synthesis stage of the *B. mori*. The differential activity of the proteases both in normal and heat shock induced larvae revealed the possible modification due to imposed thermal stress in the silkworm larvae. Although, detailed studies are required to correlate its role in thermotolerance and adaptation to heat in the domesticated silkworm, *B. mori* also has other naturally growing insects in the context of global warming that affect insect population dynamics.

ACKNOWLEDGEMENT

The authors are grateful to the University Grant Commission, New Delhi, India, for the research grant awarded in supporting this study.

REFERENCES

Aparna HS, Rajesh RK, Manjunatha HB (2010). Molecular characterisation of heat shock proteins 90 (HSP83) and 70 in tropical strains of *Bombyx mori*. Proteomics, 10: 2734-2745.

Arrigo AP (1998). Small heat shock proteins: Chaperones that act as regulators of intracellular redox state and programmed cell death. J. Biol. Chem., 379: 19-26.

Arrigo AP (2005). In search of the molecular mechanism by which small stress proteins counteract apoptosis during cellular differentiation. J. Cell Biochem., 94: 241-246.

Cupp-Enyard C (2008). Sigma's Non-specific Protease Activity Assay - Casein as a Substrate. J. Visual. Exp., 17: 899.

Evgen'ev MB, Garbuz DG, Zatsepina OG (2005). Heat shock proteins: Functions and role in adaptation to hyperthermia. Rus. J. Dev. Biol., 36: 218-224.

Goldsmith MR, Shimada T, Abe H (2004). The genetics and genomics of the silkworm, *Bombyx mori*. Ann. Rev. Entomol., 50: 71-100.

Gorman MJ, Paskewitz SM (2001). Serine proteases as mediators of mosquito immune responses. Insect Biol. Mol. Biol., 31: 257-262.

Heussen C, Dowdle EB (1980). Electrophoretic analysis of plasminogen activators in polyacrylamide gels containing sodium dodecyl sulfate and copolymerized substrates. Ann. Biochem., 102: 196.

Ikeda M, Yaginuma T, Kobayashi M, Yamashita O (1990). cDNA cloning, sequencing and temporal expression of the protease responsible for vitellin degradation in the silkworm, *Bombyx mori*. Agric. Biol. Chem., 54: 131-137.

Jolly MS (1987). Appropriate sericulture techniques, pp. 63-106.

Joy O, Gopinathan KP (1995). Heat shock response in mulberry silkworm races with different thermotolerance. J. Biosci., 20: 499-513.

Kaji K, Tomino S, Asano T (2009). A serine protease in the midgut of the silkworm, *Bombyx mori*: protein sequencing, identification of cDNA, demonstration of its synthesis as zymogen form and activation during midgut remodeling. Insect Biochem. Mol. Biol., 39: 207-217.

Kajiwara H, Ito Y, Imamaki A, Nakamura M, Mita K, Ishizaka M (2005). Protein profile of silkworm midgut of fifth-instar day-3 larvae. J. Electrophoresis, 49: 61-69.

Katsumi Y, Kihara H, Ochiai M, Ashida M (1995). A serine protease zymogen in insect plasma purification and activation by microbial cell wall components. Eur. J. Biochem., 228: 870-877.

Kim SH, Choi NS (1998). Electrophoretic analysis of protease inhibitors in fibrin zymography. Ann. Biochem., 270: 179-181.

Kregel KC (2002). Heat shock proteins: Modifying factors in physiological stress responses and acquired thermotolerance. J. Appl. Physiol., 92: 2177-2186.

Kundapur RR, Aparna HS, Manjunatha HB (2009). Comparative analysis of silk gland proteins of both heat shocked and normal silkworm larvae of NB4D2 strain by 2-DE. Int. J. Appl. Agri. Res., 4: 125-130.

Landry I, Chretien P, Lambert H, Hickey E, Weber LA (1989). Heat shock resistance conferred by expression of the human hsp27 gene in rodent cells. J. Cell Biol., 109: 7-15.

Landry J, Huot J (1995). Modulation of actin dynamics during stress and physiological stimulation by a signaling pathway involving p. 38 MAP kinase and heat shock ptotein27. Biochem. Cell Biol., 73: 703-707.

Lohmann CMF, Riddiford LM (1992). The heat shock response and heat sensitivity of *Bombyx mori*. Sericologia, 32: 533-537.

Lokeshwar BL, Selzer MG, Block NL, Gunja-Smith Z (1993). Secretion of matrix metalloproteinases and their inhibitors (tissue inhibitors of metalloproteinases) by human prostate in explant cultures: Reduced tissue inhibitors of metalloproteinase secretion by malign ant tissues. Cancer Res., 53: 4493-4498.

Lowry OH, Rosebrough NJ, Far, AL, Randall RJ (1951). Protein measurement with the Folin phenol reagent. J. Biol. Chem., 193: 265-275.

Maki N, Yamashita O (2000). The 30kP protease A responsible for 30-kDa yolk protein degradation of the silkworm, *Bombyx mori*: cDNA structure, developmental change and regulation by feeding. Dev. Genes Evol., 210: 545-558.

Manjunatha HB, Rajesh RK, Aparna HS (2010). Silkworm thermal biology: A review on heat shock response, heat shock proteins and heat acclimation in the silkworm, *Bombyx mori*. J. Insect Sci., 10: 204.

Manjunatha HB, Zamood A, Vasudha BC, Aparna HS (2005). Heat shock response and analysis of egg proteins in new bivoltine strains of *Bombyx mori*. Sericologia, 45: 403-408.

Mehlen P, Mehlen A, Godet J, Arrigo AP (1997). hsp27 as a switch between differentiation and apoptosis in murine embryonic stem cells. J. Biol. Chem., 72: 31657-31665.

Meng BO, Qian Z, Wei F, Weiwei W, Chuanqi Z, Zhuowei W, Quanhui W, Wei T, Qian W, Yanhe M, Ningzhi X,Siqi L (2009). Proteomic analysis on the temperature-dependent complexes in *Thermoanaerobacter tengcongensis*. Proteomics, 9: 3189-3200.

Naggie YC, Hu TR, Holomoan P, Bentley WE (1997). Substrate (gelatin) gel electrophoretic method for analysis of protease activity in insect (Sf-9) cells. Biotech. Tech., 11: 297-300.

Nakazawa H, Tsuneishi E, Ponnuvel KM, Furukawa S, Asaoka A, Tanaka H, Ishibashi J, Yamakawa M (2004). Antiviral activity of a serine protease from the digestive juice of *Bombyx mori* larvae against nucleopolyhydrovirus. Virology, 321: 154-162.

Neurath H, Walsh KA (1976). Role of proteolytic enzymes in biological regulation (A review). Proc. Nat. Aca. Sci., 73: 3825-3832.

Rimondia S, Percival PD, Marsh DA, Devine MR, Joseph AO, Philip S, Michael AC (1999). Effect of temperature on growth, hemagglutination and protease activity of *Porphyromonas gingivalis*. Inf. Immunol., 67: 1917-1921.

Sorensen JG, Loeschcke V (2007). Studying stress responses in the post-genomic era: its ecological and evolutionary role. J. Biosci., 32: 447-456.

Tanriverdi AS, Menderes A, Oktay G (2009). Matrix metalloproteinase-2 and -9 activities in human keloids, hypertrophic and atrophic scars: A pilot study. Cell Bio. Fun., 27: 81-87.

Tiwari PK, Joshi A, Mohan DRK (1997). Thermotolerance and heat shock response in *Musca domestica*. Curr. Sci., 72: 501-506.

Tiwari PK, Mohan DRK, Joshi A (1995). Developmental study of thermotolerance and the heat shock response in *Lucilia cuprina* (Weidemann). J. Biosci., 20: 341-354.

Tomita M, Munetsuna M, Sato T, Asachi T, Hino R, Hayashi M, Shimizu K, Namiko N, Tamura T, Yoshizato K (2003). Transgenic silkworms produce recombinant human type III procollagen in cocoons. Nature Biotech., 21: 52-56.

Truscott KC, Hoj PB, Scopes RK (1994). Purification and characterization of chaperonin 60 and chaperonin 10 from the abaerobic thermophile *Thermoanaerobacter brockii*. Eur. J. Biochem., 222: 277-284.

Ulmasov KA, Shammakov S, Karaev K, Evgen'ev MB (1992). Heat shock proteins and thermoresistance in lizards. Proc. Nat. Aca. Sci. USA, 89: 1666-1670.

Vasudha BC, Aparna HS, Manjunatha HB (2006). Impact of heat shock on heat shock proteins expression, biological and commercial traits of *Bombyx mori.* Insect Sci., 13: 243-250.

Zhang P, Aso Y, Yamamoto K, Banno Y, Wang Y, Tsuchida K, Kawaguchi Y, Fujii H (2006). Proteome analysis of silk gland proteins from the silkworm, *Bombyx mori.* Proteomics, 6: 2586-2599.

Effect of *Gongronema latifolium* crude leaf extract on some cardiac enzymes of alloxan-induced diabetic rats

E. E. Edet[1], M. I. Akpanabiatu[1]* A. E. Eno[2], I. B. Umoh[1] and E. H. Itam[1]

[1]Biochemistry Department, Faculty of Basic Medical Sciences, College of Health Sciences, University of Uyo, P. M. B. 1017, Uyo - Nigeria.

[2]Physiology Department, Faculty of Basic Medical Sciences, College of Basic Medical Sciences, University of Calabar, P. M. B. 1115, Calabar - Nigeria.

The effect of *Gongronema latifolium* on serum cardiac enzymes in alloxan induced diabetic rat models and normal control rats were determined using graded doses of 80% ethanol extract of *G. latifolium* leaf. The serum activities of aspartate aminotransferase (AST), creatine kinase (CK), creatine kinase isoenzyme (CKMB) and lactate dehydrogenase (LD) were determined in diabetic Wistar albino rats; diabetic control and normal Wistar albino rats after 2 weeks administration of *G. latifolium* leaf extract. Serum CKMB and LD activities of all the experimental groups significantly ($p \leq 0.001$) increased in diabetic rats when compared to non-diabetic rats. CK and LD decreased significantly in diabetic and non-diabetic rats treated with *G. latifolium* leaf extract when compared with the control. Equally, CKMB and LD of diabetic rats treated with *G. latifolium* leaf extract decreased significantly except CMB of group 4 and LD of group 3 animals. There were significant ($p \leq 0.001$) decreased activity of serum AST in diabetic rats when compared to the non-diabetic rats. However, AST did not change significantly in non-diabetic rats while in diabetic rats AST increased significantly ($p \leq 0.001$) when compared with their respective controls. These data suggest that the effects of *G. latifolium* leaf extract at our concentrations are not dose dependent and hepatotoxic. The extract of this plant is likely to be of biological significance in cardiovascular complication of diabetic and non-diabetic users.

Key words: Alloxan-induced diabetes, *Gongronema latifolium,* Cardiac enzymes, Myocardial infarction.

INTRODUCTION

Gongronema latifolium Benth Hook, (Asclepiadaceae) is an herbaceous shrub, with yellow flowers and the stem that yields characteristic milky exudates when cut. It is commonly grown in gardens in Calabar, Cross River State, Nigeria. It is locally called "utasi" by the Efiks, Ibibios and Quas; "utazi" by the Igbos and "arokeke" by the Yorubas in Nigeria. The Efiks and Quas in Calabar use *G. latifolium* crude leaf extract in the treatment of malaria, diabetes, hypertension, and as laxative. Also it is used as a spice and vegetable (Morebise, 2002). The use of crude leaf extract of this shrub in maintaining healthy blood glucose levels have been reported (Okafor, 1981, 1987, 1996). Scientific studies have established the hypoglycaemic, hypolipidaemic and antioxidative effects of aqueous and ethanol extracts of *G. latifolium* leaf (Ugochukwu et al., 2003; Ogundipe et al., 2003). Morebise et al. (2002) showed that the leaf extract has anti-inflammatory properties while its potential nutritional and food processing and Eleyinmi (2007) investigated preservation values of food.

Some phytochemicals such as B-sistosterol, lupenyl esters, pregnane ester, glycosides, essential oils and saponins are associated with parts of this herb (Ekundayo, 1980; Schneider et al., 1993; Morebise et al., 1998; Morebise et al., 2002). It is plausible that one or more of these phytochemicals that are found in *G. latifolium* is likely to influence cellular proteins with enzymatic activity. Some plants extract have influenced markers of myocardial infarction and clinical diagnosis of myocardial infarction is most often establish by the measurement of marker enzymes (Adam et al., 1993; Chatterjea and Shinde, 2002). Aspartate aminotransferase

*Corresponding author. E-mail: akpanabiatu@yahoo.com.

(AST), creatine phosphokinase (CK), creatine kinase isoenezyme (CKMB) and lactate dehydrogenase (LD) are often use as markers of myocardial infarction These enzymes are tightly bound to the contractile apparatus of the cardiac muscle tissue and any serious insult to the heart muscle will evoke the release of these enzymes into the serum.

A serum creatine kinase and creatine kinase isoenzyme activities have been reported in hypertension (Lott et al., 1978; Daniel et al., 1983). The present study was therefore undertaken to determine the extent to which *G. latifolium* leaf extract would influence some cardiac enzymes such as aspartate aminotransferase (AST), creatine phosphokinase (CK), creatine kinase isoenzyme (CKMB) and lactate dehydrogenase (LD) in diabetic and non-diabetic rats. Our findings may provide some useful information to support the age-long practice of using *G. latifolium* as an anti-diabetic and anti-hypertensive herb.

MATERIALS AND METHODS

Chemicals and reagents

All reagents and chemicals used in this work were of analytical grade. Alloxan monohydrate (sodium salt) was purchased from Alpha Aesar (Johnson Matthey Shore Road, Heysham, Lance, Canada). The quantitative *in-vitro* determination of serum creatine kinase isoenzyme and creatine kinase activities were carried using TECO diagnostics kits (Szasz, 1976; Gerhardt et al., 1977). Serum aspartate aminotransterase (AST) and lactate dehydrogenase (LD) activities were determined by the use Randox kit reagents (Randox Laboratories, UK). The absorbances were read using Optima Spectrophotometer SP-300 (Optima Inc. Chicago, USA).

Preparation of plant extract

Fresh leaves of *G. latifolium* were collected in Akpabuyo in Cross River State, Nigeria identified and authenticated by a botanist in the Botany Department, University of Calabar, Nigeria. The leaves were picked, sun-dried initially and finally dried in a PLUS II oven at 50°C. The dry sample was crushed to powder with a KENWOOD blender. The leaf powder was stored in a glass bottle with a plastic screw cap and kept in a refrigerator (4°C). Bulk ethanol extraction of the leaf powder was undertaken by shaking for 12 h. The extract was thereafter concentrated at 55°C using a rotary evaporator (Labo rota 3000 Resona) and a vacuum pump (Edward's). The extract was dried in an oven at 55°C until all the ethanol was removed. The dry extract was stored in a refrigerator until required for use. The concentration of the extract was determined by drying a known volume and measuring the dry weight.

Experimental animals/animal treatment

Male albino rats of the Wistar strain weighing 180 - 330 gm obtained from the animal house of Biochemistry department, Faculty of Basic Medical Sciences, University of Calabar, were used in this investigation. The experimental animals were acclimatized for three days in the experimental section of the animal house and there after were fasted for 24 h (Esmerino et al., 1998) before induction of diabetic by intraperitoneal injecting of a single dose of alloxan (150 mg/kg body weight). Diabetes was confirmed after 7 days in rats that showed fasting blood glucose (FBG) levels of ≥ 300 mg/dl accompanied with loss of body weight, polyuria and polyphagia. The mortality ranged from 5 - 10%. The 56 male albino Wistar rats were randomly selected into 8 groups consisting of diabetic (groups 5 - 8) and 4 non-diabetics (groups 1 - 4) of 7 animals each. The animals in groups I and 5 were designated normal and diabetic controls and received (1.5 ml/body weight of distilled water) as placebo for 2 wks., while the test groups 2, 3 and 4 were gavaged graded doses of 1, 1.5 and 2 ml of the extract, corresponding to 200, 300 and 400 mg/kg body weight of the *G. latifolium* extract. The same treatments were given to groups 6, 7 and 8 respectively. Experimental animals had access to feed and water *ad libitum*.

Animal sacrifice and preparation of sera

All the experimental animals were kept off extract and feed 24 h before they were weighed and sacrificed using chloroform fumes for anaesthesia. Blood samples were collected by cardiac puncture into sterile plain tubes for preparation of sera. Sera were collected from clotted blood samples by centrifugation using a bench top centrifuge (MSE, Minor, England) at 3000 g for 10 min. The separated sera were kept in a refrigerator (4°C) until required for analyses. All the enzymes were determined within 24 h of sample collection. This experiment was conducted in line with the Institution Ethical Committee as approved by the University of Calabar Graduate School.

Statistical analysis

Statistical analysis of data was by standard student's t-test method and $p < 0.05$ was regarded as significant. The group data was regarded as mean ± standard deviation (SD) of seven (7) determinations.

RESULTS

Table 1 indicates the activities of cardiac enzymes of non-diabetic and diabetic Wistar albino rats after 14 days oral administration of *G. latifolium* leaf extract in graded doses of 200, 300 and 400 mg/kg body weight. AST and CK activities were significantly lower when diabetic rats in groups 5, 6 and 8 were compared to corresponding non-diabetic animals in groups 1, 2, 4. However, the decrease in AST activity was significant ($p \leq 0.001$) in groups 6, 7 and 8 of diabetic rats when compared to that of non-diabetic rats in groups. In Table 1, the decreases in CK activity of the *G. latifolium* leaf extract treated groups: 6, 7 and 8 were significant ($p \leq 0.05$) when compared to the non-diabetic control. There was a significant increase ($p \leq 0.05$) in serum CK activity in test group 7 and a significant decrease ($p \leq 0.05$) in test group 8 when compared to the diabetic control. Serum CKMB activities were significantly ($p \leq 0.05$) higher in all diabetic rats than non-diabetic rats. Animals in test group recorded significantly ($p \leq 0.001$) lower CKMB activities while group 2 had significant ($p \leq 0.001$) higher CKMB activities, when compared with the normal control. Serum AST activity in all diabetic test groups increased (though not significantly, $p \geq 0.05$), when compared to the diabetic control. There was a non-dose dependent increase in serum CKMB activity in diabetic test groups and a non-dose dependent decrease in serum AST and CK activities fol-

Table 1. Serum activities of cardiac enzymes of diabetic and non-diabetic Wistar albino rats treated with 80% ethanol extract of *Gongronema latifolium.*

Groups/Enzyme†	Treatment	AST (U/L)	CK (U/L)	CKMB (U/L)	LD (U/L)
Group 1	Normal control	98 ± 7.60	474.19 ± 73.35	225.23 ± 55.29	1,459.80 ± 26.36
Group 2	200 mg/kgbd/wt	112.45 ± 4.84	387.98 ± 32.54 *	282.14 ± 42.30*	1164.14 ± 61.08 *
Group 3	300 mg/kgbd/wt	96.50 ± 7.01	211.68 ± 59.93 **	177.48 ± 53.73*	734.46 ± 93.21 **
Group 4	400 mg/kgbd/wt	97.07 ± 12.50	251.91 ± 28.73 **	226.67 ± 29.84	858.07 ± 37.42**
Group 5	Diabetic control	52.64 ± 15.48	340.43 ± 72.38	698.35 ± 66.93	1,603.59 ± 83.35
Group 6	200 mg/kgbd/wt	75.46 ± 20.95 *	333.37 ± 39.13	433.57 ± 32.65***	1371.51 ± 81.21*
Group 7	300 mg/kgbd/wt	77.49 ± 16.03*	495.34 ± 75.92 **	445.33 ± 28.45 ***	1581.95 ± 70.32
Group 8	400 mg/kgbd/wt	70.56 ± 20.00 *	214.38 ± 70.43 **	780.82 ± 42.78*	1459.08 ± 73.52 **

†Mean ± SD of 7 determinations. * = $p \leq 0.05$; ** = $p \leq 0.01$; *** = $p \leq 0.001$

lowing oral gavaging with graded doses of *G. latifolium* crude leaf extract for two weeks. In Table 1, serum LDH activity decreased significantly ($p \leq 0.001$) when compared to the non-diabetic control rats, while it decreased significantly in groups 6 and 8 but insignificantly different ($p \leq 0.05$) for test group 7. The significant change in LDH activities of diabetic rats when compared with the non-diabetic rats is an indication that cellular integrity of diabetic rats is more susceptible to drug metabolism.

DISCUSSION

Herbal medicinal researches have attained an incredible global level in the recent past as well as the continued application of some plant constituents in pharmaceutical industries thereby elevating the growing status of traditional herbal medicine in Africa in general and Nigeria in particular. In Calabar, Cross River State of Nigeria, the culinary and medicinal use of herbs is phenomenal as there is hardly any local meal without herbs. The use of *G. latifolium* in our local community as an anti-malarial, anti-inflammatory, laxative, spice and vegetable is common but its use in the treatment of diabetes and hypertension is one major reason for giving serious attention to research in prognosis of cardiovascular risk factors. The knowledge of toxicity or efficacy of xenobiotic(s) is function of the knowledge of drug metabolizing enzymes and marker enzyme activity. The local use of *G. latifolium* crude leaf extract in folk medicine as an anti-diabetic and anti-hypertensive remedy is one reason that has given the impetus to study its possible involvement in myocardial infarction. Experimental evidence (Kamble and Vaidya, 2002) has shown that antihypertensive drugs such as atenolol and enalapril elicit the release of cellular enzymes from injured organs such as the liver. We also believe that through a similar mechanism, enzymes may leak profusely from infarct myocardial cells.

It is well known that diagnosis of cardiac enzymes is important. Serum CK activity is a more sensitive indicator in early stage of myocardial ischemia, while peak rises in LD is roughly proportional to the extent of injury to the myocardial tissue (Chatterjea and Shinde, 2002). Also, the integrity of the cardiac apparatus in drug biotransformation and metabolism could be assessed by evaluating the levels of AST, CK, CKMB and LD in serum. The results in diabetic animals in this experiment shows a protective effect of *G. latifolium* crude leaf extract on the heart of experimental animals at a dose level of 200 mg/kg body weight. Moreover, the significantly lowered activities of CK, CKMB and LD at 200 mg/kg body weight and CK and LD at 400 mg/kg body weight scientifically suggest that the leaf extract of *G. latifolium* may have the potential of reducing the factors that produce infarction in the myocardium. This is so because the metabolism of alloxan-induced infarct myocardium may be studied by assessing the level of marker enzyme proteins in the serum. It is interesting to know that as myocardial diseases are rich sources of CKMB, so are skeletal muscular diseases good sources of creatine kinase isoenzyme (Hamm et al., 1992). Pathological value has been estimated in injured skeletal muscle. Therefore the significant reduction in CK enzyme at the dose of 400 mg/kg body weight of *G. latifolium* extract may be due to some physiological effects on muscular activity. This fact may be associated with the efficacy of *G. latifolium* crude leaf extract in the treatment of muscular pains, arthritis and inflammation (Morebise et al., 2002).

Conclusion

The findings of this investigation have revealed that the application of *G. latifolium* crude leaf extract in the treatment of hypertension may have significant effects in moderating incident of myocardial infarction.

REFERENCES

Chatterjea MN, Shinde R (2002). Serum enzymes in heart diseases. In Chatterjea and Shinde (ed) Textbook of Medical Biochemistry (5th ed.) Jappe brothers, New Delhi, Med. Publ. Ltd. pp 555-557.

Daniel LS, Anthonio V, Phyllis KR, Ronnakdo CM (1983). Cardiac isoenzyme methodology and the diagnosis of acute myocardial infarction. Am. J. Clin. Pathol. 80: 169 – 189.

Ekundayo O (1980). Constituents of *Gongronema latifolium* Benth Hook *(Asclepiadaceae).* Quart J. Crude Drug Res. 3: 127-129.
Eleyinmi AF (2007). Chemical composition and antibacterial activity of *Gongronema latifolium.* J. Zhejiang Univ. Sci. B. 8: 352 – 358.
Esmerino LA, Ranali R, Rodriguer LA Jr. (1998). Blood glucose determination of normal and alloxan diabetic rats after administration of local anaesthetics containing vasoconstrictors. Braz. Dent. J. 9(1): 33-37.
Gerhardt W, Szasz G, Gruber W (1977). Creatine kinase in serum: Further studies of adenylate kinase inhibitors. Clin. Chem. Acta 78 : 29.
Hamm CW, Ravkilde J, Gerhardt W, Jorgensen P, Penheim E, Ljungdahl L (1992). The prognostic value of serum troponin T in unstable angina. N. Engl. J. Med. 327: 146 – 150.
Kamble MM, Vaidya SM (2002). Effect of antihypertensive drugs on cardiac enzymes in hypertension with myocardial infarction in NIDDM. Indian J. Clin. Biochem. 17(2): 60 – 63.
Lott JA, Stang JM (1978). Serum enzymes and isoenzymes in the diagnosis of myocardial ischemia and necrosis. Clin. Chem. 24: 832 – 834.
Morebise O, Fafunso MA (1998). Antimicrobial and phytotoxic activities of saponin extracts from two Nigerian edible medicinal plants. Biokemistri 8(2): 69-77.
Morebise O, Fafunso MA, Makinde JM, Olayide OA, Awe E (2002). Anti-inflammatory Property of *Gongronema latifolium.* Phytother. Res. 16: 575 – 577.
Ogundipe OO, Moody JO, Akinyemi TO, Raman A (2003). Hypoglycaemic potentials of methanolic extracts of selected plant foods in alloxanized mice. Plt. Foods Hum. Nutr. 58: 1 – 7.
Okafor JC (1981). Woody plants of nutritional importance in traditional farming system of Nigerian humid tropics. Ph.D Thesis, University of Ibadan, Ibadan, Nigeria.
Okafor JC (1987). Identification and conservation of plants used in traditional medicine. Lead lecture presented at the international workshop on evaluation of traditional medicine. University of Nigeria, Nsukka.
Okafor JC, Okolo HC, Ejiofor MAN, van der Maesen LJG, van der Burgt XM, van Medenbach de Rooy JM (1994). Strategies for enhancement of utilization potential of edible woody forest species of S. E. Nigeria. The biodiversity of African plants. L.J.G. van der, Maesen; X.M. van der, Burgt eds. Proceedings of the 14th AETFAT CONGRESS. Wageningen, Netherlands pp. 684-695.
Schneider C, Rotscheidt K, Breitmaier E (1993). Four new pregnane glycosides from *Gongronema latifolium* (Asclepiadaceae). J. Prak. Chemie 335(6): 532 – 536.
Szasz G, Gruber W, Bernt E (1976). Creatine kinase in serum. Clin. Chem. 22: 650 – 656.
Ugochukwu NH, Babady NE, Cobourne MK, Gasset SR (2003). The effect of *Gongronema latifolium* extract on serum lipid profile and oxidative stress in hepatocytes of diabetic rats. J. Biosci. 28(1): 1 – 5.

Angiotensin-converting enzyme genotypes relationship with blood pressure, C-reactive protein and selected physical tests in Zulu South African cricketers

Trayana Djarova[1]*, Abigail Ramakoaba[1], Albert Basson[1], Sam Mugandani[2], Jeanne Grace[2], Jacque Cloete[3] and Gregory Watson[4]

[1]Department of Biochemistry and Microbiology, University of Zululand, KwaDlangezwa 3886, South Africa.
[2]Department of Biokinetics and Sport Science, University of Zululand, KwaDlangezwa 3886, South Africa.
[3]Department of Mathematical Sciences, University of Zululand, KwaDlangezwa 3886, South Africa.
[4]School of Biochemistry, Genetics and Microbiology, University of KwaZulu Natal, Scottsville 3209, South Africa.

Zulu cricketers (n=14) and students (n=17) as controls were genotyped (blood spots) for angiotensin converting enzyme (ACE), gene by PCR amplification followed by agarose gel electrophoresis. Systolic and diastolic blood pressure (SBP and DBP) and grip strength (kg), knee extension and flexion (Nm/kg) were measured, systolic tension time (STT) and metabolic rates (MR) were calculated. After ANOVA, the association between these parameters and I/D gene polymorphism was probed using Chi^2 maximum likelihood test and Fisher's exact test. ACE genotyping for the whole group displayed a complete absence of II genotype, 67.7% DD and 32.3% ID genotypes. The frequency of D allele was 83.8% and I allele 16.2%. In cricketers DD and ID genotypes were 50% each compared to controls-83% DD and 17% ID. No differences in grip strength and quadriceps/hamstring muscle strength between the groups were observed, but for the whole cohort 86% D allele frequency was associated with higher (greater than 43.3 kg) grip strength (p<0.037). In cricketers CRP (less than 3.0 mmol/l) was associated with 79% D allele frequency. SBP and DBP were significantly lower by 3.2 and 4.25 mmHg, whereas increased values of STT by 5.5%, and MR by 10.3% were found. Although, the number of participants in this study is small, it is concluded that in cricketers no over presentation of DD or ID genotypes was observed indicating a more balanced display of power and endurance required for the game.

Key words: Angiotensin converting enzyme (ACE) genotype, polymorphism, blood pressure, body mass index (BMI), lean body mass (LBM), fat mass, hand grip, quadriceps and hamstring muscle strength, Zulu cricketers.

INTRODUCTION

Researches have shown that genes play an important role in athletes' performance in various sports (Woods et al., 2002; Wolfarth et al., 2000; Williams and Folland, 2008; Bray et al., 2008; Ruiz et al., 2010). To date improvements in athletes' performance have been based mainly on the manipulation of physiological, physical, nutritional and psychological factors which have been referred to as nurture or environmental constrains (MacArthur and North, 2004; Davids and Baker, 2007). The general view is that there is an interactive influence of genetic and environmental factors on human physical performance. Therefore numerous studies have focused on the genetic make-up of the athletes and associations with fitness phenotypes, physical/physiological tests and molecular basis of adaptation to training (Rankinen et al., 2000; Calo and Vona, 2008; Lucia et al., 2010).

Angiotensin converting enzyme (ACE) polymorphism is the most investigated genetic variation linked to athletic status, physical performance, cardiovascular and muscle function, and trainability in athletes (Alvarez et al., 2000; Scanavini et al., 2002; Collins et al., 2004; Amir et al., 2007; Wang et al., 2008; Min et al., 2009; Ruiz et al.,

*Corresponding author. E-mail: tdjarova@pan.uzulu.ac.za or drdjarova@yahoo.co.uk.

2010).

The ACE gene is located on the long arm of chromosome 17q23.3. It is made up of 26 exons and 25 introns, stretching over 21 kb. The_polymorphism analyzed here consists of the presence (insertion, I allele) or absence (deletion, D allele) of a 287 bp Alu repeat sequence, resulting in three genotypes (homozygote DD and II and heterozygote ID) (Rigat et al., 1990; Wang et al., 2008; Wagner et al., 2006; Sipahi et al., 2006). Observations have been reported that individuals carrying ACE I allele are overrepresented in elite endurance athletes (Alvarez et al., 2000; Nazarov et al., 2001; Scanavini et al., 2002) and have been associated with higher percentage of slow twitch type I muscle fibres (Zhang et al., 2003), and have shown an enhanced response to training (Wang et al. 2008). However while the research results are not widely consistent (Scot et al., 2005; Amir et al., 2007) the general findings support the hypothesis that the I allele is associated with better performance in endurance events, while the D allele is associated with success in power events (Folland et al., 2000; Pescatello et al., 2006; Lucia et al., 2010)

The ACE gene is implicated in the rennin angiotensin aldosterone system (RAS). The local RAS is active in a variety of tissues, including lung, kidney, heart, vascular smooth muscle cells and skeletal muscle (Pieruzzi et al., 1995; Jones and Woods, 2003). Angiotensin I-converting enzyme is a key enzyme in the generation of angiotensin (AT)-II from vasoinactive angiotensin (AT)-I. Angiotensin II promotes peripheral vasoconstriction. It also stimulates aldosterone and vasopressin secretion which leads to increased blood volume and increased arterial blood pressure (Sipahi et al., 2006; Calo and Vona, 2008). Individuals who are homozygous for the D allele have been shown to have higher angiotensin-converting enzyme activity in serum and tissue than in those with the I allele (Rigat et al., 1990) but it is not known by what mechanism lower circulating levels of the enzyme could improve performance. ACE protein is also responsible for the degradation of bradykinin, respiratory drive and tissue oxygenation (Ostrander et al., 2009).

Studies involving the association of ACE gene or other muscle and metabolism related genes (example ACTN3, AMPD1, ADRB2, TNF) with blood biochemical parameters and markers of inflammation and exercise-induced oxidative stress have been rare and not studied enough (Lakka et al., 2006; Payne et al., 2007; Bray et al., 2008; Wang et al., 2008; Andonov et al., 2008; Tsianos et al., 2009; Milander et al., 2009). The association of ACTN3 and TNF gene polymorphism with C-reactive protein, uric acid and lactate in cricketers was reported (Djarova et al., 2011). Considering the role of ACE protein in the regulation of some inflammatory reactions and skeletal muscle efficiency (Woods et al., 2000), it is important to explore further the possible association of ACE polymorphism with the above mentioned markers in the same group of cricketers.

Modern cricketers are now exposed to greater physical and physiological demands. Heart rate could reach 190 beats/min and the predominant contribution from the oxygen-independent glycolysis to lactate can contribute to 60% of the total energy in multiple activities of short duration of less than 40 s which are typical for the cricket game (Noakes and Durandt, 2000).

Martens (2004) has identified the following as the estimated energy and muscular/cardiovascular fitness demands of cricket: low to moderate aerobic capacity, moderate anaerobic capacity, moderate strength and flexibility, low to moderate endurance and moderate to high speed. Fast bowling has been linked with a mesomorphic somatotype, greater percentage of type II muscle fibres and a superior phosphagenic and glycolytic metabolic pathways together with eccentric muscle strength. Speed of the ball at release was seen to determine success in bowling (Stuelcken et al., 2007). Hand grip strength was also adjudicated to be an acceptable indicator for good performance in cricket (Koley and Yadar, 2009). A shorter stature and isokinetic knee and shoulder strength were seen to be contributory to the success of batsman (Noakes and Durant, 2000; Nunes and Coetzee, 2007).

The aim of the study is to explore ACE I/D polymorphism, blood pressure and association with C-reactive protein and selected physical tests in Zulu South African cricketers.

MATERIALS AND METHODS

Experimental subjects

The participants of this study were 31 Zulu South African males (14 cricketers age 22.85±0.65 from the University of Zululand cricket team and 17 students age 22.64±0.66 as controls). All experimental subjects were volunteers and a written consent was obtained prior to the study. Experimental protocols were approved by the Ethic Committee of the Research Board of University of Zululand. The participants of the control group reported leisure physical activities once or twice weekly. Cricket players participated in regular 2 h training sessions 5-6 times weekly and played club matches in the Uthumgulu District, KZN over the weekend and inter-universities matches during the season.

Measurements of body mass index (BMI), fat percentage (Fat %), lean body mass (LBM) and fat mass (FM) were taken according to the procedures of the American College of Sports Medicine (Thompson et al., 2000). The evaluation testing procedure as suggested by Ashton and Myers, (2004) was used for the measurement of grip strength. The IsoKnee α was applied to determine the relative strength of the quadriceps (knee extension) and hamstring (knee flexion) muscles as suggested by Coetsee (1995). The speed of rotation was set at 60° for the measurement of the peak muscular strength. A warm-up routine of two to three sets of 6 (six) repetitions interspaced by 30 s rest followed by 3 (three) maximum. The subject was allowed to recover for a few minutes. Data was recorded with the subject performing maximal knee extension and knee flexion for the duration of 10 s.

The estimates of daily energy requirements and metabolic rate (MR) were done using Cunningham equation (Thompson and Manore, 1996). The equation estimates metabolic rate at rest which

Table 1. ACE genotype and allele frequency (%) in cricket players and controls.

Group	Genotype frequency in % and number in brackets)					Allele frequency in %	
	DD		ID		II	D	I
Cricket players (n=14)	50.0	(7)	50.0	(7)	Null	75.0 (10.5)	25.0 (8.5) [a]
Controls (n=17)	82.4	(14)	16.7	(3)	Null	91.2 (15.5)	8.8 (1.5) [a]
Total (31)	67.7	(21)	32.3	(10)	Null	83.8 (26)	16.2 (5.0)[a]

[a]p= 0.004 Fisher's test - two tailed based on %.

is multiplied by an activity factor (within range 1.2 to 1.9) to establish mean daily energy requirements. This estimation has been shown to be the best energy requirement prediction equation for metabolic rate in athletic population (Watson et al., 2005).

All participants were advised not to change their dietary habits and to refrain from physical exercise 24 h before blood sampling. Blood samples were collected at rest from the antecubital vein into vacutainers and analysed in the accredited Lancet laboratory at Bay Hospital, Richards Bay according to the South African standards of good laboratory practice. The Dimension Xpanda (Siemens, Germany) equipment was used for the determination of C-reactive protein (CRP, range 0-8 mg/L), uric acid (UA, range 0.26-0.45 mmol/L and lactate (LA, range 0.63-2.4 mmol/L).

Genotyping

Blood spots were collected on FTA® Classic cards according to the manufacturer's instructions (Whatman International, UK). Samples were prepared by punching 1.2 discs from the cards and washing with FTA® purification reagent and TE (10 mM Tris-HCl, 0.1 mM EDTA, pH 8.0) according to the manufacturer's instructions. PCR was then performed directly from the dried disc. The detection of the insertion (I) and deletion (D) alleles of the *ACE* gene was performed by a modified method of Alvarez and Coto (1998). The primer sequences were ACE F (forward): 5´-CTGGAGACCACTCCCATCCTTTCT -3´ and ACE R (reverse): 5´-GATGTGGCCATCACATTCGTCAGAT -3´. PCR reactions were performed using the SensiMix™ dT kit according to the manufacturer's instructions (Quantace, UK). The final reaction mixtures contained 1× SensiMix (with a final Mg^{2+} concentration of 3 mM) and 200 nM of each of each primer. 20 µl of the PCR mix was added to a single dried disc in a thin-walled 200 µl PCR tube. All amplifications were performed in a Rotor-Gene 6000 (Corbett Research, Australia) using the following conditions: activation step 95℃ for 10 min. followed by 40 cycles of 95℃ for 10 s, 60℃ for 20 s 72℃ for 20 s. 1 µl 6× loading buffer was added to 5 µl of each PCR reaction which was then loaded and analysed in a 2% (w/v) agarose 1× TBE gel. All genotypes were determined in duplicate.

Statistical analysis

The Student's *t*-test was used to analyze the statistical difference in the blood biomarkers and physical characteristics between the cricket players and the control group. The results are presented as mean ± SEM. Statistical significance was accepted at p<0.05. Statistical analysis for the genotype associations was done using GenStat Discovery Edition 3. The distribution of some variables was skewed; hence these variables were transformed for the Analysis of Variance (Unbalanced design). For the association tests, CRP levels were categorized as less than 3 mg/L (low) and less than 3 mg/l (high) according to Pearson et al. (2003). Other variables were categorized according to their median (M). After ANOVA the association was examined using Chi^2 maximum likelihood test and Fisher's exact test.

RESULTS

ACE genotyping and allele frequencies are shown in Table 1. ACE genotyping showed a complete absence of II genotype (Figure 1). For the whole group 67.7% DD and 32.3% ID genotypes were observed (Figure 2) In cricketers, DD and ID genotypes were 50% each compared to controls – 83% DD and 17% ID (Figure 2). The total frequency of 83.8% D allele for the cohort was significantly higher (p=0.004) compared to 16.2% I allele. It was also found that in cricketers 25% ACE I allele frequency was higher (p=0.004) than 8.8% in controls, and 75% D allele frequency was lower (p=0.004) compared to 91.2%, respectively (Figure 3 and Table 1).

The blood pressure results are shown in Table 2. In cricket players SBP was lower by 3.2 mmHg (p<0.05) and DBP by 4.25 mmHg (p<0.001), where the values of STT were increased by 5.5% (p<0.05) compared to controls. No differences in heart rate and pulse pressure and no associations between blood pressure and allele frequencies were noted.

Cricketers have shown higher (p<0.05) basic metabolic rate and increased values (p<0.001) of metabolic rate by 10.3% and energy requirements by 14% (Table 3).

C-reactive protein (Table 4) in controls is much higher (p<0.001) than in cricketers, but still within the reference range of 0-8 mg/L accepted by Lancet Laboratory, South Africa. The results in cricketers have shown that 79% D allele frequency was associated (p<0.001) with lower CRP levels (<3.0 mg/L). Uric acid (<0.30 mmol/L) was associated (p=0.001) with 43% D allele frequency.

BMI, LBM and FM were higher (p<0.001) in cricket players (Table 5). High D allele frequency (91-94%) were associated with BMI and FM in cricketers (p=0.001) and in controls (p=0.029), where LBM has shown an association with 71% D allele (p<0.041) and with 94% D allele (p<0.029) respectively.

No differences in grip strength and the strength of the quadriceps (knee extension) and hamstring (knee flexion) muscles between the groups were observed (Table 6).

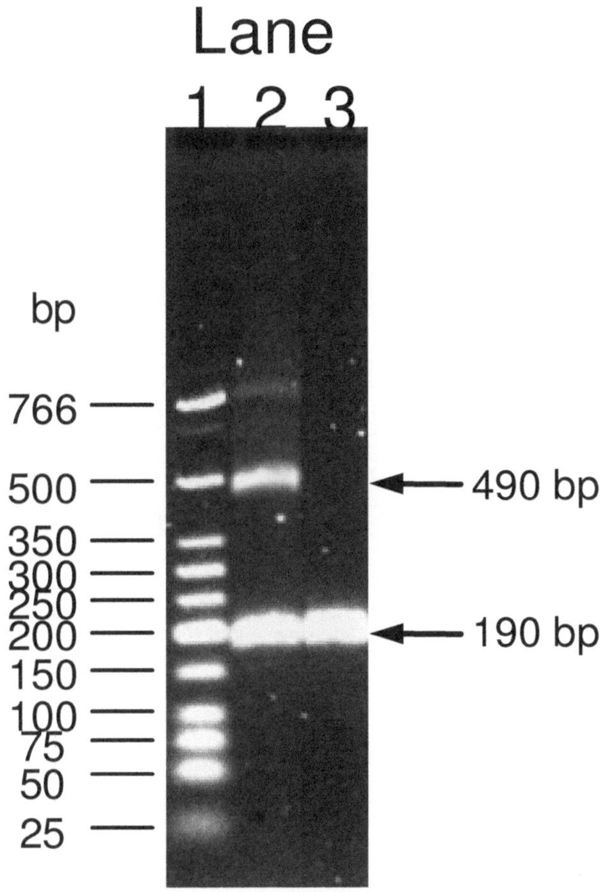

Figure 1. Analysis of ACE ID and DD genotypes. Amplified fragments were resolved in 2% (w/v) agarose, 1 ×TBE gels. Lane 1 - Low molecular weight DNA ladder (New England Biolabs); Lane 2 - ID genotype; Lane 3 - DD genotype. II genotype is absent.

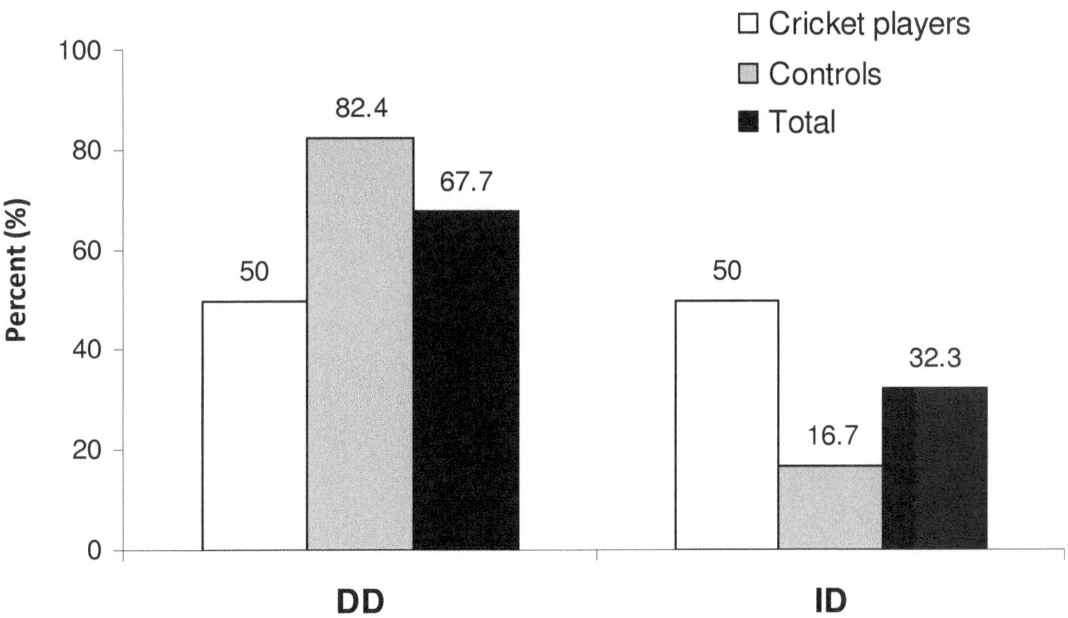

Figure 2. ACE DD and ID genotype frequencies (%) in cricketers, controls and the whole group.

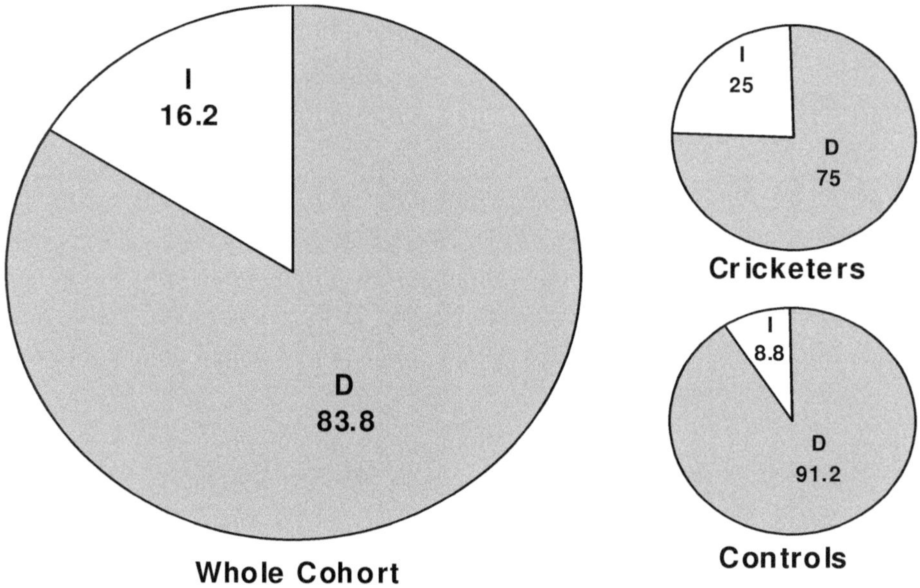

Figure 3. D and I allele frequencies (%) for the whole cohort, cricketers and controls.

Table 2. Systolic and diastolic blood pressure (mmHg) pulse pressure, heart rate (beats/min) and systolic tension time in cricket players and control group (mean ± SEM).

Parameters	Cricket players	Control group
Systolic blood pressure mmHg	120.81 ± 2.03	123.75 ± 1.07*
Diastolic blood pressure mmHg	76.88 ± 1.4	81.13 ± 1.31**
Pulse pressure mmHg	43.00 ± 1.29	42.00 ± 2.01
Heart rate beats/min	60.75 ± 0.85	59.80 ± 1.11
Systolic tension time (SBP x HR)	7654.27 ± 147.59	7253.60 ± 124.48*

*p < 0.05; *p > 0.0001.

Table 3. Basic metabolic rate, metabolic rate and energy requirements (kilojoules) in cricket players and control group (mean ± SEM).

Parameters	Cricket players	Control group
Basic metabolic rate	1502.40 ± 25.91	1480.31 ± 13.34*
Metabolic rate	2246.99 ± 102.24	2035.40 ± 18.34**
Energy requirements (kilojoules per food intake)	9737.41 ± 76.79	8520.18 ± 76.79**

*p < 0.05; *p > 0.0001.

Table 4. C-reactive protein (CRP), uric acid (UA) and lactate (LA) blood levels at rest in cricket players and control group (mean ± SEM) and ACE association tests.

Biomarkers	Cricket players	Control group
C-reactive protein (mg/L)	1.81 ± 0.37 [a,]	5.81 ± 0.51** NS
Uric acid (mmol/L)	0.31 ± 0.008 [b]	0.29 ± 0.007* NS
Lactate (mmol/L)	1.55 ± 0.08 NS	1.95 ±. 0.11** NS

Student's *t*-test: *p<0.05 control group vs. cricket players; **p<0.001 control group vs. cricket players.
Association tests: [a]p=0.001 CRP <3 mg/L in cricketers – ACE D allele frequency (79%), I allele (21%); [b]p=0.001 UA <0.30 mmol/L in cricketers – ACE D allele frequency (43%), I allele (57%); NS – no significant association tests.

Table 5. Physical characteristics of cricket players and control group (mean ± SEM) and ACE association tests.

Physical characteristics	Cricket players	Control group
Weight (kg)	68.68 ± 2.54	61.00 ± 1.61 **
Stature (cm)	175.08 ± 1.25	170.58 ± 0.33 **
Body mass index-BMI (kg/m^2)	22.40 ± 0.81 [a,]	20.79 ± 0.36 **, [C]
Lean body mass-LBM (kg)	61.81 ± 2.01 [b]	55.41 ± 1.49 **,[C]
Fat mass-FM (kg)	6.87 ± 0.54 [a,]	5.59 ± 0.22 **,[C]
Fat %	9.84 ± 0.39 NS	9.13 ± 0.32 * NS

Student's *t*-test: *p< 0.05 control group vs. cricket players; **p<0.001 control group vs. cricket players.
Association tests for BMI, LBM and FM: [a]p=0.001 BMI below 22.4 kg/m^2 and FM below 6.9 kg – ACE D allele frequency (91%), I allele frequency (9%) in cricketers; [b]p=0.041 LBM below 61.8 kg – ACE D allele frequency (71%), I allele frequency (29%) in cricketers; [c]p=0.029 BMI below 22.4 kg/m^2, LBM below 55.4 kg and FM below 5.6 kg – ACE D allele frequency (94%), I allele frequency (6%) in controls; NS – no significant association tests.

Table 6. Grip strength (kg), quadriceps strength (knee extension - Nm/kg) and hamstring strength (knee flexion - Nm/kg) of cricket players and control group (mean ± SEM) and ACE association tests.

Physical characteristics	Cricket players	Control group
Grip strength – L (kg)	44.00 ± 2.12 NS	42.57 ± 1.97 NS
Grip strength – R (kg)	45.79 ± 2.25 NS	45.11 ± 1.84 NS
Knee extension – L (N/kg)	3.73 ± 0.11 [a,]	3.71 ± 0.13 [e,]
Knee extension – R (N/kg)	3.63 ± 0.08 [b]	3.59 ± 0.12 [e]
Knee flexion –L (N/kg)	2.04 ± 0.08 [c]	2.08 ± 0.88 [f]
Knee flexion –R (N/kg)	2.00 ± 0.02 [d]	2.00 ± 0.08 [f]

L = left; R = right. Association tests: NS – no significant association tests for grip strength (L) and (R) per group of cricketers and controls. p=0.037 for the whole cohort grip strength (L) and (R) above 44.3 kg – ACE D allele frequency (86%) and I allele frequency (14%). [a]p=0.010 Knee extension (L) .above 3.7 Nm/kg ACE D allele frequency (86%) and I allele frequency (14%) in cricketers. [b]p=0.014 Knee extension (R) above 3.6 Nm/kg – ACE D allele frequency (79%); I allele frequency (21%) in cricketers. [c]p=0.014 Knee flexion (L) above 2.0 Nm/kg – ACE D allele frequency (78%) and I allele frequency (22%) in cricketers. [d]p=0.014 Knee flexion ® above 2.1 (Nm/kg) –ACE D allele frequency (83%) and I allele frequency (175) in cricketers. [e]p=0.001 Knee extension (L) above 3.7 Nm/kg and (R) above 3.6 N/kg – ACE D allele frequency (83%) and I allele frequency (17%) in controls.[f]p= 0.001 Knee flexion (L) and (R) above 2.0 Nm/kg – D allele frequency 87% and I allele frequency (13%) in controls

Knee extension L (>3.73 Nm/kg) and R (>3.63 Nm/kg) was associated with D allele frequency of 86% (p=0.010) and 79% (p=0.014). Knee flexion L (>2.04 Nm/kg) and R (>2.0 Nm/kg) was associated (p=0.014) with D allele frequency of 78% and 83%. For the whole cohort (Table 7) 86% D allele frequency was associated (p=0.037) with grip strength L (>43.3 kg) and R (R>45. 5 kg).

DISCUSSION

In our study a complete absence of ACE II genotype was established for the first time in Zulu South Africans. The genotype distribution for the whole cohort was skewed (67.7% DD and 32.3% ID). Low frequency of II genotype was reported in African Americans, Kenyans, Jamaicans (Scott et al., 2005), Nigerians (Woods, 2009) and Xhosa South Africans (Payne et al., 2007). The ACE distribution in Caucasian Europeans (Woods, 2009) was found to be in ratio 1:2:1 (e.g. 26% DD, 50%ID, and 24% II in British males).

Collins et al. (2004) tested a mixed group of South African-born athletes participating in Ironman triathlons and observed genotype frequencies of 24.3% DD, 54% ID and 21.6% II compared to 32.5% DD, 50.6% ID and 16.95% II in controls, pointing out that significantly higher 51.5 % ACE I allele frequency was found in the fastest South African finishers. The frequency of I allele was higher in Lithuanian elite athletes than in controls (Gineviciene et al., 2010). On the other hand, a large study of East African distance runners did not find any association between ACE genotypes and elite endurance athletic status (Scott et al., 2010). In the present study no over presentation of DD and ID genotypes was displayed in Zulu cricketers.

Our findings of high D allele frequency in Zulu South Africans are in line with the trend reported in Afro-Caribbean people (Berley et al., 1996), Nigerians (Woods, 2009) and elite Taekwondo athletes of Turkish and Azerbaijan origin (Gunay et al., 2010). The excess of D allele was represented more in athletes participating in power-oriented and short-distance/high intensity events (Myerson et al., 1999; Woods et al., 2000; Nazarov et al., 2001; Cerit et al., 2006; Charbonneau et al., 2008).

Previously, we used the same cohort to investigate the

Table 7. ACE D and I allele frequency (%) association with left (L) and right (R) grip strength in the whole cohort.

Physical test		Whole cohort		
		D (%)	I (%)	P
Grip	< M	92	8	-
Strength (L)	> M	81	19	-
Total	> M	86	14	0.037
Grip	< M	92	8	-
Strength (R)	> M	81	19	-
Total	> M	86	4	0.037

association of ACTN3 gene polymorphism and CRP, uric acid and lactate. Therefore in this study after performing ACE genotyping we conducted the association tests using the same results of the above mentioned biomarkers.

We found that the high frequency of power linked ACE D allele is also associated with lower CRP and uric acid levels in cricketers. It is important to mention that this is in concordance with our previous findings of strong association in the same cohort between these biomarkers and another power-related R allele of the ACTN3 gene (Djarova et al., 2011). C-reactive protein might be involved via cytokines in triggering metabolic signaling pathways in the exercising muscles that could be under genetic control by both genes.

It is important to emphasize that associations of the same trend as the above mentioned between high ACE D allele and high ACTN3 R allele frequencies, and BMI, LBM and FM were also found. ACE I/D polymorphism associations with BMI and body fat have been reported (Thompson et al., 2007) suggesting that it may affect adherence to exercise training.

In sports events like cricket requiring short power/sprint bursts, considering the fact that both ACE D and ACTN3 alleles have shown similarity in the association tests is a finding of interest that needs further studies. The other similarity found was the complete absence of ACE II genotype and ACTN3 XX genotype.

When comparing cricketers to control subjects we established no differences in heart rate, lower BP and higher SST, metabolic rate and energy requirements. The I/D polymorphism may play a role in enhanced performance but this is not mediated by differences in the heart rate/VO_2 relationship to training (Woods, 2009). Lower blood pressure and higher systolic tension time at rest indicate better cardiac efficiency in cricketers. Despite the interaction between ACE genotypes and serum ACE activity and the fact that D allele has been related to higher circulating/tissue ACE levels and enhanced performance, no associations between I and D alleles and BP have been reported (Bloem et al., 1996; Ostrander et al., 2009). Significantly higher estimated energy requirements were noted in cricketers. This corresponds to the findings of Noakes and Durandt (2000) that the energy demands of different cricket activities varied from 760 kJ/h in fielding to 1064 kJ/h in bowling and 1368 kJ/h in batting.

Strength, flexibility and speed parameters are among many factors contributing to the success in cricket (Nunes and Coetsee, 2007). The high D allele frequency association with grip strength that was established in the whole cohort in our study is in accordance with the findings that the D allele is related to the power/sprint output (Ruiz et al., 2010). Associations between the higher values of quadriceps/hamstring strength in cricketers and ACE D allele frequency has been observed for the first time. In batting and especially in fast bowling the trunk must flex, extend and rotate within a short period. The knee circumvents through flexion, rotation and extension. The bowling arm circumvents through extension, abduction, external rotation, thrusting flexion and internal rotation (Stuelcken et al., 2007). The average running sprint between the wickets was found to be 18.7 km/h which reflects high intensity work bouts (Christie and King, 2008).

The interpretation of association studies has always been controversial especially when the limitation is the small sample. Genetic studies need large population samples, but it is difficult to reconcile this premise with the scarce number of world-class champions or a given ethnicity and sport event (Ruiz et al., 2010). The analysis of a single sport discipline and association with I/D ACE gene polymorphism has been done in groups of athletes from 25 elite climbers, 37 swimmers and 27 up to 291 runners (Woods, 2009).

The unique demands of cricket may require specific physical characteristics and genetic traits play a substantial role. The perspective is to consider individual genetic endowment and develop training programmes that allow it to be optimized (Ostrander et al., 2009; Djarova et al., 2011). This study also might provide insight in talent identification and nurturing of young South African cricketers of various ethnicities.

Although, the number of participants in this study is small, it is concluded that ACE I/D genotyping has shown a complete absence of II genotype. Zulu cricket players display a balanced DD and ID genotypes distribution in conjugation with significantly higher D allele frequencies associated with physical tests and beneficial differences in blood pressure and systolic tension time compared to controls. This could be considered as a competitive advantage in the cricket training and performance.

REFERENCES

Alvarez R, Terrados N, Ortolano R, Iglesias-Cubero G, Reguero JR, Batalia A, Cortina R, Fernandez-Garcia B, Rodriguez C, Braga S, Alvarez V, Coto E (2000). Genetic variation in the rennin-angiotensin system and athletic performance. Eur. J. Appl. Physiol., 82: 117-120.

Alvarez V, Coto E (1998). Angiotensin-converting enzyme and angiotensin II receptor 1 polymorphisms: association with early coronary disease. Cardiovasc. Res., 40: 375-379.

Amir O, Amir R, Yamin R, Attias E, Enion N, Sagiv M, Meckel Y (2007). The ACE deletion allele associated with Israeli elite endurance athletes. Exp. Physiol., 92: 881-886.

Andonov S, Saraeva R, Andonova S, Kaneva R, Gigova V, Stefanov L, Kremenski I, Atanassov P (2008). Polymorphism of ACTN3, ACE and AMPD1 genes and physical performance in Bulgarian sub-elite athletes. European Human Genetics Conference, Barcelona, 31 May–3 June, Supplement 2. Eur. J. Hum. Genet., pp. 288-289.

Ashton LA, Myers S (2004). Serial grip strength testing-Its role in assessment of wrist and hand disability. Int. J. Surg., 5: 1528-1642.

Bloem LJ, Manatunga AK, Prat JH (1996). Racial differences in the relationship of an angiotensin I–converting enzyme gene polymorphism to serum angiotensin I–converting enzyme activity. Hypertension, 27: 62-66.

Bray MS, Hagbero JM, Perusse L, Rankinen T, Roth SM, Wolfarth B, Bouchard C (2008). The human gene map for performance and health-related fitness phenotypes: The 2006-2007 update. Med. Sci. Sports Exerc., 41(1): 34-72.

Calo CM, Vona G (2008). Gene polymorphisms and elite athletic performance. J. Anthropol. Sci., 86: 113-131.

Cerit M, Colakoglu M, Erdogan M, Berdeli A, Cam FS (2006). Relationship between ace genotype and short duration aerobic performance development. Eur. J. Appl. Physiol., 98: 461-465.

Charbonneau DE, Hanson ED, Ludlow AT, Delmonico MJ, Hurley BF, Roth SM (2008). ACE genotype and the muscle hypertrophic and strength responses to strength training. Med. Sci. Sports Exerc., 40: 677-683.

Christie CJ, King GM (2008). Heart rate and perceived strain during batting in a warm and cold environment. Int. J. Fitness, 4(1): 33-38.

Coetsee MF (1995). Isoknee α Instruction Manual, Isokinetic Equipment cc Publisher, South Africa, pp. 1-163.

Collins M, Xenophontos SL, Cariolou MA, Mokone GG, Hudson DE, Anastasiades L, Noakes TD (2004). The ACE gene and endurance performance during the South African Ironman Triathlons. Med. Sci. Sports Exerc., 36: 1314-1320.

Davids K, Baker J (2007). Genes environment and sports performance. Sports Med., 37: 901-980.

Djarova T, Watson G, Basson A, Grace J, Cloete J, Ramakoaba A (2011). ACTN3 and TNF gene polymorphism association with C-reactive protein uric acid, lactate and physical characteristics in young African cricket players. Afr. J. Biochem. Res., 4: 22-27.

Folland J, Leach B, Little T, Howker K, Myerson S, Montgomery H (2000). Angiotensin converting enzyme genotype affects the response of human skeletal muscle to functional overload. Exp. Physiol., 85: 75-579.

Gineviciene V, Pranckeviciene E, Millasius K, Kuchinskas V (2010). Relating fitness phenotypes to genotypes in Lithuanian elite athletes. Acta Medica Lithuanica, 17: 1-10.

Gunay M, Ulkuer MK, Celenk C, Bezci S, Gokdemiir K, Gevat C, Kesici T (2010). Angiotensin-converting enzyme polymorphism in elite Taekwondo athletes of Turkish and Azzerbaijan Taekwondo teams. Ovidia University Annals, Series Physical Education and Sports/Science, Movement Health, 10: 165-168.

Jones A, Woods DR (2003). Skeletal muscle RAS and exercise performance. Int. J. Biochem. Cell Biol., 35: 855-866.

Koley S, Yadav MK (2009). An association of hand grip strength with some anthropometric variables in Indian cricket players. Facta Universitatis, Phy. Edu. Sport, 7: 113-123.

Lakka HM, Lakka TA, Rankinen T, Rice T, Rao DC, Leon AS, Skinner JS, Bouchard C (2006). The TNF-α G308A polymorphism is associated with C-reactive protein levels: The HERITAGE Family Study. Vascul. Pharmacol., 44: 377-383.

Lucia A, Moran M, Zihong, H, Ruitz R (2010). Elite Athletes: Are the Genes the Champions. Int. J. Sports Physiol. Perform., 5: 93-102.

MacArthur DG, North KN (2004). A gene for speed? The evolution function of alpha-actinin 3 gene. Bio. Essays, 26: 785-789.

Martens R (2004). Successful Coaching (3rd edition). Champaign Illinois, USA, Human Kinetics.

Milander L, Stein DJ, Collins M (2009). The interleukin-6, serotonin transporter, and monoamine oxidase A genes and endurance performance during the South African Ironman Triathlon. Appl. Physiol. Nutr. Metab., 34: 858-865.

Min S-K, Takahashi K, Ishigami H, Hiranuma K, Mizuno M, Ishii T, Kim C-S, Nakazato K (2009). Is there a gender difference between ACE gene and race distance? Appl. Physiol. Nutr. Metab., 34: 926-932.

Myerson S, Hemingway H, Budget R, Martin J, Humphries S Montgomery H (1999). Human angitensin I-converting enzyme gene and endurance performance. J. Appl. Physiol., 87: 1313-1316.

Nazarov IB, Woods DR, Montgomery VA, Shneider OV, Kasakov VT, Tomilin NV, Rogozkin VA (2001). The angiotensin converting enzyme I/D polymorphism in Russian athletes. Eur. J. Hum. Genet., 9: 797-801.

Noakes TD, Durandt JJ (2000). Physiological requirement of cricket. J. Sports Sci., 18: 919-929.

Nunes T, Coetsee B (2007). The contribution of isokinetic strength parameters to the performance of cricket batsmen. Isokinetic Exerc. Sci., 15: 233-244.

Ostrander EA, Huson HJ, Ostrander GK (2009). Genetics of athletic performance. Annu. Rev. Genomics Hum. Genet., 10: 407-429.

Payne JR, Dhamrait SS, Gohlke P, Cooper J, Scott, RA, Pitsiladis YP, Humphries SE, Rayner B, Montgomery HE (2007). The impact of ACE genotype on serum ACE activity in a Black South African male population. Ann. Hum. Genet., 71: 1-7.

Pescatello LS, Kostek MA, Gordish-Dressman H, Thompson PD, Seip RI, Price TB, Angelopoulos TJ, Clarkson PM, Gordon PM, Moyna NM, Visich PS, Zoeller RF, Devaney JM, Hoffman EP (2006). ACE ID genotype and the muscle strength and size response to unilateral training. Med. Sci. Sports Exerc., 38: 1074-1081.

Pieruzzi F, Abassi A, Keiser HR (1995). Expression of rennin-angitensin system compounds in heart, kidneys and lungs of rats with experimental heart failure. Circulation, 92: 3105-3112.

Rankinen T, Perusse L, Gagnon J, Changon YC, Keon AC, Skinner JS, Wilmore JH, Rao DC, Bouchard C (2000). Angiotensin-converting enzyme ID polymorphism and fitness phenotype in the HERITAGE family study. J. Appl. Physiol., 88: 1029-1035.

Rigat B, Hubert C, Athenc-Gelas F, Cambien F, Corvol P, Soubrier F (1990). An insertion/deletion polymorphism in the angiotension I-converting enzyme gene accounting for half the variance of serum levels. J. Clin. Invest., 86: 1343-1346.

Ruiz JR, Arieta D, Buxens A, Arieda M, Gomez-Galego F, Santiago C, Yvert T, Moran M, Lucia A (2010). Can we identify a power-oriented polygenic profile? J. Appl. Physiol., 108: 561-566.

Scanavini D, Bernardi F, Castoldi E, Conconi F, Mazzoni G (2002). Increased frequency of the homozygous II Ace genotype in Italian Olympic endurance athletes. Eur. J. Hum. Genet., 10: 576-577.

Scott RA, Irving R, Irwin L, Morrison E, Charlon V, Austin K, Tladi D, Deason M, Headley S, Kolkhorst FW, Yang N, North K, Pitsiladis Y (2010). ACTN3 and ACE genotypes in elite Jamaican and US sprinters. Med. Sci. Sports Exerc., 42(1): 107-112.

Scott RA, Moran C, Wilson RH, Onywera V, Boit MK, Goodwin WH, Gohike P, Montgomery H, Pitsiladis YP (2005). No association between angiotensin converting enzyme (ACE) gene variation and endurance athlete status in Kenyans. Comp. Biochem. Physiol. Mol. Intregr. Physiol., 141: 169-175.

Sipahi T, Budak M, Sen S, Ay A, Seneqr S (2006). Association between

Ace gene insertion (I) / deletion (D) polymorphism and primary hypertension in Turkish patients of Trakya region. Botechnol. Biotechnol. Eq., 20: 104-108.
Stuelcken M, Pyne D, Sinclair P (2007). Anthropometric characteristics of elite cricket fast bowlers. J. Sports Sci., 25: 1587-1597.
Thompson J, Manore MM (1996). Predicted and measured resting metabolic rate of male and female endurance athletes. J. Am. Diet Assoc., 96: 30-34.
Thompson J, Raitt J, Hutchings, I, Bjargo E, Loset A, Grocott M, Montgomery H, Caudwell E (2007). Angiotenzin-converting enzyme genotype and successful ascent to extreme high altitude. Extreme Everest Research Group. High Alt. Med. Biol., 8: 278-283.
Thompson WR, Gordon NF, Pescatello LS (2000). Guidelines for exercise training and prescription. American College of Sports Medicine. 6th Ed. Philadelphia, PA. Williams and Wilkins.
Tsianos GI, Evangelou E, Boot A, Zillikens MC, van Meurs JNJ, Uitterlinden AG, Ioannidis JPA (2009). Associations of polymorphism of eight muscle- or metabolism-related genes with performance of Mount Olympus marathon runners. J. Appl. Physiol., 108: 569-574.
Wagner H, Thaller S, Dahse R, Sust M (2006). Biomechanical muscle properties and angiotensin-converting enzyme gene polymorphism: a model-based study. Eur. J. Appl. Physiol., 98: 507-515.
Wang P, Fedoruk MN, Rupert JL (2008). Keeping pace with ACE. Sports Med., 38: 1065-1079.
Watson TA, Callister R, Taylor RD, Sibbritt DW, MacDonald-Wicks LK, Garg ML (2005). Antioxidant restriction and oxidative stress in short-duration exhaustive exercise. Med. Sci. Sports Exerc., 37: 63-71
Williams AG, Folland JP (2008). Similarity of polygenic profiles limits the potential for elite human physical performance. J. Physiol., 1(586): 113-121.
Wolfarth B, Rivera MA, Oppert JM, Bowlay MR, Dionne FT (2000). A polymorphism of the alpha 2a – adrenoreceptor gene and endurance athlete status. Med. Sci. Sports Exerc., 32: 1709-1712
Woods D (2009). Angiotensin-converting enzyme, rennin-angiotensin system and human performance. In Collins M (Ed) Genetics and Sports. Med. Sport Sci. Basel, Karger, 54: 72-87.
Woods DR, Brull D, Montgomery HE (2000). Endurance and the ACE I/D polymorphism. Sci. Prog., 84: 317-336.
Woods DR, World M, Rayson MP, Williams AG, Jubb M, Jamshidi Y, Hayward M, Mary DASG, Humphries SF, Montgomery HF (2002). Ednurance enhancement related to the human angiotensin I-converting enzyme I-D polymorphism is not due to differences in the cardiorespiratory response to training. Eur. J. Appl. Physiol., 86: 240-244.
Zhang B, Tanaka H, Shono N, Miura S, Kkiyomaga A, Shindo M, Saku K (2003). The lallele of the angiotensin-converting enzyeme gene is associated with an increased percentage of slow twitch type I fibres in human skeletal muscle. Clin. Genet., 63: 139-144.

Changes in enzymes activities of soil samples exposed to electromagnetic radiations (EMR) from mobile phone

C. I. Iheme, R. N. Nwaoguikpe, S. E. Abanobi, C. U. Igwe, E. C. Udenze, A. U. Ezirim, C. P. Ihedimbu and D. I. Ukairo

Department of Biochemistry, Federal University of Technology, P.M.B. 1526, Owerri, Imo State, Nigeria.

This study aimed at investigating the changes in activities of dehydrogenase, catalase, alkaline phosphatase, acid phosphatase and alkaline protease of soil samples exposed to electromagnetic radiations (EMR) from mobile phone for the periods of 0, 30, 60 and 90 days. EMR-unexposed soil samples served as the source of control enzyme activities. The mean enzyme activities from the EMR-exposed soil were significantly ($p<0.05$) lower than those from the unexposed samples. The overall percentage changes in enzymes activities of the EMR-unexposed and exposed soil samples for dehydrogenase, catalase, alkaline phosphatase, acid phosphatase and alkaline protease relative to their respective starting values were 124.42 and -65.15%, 138.01 and -13.87%, 94.09 and 19.70%, 101.01 and -41.00%, and 162.55 and -21.71% respectively. The results show that EMR from mobile phones elicited significant negative impact on soil enzymes activities.

Key words: Electromagnetic radiations, mobile phones, enzymes.

INTRODUCTION

The use of mobile phones in recent years has raised and continued to raise tremendous safety questions. This is because users are exposed to electromagnetic radiations, the effects of which on the body depend on their frequencies, durations and powers.

The radiations generated by mobile cell phones are non-ionizing radiations (NIR). There are wide ranges of data documenting the ability of non-ionizing radiations (NIR) to affect living cells, including changes in the biochemical and molecular mechanisms of cells both *in vitro* and *in vivo* (Barnes, 1996). Changes also occur in cell metabolism and proliferation, inducing potentially damaging effects in various cell components ranging from the cytoplasmic membrane, where the distribution of proteins is modified (Bersani et al., 1997) to the cytoplasm itself and the nucleus, where the activities of intracellular enzymes and molecules regulating cell growth are altered (Hill, 1998).

Contrary to what the Telecommunication Industries proclaim, there are vast scientific, epidemiological and medical data that affirm that exposure to electromagnetic radiations from mobile phones and their towers, even at low levels, can have profound adverse effects on biological systems.

Experiments have revealed the biological effects that occur in activities of isolated enzymes, cell cultures and animals after exposures to low-intensity electromagnetic radiations. Mobile phone radiation exposure can cause cell damage via reactive oxygen species formation, and cell death (Oral et al., 2006; Sokolovic et al., 2008). Exposure of acetylcholinesterase, an important central nervous system enzyme, to mobile phone EMR altered the structural and biochemical characteristics of the enzyme, resulting to a significant change of its activities (Barteri et al., 2005). Three different cell types (rat hepatoma cells, egg cells of the Chinese hamster, human melanoma cells) were exposed for 1 h to a 450 MHz fields with a 16 Hz amplitude modulation and a power flux density of 10 W/m^2. The exposure raised ornithine decarboxylase (ODC) activity by a little more than 50%. The increased ODC activity remained fairly constant for several hours after the exposure. Similar fields with a 60 Hz and a 100 Hz modulation had no effects (Byus et al., 1988).

The mobile phones in their receiving and sending processes use (mostly) microwave frequency of 945 MHz (Bakr, 2004), emitting radiation that may be absorbed by various organs of the body depending on their location (Ozguner et al., 2005; Oktem et al., 2005). Mobile phones also emit low-frequency magnetic field pulse generated by battery currents in the phone that are too weak to produce non-thermal effects (Hyland, 2000). Recent studies have shown that the intense radiations from mobile phone towers adversely impact on every biological organism within 1 square kilometer (Bakr, 2004).

Mobile phone towers are especially dangerous because they emit electromagnetic radiations at a frequency range of 900 to 1900 MHz. Radiofrequency of this range is considered to be within high frequency range (HF-EMF) (Scehihr, 2006). However, the electromagnetic radiations emitted by mobile phones and base stations penetrate the living bodies to a distance that decreases with increasing frequency (Stewart, 2000).

It has been reported that at 900-1900 MHz, EMR does not penetrate deeply into the body; instead, it is absorbed by the skin and the underlying tissues. The heat that is generated in the tissue is then channeled into blood circulatory system (Otto and von Muhlendahl, 2007). However, the same cannot be said for unicellular organisms or even enzymes where it is expected to penetrate deep into the nucleus where DNA is located.

The importance of microorganisms in the soil cannot be over emphasized. For example, the main role of mould and fungi in the soil is to breakdown the remains of plant materials using their appropriate extracellular enzymes such as pectinases, celluloses etc., and these are further broken down through the activities of bacterial enzymes. Through the activities of these microbial enzymes, carbon, nitrogen and other minerals are released to the soil for plants utilization. Therefore, such factors like temperature, pH, substrate concentrations etc that affect the activities of these enzymes cum their host microorganisms in soil, for instance, will negate the availability of these minerals and by extension soil fertility. Also, bacterial and fungal communities occupy overlapping niches in soil. Disturbing these communities, for instance, through the denaturation or reduction of their enzymes activities, may alter the balance existing between them. The resulting imbalance may affect the influence of bacteria and fungi on their niches, and consequently, to the functional ecosystem.

Duration of EMR exposure for instance, seems to be a major determinant of EMR effect on the activities of enzymes. The time of exposure and power density are correlated in a way that decrease in power density (PD) could be compensated by increase in duration of exposure.

MATERIALS AND METHODS

Sampling and sample preparation

Soil samples were collected from the surface to a depth of about 30 cm into thoroughly washed and air-dried plastic containers from five different sites (at a distance of 2m from each other) on a five-year fallowed garden farm of Federal University of Technology, Owerri, Nigeria, in the month of August, 2013 and then homogenized (Kenwood, UK). Five kilograms (5 kg) of the homogenized soil sample were then weighed into each of two experimental tanks, measuring120 × 120 × 100 cm each.

Experimental set up

A commercially available Nokia 2700 mobile phone (90 0MHz band with modulated voice and Specific Absorption Rate of 0.927 W/kg) was used for the generation of EMR for this study. The mobile phone was sandwiched at a depth of 5cm in the 5 kg soil sample of experimental tank one. During exposure, the mobile phone was kept on talking mode for 4 h per day, in vibration and the phone battery was also kept charged. The unexposed sample (with no phone) was placed away (at a distance of 150 m) from the exposed sample (with phone) and at the same atmospheric conditions.

Both the exposed and the unexposed soil samples were irrigated continuously to maintain 75% humidity and a room temperature of 28±2°C for periods of 0, 30, 60 and 90 days. The mobile phone was temporarily removed (2 min) from test experimental tank during every irrigation exercise. The exposed and unexposed sample set-ups were kept away from any instrument or machine to avoid interference from any form of unwanted EMR. At the end of each period, the soil enzyme activities were assayed.

Soil enzyme activities assay

Dehydrogenases (EC 1.1.1) activity

Soil dehydrogenase activity was assayed based on the estimation of the rate of reduction of triphenyltetrazolium chloride (TTC) to triphenylformazan (TPF) after incubation at 30°C for 24 h. Field-moist soil sample (5 g) was weighed into five test tubes and mixed with 5 ml of TTC solution. The tubes were sealed with rubber stoppers and incubated for 24 h at 30°C. After the incubation, 40 ml acetone was added to each tube, and the tubes were shaken thoroughly and further incubated at room temperature for 2 h in the dark (shaking the tubes at intervals). The soil suspension (15 ml) was then filtered and the optical density of the clear supernatant was measured colorimetrically against the blank at 546 nm (Thalmann, 1968).

Phosphatases

The method used was based on the determination of p-nitro-phenol released after the incubation of the soil samples with p -nitrophenyl phosphate (ρNP) for 1h at 37°C. Five grams (5 g) of the moist soil sample were placed in Erlenmeyer flasks (50 ml) and treated with 0.25 ml of toluene, 4 ml of buffer solutions (pH 6.5 for the assay of acid phosphatases and pH 11 for the assay of alkaline phosphatases), and 1 ml of p-nitrophenyl phosphate solutions made in the same buffer. After stoppering the flasks, the contents were mixed and incubated for 1 h at 37°C. After the incubation, 1 ml of $CaCl_2$ (0.5 M) and 4 ml of NaOH (0.5 M) were added. The contents were mixed and the soil suspension filtered through a Whatman filter paper grade 2. For the standard, 1ml of pNP solution was added after the addition of $CaCl_2$ (0.5 M) and 4 ml of NaOH (0.5 M) immediately before filtration of the soil suspension. Absorbance of the solutions were read at 400 nm (Eivazi and Tabatabai, 1977).

Proteases

Soil alkaline protease activity was assayed using Folin-Ciocalteu reagent based on the determination of amino acid released after the incubation of the soil samples with sodium caseinate for 30 min. The soil suspension was subjected to centrifugation (5,000 rpm for 10 min), and the supernatant was used for alkaline protease activities determination according to (Nigam and Ayyagari, 2007).1 ml of supernatant was mixed with 1 ml of casein (1%) [prepared in 0.1 M citrate buffer of pH 5 and 0.1 M glycine-NaOH buffer of pH 10] followed by incubation at 50°C for 30 min. Then 5 ml of 5% trichloroacetic acid (TCA) was added to precipitate the undigested protein. The solution was then centrifuged at 10,000 rpm for 10 min, and the supernatant subjected to estimation for the amount of amino acids released by Lowry's method (Lowry et al., 1951). International unit of protease was calculated as:IU= Net amount of amino acid released (µg) × dilution factor/ 181× 30, where 30=incubation period, 181=amount of protein present in the sample.

Catalase (EC1.11.1.6) activity

The method used was based on the volumetric determination of oxygen liberated after incubation of soil sample with hydrogen peroxide for 3 min at room temperature (Isamah et al., 2000; Cohen et al., 1970). Five grams (5 g) of moist soil samples were weighed into five test tubes, mixed in 20 ml of distilled water, and centrifuged at 3500 rpm for 10 min at 4°C. To the supernatant, 5ml of cold 6mM H_2O_2 was added and the reaction stopped after 3 min by adding 0.25 ml of 6 N H_2SO_4 rapidly with thorough mixing. Excess potassium permanganate (10 ml) was added, shaken gently and absorbance read at 480 nm for 30 s. The blank was prepared as the test but with the replacement of supernatant with distilled water.

A standard was prepared by mixing 10 ml of potassium permanganate with 5.5 ml of potassium phosphate buffer (pH 6.5) and 0.25 ml of 6N H_2SO_4 and the absorbance read at 480 nm. Enzyme activity was calculated using the following formula: $K/_{0.00693}$.

Where the first-order rate constant $(K) = \text{Log}\left(\frac{S_o}{S_t}\right) \times \left(\frac{2.3}{t}\right)$.

Where, S_0 (Initial substrate concentration) = Absorbance of standard minus Absorbance of blank; S_t (final substrate concentration) = Absorbance of standard minus Absorbance of sample; t = reaction time (min).

Statistical analysis

All analyses were done in triplicate and the data generated were analyzed using ANOVA and Duncan's test with the aid of Statistical Package for Social Science (SPSS). Values for $p<0.05$ were considered statistically significant; data were presented as mean± standard deviation.

RESULTS AND DISCUSSION

The effects of non-ionizing radiation, emitted by mobile phones, in living cells have been widely reported, however, the mechanism of action has not been fully documented, though it is thought to be through the action of reactive oxygen species. Living tissues are 70-90% water by weight, non-ionizing radiation therefore, would be expected to induce cellular generation of free radicals by excitation of water molecules. The dividing line between radiations that excites electrons and radiation that forms ions is often assumed to be equal to the energy of ionization of water (1216 kJ/mol). Radiation that carries less energy can excite the water molecule and is therefore called non-ionizing radiation. Radiation that carries more energy than 1216 kJ/mol can remove an electron from a water molecule, and is called ionizing radiation (Daniel et al., 2008). The capability of non-ionizing radiation to induce cellular damage has been reported by Barnes (1996) and other scientists; whether these reported effects were as a result of free radical formation, thermal or athermal action, remained an interesting area for further investigation.

EMR is also believed to exert its biological effects through: non-thermal action, thermal action, and/or a combination of the two. EMR emitted by mobile phone is at non-thermal power density level, so far, no common ground exist on non-thermal exposure levels of EMR in literatures. Thermal effect results from the conversion of the EMR energy into heat energy in the living systems. Polar molecules in cells exist in the form of water, DNA and proteins, and these molecules respond to electromagnetic radiations by rotating. This rotation creates an angular momentum which results in friction with neighboring molecules, thereby developing a linear momentum (vibrational energy) (Saifuddin et al., 2009), through this means, radiation energy is converted into thermal energy.

It can therefore be said that the effect resulting from vibrational energy is ther-mal effect which occurs in a biosystem due to penetration of electromagnetic radiation into biological materials and subsequently heating up the intra- and extra- cellular fluids by transfer of vibrational energy (Tahir et al., 2009). It must be noted that EMR thermal effects is different from conventional heating effect. This is because dipolar polarization and rotation of molecules in an attempt to align the dipoles with applied electromgnetic field produces effects which cannot be achieved by conventional heating (Zelentsova et al., 2006).

The non-thermal effect of EMR has been a matter of debate in scientific community. A non-thermal effect has been suggested to result from a direct stabilizing interaction of electric field with specific (polar) molecules in reaction medium with no rise in temperature (Herrero et al., 2008). The interest in non-thermal effect is predicated on the established fact that thermal effects alone

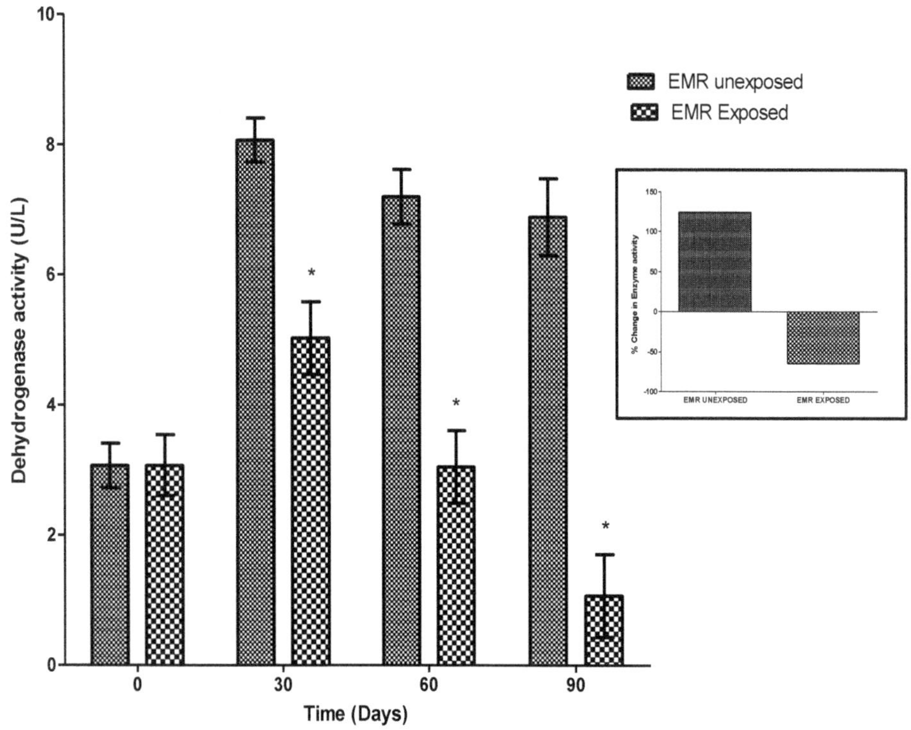

Figure 1. Dehydrogenase activity (U/L) of Electromagnetic radiation (EMR) exposed and unexposed soil samples. %Δ, *Values are significant ($p < 0.05$) in comparison with their respective EMR unexposed control. Insert chart; overall percentage change in dehydrogenase activities of EMR unexposed and exposed soil samples.

cannot explain the manner in which the EMR affects biological systems. As a result of pausity of information on the exact mechanism involved in *'nonthermal* EMR effects' (*athermal* effects), their existence has been a subject of controversy. The influence of EMR on the soil enzymes may also result from its effect on the proteinaceous nature of the enzymes that changes the secondary and tertiary protein structure of enzymes and thus, denaturation of enzymes. These changes in enzyme activities may be correlated with the probability of electroporation (Reina et al., 1998) and reported irreversible cell membrane break down of microbial hosts.

The concept of resonant absorption and resonant interactions has been suggested as another possible explanations for the marked sensitivities of living systems to EMR (Cosic, 1997). In the present study, changes in enzymes activities of soil samples exposed to electromagnetic radiations (EMR) from mobile phone were assessed.

Figure 1 shows the dehydrogenase (DHG) activity of the electromagnetic radiation (EMR) of exposed and unexposed soil samples. It indicates significant ($p < 0.05$) reductions in DHG activity of EMR exposed soil samples at exposure periods of 30, 60 and 90 days, giving an overall percentage change in DHG activity between 0 to 90 days of 124.42% and -65.15% for the EMR unexposed and exposed soil samples respectively. Dehydrogenases are used as an indicator of overall soil microbial activity because they occur intracellularly in all living microbial cells. They are tightly linked with microbial redox processes (Moeskops et al., 2010). Most importantly, dehydrogenases do not accumulate extracellular in the soil. They play vital roles in the biological oxidation of soil organic matters by transfer of hydrogen from organic substrates to inorganic acceptors (Zhang et al., 2010). DHG activity, therefore, reflects metabolic ability of the soil and it is usually proportional to the biomass of the microorganisms in soil (Salazar et al., 2011). Stress on microbial biomass is reflected on dehydrogenase activities.

Our results show that between 0 to 30 days of EMR exposure to the soil samples, the activities of catalase were enhanced (Figure 2). This finding is consistent with the induction of catalase and peroxidase activities following microwave pretreatment of wheat seedlings (Chen et al., 2008). Generally, the catalase activity of the EMR exposed soil samples decreased significantly ($p<0.05$) with increase in duration of exposure, and in comparison with their respective EMR unexposed controls. The overall percentage change in catalase activities between 0 to 90 days of EMR exposed soil sample was -13.87% as compared to 138.01% from the EMR unexposed soil samples, thus indicating the negative impact of the treatment on the enzyme activities. Catalase, also known as hydrogen peroxidase oxido-reductase is an

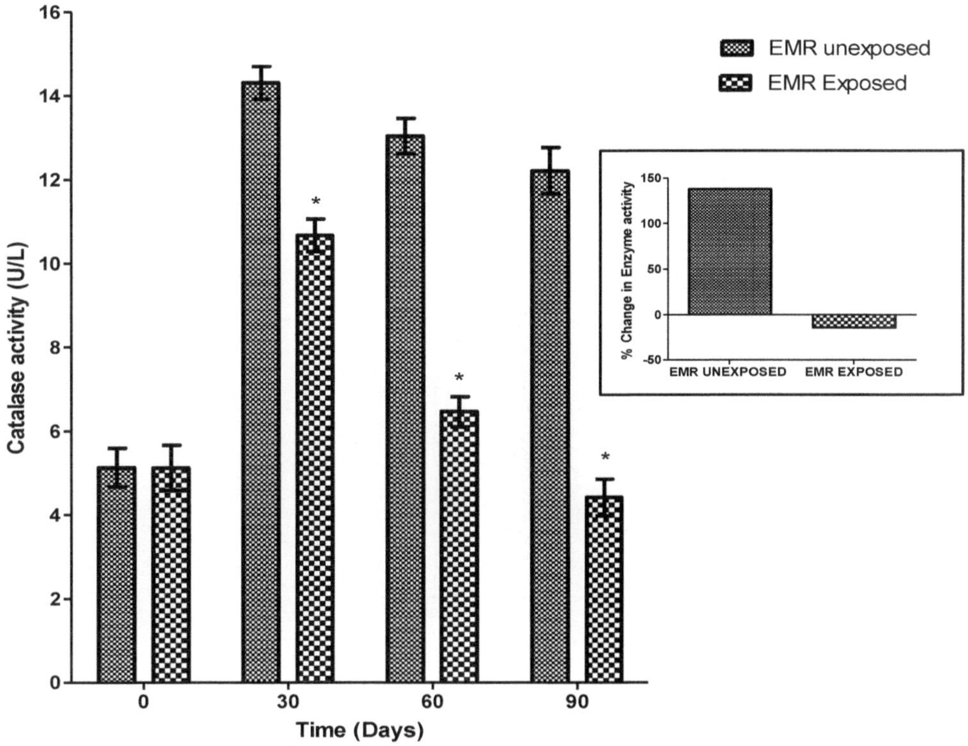

Figure 2. Catalase activity (U/L) of Electromagnetic radiation (EMR) exposed and unexposed soil samples. %Δ, *Values are significant ($p < 0.05$) in comparison with their respective EMR unexposed control. Insert chart; overall percentage change in catalase activities of EMR unexposed and exposed soil samples.

enzyme that has a detoxifying function in cells, catalyzing the reaction: $H_2O_2 \longrightarrow H_2O + \frac{1}{2}O_2$. All aerobic and most facultative anaerobic bacteria, but not obligate anaerobic bacteria, exhibit catalase activity. Catalase activity has been shown to be very stable in soil and significantly correlates with the content of organic carbon and decreases with soil depth (Ladd, 1978). However, no relation has been detected between catalase activity and soil biomass. Furthermore, storage of moist or air-dried soils at room temperature for up to a period 4 months had not been found to have effect on catalase activity.

Figures 3 and 4 show the effect of EMR exposure on alkaline phosphatase (ALP) and acid phosphatase (ACP) activities of soil samples respectively. ALP and ACP are both phosphatases, which are known to catalyse the hydrolysis of organic phosphomonoester to inorganic phosphorus. Based on their optimum pH, the enzymes are classified as acid, neutral and alkaline phosphatases. Acid (pH optimum: 4 – 6.5) and alkaline phosphatase (pH optimum: 9-10) have been found in soils (Speir and Ross, 1978). Acid phosphatase is predominant in acidic soils, while alkaline phosphatase prevails more in alkaline soils (34). Both ALP and ACP activities were significantly ($p<0.05$) lower in the soil samples exposed to the EMR. The observed reductions in the phosphatase activities decreased with increase in duration of exposure giving an overall percentage activity decrease of -19.70 and -41.00% in comparison with the increased activity observed in the unexposed soil samples (94.09% and 101.01%) for ALP and ACP respectively. Generally, ACP recorded minimal activities than ALP in both the EMR exposed and unexposed samples. This is because the soil samples were basic with an average pH value of 9.40.

Alkaline proteases are degradative enzymes that catalyze the partial or the total hydrolysis of proteins. These enzymes are mainly produced by bacteria and fungi. Microbial proteases are predominantly extracellular and can be secreted in the fermentation medium by several species of bacteria, bacillus and fungi for example *notatum* (Ellaiah et al., 2002; Raju et al., 1994; Haq et al., 2006). In this study, EMR inhibited alkaline protease activities by -21.71% and this is consistent with the report of Dholiya et al. (2012). However, between 0-30 days, the activities of the protease were enhanced in the EMR-exposed samples. This agrees with the findings of Afzal and Mansoor (2012) who reported increased activities of proteases following EMR exposure to radicles for 72 h. While the alkaline protease activity remained high in the EMR unexposed samples, that of the exposed samples decreased significantly ($p<0.05$) with increased duration of exposure. Thus, the overall percentage changes

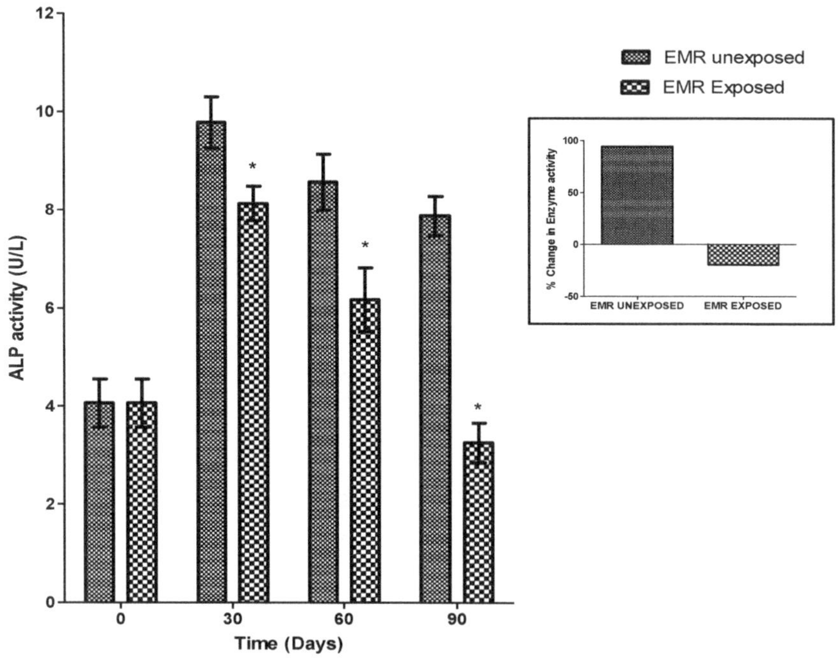

Figure 3. Alkaline phosphatase (ALP) activity (U/L) of Electromagnetic radiation (EMR) exposed and unexposed soil samples. %Δ, *Values are significant ($p < 0.05$) in comparison with their respective EMR unexposed control. Insert chart; overall percentage change in ALP activities of EMR unexposed and exposed soil samples.

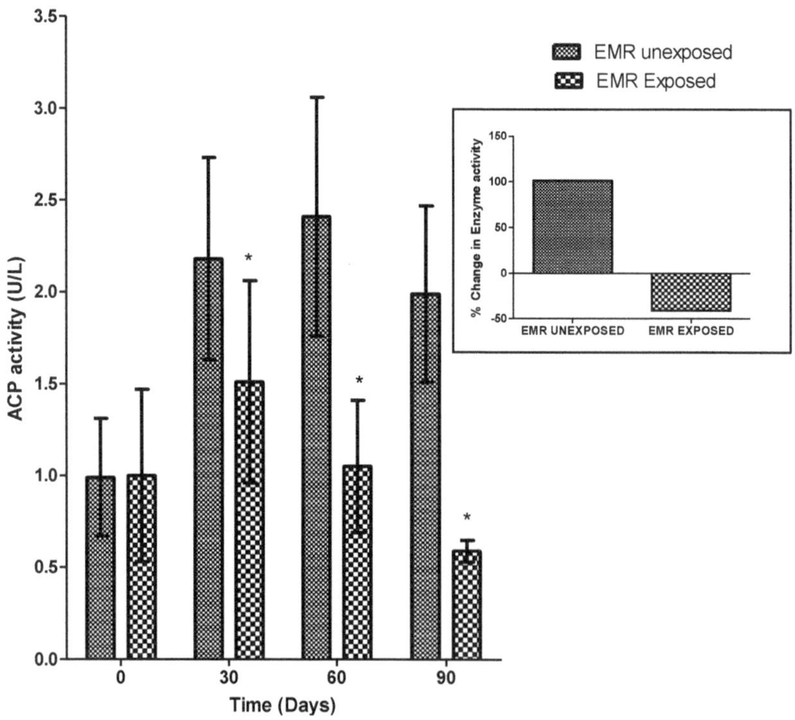

Figure 4. Acid phosphatase (ACP) activity (U/L) of Electromagnetic radiation (EMR) exposed and unexposed soil samples. *Values are significant ($p < 0.05$) in comparison with their respective EMR unexposed control. Insert chart; overall percentage change in ACP activities of EMR unexposed and exposed soil samples.

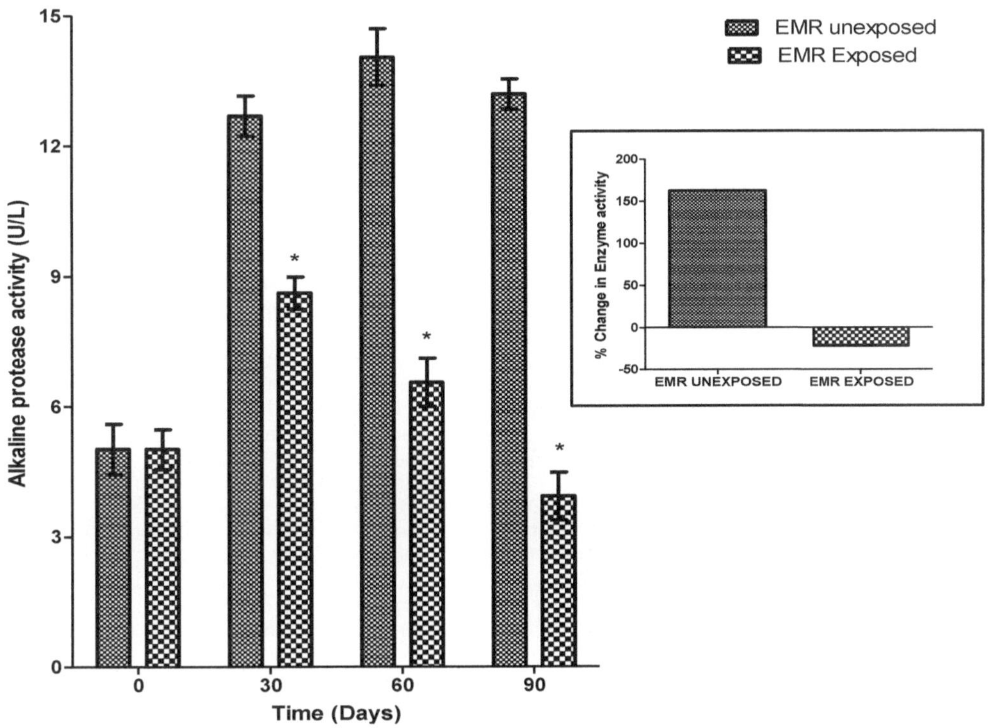

Figure 5. Alkaline protease activity (U/L) of Electromagnetic radiation (EMR) exposed and unexposed soil samples. *Values are significant ($p < 0.05$) in comparison with their respective EMR unexposed control. Insert chart; overall percentage change in alkaline protease activities of EMR unexposed and exposed soil samples.

in activities of the enzymes between 0 to 90 days were found to be -21.71 and 162.55% for the exposed and unexposed samples respectively (Figure 5).

Conclusion

This study indicated that EMR had significant negative impacts on soil enzyme activities at the exposure periods of 30, 60, and 90 days, with dehydrogenase being the most affected and catalase the least affected vis-a-vis their corresponding enzymes activities in the unexposed soil samples. It can be suggested that the EMR effects might be the result of micro-thermal heating that was importantly different from conventional heating.

Based on the observed effects, it is therefore concluded that chronic ex-posure to electromagnetic radiations from mobile phones can inhibit enzyme activities in soil. The findings have further paved ways for studies on the impact of EMR on soil and its biomass around telecommunication masks.

Conflict of interests

The authors do hereby declare that no conflict of interest exist.

ACKNOWLEDGEMENT

The authors appreciate Mr. Uche I. Arukwe, a technical staff in the Department of Biochemistry, Federal University of Technology Owerri, for his technical assistance during the course of the research.

REFERENCES

Afzal M, Mansoor S (2012). Effect of mobile phone radiations on morphological and biochemical parameters of mung beans (Vigna radiata) and wheat (Triticum aestivum) seedlings. Asia J. Agric. Sci. 4(2):149-152.

Bakr AD (2004). Study of Microwave Effects on Biophysical and Histological Properties of Rat Brain, Ph.D. Thesis, Physics Department, Faculty of Science, Benha Branch, Zagazig University, p. 21.

Barnes FS (1996). The effect of ELF on chemical reaction rate in biological systems. In S. Ueno (ed.), biological effects of magnetic and electromagnetic fields, New York, Plenum Press, pp. 37-44.

Barteri M, Pala A, Rotella S (2005). Structural and kinetic effects of mobile phone microwaves on acetylcholinesterase activity. Biophys. Chem. 113:245-253.

Bersani T, Marinelli I, Ogmbene A, Matteueci A, Cecchi S, Sahti S, Squarzohi S, Maraldi NM (1997). Intermembrane protein distribution in cell cultures is affected by 50Hz Ru/sed magnetic fields. Bioelectromagnetic 18:463-496.

Byus CV, Kartun K, Pieper S, Adey WR (1988). Increased ornithine decarboxylase activity in cultured cells exposed to low energy

modulated microwave fields and phorbol ester tumor Promoters. Cancer Res. 48:422-426.
Chen YP, Jia JF, Han XL (2008). Weak microwave can alleviate water deficit induced by osmotic stress in wheat seedlings. Planta 229(2): 291-298.
Cohen G, Dembiec D, Marcus J (1970). Measurement of Catalase Activity in Tissues Extracts. Anal. Biochem. 34:30-38.

Cosic I (1997). The Resonant Recognition Model of Macromolecular Bioactivity: Theory and Applications. Basel, Birkhauser Verlas, p. 8.

Daniel RM, Danson MJ, Eisenthal R, Lee CK, Peterson ME (2008). The effect of temperature on enzyme activity: New insights and their implications. Extremophiles 12:51-59.

Dholiya K, Patel D, Kothari V (2012). Effect of low power microwave on microbial growth, enzyme activity, and aflatoxin production. Res. Biotechnol. 3(4):28-34.
Eivazi F, Tabatabai MA (1977). Phosphatases in soil. Soil Biol. Biochem. 9:167-172.
Ellaiah P, Adinarayana K, Bhavani Y, Padmaja P, Srinivasulu B (2002). Optimization of process parameters for glucoamylase production under solid state fermentation by a newly isolated Aspergillus species. Process Biochem. 38:615-620.

Haq IU, Mukhtar H, Umber H (2006). Production of protease by Penicillium chrysogenum through optimization of environmental conditions. J. Agric. Sci. 2(1):23-25.
Herrero MA, Kremsner JM, Kappe CO (2008). Nonthermal microwave effects revisited: on the importance of internal temperature monitoring and agitation in microwave chemistry. J. Org. Chem. 73(1):36-47.

Hill SM (1998). Receptor crosstalk: Communication through cell signaling pathways. Bioelectromagnetics 16:207-210.
Hyland GJ (2000). Physics and biology of mobile telephony. Lancet 356:1833-1836.
Isamah GK, Asagba SO, Thomas AE (2000). Lipid peroxidation, O-diphenolase, superoxide dismutase and catalase profile along the three physiological regions of Dioscorea ratundata. Poir CV Omi. Food Chem. 69:1-4.

Ladd JN (1978). Origin and range of enzymes in soil. In Soil Enzymes (Burns, R. G. Ed.), Academic Press, New York. pp. 51-96.
Lowry, O.H., Rosenbrough, H.I., Fair, A. L. and Randall, R. I. (1951). Protein measurement with folin phenol reagent. J. Biol. Chem. 193: 265-275.
Moeskops B, Buchan D, Sleutel S, Herawaty L, Husen E, Saraswati R, Setyorini D, De Neve S (2010). Soil Microbial Communities and Activities Under Intensive Organic and Conventional Vegetable Farming In West Java, Indonesia. Appl. Soil Ecol. 45:112-120.

Nigam A, Ayyagari A (2007). Lab Manual in Biochemistry, Immunology and Biotechnology, New Delhi: Tata-McGraw Hill Education.
Oktem F, Ozguner F, Mollaoglu H, Koyu A, Uz E (2005). Oxidative damage in the kidney induced by 900-MHz-emitted mobile phone: Protection by Melatonin. Arch. Med. Res. 4:350-535.

Oral B, Mehmet G, Fehmi O, Nermin K, Tamer M, Selcuk C, Gokhan C (2006). Endometrial apoptosis induced by a 900-MHz mobile phone: Preventive effects of vitamins E and C. Adv. Ther. 6:957-973.

Otto M, von Muhlendahl KE (2007). Electromagnetic fields (EMF): Do they play a role in children's environmental health (CEH)? Int. J. Hyg. Environm. Health 210:635-644.

Ozguner F, Altinbas A, Ozaydin M, Dogan A, Vural H, Kisioglu AN, Cesur G, Yildirim NG (2005). Mobile phone-induced myocardial oxidative stress: Protection by a novel antioxidant agent caffeic acid phenethyl ester. Toxicol. Ind. Health 9:223-230.

Raju K, Jaya R, Ayyanna C (1994). Hydrolysis of casein by Bajara protease importance. Biotechnol. Coming Decadea 181:55-70.
Reina LD, Jin ZT, Zhang QH, Yousef AE (1998). Inactivation of Listeria monucytogenes in milk by pulsedelectric field. J. Food Prot. 61(9): 1203-1206.
Saifuddin N, Wong CW, Yasumira AA (2009). Rapid biosynthesis of silver nanoparticles using culture supernatant of bacteria with microwave irradiation. J. Chem. 6(1):61-70.
Salazar S, Sanchez L, Alvarez J, Valverde A, Galindo P, Igual J, Peix A, Santa- Regina I (2011). Correlation Among Soil Enzyme Activities Under Different Forest System Management Practices. Ecol. Eng. 37: 1123-1131.
Scehihr C (2006). Preliminary opinion on possible effect of electromagnetic fields (EMF) on human health, H. a. C. P. D.-G. European Commission, ed.
Sokolovic D, Djindjic B, Nikolic J, Bjelakovic G, Pavlovic D, Kocic G, Krstic D, Cvetkovic T, Pavlovic V (2008). Melatonin reduces oxidative stress induced by chronic exposure of microwave radiation from mobile phones in rat brain. J. Radiat. Res. Tokyo 49(6):579-586.

Speir TW, Ross DJ (1978). Soil phosphatase and sulphatase. In: Burns, R.G. (ed) Soil enzymes. Academic, London, pp. 197-250.
Stewart W (2000). Mobile Phone and Health; Report Of Independent Expert Group On Mobile Phones, London, p. 38.
Tahir A, Mateen B, Univerdi S, KaraGoban O, Zengin M (2009). Simple method to study the mechanism of thermal and non thermal bactericidal action of microwave radiations on different bacterial species. J. Bacteriol. Res. 1(5):58-63.
Thalmann A (1968). The method of determining dehydrogenase activities in the soil using triphenyltetrazolium chloride (TTC). Landwirtsch Forsch 21:249-258.
Zelentsova NV, Zelentsov SV, Semchikov YD (2006). On the mechanism of microwave initiated reactions. Cheminform 37(8).

Zhang N, He X, Gao YB, Li YH, Wang HT, Ma D, Zhang R, Yang S (2010). Pedogenic Carbonate and Soil Dehydrogenase Activity in Response to Soil Organic Matter in Artemisia ordosica Community. Pedosphere 20:229-235.

Evaluation of different matrices for production of alkaline protease from *Bacillus subtilis* –K 30 by entrapment technique

K. Suresh Babu Naidu[1]*, K. Lakshmi Devi[2] and K. Jamila Adam[1]

[1]Department of Biomedical and Clinical Technology, Durban University of Technology, Durban-4000, South Africa.
[2]Faculty of Biochemistry, SriKrishna Devaraya University, Anantapur-515 003, Andhra Pradesh, India.

The present investigation evaluates the suitability of different matrices such as calcium alginate, polyacrylamide, agar-agar and gelatin for production of alkaline protease from *Bacillus subtilis*-K 30 using immobilization approach. Calcium alginate was found to be an effective and suitable matrix for higher alkaline protease productivity compared to other matrices studied. All the matrices were selected for repeated batch fermentation. The average specific volumetric productivities with calcium alginate was 14.83 U/ml/h which is 42.83% higher production over the convention free cell fermentation. Similarly, the specific volumetric productivity by repeated batch fermentation was 12.29 U/ml/h with polyacrylamide, 12.22 U/ml/h with agar-agar and 12.09 U/ml/h with gelatin. From the results, it is concluded that the immobilized cells of *B. subtilis* K-30 in calcium alginate are more efficient for the production of alkaline protease with repeated batch fermentation.

Key words: *Bacillus subtilis* K-30, Alkaline protease, immobilized cells, repeated batch fermentation, matrices.

INTRODUCTION

Proteases constitute one of the most important groups of industrial enzymes that execute a wide variety of functions and have various important biotechnological applications (Mohen et al., 2005). These enzymes occupy a pivotal position due to their applications in detergent industry, leather industry, food industry; pharmaceutical industry, silk industry and bio-remediation processes (Anwar and Saleemuddin, 1998; Gupta et al., 2005; Prakasham et al., 2005; Naidu and Devi, 2010). Although, a variety of proteolytic fungi and bacteria are known; only few microbial strains provide high activities with commercial success (Berla and Suseela, 2002). Moreover, the increasing demand for protease as a detergent supplement justifies a study aiming at a search for proteolytic enzymes from various microbial sources. Commercial proteases are mostly from various sources and it was reported that about 35% of the total microbial enzymes used in detergent industry are the proteases from bacteria (Ferrero et al., 1996). *Bacillus* spp. are extensively exploited for protease production (Karl-Heinz, 2004) due to lack of pathogenicity and ability to grow in simple medium (Joshi, 2010).

At present, the use and applications of alkaline proteases have increased remarkably with large proportions of commercially available alkaline proteases derived from *Bacillus* strains (Sen and Satyanarayana, 1993; Prakash et al., 2005; El Enshasy et al., 2008). Modification of processes using immobilized biocatalysts has recently gained much attention from biotechnologists (Kumari et al., 2009). Use of immobilized or whole cells is advantageous because such biocatalyst offer better operational stability, ability to separate from the bulk liquid for possible reuse, continuous operation and decreasing contamination from the product stream during continuous fermentations without loss in biomass (Helmo et al., 1985, Beshay, 2003; Ahmed and Abdel-Fattah, 2010). Moreover, immobilized cells show modifications in

*Corresponding author. E-mail: dr.naidu@hotmail.com.

Abbreviations: DCW, Dry cell weight; **TEMED,** tetra methyl ethylene diamine.

increased mechanical, thermal or chemical resistance in accordance to the method and materials used for their entrapment. As a result, the same cells can be used multiple times with no apparent decrease of their overall biological activity compared to suspended cells (Lukasz et al., 2011; Adinarayana et al., 2004). In the present study, we report immobilization of *B. subtilis* K-30 cells originally isolated from soil sample for higher alkaline protease production using different entrapment techniques with matrices such as calcium alginate, polyacrylamide, agar-agar and gelatin gel. The reusability of immobilized cells for alkaline protease production under repeated batch fermentation conditions was also investigated.

MATERIALS AND METHODS

Microorganism

B. subtilis K-30 was used throughout this study. It was isolated using rice bran as a carbon source in our laboratory (Naidu and Devi, 2005). It was maintained on nutrient agar slants at 4°C and was sub cultured every two weeks.

Inoculum preparation

The inoculum was prepared as previously reported (Adinarayana et al., 2005) with modifications. Briefly, 5 ml of sterile distilled water was added to 24 h old slant of *B. subtilis* K-30. The cells were scraped from the slant into sterile distilled water and the resulted cell suspension at 10% level was aseptically transferred into 250 ml Erlenmeyer flasks containing 45 ml of sterile inoculum medium. The composition of inoculum medium (g/L): glucose, 2.0; casein, 0.5; peptone, 0.5; yeast extract, 0.5; and salt solution, 50 ml (salt solution containing {g/L}: KH_2PO_4, 5.0; $MgSO_4.7H_20$, 5.0, and $FeS0_4.7H_20$, 0.1) and pH was maintained at 7.0. The flask was kept in a shaker at 150 rpm at 37°C.The content of the flask was centrifuged at 3000 rpm for 10 min and the supernatant was decanted. The cell pellet was washed thoroughly with sterile 20.0 g/L potassium chloride solution followed by sodium chloride solution (20.0 g/L) and sterile distilled water. Finally, the cell mass was suspended in sterile sodium chloride solution (9.0 g/L). This cell suspension was used as inoculum source for all immobilization studies.

Immobilization of whole cell in alginate, polyacrylamide, agar-agar and gelatin

The alginate entrapment of cells was performed according to the method of Johnsen and Flink (1986). Briefly, 3% sodium alginate solution was prepared under sterile conditions. Both alginate slurry and cell suspension [(0.03 g dry cell weight, (DCW)] were mixed and stirred for 10 min. The resultant slurry was added drop wise into 0.2 M calcium chloride solution with the aid of a syringe of 1.0 mm and left for curing at 4°C with mild stirring at 60 rpm for 1 h. The beads were then washed for 3 to 4 times with sterile distilled water. When the beads were not used, they were preserved in 0.9% sodium chloride solution at 4°C. All operations were carried out aseptically in a laminar air flow unit. Immobilization in polyacrylamide was done by the method described by Adinarayana et al. (2005) by adding 0.03 g cells to 10 ml chilled sterile distilled water.

To another 10 ml of 0.2 M sterile phosphate buffer(pH 7.0), the following chemicals were added: 2.85g acrylamide (Sigma-Aldrich), 0.15 g bisacrylamide (Sigma-Aldrich), 10 mg ammonium persulphate (Sigma-Aldrich), and 1 ml tetra methyl ethylene diamine (TEMED), (Sigma-Aldrich). The cell suspension and the above phosphate buffer mixture was mixed well and poured into sterile flat bottom 10 cm-diameter petri plates. After polymerization, the acrylamide gel was cut into equal size cubes (4 mm^3), transferred to 0.2 M phosphate buffer (pH 7.0), and kept in the refrigerator for 1 h for curing with constant stirring. Whole cell immobilization of *B.subtilis* K-30 was carried out in accordance with the method described by Veelken and Pape (1982). Encapsulation in agar was done by adding cell suspension (2 ml equivalent to 0.03 g DCW) into the molten agar-agar. The solidified agar block was cut into equal size cubes (4 mm^3), added to sterile 0.1M phosphate buffer (pH 7.0), and kept in the refrigerator (overnight) for 1 h for curing. Five milliliters (0.06% DCW) of cell suspension was added to 15 ml of 20% sterile gelatin, maintained at 45°C and poured into sterile petriplate. The gel was over-layered with 10 ml of 5% glutaraldehyde for covalent crosslinking and hardening at 30°C. The resulting block was cut into small cubes (4 mm^3) and the cubes were washed thoroughly with sterile distilled water for complete removal of excess glutaraldehyde.

Production of alkaline protease by batch process with immobilized cells

The immobilized beads, prepared by using all the four matrices along with control were transferred into 50 ml of production medium in 250 ml Erlenmeyer flasks. The composition of production medium was (g/L) glucose, 5; peptone, 7.5; and 5% salt solution ($MgSO_4 \cdot 7H_2O$, 5 g/L; KH_2PO_4, 5 g/L; and $FeSO_4 \cdot 7H_2O$, 0.1 g/L) with a pH of 9.0. The flasks were incubated at 37°C for 48 h. Samples were withdrawn at regular intervals of 6 h and assayed for alkaline protease activity.

Alkaline protease assay

The culture broth was centrifuged at 3000 rpm for 20 min and supernatant served as the crude enzyme source. Protease activity was assessed by the modified procedure based on the method of Tsuchida et al. (1986) using 2.0% casein in 0.2 M carbonate buffer (pH 10.0) as substrate. One unit of enzyme activity is defined as the amount of enzyme that released 1 µg of tyrosine/mL/min. All experiments were carried out in triplicate.

RESULTS AND DISCUSSION

Application of immobilized cells or whole cells is advantageous because such biocatalysts display better operational stability, higher efficiency of catalysis and also their suitability. Immobilization is a strategy for protecting cells from shear forces.

Production of alkaline protease by immobilized cells in calcium alginate

Immobilization of *B.subtilis* K-30 cells using different entrapment techniques with matrices such as calcium alginate, polyacrylamide, agar-agar and gelatin gel was investigated. The enzyme production with immobilized

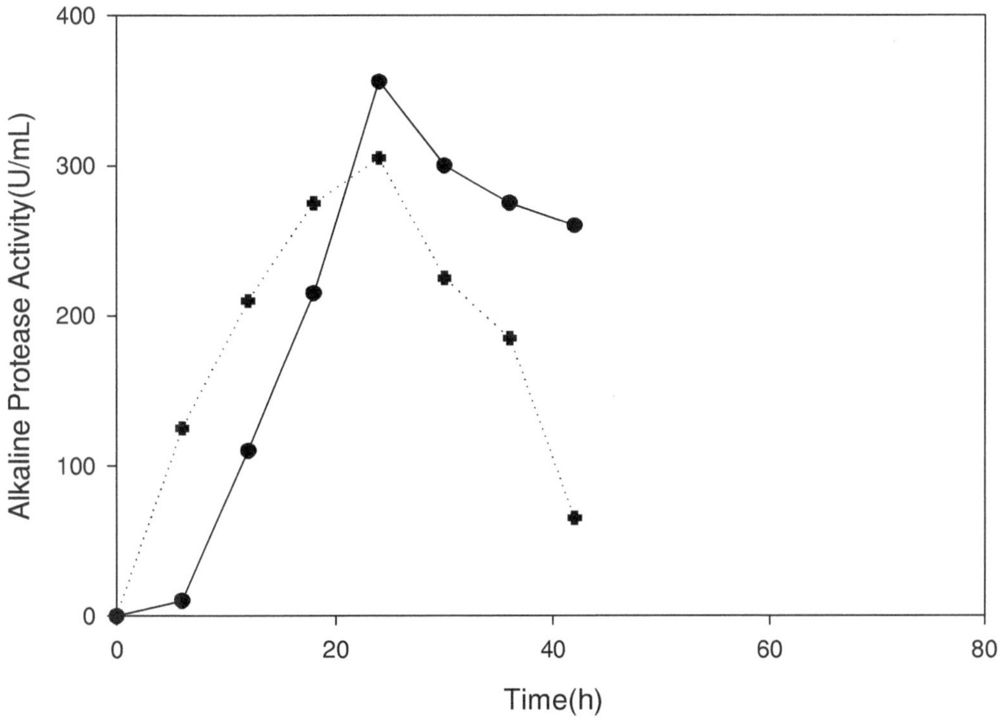

Figure 1. Time course profile of alkaline protease production by free cells and immobilized culture of *B. subtilis* K-30 in calcium alginate.

cells in calcium alginate started at 6 h and reached the maximum level (356 U/ml) by 24 h (Figure 1). On further incubation, enzyme production gradually decreased, while maximum enzyme titer was observed at 48 h in the case of free cells. It is evident that the alkaline protease production was higher with immobilized cells (356 U/ml) compared to free cells (305 U/ml). Beshay (2003) studied the effect of alginate concentration on immobilization of alkaline protease producing bacterial strain, *Teredinobacter turnirae.* The production of alkaline protease improved significantly with increasing alginate concentration and reached a maximum enzyme yield of 8000 U/ml at 25 g/L alginate concentration, which was about 176.8% higher than that obtained by free cells (2890 U/ml). However, at low alginate concentration (20 g/L), the beads are relatively soft and showed rapid leakage of cells from the beads. Another explanation for decrease in protease production with further increase in alginate concentration is low rate of substrate mass transfer and lower porosity of gel beads as judged by earlier reports (Alvaro and Maria, 1996; Fumi et al., 1992). Ramakrishna et al. (1992) reported the immobilization of *Bacillus cereus* in calcium alginate and employed packed bed reactor and fluidized bed-reactors to continuously synthesize thermostable α-amylase. They spun alginate fibres by pulstration technique to reduce the diffusional resistances in the gel matrix, and thereby a 24-fold increase in the productivity compared to batch fermentation with free cells was attained.

Production of alkaline protease by immobilised cells in polyacrylamide

Polyacrylamide was successfully used for immobilization of microbial cells, sub cellular organelles and many enzyme systems (Kierstan and Bucke, 1977; Kim et al., 1994). A gradual increase in alkaline protease production was noticed from 6 h onwards to 24 h (Figure 2). On further incubation, there was a decline in alkaline protease titer with immobilized cells in polyacrylamide. Alkaline protease titer was 295 U/ml at 24 h which was a low titer compared to alginate matrix.

Production of alkaline protease by immobilized cells in agar-agar

Protease production increased gradually from 6 h onwards and reached maximum level by 24 h (294 U/ml). Further, alkaline protease production with immobilized cells in agar-agar was comparable with polyacrylamide, whereas it was less than alginate matrix (Figure 3). The cell leakage from the matrix was gradually increased with increase of fermentation time. The operational stability of *Bacillus circulans* ATCC 21783 immobilized on agar-agar for cyclodextrin glucanotransferase production was studied by Anna et al. (2003) by repeated batch cultivation for 24 h in a fluidized bed reactor and found to be effective. In another report, an acid protease from

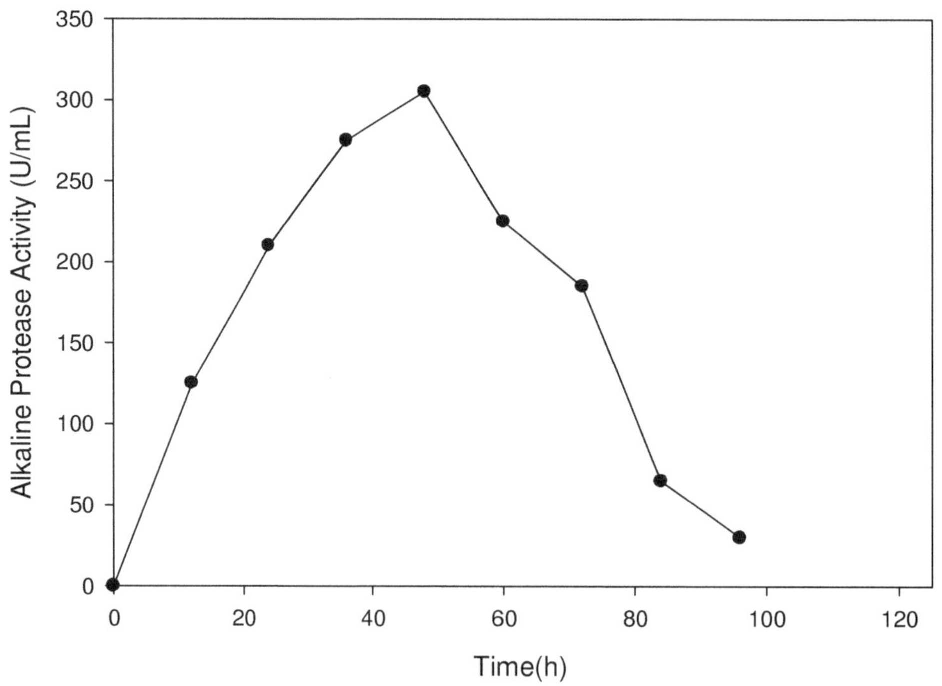

Figure 2. Time course profile of alkaline protease production by free cells of *B.subtilis* K-30.

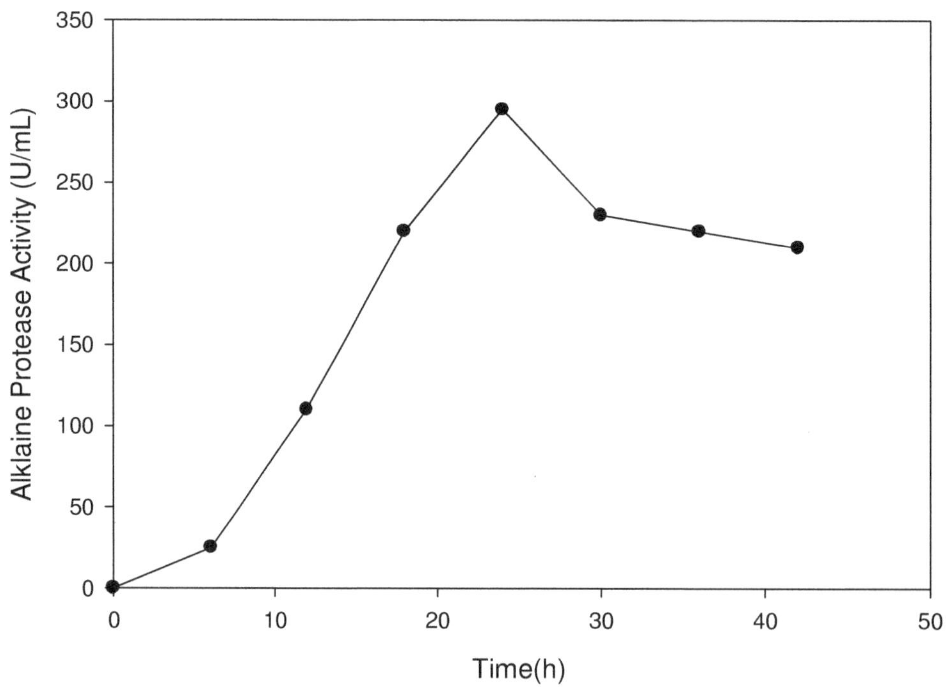

Figure 3. Time course profile of alkaline protease production by *B.subtilis* K-30 in polyacrylamide.

Aspergillus saitoi was immobilized on agar beads of approximately 100 µm diameters, the hydrolysis of gelatin by immobilized acid protease was found to be 6.8 x 10^{-7} M per second (Gregg and Roger, 1981) Only a detectable level of alkaline protease titer was observed after 6 h of fermentation and reached the maximum level (290 U/ml) by 24 h in gelatin (Figure 4).The alkaline protease titer obtained with this carrier was low compared to that in alginate matrix and comparable to that of free cells, polyacrylamide and agar-agar. *B. subtilis* PE-11

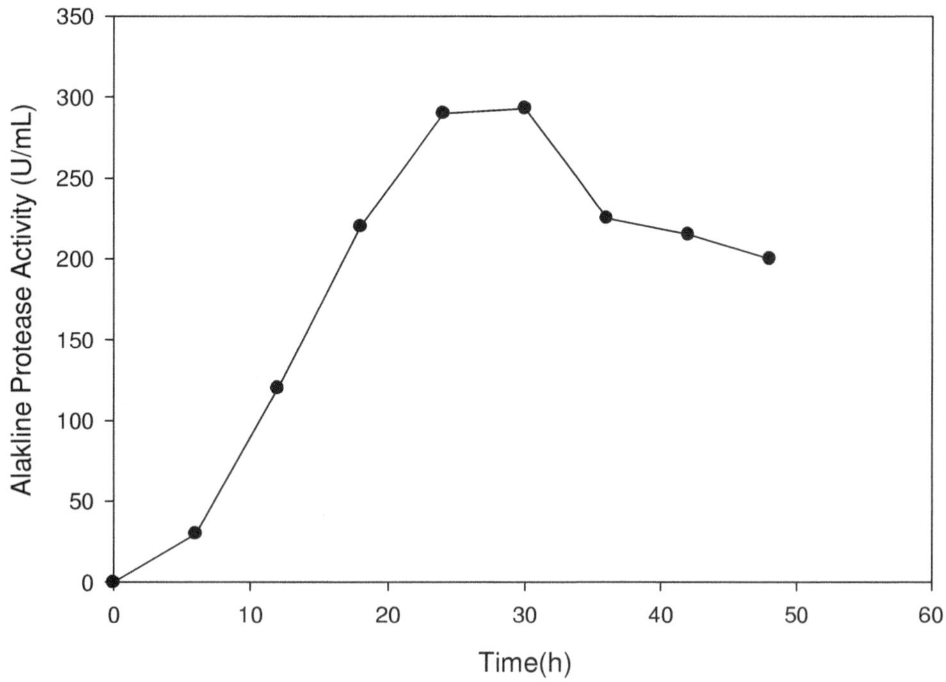

Figure 4. Time course profile of alkaline protease production by *B.subtilis* K-30 in gelatin

Table 1. Comparison of alkaline protease production with cells immobilized in different matrices.

Matrix	Fermentation period for each batch (h)	Number of batches	Total fermentation time (h)	Total alkaline protease titer (U/ml)	Average alkaline protease activity per batch (U/ml)	Specific Volumetric productivity U/ml/h)
Calcium alginate	24	9	216	3204	356	14.83
Polyacrylamide	24	6	144	1170	295	12.29
Agar-agar	24	6	144	1760	294	12.22
Gelatin	24	6	144	1740	290	12.09
Free cells	48	1	48	305	305	6.37

isolated from industrial effluents upon immobilization in gelatin retained 79% of initial activity after 6 cycles of repeated batch fermentation (Adinarayana et al., 2004). Gelatin is employed as a carrier material for the immobilization of *Kluyveromyces fragilis* for β-galactosidase production and Escherichia *coli* for penicillin acylase production (Ramakrishna and Prakasham, 1999).

Comparison of alkaline protease production by immobilized cells in different matrices by entrapment technique

The results show that the average specific volumetric productivity with alginate matrix was 14.83 U/ml/h which is 42.83% higher production over the conventional free cell fermentation (Table 1). Similarly, low level of alkaline protease was observed with polyacrylamide (12.29 U/ml/h), agar (12.22 U/ml/h) and gelatin (12.09 U/ml/h). The maximum enzyme production in immobilized cells required only 24 h where as the free cells required 48 h. From the results, it is concluded that the immobilized cells of *B. subtilis* K-30 in alginate matrix are more efficient for the production of alkaline protease with repeated batch fermentation.

Conclusion

Alkaline protease production by immobilized cells is superior to that of free cells because it leads to higher volumetric activities within the same time of fermentation. The natural polymers such as agar, agarose, pectin and gelatin are employed for cell immobilization (Ramakrishna and Prakasham, 1999). In the present study, we have determined the optimum parameters for maximum production of alkaline protease by the newly isolated strain by various matrices by entrapment technique using alginate, polyacrylamide, agar-agar and gelatin. Specific advantages of this technique include long life-term stability, reusability, and possibility of regeneration to be adaptable also to scale-up the obtained data. Calcium alginate was found to be a promising matrix to immobilize *B.subtilis* K-30 cells for optimum protease production. In addition, experiments with repeated batch of bacterial growth by introducing fresh nutrients at every 24 h leads to increased enzyme activity which is 1.7 times higher than obtained from free cells.

REFERENCES

Adinarayana K, Bapi Raju KVVSN, Ellaiah P (2004). Investigation on alkaline protease production with *B. subtilis* PE-11 immobilized in calcium alginate gel beads. Proc. Biochem., 39: 1331-1339.

Adinarayana K, Jyothi B, Ellaiah P (2005). Production of alkaline protease with immobilized cells of *Bacillus subtilis* PE-11 in various matrices by entrapment technique. AAPS Pharm. Sci. Tech., 06(03): E 391-E397.DOI: 10.1208/pt060348.

Ahmed SA, Abdel-Fattah AF (2010). Production of *Bacillus licheniformis* ATCC 21415 alkaline protease in batch, repeated batch and continuous culture. Mal. J. Microbiol., 6: 156-160.

Alvaro A, Maria HAS (1996). Aerobic immobilized cells in alginate gel particles of variable density. Appl. Biochem. Biotechnol., 57-58: 543-550.

Anna V, Nigar B, Venko B (2003). Cyclodextrin glucanotransferase production by free and agar gel immobilized cells of *B. circulans* ATCC 21783. Proc. Biochem., 38: 1585-1591.

Anwar A, Saleemuddin M (1998). Alkaline proteases: A Review. Bioresour. Technol., 6: 175-183.

Berla TE, Suseela RG (2002). Purification and characterization of alkaline protease from *Alcaligenes faecalis*. Biotechnol. Appl. Biochem., 35: 149-54.

Beshay U (2003). Production of alkaline protease by *Teredinobacter turnirae* cells immobilized in ca- alginate beads. Afr. J. Biotechnol., 2: 60-65.

El Enshasy H, Abuoul-Enein A, Helmy S, El Azaly Y (2008). Optimization of the industrial production of alkaline protease by *Bacillus licheniformis* in different production scales. Aust. J. Basic Appl. Sci., 2: 583-593.

Ferrero MA, Castro GR, Abate CM, Baigori MD, Sineritz F (1996). Thermostable alkaline protease of Bacillus licheniformis MIR29: isolation, production and characterization. Appl. Microbial. Biotechnol., 45: 327-332.

Fumi MD, Silva A, Battistotti G, Golagrande O (1992). Living immobilized *Acetobacter* in Ca-alginate in vinegar production: preliminary study on optimum condition for immobilization. Biotechnol. Letts., 14: 605-608.

Gregg AC, Robert B (1981). The hydrolysis of gelatin by an immobilized acid protease. I. Immobilization and hydrolysis kinetics. Am. J. Enol. Vitic., 32: 14-17.

Gupta A, Roy I, Khare SK, Gupta MN (2005). Purification and characterization of a solvent stable protease from *Pseudomonas aeruginosa* PseA. J. Chromatogr., 1069: 155-161.

Helmo K, Winther NM, Emborg C (1985). Protease productivity in chemostat fermentation with retention of biomass on suspended particles. Enz. Microbiol. Technol., 7: 443-444.

Johnsen A, Flink JM (1986). Influence of alginate properties and gel reinforcement on fermentation characteristics of immobilized yeast cells. Enz. Microbiol. Technol., 8: 737-748.

Joshi BH (2010). Purification and characterization of a novel protease from *Bacillus firmus* Tap5 isolated from tannery industry. J. Appl. Sci. Res., 6: 1068-1076.

Karl-Heinz M (2004). Detergent proteases. Curr. Opin. Biotechnol., 15: 330-334.

Kumari D, Sharm N., Pandove G, Achal V (2009). Alkaline protease production by immobilized cells of *Bacillus pumilis* MTCC 2296 in various matrices (*http://lsj.edu.cn*). Life Sci. J., 6: 8-10.

Kim DM, Kim GJ, Kim HS (1994). Enhancement of operational stability of immobilized whole cell D-Hydantoinase. Biotech. Letts., 16: 11-16.

Kierstan M, Bucke C (1977). The immobilization of microbial cells, subcellular organelles, and enzymes in calcium alginate gels. Biotechnol. BioEng., 19: 387-397.

Lukasz L, Ewa K, Andrezej O (2011). The influence of cell immobilization by biofilm forming on the biodegradation capabilities of bacterial consortia. World J. Microbial. Biotechnol., 27: 1183-1188.

Mohen FN, Dileep D, Deepthi D (2005). Potential application of protease isolated from *Pseudomonas auriginosa* PD100. Biotechnol. Ind., 8: 197-203.

Naidu KSB, Devi KL (2010). Purification and properties of thermostable alkaline protease from *Bacillus subtilis* K-30. J. Pure Appl. Microbiol., 4: 83-90.

Naidu KSB, Devi KL (2005). Optimization of thermostable alkaline protease production from species of Bacillus using rice bran. Afr. J. Biotechnol., 4: 724-726.

Prakash M, Banik RM, Koch-Brandt C (2005). Purification and characterization of *Bacillus cereus* protease suitable for detergent industry. Appl. Biochem. Biotechnol., 127: 143-145.

Prakasham RS, SubbaRao CH, Sreenivas Rao R, Rajesham S, Sarma PN (2005). Optimization of alkaline protease production by *Bacillus* sp using Taguchi methodology. Appl. Biochem. Biotechnol., 120:133-144.

Ramakrishna SV, Prakasham RS (1999). Microbial fermentations with immobilized cells. Curr. Sci., 77: 87-100.

Ramakrishna SV, Jamuna R, Emery AN (1992). Production of ethanol by immobilized yeast cells. Appl. Biochem. Biotechnol., 37: 275-282.

Sen S, Satyanarayana T (1993). Optimization of alkaline protease production by thermophilic *Bacillus licheniformis* S-40. Indian J. Microbiol., 33: 43-47.

Tsuchida O, Yamagota Y, Ishizuka J (1986). An alkaline protease of an alkalophilic *Bacillus* sp. Curr. Microbiol., 14: 7-12.

Veelken M, Pape H (1982). Production of tylosin and nikkomycin by immobilized *Streptomyces* cells. Eur. J. Appl. Microbiol. Biotechnol., 15: 206-210.

Effect of combined use of *Ocimum gratissimum* and *Vernonia amygdalina* extract on the activity of angiotensin converting enzyme, hypolipidemic and antioxidant parameters in streptozotocin-induced diabetic rats

Abdulazeez Mansurah A.[1,2], Ibrahim Kassim[1], Bulus Kenpia[1], Babvoshia Hope B.[1] and Abdullahi, Yusuf[1]

[1]Department of Biochemistry, Ahmadu Bello University, Zaria, Kaduna State, Nigeria.
[2]Natural Product Research and Development Centre, Faculty of Pharmacy, Chiang Mai University, Thailand.

This study was aimed at investigating the angiotensin converting enzyme (ACE) inhibitory effect, hypolidemic and antioxidant properties of combined extract of *Occimum gratissimum* and *Vernonia amygdalina* in streptozotocin-induced diabetic rats. Thirty (30) apparently healthy male Wistar rats were grouped into: control, diabetic, diabetic + insulin, diabetic + 100 mg/kg and diabetic + 250 mg/kg extract groups. All rats except those in the control group were made diabetic using a single intraperitoneal injection of 60 mg/kg streptozotocin. Administration of extract began four days after induction of diabetes and lasted for four weeks. Results show a significant (P<0.05) increase in serum and liver triacyglycerol, total cholesterol and low density lipoprotein levels upon induction of diabetes, which reduced significantly (P<0.05) in rats given insulin and the extract. HDL-C levels remained unchanged in both liver and serum of diabetic rats when compared to control rats. The serum levels of sulfuroxide dismutase (SOD) and catalase increased significantly (P<0.05) in rats given insulin and the extract, while thiobarbituric acid reactive substances decreased significantly (P<0.05) compared to diabetic rats. The angiotensin converting enzyme (ACE) activity in lungs, kidney and heart of diabetic rats were significantly (P<0.05) higher than that of control rats. Also kidney and liver functions were affected on induction of diabetes with a significant increase in levels of urea and aspartate aminotransferase (AST). All of the above factors reduced significantly (P<0.05) when rats were given either insulin or the extract. In conclusion, this study has shown that a combination of the aqueous leaf extracts of *Vernonia amygdalina* and *Ocimum gratissimum* inhibits ACE activity, lowers blood glucose and lipid levels; and is a good antioxidant.

Key words: Angiotensin converting enzyme (ACE), *Ocimum gratissimum* and *Vernonia amygdalina,* diabetes.

INTRODUCTION

Diabetes is a serious metabolic disorder with micro and macrovascular complications that results in significant morbidity and mortality (Rang et al., 1991). According to the World health organization (WHO), there are over 150 million diabetics worldwide and this is likely to increase to about 300 million by the year 2023, in spite of major inroads in understanding the pathophysiology and treatment of the disease (WHO, 1999). Experimental diabetes in animals has provided a considerable insight into the physiologic and biochemical derangements of the diabetic state (Magarinos and McEwen, 2000; Srinivasan and Ramarao, 2007).

Studies have shown that individuals with diabetes have a higher incidence of liver and kidney function abnormalities, as well as formation of free radicals due to glucose oxidation, non-enzymatic glycosylation of proteins and subsequent oxidative degradation of glycated proteins, leading to a decline in antioxidant defence mechanisms and damage of cellular organelles and enzymes, increased lipid peroxidation and development of insulin resistance (Kangralkar et al., 2010; Arora, 2010).

Diabetes mellitus is a complex metabolic disease caused by impairment of insulin signaling path ways and usually results from pancreatic β-cells deficiency and/ or a deficiency of insulin. It is associated with profound alteration in serum lipid and lipoprotein profile with an increase risk in coronary heart disease (Betteridge, 1997; Arora, 2010).

Studies have also demonstrated the possible role of renin angiotensin aldosterone system (RAAS) in the pathogenesis of glomerular injury in diabetes mellitus, with conflicting results on activity of ACE, the major enzyme regulating angiotensin II production (Schernthaner et al., 1984). Also, the ability of ACE inhibitors and some direct renin inhibitors to prevent or show the progression of some complications of diabetes further suggest a pathologic role for RAAS in diabetes mellitus (Schernthaner et al., 1984).

Diabetes without proper treatments can lead to complications such as hypoglycaemia, diabetic ketoacidosis or non-ketotic hyperosmolar coma. More serious complications associated with long term diabetes include cardiovascular diseases, chronic renal failure, renal damage and neuropathy with damage to extremities. Adequate treatment is therefore important. However, population increase, inadequate drug supply, exorbitant cost of treatment and side-effects of several conventional drugs have increased the dependence on plant materials as source of medicine for a variety of ailments, many of which are yet to be scientifically validated (Yaro et al., 2007).

Ocimum gratissimum (basil) and *Vernonia amygdalina* (commonly known as bitter leaf) have beenreportedto have antidiabetic properties (Mohammed et al., 2007). However, very little have been documented on the combined use of both plants with respect to diabetes. Therefore, this work was designed to investigate the effects of the combined use of both plants on blood glucose, activity of ACE and some bio-chemical para-meters in an animal model of diabetes.

MATERIALS AND METHODS

Plant material

The leaves of *O. gratissimum* and *V. amygdalina* were obtained from Ibadan, Oyo state, Nigeria, in the month of March, 2010. They were authenticated by the Botanist at the department of Biological sciences, Faculty of Sciences, Ahmadu Bello University Zaria, Kaduna State, Nigeria.

Preparation of extract

The fresh leaves were air dried and made into powder by grinding. 100 g of the powdered leaves was soaked in 500 ml of distilled water for 48 h. The extract was sieved using a muslin cloth and then filtered under suction pressure with a Whatman's filter paper. All extracts were then concentrated under reduced pressure using a rotary evaporator (Buchi, Switzerland), lyophilized (Christ Alpha 1-2 LD, Germany) and stored at 4°C until needed.

Animals

Thirty (30) apparently healthy male Wistar rats (*Rattus norvegicus*) weighing between 150 to 250 g were obtained from Department of Anatomy, Faculty of Medicine, Ahmadu Bello University, Zaria, Nigeria. The animals were allowed to acclimatize for two weeks before commencement of experiment. They were allowed free access to water and feed (Vital Agricultural feeds, Zaria, Kaduna State) throughout the period of the experiment. All experimental protocol were assessed and approved by the Animal Care and Use Committee of Ahmadu Bello University, Zaria, Kaduna State, Nigeria.

Experimental design

Diabetes was induced by single intraperitoneal dose of 60 mg/kg of streptozotocin (STZ) dissolved in 0.1 ml fresh cold citrate buffer pH 4.5 into 12 h-fasted rats. On the third day of STZ-injection, the rats were fasted for 6 h and blood glucose determined (Elias et al., 1994). Rats given the extracts and distilled water were treated by oral intubation for four weeks after induction of diabetes.
The rats were divided into five groups of six rats each:

Group 1: Control group; rats were normoglyceamic and given distilled water
Group 2: Diabetic group; rats were diabetic, but not treated.
Group 3: Diabetic + insulin; rats were diabetic and treated with insulin (5 iu/100 g).
Group 4: Diabetic + 100 mg/kg extract.
Group 5: Diabetic + 250 mg/kg extract.

Acute toxicity studies

Acute oral toxicity studies of the combined extracts of *O. gratissimum* and *V. amygdalina* were carried out as described by Lorke (1983) with oral administration of increasing doses of the extracts from 10 mg/kg to 5000 mg/kg, while animals were observed for behavioural changes, toxicity and mortality for 24 h.

Determination of fasting blood glucose

Blood samples were collected from the tail of the rats and blood glucose levels determined by the glucose oxidase method using the ONE TOUCH Basic. The results were expressed as mg/dl (Rheney and Kirk, 2000).

Sample collection

At the end of the experiment, rats were sacrificed under anaesthesia and blood collected by cardiac puncture. Blood samples were collected into ethylenediaminetetraacetic acid (EDTA) bottles and centrifuged to obtain serum. Lungs, kidneys and hearts were rapidly removed, cleaned of fatty and connective tissues, blotted dry, weighed and stored at 4°C until use.

Determination of antioxidant activity of the extracts

Thiobarbituric acid reactive substances were determined as described by Buege and Aust (1978). Superoxide dismutase activity was determined using the McCord and Fridovich method (1988) and catalase was measured as described by Machly and Chance (1959).

Determination of ACE Activity

The ACE activity was determined as described by Cushman and Cheung (1971). Organs were homogenised in cold Trizma-HCl buffer (pH = 7.8) (Sigma) and centrifuged at 4°C for 15 min and 5000 x *g*. The supernatants were used for the assay. Briefly, the sample (0.2 mL) was added to ACE solution (50 µL) and the reaction started by adding 0.2 mL of 5 mmol/L hippuric histidyl leucine. This was incubated at 37°C for 15 min. The reaction was terminated by adding 0.25 mL of 1.0 N hydrochloric acid and then 2.0mL ethyl acetate to extract the hippuric acid formed by the action of ACE. This was centrifuged at 3600 x *g* for 2 min, and 1 mL of upper layer transferred into a microcentrifuge tube and heated by dry bath at 100°C for 15 min to remove ethylacetate by evaporation. The resulting hippuric acid was dissolved in 3.0 mL of distilled water, and the absorbance read at 228 nm using a spectrophotometer (Jenway 6100, Dunmow, Essex, U.K.). The activity was determined by:

$$\text{Units/ml enzyme} = \frac{(\text{A228nm Test} - \text{A228 nm Blank})\,(2)\,(3)}{(a)(b)(c)(d)}$$

Where; 2 = Conversion factor since the hippuric acid detected was 1/2 of the total amount produced in the assay: 2 mL of ethyl acetate was added and 1 mL of the organic layer containing the product, hippuric acid, was removed; 3 = total volume of hippuric acid solution; a = millimolar extinction coefficient of hippuric acid at 228 nm; b = time (in minutes) of the assay as per the unit definition; c = extraction efficiency of ethyl acetate, d = volume (in milliliter) of enzyme used.

Activity in tissues was expressed in units which corresponded to 1 µmol of hippuric acid released by enzymatic hydrolysis of HHL per minute per milligram of tissue.

Liver lipid extraction

Liver lipids were extracted as described by Folch et al. (1957). Briefly, 1 g of each tissue sample was homogenized in 20 ml chloroform/methanol, agitated for 15 to 20 min in an orbital shaker and centrifuged (3500x*g*) to recover the liquid phase. Water (0.2 ml) was added to the recovered solvent vortexed at 2000 rpm for some seconds and centrifuged at 3500 x*g* to separate the two phases. The lower phase containing liver lipids was collected for lipid profile determination.

Determination of biochemical parameters

The biochemical analysis of serum and liver samples was performed using reagents kits (Randox Kits). Biochemical parameters measured were aspartate aminotransferase, alanine aminotransferase, creatinine, cholesterol, triacylglycerol, total cholesterol, HDL-cholesterol and LDL-cholesterol.

Statistical analysis

The results were analyzed for statistical significance by one way analysis of variance (ANOVA) using the SPSS 17 statistical program and Post Hoc Test (LSD) between groups. All data were expressed as mean ± SD. P values < 0.05 were considered significant.

RESULTS

Acute toxicity

In the acute toxicity study, a single oral administration of combined extract of *Ocimum gratissimum* and *Vernonia amygdalina* at doses from 10 to 5000 mg/kg did not produce any apparent toxic symptom or mortality after the 24 h observation period.

Effect of *O. gratissimum* and *V. amygdalina* on blood glucose of streptozotocin-induced diabetic rats

The antidiabetic effect of the extract of *O. gratissimum* and *V. amygdalina* on fasting blood sugar levels of diabetic rats is shown on Table 1. The administration of streptozotocin (60 mg/kg) led to over 100% increase in blood glucose levels when compared to control rats (130.16 ± 17.16 mg/kg). Four weeks after daily treatment with insulin and the extract at 100 and 250 mg/kg, fasting blood glucose levels significantly ($P < 0.05$) reduced in diabetic rats. The glucose level in diabetic untreated rats increased from 130.16 ± 17.16 to 178.33 ± 27.55 mg/dl, while that of rats in the control group did not increase significantly. Treatment of rats with insulin significantly ($P < 0.05$) decreased blood glucose levels by 52% (from 140.00 ± 10.95 to 67.21 ± 8.75 mg/dl); the 100 mg/kg extract decreased glucose level by 21.93% (from 114.00 ± 10.67 to 89.33 ± 7.71 mg/dl) and the 250 mg/kg extract decreased blood glucose by 47.23% (from 148.50 ± 24.59 to 78.83 ± 9.75 mg/dl).

Effect of *O. gratissimum* and *V. amygdalina* on serum and liver lipid profile of streptozotocin-induced diabetic rats

The effect of the extract on diabetes-induced hyperlipidemia was also studied. It was observed that induction of diabetes increased triacylglycerols (TAG) from 169.89 ± 6.08 to 262.9 ± 36.63 mg/dl), TC (259.00 ± 8.11 mg/dl) and LDL-C (from 223.66 ± 46.50 to 383.24 ± 39.59 mg/dl)

Table 1. Fasting blood glucose levels of rats after before and after four weeks treatment.

FBG	Control	Diabetic	Diabetic + Insulin	Diabetic + 100 mg/kg extract	Diabetic + 250 mg/kg extract
Initial (mg/dl)	63.00 ± 9.95[a]	130.16 ± 17.16[b]	140.00 ± 10.95[a]	114.00 ± 10.67[c]	148.50 ± 24.59[e]
Final (mg/dl)	70.20 ± 6.87[a]	178.33 ± 27.55[c]	67.21 ± 8.75[b]	89.33 ± 7.71[d]	78.83 ± 9.75[f]

[a, b, c, d, e, f] Mean values within the same column with different superscript letters are significantly different (P<0.05)

Table 2. Effect of combined use of *Ocimum gratissimum* and *Vernonia amygdalina* on serum lipid profile of streptozotocin-induced diabetic rats.

Parameter (mg/dl)	Control	Diabetic	Diabetic + insulin	Diabetic + 100 mg/kg extract	Diabetic + 250 mg/kg extract
TAG	169.89 ± 6.08[a]	262.9 ± 36.63[b]	169.98 ± 12.16[a]	201.03 ± 64.76[a,b]	165.58 ± 24.80[a]
TC	196.70 ± 25.78[a]	259.00 ± 8.11[b]	184.26 ±3.17[a]	185.39 ±52.36[a]	134.38 ±48.23[a]
HDL-C	23.09 ± 10.79[a]	18.87 ± 3.94[a]	30.36 ± 10.85[a]	20.72 ± 6.02[a]	27.96 ± 18.26[a]
LDL-C	223.66 ± 46.50[a]	383.24 ± 39.59[b]	240.52 ± 14.38[a]	248.79 ± 84.60[a]	195.48 ± 47.43[a]

Values are mean ± SD. [a, b]Mean values with different superscript along the same row are significantly different (P<0.05). TAG, Triacylglycerol; TC, total cholesterol; HDL-C, high density lipoprotein cholesterol; LDL-C, low density lipoprotein cholesterol.

Table 3. Effect of combined use of *Ocimum gratissimum* and *Vernonia amygdalina* on liver lipid profile of streptozotocin-induced diabetic rats.

Parameter (mg/dl)	Control	Diabetic	Diabetic + insulin	Diabetic + 100 mg/kg extract	Diabetic + 250 mg/kg extract
TAG	216.21 ± 45.88[a,c]	251.98 ± 52.76[a]	127.97 ± 35.55[b]	158.72 ± 29.76[b,c]	209.51 ± 58.70[a,b]
TC	177.78 ± 29.92[a]	359.26 ± 24.32[b]	161.11 ± 41.07[a]	169.45 ± 33.17[a]	166.73 ± 45.05[a]
HDL-C	46.00 ± 9.09[ab]	55.01 ± 12.60[ab]	36.37 ± 12.17[ab]	38.62 ± 8.55[ab]	37.86 ± 15.39[ab]
LDL-C	207.00 ± 76.60[a]	418.79 ± 53.66[b]	183.13 ± 39.12[a]	174.48 ± 72.83[a]	207.11 ± 65.90[a]

Values are mean ± SD. [a, b]Mean values with different superscript along the same row are significantly different (P<0.05). TAG, Triacylglycerol; TC, total cholesterol; HDL-C, high density lipoprotein cholesterol; LDL-C, low density lipoprotein cholesterol.

levels significantly ($P < 0.05$) compared to rats in the control group. The levels of HDL-C in diabetic rats (18.87 ± 3.94 mg/dl), though lower than control rats (23.09 ± 10.79 mg/dl) was not significantly ($P > 0.05$) different. Treatment of diabetic rats with insulin significantly ($P < 0.05$) reduced TAG (169.98 ± 12.16 mg/dl), LDL-C (240.52 ± 14.38 mg/dl), and TC (184.26 ±3.17 mg/dl). The reduction of TAG by the extract was dose-dependent, with rats given 250 mg/kg of the extract having significantly lower TAG (165.58 ± 24.80 mg/dl) than those given 100 mg/kg (201.03 ± 64.76 mg/kg). Also, LDL-C and TC decreased significantly in rats treated with both doses of the extract. There was no significant difference ($P > 0.05$) in HDL-C levels of diabetic rats and rats given insulin and the extract at both doses (Table 2).

The effect of combined use of *Ocimum gratissimum* and *Vernonia amygdalina* on liver lipid profile of streptozotocin-induced diabetic rats is shown on Table 3. From the results, the levels of TC (359.26 ± 24.32 mg/dl) and LDL-C (418.79 ± 53.66 mg/dl) increased significantly ($P<0.05$) in diabetic rats compared to rats in the control group. Treatment of diabetic rats with insulin, 100 and 250 mg/kg of the extract significantly decreased TC (161.11 ± 41.07, 169.45 ± 33.17 and 166.73 ± 45.05 mg/dl) to levels not significantly ($P > 0.05$) different from rats in the control group (177.78 ± 29.92 mg/dl). The TAG (251.98 ± 52.76 mg/dl) level of diabetic rats was not significantly ($P < 0.05$) different from control rats, but it reduced significantly ($P < 0.05$) when rats were given insulin (127.97 ± 35.55 mg/dl) and the extract at 250 mg/kg (158.72 ± 29.76 mg/dl). However, there was no significant ($P > 0.05$) difference in HDL level in untreated diabetic rats (55.01 ± 12.60 mg/dl) and those given insulin (36.37 ± 12.17 mg/dl), 100 mg/kg (38.62 ± 8.55 mg/dl)

Table 4. Effect of *Ocimum gratissimum* and *Vernonia amygdalina* on some oxidative stress parameters of streptozotocin-induced diabetic rats.

Paramater	Control	Diabetic	Diabetic + insulin	Diabetic + 100 mg/kg extract	Diabetic + 250 mg/kg extract
TBARS (nmol/mg protein)	0.672 ± 0.24[a]	2.04 ± 1.21[b]	0.804 ± 0.25[a]	0.731 ± 0.01[a]	0.787 ± 0.23[a]
SOD (u/ml)	88.23 ± 1.12[a]	33.92 ± 0.26[b]	57.02 ± 1.03[c]	58.52 ± 2.12[c]	68.02 ± 0.60[d]
Catalase (μmol/min)	68.35 ± 0.45[a]	22.92 ± 3.02[b]	56.30 ± 2.73[c]	53.50 ± 1.03[c]	66.25 ± 1.85[a]

Values are mean ± SD. [a, b, c, d]Mean values with different superscript along the same row are significantly different (P<0.05). TBARS, Thiobarbituric acid reactive substances; SOD, superoxide dismutase.

Table 5. Effect of *Ocimum gratissimum* and *Vernonia amygdalina* on some liver and kidney function parameters of streptozotocin-induced diabetic rats.

Paramater	Control	Diabetic	Diabetic ± insulin	Diabetic ± 100 mg/kg extract	Diabetic ± 250 mg/kg extract
AST (IU/L)	12.49 ± 3.91[a]	22.16 ± 2.57[b]	12.00 ± 1.96[a]	12.90 ± 1.93[a]	12.30 ± 1.70[a]
ALT (IU/L)	52.50 ± 10.08[a]	63.90 ± 15.50[a]	51.00 ± 0.08[a]	57.55 ± 2.64[a]	52.45 ± 7.71[a]
Creatinine (mg/dl)	13.00 ± 3.83[a,b]	16.26 ± 2.23[b]	12.33 ± 1.95[a,b]	11.32 ± 2.75[a]	10.22 ± 0.70[a]
Urea (mg/dl)	28.57 ± 7.15[a]	45.23 ± 10.91[b]	30.95 ± 4.12[a]	30.95 ± 3.35[a]	28.90 ± 2.72[a]

Values are mean ± SD. [a, b]Mean values with different superscript along the same row are significantly different (P<0.05). AST, Alanine aminotransferase; AST, aspartate aminotransferase.

and 250 mg/kg (37.86 ± 15.39 mg/dl).

Effect of combined use of *Ocimum gratissimum* and *Vernonia amygdalina* on some oxidative stress parameters of streptozotocin-induced diabetic rats

The induction of diabetes significantly ($P < 0.05$) increased TBARS from 0.672 ± 0.24 nmol/mg protein in control rats to 2.040 ± 1.21 nmol/mg protein in diabetic rats. Administration of insulin, 100 and 250 mg/kg of the extract significantly ($P < 0.05$) decreased TBARS in serum of diabetic rats. There was no significant difference ($P > 0.05$) in TBARS levels of rats in the control group (0.672 ± 0.24 nmol/mg protein), rats treated with insulin (0.804 ± 0.25 nmol/mg protein), 100 (0.731 ± 0.01 nmol/mg protein) and 250 mg/kg (0.782 ± 0.23 nmol/mg protein) of the extract.

The activity of SOD in rats within the control group (88.23 ± 1.12 U/ml) was significantly higher ($P < 0.05$) than diabetic rats (33.92 ± 0.26 U/ml). The SOD activity increased significantly when rats were given insulin (57.02 ± 1.03 U/ml) and the extract at 100 mg/kg (58.52 ± 2.12 U/ml) and 250 mg/kg (68.02 ± 0.60 U/ml). The effect of the extract on SOD was dose-dependent as serum SOD activity of rats given 250 mg/kg extract was significantly higher than that of rats given 100 mg/kg.

Diabetic rats had significantly lower ($P < 0.05$) serum catalase activity (22.92 ± 3.02 μmol/min) compared to rats in the control group (68.35 ± 0.45 μmol/min). There was no significant ($P > 0.05$) difference in serum catalase activity of diabetic rats given insulin (56.30 ± 2.73 μmol/min) and 100 mg/kg of the extract (53.50 ± 1.03 μmol/min), although both groups had significantly ($P < 0.05$) higher serum catalase activity when compared to diabetic rats. The serum catalase activity of rats given 250 mg/kg extract (66.25 ± 1.85 μmol/min) was significantly ($P < 0.05$) higher than that of diabetic rats given the extract at 100 mg/kg (53.50 ± 1.03 μmol/min), showing a dose-dependent effect (Table 4).

Effect of *O. gratissimum* and *V. amygdalina* on some liver and kidney parameters of streptozotocin-induced diabetic rats

From the Table 5, diabetes significantly ($P < 0.05$) increased the serum levels of AST (22.16 ± 2.57 U/L) and urea (45.23 ±10.91 U/L) than rats in the control group. This decreased significantly in rats given insulin and both doses of the extract. The serum AST levels of rats treated with insulin (12.00 ± 1.96 U/L), the extract at 100 mg/kg (12.90 ± 1.93 U/L) and 250 mg/kg (12.30 ± 1.70 U/L) were not significantly different from each other but were significantly ($P < 0.05$) lower than AST levels in serum of untreated diabetic rats. There was no significant ($P > 0.05$) difference in serum ALT of untreated diabetic rats (63.90 ± 15.50 U/L), rats treated with insulin (51.00 ± 0.08 U/L), those given the extract at 100 mg/kg (57.55 ± 2.64 U/L) and 250 mg /kg (52.45 ±7.71 U/L) when

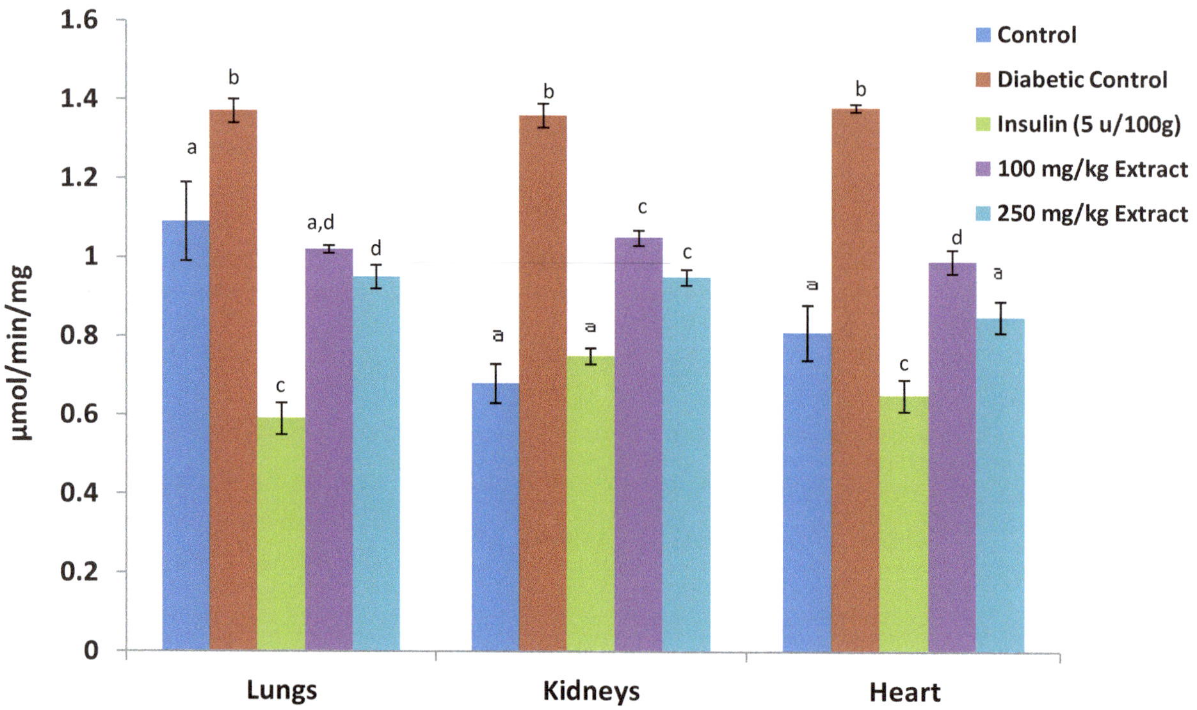

Figure 1. Effect of *Ocimum gratissimum* and *Vernonia amygdalina* on the activity of angiotensin converting enzyme in streptozotocin-induced diabetic rats. [a, b, c, d]Mean values within the same group having different superscript letters are significantly different ($P < 0.05$).

compared to control rats (52.50 ±10.08 U/L).

The serum creatinine level of untreated diabetic rats (16.26 ± 2.23 mg/dl) was not significantly (P<0.05) different from control rats (13.00 ± 3.83 mg/dl) and rats given insulin. However, it decreased significantly (P<0.05) when rats were given the extract at 100 mg/kg (11.32 ± 2.75) and 250 mg/kg of the extract (10.22 ± 0.70 mg/dl) compared to the untreated diabetic rats. Also, serum urea levels decreased significantly ($P < 0.05$) in diabetic rats given insulin (30.95 ± 4.12 mg/dl), the extract at 100 mg/kg (30.95 ± 3.35 mg/dl) and 250 mg/kg (28.90 ± 2.72 mg/dl) when compared to untreated diabetic rats (45.23 ± 10.91 mg/dl).

Effect of combined extract of *V. amygdalina* and *O. gratissimum* on tissue ACE activity of streptozotocin-induced diabetic rats

The effect of the extract on tissue ACE activity was also studied, and from the result obtained (Figure 1), induction of diabetes significantly increased ($P < 0.05$) lung ACE activity (1.37 ± 0.03 µmol/min ml) compared to rats in control group (1.09 ± 0.10 µmol/min ml). This decreased significantly ($P < 0.05$) when rats were given insulin (0.59 ± 0.04 µmol/min ml), 250 mg/kg (0.95 ± 0.03 µmol/min ml) and 100 mg/kg (1.02 ± 0.01 µmol/min ml) of the extract.

The ACE activity in the kidneys (1.36 ± 0.03 µmol/min ml) and heart (1.38 ± 0.01 µmol/min ml) of untreated diabetic rats increased significantly ($P < 0.05$) when compared to the kidneys (0.68 ± 0.04 µmol/min ml) and heart (0.81 ± 0.04 µmol/min ml) of control rats. Treatment of diabetic rats with insulin significantly ($P < 0.05$) reduced kidney (1.05 ± 0.02 µmol/min ml) and heart (0.65 ± 0.02 µmol/min ml) ACE activity than rats treated with the extract at both doses. The decrease in ACE activity in heart of rats given the extract at 100 (0.99 ± 0.01 µmol/min ml) and 250 mg/kg (0.85 ± 0.02 µmol/min ml) was dose-dependent.

DISCUSSION

The measurement of blood glucose is the most important biological marker used in the diagnosis and monitoring of diabetes mellitus both in clinical and experimental settings (Mayfield, 1998). The significant increase in blood glucose level upon treatment with streptozotocin is evident that diabetes was induced in the rats. The diabetic effect of streptozotocin has been attributed to a specific cytotoxic action mediated by hydroxyl radical generation in pancreatic β-cell, which damages a large number of β-cells resulting in a decrease in endogenous insulin release (Szkudelski, 2001). These result in elevated blood glucose within a short period of time after streptozotocin administration (Milagro and Martinez, 2000; Haidari et al., 2012).

In the present study, it was observed that the combined extract of *V. amygdalina* and *O. gratissimum* reversed this (Table 1). Our result is in consonance with reports by Asuquo et al. (2010) and Okokon et al. (2013) who demonstrated the hypoglycemic effect of the combined extract. Although the mechanism of action of the extract was not explored, studies have shown that the causes and sites of intervention in the biochemical process of diabetes is diverse (Larner, 1985). One of such is that blood glucose lowering drugs may act through stimulation of synthesis and/or release of insulin from the beta-cells of the pancreatic islets. They may also increase sensitivity of receptors to insulin, insulinase inhibiting effect, and stimulation of peripheral tissues uptake of glucose cannot be ruled out (Szkudelski, 2001). Also, it has been reported that some herbs or plants reduce absorption of carbohydrates in the digestive system causing progressive entry of glucose into the blood and prevent sudden increase in blood glucose after food intake (Jenkins et al., 1980).

In other studies, the hypoglycaemic and hypolipidemic activities of some plants have been attributed to their fiber and antioxidant contents (Lemhadri et al., 2007). Recently, Ong et al. (2011) demonstrated that *V. amygdalina* decreased blood glucose by increasing GLUT 4 translocation and inhibiting hepatic glucose-6-phosphatase. Thus, it is possible that the two extracts may have exerted their hypoglycemic effects by utilizing one or more of the above mechanisms. A combination of these mechanisms could have resulted in the significant hypoglycemic activity observed in this study, which is likely to be sustained and better than that of a single extract.

The high concentration of TAG, TC and LDL-C observed in diabetic rats compared to control rats in this study is consistent with reports of several studies (Pari and Latha, 2002; Akah et al., 2004; Nwanjo and Oze, 2007; Ayinla et al., 2011) demonstrating that a rise in glucose level on induction of diabetes, results in a corresponding increase in plasma lipids. Hyperlipidemia is a recognized complication of diabetes mellitus characterized by elevated levels of cholesterol, triglycerides, phospholipids and other lipoproteins (Segal et al., 1984; Ayinla et al., 2011). It has been reported that elevated serum lipids in diabetes is due to the increased mobilezation of free fatty acids from peripheral fat depots as a result of inhibition of the hormone sensitive lipase (Sharma et al., 2010). The excess fatty acids produced are converted into phospholipids and cholesterol, which together with excess triacylglycerols formed at the same time in the liver are discharged into the blood in form of lipoproteins. Thus, the marked hyperlipidemia observed in diabetic rats may be regarded as a consequence of uninhibited actions of lipolytic hormones in fats depots (Goodman and Gilman, 1985). Treatment of diabetic rats with the combined extract of *V. amygdalina* and *O. gratissimum* caused a significant ($P < 0.05$) decrease in serum and liver lipids, showing its hypolipidemic effect.

The results of this study support earlier reports (Akah et al., 2004; Nwanjo, 2005; Nwanjo and Oze, 2007; Akah et al., 2009; Ekeocha et al., 2012; Ayinla et al., 2011) and could be related to the presence of alkaloids, saponins, flavonoids and polyphenols (Ayoola et al, 2008; Akinmoladun et al., 2007) known to reduce serum lipid level in animals (Ezekwe and Obioha, 2001).

Lipid peroxidation is a characteristic of diabetes. The increase of free radicals in diabetic condition is suggested to be due to increased lipid peroxidation (Kakkar et al., 1995; Baynes and Thorpe, 1999) and damage of antioxidant defence system as a result of increased production and/ or decreased destruction by non-enzymic and enzymic catalase (CAT), reduced glutathione (GSH), and superoxide dismutase (SOD) (Moussa, 2008). Elevated lipid peroxide commonly based on concentration of TBARS and decreased activity of antioxidant molecules in diabetic rats could probably be associated with oxidative stress and decreased antioxidant defence potential (Koshio et al., 1994; Rosa et al., 2003). Results of the present study is consistent with previous findings (Chis et al., 2009; Bhaskar and Kumar, 2012) demonstrating a significant ($P < 0.05$) increase in TBARS level and concomitant decrease in SOD and catalase in diabetic rats compared to control rats. SOD protects tissues against oxygen free radicals by converting the superoxide radical into hydrogen peroxide and molecular oxygen, while catalase catalyses the detoxification of hydrogen peroxide, preventing damage to cell membranes and other biological structures (Chis et al., 2009). Administration of combined extract of *V. amygdalina* and *O. gratissimum* reduced TBARS to normal levels, while SOD and catalase increased in a dose-dependent manner. Our result supports several studies demonstrating the antioxidant properties of the extracts (Nwanjo, 2006; Akinmoladun et al., 2007; Ayoola et al., 2008; Owolabi et al., 2008; Ong et al., 2011), which has been attributed to the presence of flavonoid and polyphenols in both plants known to possess antioxidant activities (Akah and Okafor, 1992; Ong et al., 2011).

Liver function tests (LFTs) are commonly used in clinical practice to screen for liver disease, monitor the progression of a known disease and determine the effects of potentially hepatotoxic drugs (Harris, 2005). The significant ($P<0.05$) increase in the serum AST observed in diabetic rats (Table 5) is consistent with studies by Nwanjo (2007), indicating liver damage as a result of an increase in hepatic glucose output and a decline in heaptic insulin sensitivity (Harris, 2005). The decrease in AST in rats given the combined extract indicates the possible hepatoprotective effect of the plant extracts (Iwalokun et al., 2006).

With respect to kidney function, the high serum urea

levels in diabetic rats compared to control is consistent with studies by Atangwho et al. (2007) and proves that diabetes affects kidney function. This has been shown to be due to the stimulation of gluconeogenesis as alternative glucose supply route as a result of insulin deficiency. Gluconeogensis is sustained by increased proteolysis which releases glucogenic amino acids that are deaminated in the liver resulting in high urea levels (Robinson and Johnston, 1997). The significant reduction in urea after treatment can be attributed to the ability of the extract to reduce glucose concentration and thus increase insulin effect causing a decline in proteolysis (Robinson and Johnston, 1997).

Results of the present study, which clearly demonstrates that lung, heart and kidney ACE activity is elevated in STZ-induced diabetic rats, supports earlier reports (Erman et al., 1998; Hosseini et al., 2007) that ACE levels increase in serum of diabetic rats. It is well established that the renin-angiotensin aldosterone system (RAAS) is involved in diabetic nephropathy, due to high ACE levels, abnormal aldosterone to renin ratio and altered angiotensin II sensitivity (Lely et al., 2007).

Although it has been speculated that diffuse vascular damage (Lieberman, 1975), and increased shedding from the endothelium in diabetes mellitus may release ACE in the blood stream (Van dyk et al., 1994), the precise underlying mechanism for the increased ACE activity in diabetes mellitus remains to be clarified. The significant ($P<0.05$) decrease in lung, heart and kidney ACE activity in rats given 250 mg/kg aqueous extracts compared to diabetic rats, may be attributed to its ability to reduce blood sugar levels.

Also, the extract may be a good ACE inhibitor, as it has been shown that blocking the renin-angiotensin system with an ACE inhibitor slows the progression of diabetic nephropathy by a mechanism independent of their antihypertensive property (Lewis et al., 1993; Nielsen et al., 2009).

In conclusion, this study has demonstrated that a combination of the aqueous leaf extract of *V. amygdalina* and *O. gratissimum* induces significant reductions in the blood glucose, lipoprotein and ACE levels of streptozotocin-induced diabetic rats. It also exhibits antioxidant properties and could prevent impairment of the kidneys and liver in the diabetic state.

REFERENCES

Akah PA, Okafor CI (1992). Blood sugar lowering effect of *Vernonia amygdalina* (del) in an experimental rabbit model. Phytother. Res. 6: 171-173.

Akah PA, Alemji JA, Salawu OA, Okoye TC, Offiah NV (2009). Effects of *Vernonia amygdalina* on biochemical and haematological parameters in diabetic rats. Asian J. Med. Sci. 1(3): 108-113.

Akah PA, Njoku O, Nwanguma A, Akunyuli D (2004). Effect of aqueous leaf extract of *Vernonia amygdalina* on blood glucose and triglyceride levels of alloxan-induced diabetic rats (Rattus rattus). Anim. Res. Int.1: 90-94.

Akinmoladun AC, Ibukun EO, Afor E, Obuotor EM, Farombi EO (2007). Phytochemical constituent and antioxidant activity of extract from the leaves of *Ocimum gratissimum*. Sci. Res. Essay 2 (5):163-166.

Arora S (2010). Renal function in diabetic nephropathy. World J Diabetes. 1(2): 48-56

Asuquo OR, Igiri AO, Akpan JE,Akpaso MI (2010). Cardioprotective potential of*Vernonia Amygdalina*and *Ocimum Gratissimum*against streptozotocin (Stz) - induced diabetes in wistar rats. Int. J. Trop. Med. 7 (1): 3

Atangwho IJ, Ebong PE, Eteng MU, Eyong EU, Obi AU (2007). Effect of *Vernonia amygdalina* Del leaf on kidney function of diabetic rats. Int. J. Pharmacol. 3(2):143-148

Ayinla MT, Dada SO, Shittu ST, Olayaki LA, Akiode AO, Ojulari SL (2011). Anti-hyperlipidemic effect of aqueous leaf extract of *Ocimum gratissimum* in alloxan induced diabetic rats. Int. J. Med. Med. Sci. 3(12):360-363

Ayoola GA, Coker HAB, Adesegun SA, Adepoju-Bello AA, Obaweva K, Ezennia EC, Atangbayila TO (2008). Phytochemical screening and antioxidant activities of some selected medicinal plants used for malaria therapy in South western Nigeria. Trop. J. Pharm. Res. 7: 1019-1024.

Bhaskar A, KumarA (2012). Antihyperglycemic, antioxidant and hypolipidemic effect of*Punica granatum* L flower extract in streptozotocin induced diabetic rats. Asian Pac. J. Trop Biomed. S1764-S1769

Baynes JW, Thorpe SR (1999). Role of Oxidative stress in diabetic complications. Diabetes 48:1-9.

Betteridge J (1997). Lipid disorder in diabetes mellitus In: Text book of diabetes. Witham SG (Ed.) London: Blackwell Science

Buege JA, Aust SD (1978). Microsomal lipid peroxidation. Methods Enzymol. 52:302-310.

Chis IC, Ungureanu MI, Marton A, Simedrea R, Muresan A, Postescu I, Decea N (2009). Antioxidant effects of a grape seed extractin a rat model of diabetes mellitus. Diab. Vasc. Dis. Res. 6(3): 200-204.

Cushman DW, Cheung HS (1971). Spectrophotometric assay and properties of the angiotensin converting enzyme of rabbit lung. Biochem. Pharm. 20: 1637-1648.

Ekeocha PC, Fasola TR, Ekeocha, AH (2012). The effect of *Vernonia amygdalina* on alloxan induced diabetic albino rats. Afr. J. Food Sci. Technol. 3(3):73-77.

Elias D, Prigoin H, Polak N, Kapoport ML, Cohen AW (1994). Autoimmune diabetes included by the B-cell toxin size. Diabetes 43:992-998.

Erman A, Chen-gal B, David I, Giller S, Boner G, Van Dijk DJ (1998). Insulin treatment reduces the increased serum and lung angiotensin converting enzyme activity in streptozotocin-induced diabetic rats. Scand. J. Clin. Lab. Inv. 58: 81-87.

Ezekwe CI, Obioha O (2001). Biochemical effects of *Vernonia amygdalina* on rat liver microsome. Niger J. Biochem. Mol. Biol. 16: 1745-1798.

Folch J, Lees M, Stanley GH (1957). A simple method for the isolation and purification of total lipids from animal tissues. J. Biol. Chem. 226:497–509.

Goodman LS, Gilman AG (1985). The pharmacological basis of therapeutics. Macmilliian Publishing Company, New York.

Haidari F, Shahi MM, Zarei M, Rafiei H, Omidian K (2012). Effect of green tea extract on body weight, serum glucose and lipid profile in streptozotocin-induced diabetic rats. A dose response study. Saudi Med. J. 33(2):128-33.

Harris EH (2005) Elevated liver function test in type 2 diabetes. Clin. Diabetes 23(3): 115-119.

Hosseini M, Shafiee SM, Baluchnejadmorjarad T (2007). Garlic extract reduces serum angiotensin converting enzyme activity in nondiabetic and streptozotocin-induced diabetic rats. Pathophysiology 14: 109-112.

Iwalokun BI, Alibi-Sofunde JA, Odunala T, Magbagbeola OA, Akinwande AI (2006). Hepatoprotective and antioxidant activities of *Vernonia amygdalina* on acetaminophen-induced hepatic damage in mice. J. Med. Food. 9(4): 526-530.

Jenkins DJ, Wolever TM, Nineham R (1980). Improved glucose tolerance four hours after taking guar with glucose. Diabetologia

19(1): 21-4.

Kakkar R, Kalra J, Mantha SV, Prasad K (1995). Lipid peroxidation and activity of antioxidant enzymes in diabetic rats. Mol. Cell. Biochem. 151: 113–119.

Kangralkar VA, Shivraj DP, Bandivadekar R M (2010). Oxidative stress and diabetes: A review. Int. J. Pharm. App. 1(1): 38-45

Koshio S, Ackmar RG, Lall SP (1994). Effects of oxidized herring and Canola oil diets on growth, survival and flavour of Atlantic salmon, Salmon Salar. J. Agric. Food Chem. 42:1164-1169.

Larner J (1985) Insulin and oral hypoglycemic drugs: Glucagon. In: Gilman, A. G., Goodman, L. S. and Gilman, A. (eds.) Pharmacological Basis of Therapeutics, 7th Edition, Macmillan, London. pp. 1490 – 1516.

Lely LA, Luik PT, Navis G (2007). Angiotensin I-Converting Enzyme: A Pathogenetic Role in Diabetic Renal. Curr. Diabetes Rev. 3: 41-52

Lemhadri A, Eddouks M, Sulpice T, Burcelin R (2007). Anti-hyperglycaemic and anti-obesity Effects of Capparisspinosa and Chamaemelumnobile aqueous extracts in HFD mice. Am. J. Pharmacol. Toxicol. 2(3): 106-110.

Lewis EJ, Hunsicker LG, Bain PR, Rohde RD (1993). The effect of angiotensin converting enzyme on diabetic nephropathy. N. Engl. J. Med. 329 (20): 1456-1462.

Lieberman J (1975). Elevation of serum angiotensin converting enzyme (ACE) level in sarcoidosis. Am. J. Med. 59:356-72.

Lorke D (1983). A new approach to practical acute toxicity testing. Arch. Toxicol. 54: 275-287.

Machly AC, Chance B (1959) The assay of catalase and peroxidase. In: Methods of Biochemical Analysis. Glick, D. (Ed.) Inter Science Publishers Inc., New York. 1:357-425.

Mayfield J (1998). Diagnosis and classification of diabetes mellitus: New criteria. Am. Family Phys. 58(6): 1355-1362.

Magarinos AM, McEwen BS (2000). Experimental diabetes in rats causes hippocampal dendritic and synaptic reorganization and increased glucocorticoid reactivity to stress. Proc. Natl. Acad. Sci. USA 97 (20): 11056–11061.

McCord JM, Fridovich I (1988) Superoxide dismutase: the first twenty years (1968–1988). Free Radic. Biol. Med. 5: 363–369.

Milagro FI, Martinez JA (2000). Effect of the oral administration of a beta 3-adrenergic agonist on lipid metabolism in alloxan-diabetic rats. J. Pharm. Pharmacol. 52(7):851-856.

Mohammed YT, Okasha MA, Magaji RA, Yaro AH (2007). Effects of aqueous leaves extract of *Ocimum gratissimum* on blood glucose levels of streptozotocin-induced diabetic wistar rats. Afr. J. Biotechnol. 6 (18):2087-2090.

Moussa SA (2008). Oxidative stress in Diabetes Mellitus. Romanian J. Biophys. 18(3):225-236.

Nielsen SE, Takeshi S, Tarnow L, Lajer M, Schjoedt KJ, Strup AS, Baba T, Hans-Henrik P, Rossing P (2009). Tubular and Glomerular Injury in Diabetes and the Impact of ACE Inhibition. Diabetes Care 32(9): 1684-1688.

Nwanjo HU (2005). Efficacy of aqueous leaf extract of *Vernonia amygdalina* on plasma lipoprotein and oxidative status in diabetic rat models. Niger. J. Physiol. Sci. 20(1-2):39-42.

Nwanjo HU (2006). Efficacy of aqueous leaf extract of *Vernonia amygdalina* on plasma lipoprotein and oxidative status in diabetic rat models. Niger. J. Phys. Sci. 20 (1-2): 39-42

Nwanjo HU (2007). Studies on the effect of aqueous extract of *Phyllanthus niruri* on plasma glucose level and some hepatospecific markers in diabetic Wistar rats. Int. J. Lab. Med. 2(2): 1-18.

Nwanjo HU, Oze GO (2007). Hypolipidaemic and antioxidant properties of *Ocimum gratissimum*on diabetic rats. Plant Prod. Res. J. 11:1-4

Okokon JE, Umoh UF, Ekpo BAJ, Etim EI (2013). Antidiabetic study of combined extracts of *Vernonia amygdalina, Ocimum gratissimum*, and *Gongronema latifolium* on alloxan-induced diabetic rats. J. Nat. Pharm. 4(1): 28-31.

Ong KW, Hsu A, Song L, Huang D, Tan BK (2011). Polyphenols-rich Vernonia amygdalina shows anti-diabetic effects in streptozotocin-induced diabetic rats. J. Ethnopharmacol. 133(2):598-607

Owolabi MA, Jaja SI, Oyekanmi OO, Olatunji J (2008). Evaluation of the antioxidant activity and lipid peroxidation of the leaves of *Vernonia amygdalina*. J. Compl. Integr. Med. 5:1.

Pari L, Latha M (2002). Effect of *Cassia Auriculata* flowers on blood sugar levels, serum and tissue lipids in streptozotocin diabetic rats. Singapore Med. J. 43(12): 617-621

Rang HP, Dale MM, Ritters, JM (1991). The endocrine pancrease and the control of blood glucose: In: Barbara Simmons and Susan Beasley (Ed.) Pharmacology, United Kingdom: Longman Group Ltd, pp. 403-410.

Rheney CC, Kirk KK (2000). Performance of three blood glucose meters. Ann. Pharmacother. 34(3):317-321

Robinson G, Johnston DE (1997). Metabolic disorder: Diabetes. In: Mechanisms of disease. An introduction to clinical science. 1st Edn., Cambridge: Cambridge University Press

Rosa EVC, Valgas C, Souza-Sierra MM, Correa AXR, Radetski CM (2003). Biomass growth, micronucleus induction, and antioxidant stress enzyme responses in *Vicia faba* exposed to cadmium in solution. Environ. Toxicol. Chem. 22(3):645 - 649.

Schernthaner G, Schwarzer CH, Kuzmits R, Muller MM, Klemen U, Freyler H (1984). Increased angiotensin-converting enzyme activities in diabetes mellitus: analysis of diabetes type, state of metabolic control and occurrence of diabetic vascular disease. J. Clin. Pathol. 37:307-312

Segal RA, Huang B, Ramanis, Z, Luck DJL (1984). Mutant strains of *Chlamydomonas reinhardtii* that move backwards only. J. Cell Biol. 98: 2026-2034.

Sharma SB, Gupta S, Ac R, Singh UR, Rajpoot R, Shukla SK (2010). Antidiabetogenic action of *Morus rubra* L. leaf extract in streptozotocin-induced diabetic rats. J. Pharm. Pharmacol. 62(2):247-55.

Srinivasan K, Ramarao P (2007). Animal models in type 2 diabetes research: An overview. Indian J. Med. Res. 125: 451-472

Szkudelski T (2001). The mechanism of alloxan and streptozotocin action in B cells of the rat pancreas. Physiol. Res. 50(6): 537-46.

Van Dyk DJ, Erman A, Erman T, Chen-Gal B, Sulkes J, Boner G. (1994). Increased serum angiotensin converting enzyme activity in type I insulin-dependent diabetes mellitus: its relation to metabolic control and diabetic complications. Eur. J. Clin. Invest. 24: 463-467.

World Health Organization (WHO) (1999). Definition, Diagnosis and classification of Diabetes Mellitus and its complication www.who.org

Yaro AH, Anuka JA, Salawu OA, Magaji MG (2007). Anticonvulsant activities of methanol extract of *Chrysanthellum indicum* linn. Vatke in mice and chicks. Niger. J. Pharm. Sci. 6(2): 22-27.

Haematology, plasma enzymes and organ indices of *Clarias gariepinus* after intramuscular injection with aqueous leaves extracts of *Lepidagathis alopecuroides*

[1]Gabriel, U. U., Obomanu, F. G[2]. and Edori, O. S.[3]

[1]Department of Fisheries and Aquatic Environment, Rivers State University of Science of Science and Technology, P.M.B. 5080, Port Harcourt, Nigeria.
[2]Department of Chemistry, Rivers State University of Science of Science and Technology, P.M.B. 5080, Port Harcourt, Nigeria.
[3]School of Remedial and General Studies, Rivers State College of Health Science and Technology, P.M.B. 5039, Port Harcourt, Nigeria.

Tank-raised *Clarias gariepinus* (mean total length, 30.05 ± 4.05cm,SD; mean weights, 250.04 ± 50.22 gSD) were injected intra-muscularly with two millilitres per kilogram of each of five concentrations (2.00, 2.00, 6.00, 8.00 and 10.00 ppm) of aqueous extracts of leaves of *Lepidagathis alopecuroides* above the lateral line. The fish in the control group were injected with same dose of distilled water. There were four fish per treatment level. On the 14th day the fish was assessed for organ indices (hepatosomatic index, HIS; cardiosomatic index, CSI; renatosomatic index, RSI and spleenosomatic index, SSI) and Fulton's condition, K . Blood samples were analysed for packed cell volume, PCV; haemoglobin, Hb; red blood cell, RBC white blood cell, WBC and platelets. The plasma was analysed for the transaminases (aspartate transaminase, AST; alanine transaminase, AST), alkaline phosphostase, ALP and lactate dehydrogenase, LDH. The values of PCV ($p < 0.05$), HB and RBC ($p > 0.05$) in the control were higher than those in the treated group. However, the decline was not concentration-dependent. The increase in the number of WBC in the exposed fish was twofold and above ($p < 0.05$) that in the control, 66.67 ± 23.09x10^{6}cells/l with highest value (360.00 ± 33.24x10^{6}cells/l) recorded at 10.00 ppm. The response pattern of platelets to the extracts varied very widely among the treated group ($p < 0.05$) with fish injected with 2.00 ppm having the highest value (100.00 ± 33.20x10^{9}cells/l), which was about twofold that at other concentrations. Generally, the activities of AST, ALT, ALP and LDH were inhibited below the control value without any direct relationship to toxicant concentration. Nevertheless, AST generally showed a decline in activity from that of the control except at 4.00 ppm (117.75 ± 9.80 IU/L); ALP declined with a slight increase at 10.00 ppm above the control. There was no difference ($p > 0.05$) between the organ indices in control and treated group except in the SSI. K and HSI showed variable responses relative to extracts concentration, while the heart showed a very slight decline in size. RSI and SSI had a reverse pattern. Exposure of *C. gariepinus* to *L. alopercuroides* in the open waters during fishing activity may impact negatively on the physiology of the fish as manifested in changes in some of the blood parameters, plasma enzymes and organ indices of the fish.

Key words: *Clarias gariepinus, Leidagagthis.alopercuroides,* plasma enzymes, blood parameters, organ indices, condition.

INTRODUCTION

In recent years there is preference for safe and environmentally friendly piscicides of plant origin than synthetic piscicides for catching fish and clearing ponds. This is because ichthyotoxins are less expensive, biodegradable, readily available, easy to handle and safe for mankind and the environment (Singh et al., 1996). Obnoxious fish-

*Corresponding author. E-mail: ugwemg@yahoo.com.

ing activities in the form of sublethal pollution have negative or destructive influence on the aquatic environment and may result in chronic stress conditions that may have adverse effect on aquatic species (Mason, 1991). Fish physiology (haematological and biochemical parameters) are suitable tools for assessing environmental influences and stress effects of anthropogenic origin on the condition and health of aquatic vertebrates (Cêlik, 2004). Since there is a close association between the circulatory system of fish and the external environment (Cech et al., 1996; Bonga, 1997), the effect of external stressors and toxic substances on exposed fish could be made mani-fest through clinical diagnosis of fish physiology.

Plant parts have been shown to cause death of fish and changes in biochemical responses of *Channa punctatus* (Tiwari and Singh, 2004), haematological and histopathological effects on *Clarias gariepinus* (Fafioye et al, 2004; Omoniyi et al., 2002). The leafy parts of *Lepidagathis alepecuroides* is used to immobilize and kill mudskippers (*Periophthalmus papillio*) and other fish species in Rivers and Cross River States of Nigeria (Obomanu et al., 2007) but the disruptive internal changes leading to death of the poisoned fish is yet to be examined. It has been shown that before death resulting from animal intoxication, various physiological and biochemical changes take place in the internal environment of fish (Adeyemo 2005; Fafioye et al., 2004; Obomanu et al., 2007).

The plant, *L. alopecuroides* have been in use for a long time but the effects of direct use in water and mudflats, its mode of action in fish and the resultant fish kill have not been examined. Therefore this study was carried out to assesess some haematologic, biochemical changes, condition and organ indices of the commercial fish, *Clarias gariepinus* injected intramuscularly with different sublethal concentrations of aqueous extracts of *L. alopecuroides* under laboratory conditions.

MATERIALS AND METHODS

Fresh leaves of *L. alopecuroides* were collected from the wild and air dried for three weeks in the Department of Chemistry, Rivers State University of Science and Technology, Port Harcourt. The dried leaves were ground into power with an electric blender and sieved with a test sieve into a fine power and then stored in a dry airtight container. Tank-raised *C. gariepinus* (mean total length, 30.05 ± 4.05 cmSD; mean weights, 250.04 ± 50.22 gSD) were obtained from a private farm and transported in aerated aquaria to the same laboratory. They were acclimated individually in bore hole water (characteristics: dissolved oxygen -4.50 ± 0.05 mg/l, pH- 7.5 ± 1.3, conductivity -410 ± 20.4 µS/cm, total dissolved solid- 400 ± 10.25 ppm) in circular plastic aqaarium (volume, 30*l*, effective water volume, 10*l*) for 7days and fed with a 35% crude protein diet at 1% biomass.

The aquaria were washed daily with a piece of foam to remove uneaten food and faecal matters during the accli-mation and experiment. A stock solution (100 mg/l) of the plant powder was prepared from which five concentrations (2.00, 2.00, 6.00, 8.00 and 10.00 ppm) were prepared. Each fish was injected with two millilitres per kilogram of each of the extracts concentrations above the lateral line. Fish in the control group was also injected with same dosage of distilled water. Blood was collected from each fish on the 14th day after injection with a syringe and 21G hypodermic needle from the kidney between the urino-genital organ and anal fin. Blood samples for haematological studies were pre-served in EDTA embedded bottles and that for enzymes (AST, ALT, ALP and LDH) analysis in heparinised bottles. The fishes were then killed by a blown on the head and the organs (kidney, liver, heart and spleen) excised and weighed for organ indices assessment.

Packed cell volume (PCV) was determined using Hawsley micropillary tubes and centrifuged for 5 min (Abudu and Safola, 1994). Red blood cells (RBC), white blood cells (WBC), haemoglobin (Hb) and platelets were analysed according to the methods of Blaxhal and Daisley (1973). The heparinised blood samples were centrifuged at 300 rpm for 10 minutes and the plasma collected for analysis. AST and ALT were analysed according to the method of Reitman and Frankel (1957); ALP according to Bessey et al. (1946) and LDH, according to the method of the Huckabee (1961). The organ indices were calculated according to the methods of (Jenkins, 2004; Adams et al., 1993). Data obtained were subjected to statistical analysis using one way analysis of variance at 95% probability and means separated by Duncan's multiple range test at 95% probability (Wahua, 1999).

RESULTS

The values of PCV ($p < 0.05$), HB and RBC ($p > 0.05$) in the control were higher than those in the treatment levels. However, the decline was not concentration dependent (Table 1). The increase in the number of WBC in the exposed fish was twofold and above ($p < 0.05$) that in the control, 66.67 ± 23.09x10^9cells/l with highest value (360.00 ± 33.24x10^9cells/l) recorded at 10.00 ppm. The response pattern of platelets to the extracts varied very widely among the treated group ($p < 0.05$), with fish injected with 2.00 mg/l extract having the highest value (100.00 ± 33.20x10^9cells/l), which was about twofold that at the other concentrations.

Generally, the activities of AST ($p > 0.05$), ALT ($p > 0.05$), ALP ($p > 0.05$) and LDH ($p < 0.05$) were inhibited below the control value without any direct relationship to toxicant concentration. Never- theless, AST generally showed a decline in activity from that of the control, 108.6743.09IU/L except at 4.00ppm (117.75 ± 9.82 IU/L). ALT and ALP activities declined below their respective control values (ALT, 18.33 ± 10.97 IU/L; ALP, 43.33 ± 8.74 IU/L) with a slight increase at 10.00 ppm above the control. LDH appeared to be most impacted upon by extracts with a one-sixth reduction at 10.00 ppm (Table 2). There was no difference ($p > 0.05$) in all the organ indices between the control and treated group except in the spleenosomatic index (Table 3).

Condition factor (K) and hepatosomatic index (HSI) showed variable response relative to extract concentration, while the heart showed a very slight decline in size (Table 3). Renatosomatic and spleenosomatic indices (RSI and SSI) had a reverse pattern (Table 3).

Table 1. Haematological variables of *Clarias gariepinus* 14days from the day of injection with aqueous extracts of *Lepidagathis alepecuroides* (mean ± SD).

Conc. of *L. alepecuroides* (ppm)	Haematological variables				
	PVC (%)	Hb (g/dl)	RBC ($x10^6$cells/L)	WBC ($x10^9$cells/L)	Platelets ($x10^9$cells/L)
0.00	26.00 ± 14.93[a]	8.60 ± 54.76[a]	2.70 ± 1.56[a]	66.67 ± 23.09[d]	53.33 ± 23.09[b]
2.00	8.00 ± 4.55[b]	2.65 ± 1.49[b]	0.90 ± 0.48[b]	130.00 ± 38.30[c]	100.00 ± 43.20[a]
4.00	13.25 ± 14.66[ab]	4.60 ± 4.85[ab]	1.40 ± 1.49[ab]	165.00 ±156.95[bc]	50.00 ± 20.00[b]
6.00	15.50 ± 9.85[ab]	5.10 ± 3.24[ab]	1.50 ± 0.87[ab]	240.00±374.17[ab]	40.00 ± 28.28[b]
8.00	8.75 ± 5.56[b]	3.05 ± 1.81[b]	1.03 ± 0.56[b]	285.00±288.62[a]	35.00 ± 19.15[b]
10.00	13.25 ± 6.18[ab]	4.53 ± 2.06[ab]	1.40 ± 0.63[ab]	360.00±233.24[a]	55.00 ± 30.00[b]

Means with the same superscript are not significantly different ($p < 0.05$).

Table 2. AST, ALT, ALP and LDH in C. *gariepinus* 14 days from the day of injection with aqueous extracts of *L. alepecuroides* (means ± SD).

Conc of L. alepel (ppm)	AST (IU/L)	ALT (IU/L)	ALP (IU/L)	LDH (IU/L)
0.00	108.67 ± 43.09[a]	18.33 ± 10.97[a]	43.33 ± 8.74[a]	66.67 ± 23.09[ad]
2.00	76.75 ± 42.77[a]	17.25 ± 12.12[a]	37.25 ± 6.65[a]	125.00 ± 21.21[a]
4.00	17.75 ± 98.02[a]	13.50 ± 3.79[a]	34.25 ± 4.72[a]	57.75 ± 16.97[c]
6.00	91.25 ± 50.99[a]	13.25 ± 10.18[a]	41.25 ± 6.18[a]	81.25 ± 19.96[bc]
8.00	88.75 ± 34.26[a]	16.75 ± 8.26[a]	38.50 ± 6.03[a]	88.50 ± 11.47[ab]
10.00	99.25 ± 42.57[a]	22.00 ± 13.64[a]	46.00 ± 15.14[a]	10.00 ± 37.05[ab]

Means with the same superscript in the same column are not significantly different ($P < 0.05$).

Table 3. Organosomatic indices of *C. gariepinus* injected with *L. alepecuroides* 14 days after injection (mean ± SD).

Conc of *L. alepecuroides* (ppm)	Fulton's Condition (K)	Hepatosomatic index (HSI)	Cardiosomatic index (CSI)	Renatosomatic index (RSI)	Splenosomatic index (SSI)
0.00	0.78 ± 0.05[a]	0.67 ± 0.08[a]	0.15 ± 0.01[a]	0.59 ± 0.52[a]	0.03 ± 0.02[b]
2.00	0.73 ± 0.05[a]	0.83 ± 0.035[a]	0.15 ± 0.01[a]	0.82 ± 0.52[a]	0.06 ± 0.02[ab]
4.00	0.84 ± 0.09[a]	0.71 ± 0.21[a]	0.15 ± 0.02[a]	0.72 ± 0.08	0.09 ± 0.03[a]
6.00	0.72 ± 0.09[a]	0.59 ± 0.44[a]	0.14 ± 0.06[a]	0.73 ± 0.12[a]	0.07 ± 0.03[ab]
8.00	0.79 ± 0.05[a]	0.71 ± 0.23[a]	0.14 ± 0.02[a]	0.69 ± 0.09[a]	0.06 ± 0.02[ab]
10.00	0.76 ± 0.03[a]	0.60 ± 0.16[a]	0.14 ± 0.02[a]	0.65 ± 0.12[a]	0.05 ± 0.05[b]

Means with the same superscript in the same column are not significantly different ($P < 0.05$).

DISCUSSION

According to Barton (2002), stressors evoke non-specific responses in fish which enables the fish to cope with the disturbance and maintenance of its homeostatic state. If severe or long lasting, the response then becomes maladaptive and threatens the fish health and wellbeing. Therefore, in the presence of stressors (contaminants /pollutants), blood parameters and blood chemistry can be employed as standard laboratory test to determine diseased conditions and metabolic disturbances in fish (Celik, 2004).

The decrease in haemoglobin concentration with increase in the concentration of the plant extract is similar to those reported in *C. gariepinus* exposed to cassava effluents and tobacco (*Nicotiana tobaccum)* leaf extracts (Adeyemo, 2005, Omoniyi et al., 2002). This pattern of response may be attributed to haemolysis which results in haemodilution, a means of diluting the haemoconcentration of the extracts thus reducing the effect of the toxicant/pollutant in its system (Smith et al., 1979; Sampath, 1993). Besides, it may result from either an increase in the rate of Hb destruction or decrease in its production/synthesis (Reddy and Bashamohideen, 1989). Decrease in PCV shows the extent of the shrink-age of cell size and decrease in the number of cells (Ahmad et al., 1995). Decrease in the Hb levels may impair oxygen supply to various tissues resulting in slow metabolic rate and low energy production (Ahmad et al., 1995; Atamanalp and Yanik, 2003). Prolonged reductions also lead to blood dyscrasis and degeneration of the erythrocytes being described as a pathological condition in fishes exposed to toxicants (Buckley et al., 1976).

Adeyemo (2005) and Gabriel et al. (2007) respectively recorded significant changes in the WBC of *C. gariepinus* exposed to cassava mill effluents and refined petroleum oil (kerosene). Similar changes in was also recorded by Shakoori et al., 1996) in *Cyprinus idella* treated with fenvalerate, a pyrethroid pesticide. The above two to fivefold increase in WBC (leukocytosis) may have resulted from the excitation of the defense mechanism of the fish to counter the effect of the toxicant. Platelets showed variable responses to different levels of the toxicant which may be a mechanistic response of the cells to fight or combat the effect of the pollutant or the toxicant on the fish. Alteration in the rate of production of platelets by the head kidney and bone marrow due to toxicant effects may be responsible for the variations in the number of platelets in the exposed fish.

The general decrease in the activities of ALT and AST in the exposed fish corroborates the findings of Luskova et al. (2002) in *Cyprinus carpio* exposed to diazinon but contradicts that of Tiwari and Singh (2004) in *Channa punctatus* treated with sublethal levels of alcoholic extracts of *Nerium indicum.* ALT and AST are non plasma specific enzymes that are localized in tissues cells of liver, heart, gills, kidneys, muscles and other organs and their presence in the blood (plasma) may give specific information about organ dysfunction (Wells, 1986; Gabriel and George, 2005). A decrease in the transaminases suggests that there was no tissue damage (Luskova et al., 2002; Ayalogu et al., 2001), the parenchymatous tissues and skeletal muscles being intact, but rather a decrease in the rate of transfer of the amine groups which eventually affects the rate of protein and carbohydrate synthesis in the fish (Watts and Watts, 1974). This is a direct effect of a reduction in the conversion of α-ketoglutarate and alanine to pyruvic and glutamic acid, respectively. The decline in ALP in exposed fish in this study may be due to the fall in the rate of synthesis of glycogen resulting from the low metabolic demands (Shaffi, 1979) and a decrease in metabolic transport (Begum, 2004; Edquist et al., 1992). The decrease may also indicate that there was no kidney damage but a reduction in the hydrolytic action on a number of phosphormonoesters of organic origin such as glucose (Edquist et al., 1992). A reduction in the concentration of LDH in the plasma of the experimental fish infers a decrease in the glycolytic process due to lower metabolic rate (Luskova et al., 2002), a shift towards anaerobic respiration (Tiwari and Singh, 2004), possibly due to a hypoxic internal environment.

Reduction in organ indices and fish wellbeing (condition) depending on its severity may limit physiological systems, reduce growth rate and impair reproducetion (Adams et al., 1996; Jenkins, 2004). The minimal change ($p > 0.05$) in fish condition may be due to the fact that the concentrations of the toxicant were below threshold values or before sampling the fish was able to efficiently detoxify and excrete the wastes, thereby maintaining its physiological integrity without any negative effects on its wellbeing. A slight increase in the HSI of exposed fish indicates that the liver cells were affected possibly causing an increase in the rate of production of endoplasmic reticulum for the synthesis of protein in liver tissue (Anderson et al., 1988). The liver is responsible for enzymatic decontamination process, vitellogenin production and storage of glycogen as energy reserves. Therefore, in the presence of stressors, these qualities are altered resulting in deleterious effect on the fish (Jenkins 2004; Adams et al., 1996). Minimal changes recorded in the cardiosomatic index of fish exposed to extracts of *L. alepecuroides* suggest that the extracts did not interfere with the condition of the heart. The kidney (renatosomatic index) and spleen (spleenosomatic index) increased in this study. These organs are considered as the haematopoietic (blood producing) tissues in fish (Singh and Singh, 1982; Jenkins, 2004). Both organs are also involved in blood filtration, the development of new blood cells and the immunological interactions. Hence, increase in their size may indicate a pathological response to combat the effect of the toxicant enhancing their ability in the destruction of blood cells (Jenkins, 2004).

Exposure of *C. gariepinus* to *L. alopercuroides* under laboratory conditions moderately affected some aspects of its physiology, a condition which may be worse in field applications where lethal doses of the plant materials are employed in stupefying and catching fish in many parts of the Niger Delta region of Nigeria. Hence, the practice should be discouraged.

REFERENCES

Abudu JO, Sofola DA (1994). Relationship between red cell mass and packed Cell volume in Nigeria primigraviadae. Niger. J. Physiol. Sci. 10(1/2): 13-21.

Adams SM, Ham KD, Greeley MS, LeHew RF, Hinton DE, Saylor CF (1996). Down stream gradients in bioindicator responses: point source contamination effects on fish health. Can. J. Fish. Aquat. Sci. 53: 2177-2187.

Adeyemo OK (2005). Haematological and histopathological effects of cassava mill effluent in *Clarias gariepinus*. Afri. J. Biomed. Res. 8(3): 179-183.

Ahmad F, Ali SS, Shakoori AR (1995). Sublethal effects of danitor (fenpropathrin), a synthetic pyrethroid on fresh water Chinese grass carp, Ctenopharyngodon idella. Folia Biol. (Krakow) 43: 151-159.

Atamanalp M, Yanik T (2003). Alterations in hematological parameters of rainbow trout (*Oncorhynchus mykiss*) exposed to mancozeb. Turk. J. Vet. Anim. Sci. 27: 1213-1217.

Ayalogu OE, Igboh NM, Dede EB (2001). Biochemical changes in the serum and liver of albino rats exposed to petroleum samples (gasoline, kerosene and crude petroleum). J. Appl. Sci. Environ.

Manag. 5(1): 97-100.
Anderson T, Forlin L, Hardig J, Larsson A (1988). Physiological disturbances in fish living in coastal water polluted with bleached kraft mill effluents. Can. J. Fish Aquat. Sci. 45: 1525-1536.
Barton AB (2002). Stress in Fishes: a diversity of responses with particular reference to changes in circulating corticosteroids. Integr. Comp. Biol. 42: 517-525.
Begum G (2004). Carbofuran insecticide induced biochemical alternations in liver and muscle tissues of fish *Clarias batrachus* (Linn.) and recovery response. Aquat. Toxicol. 66(1): 83-91.
Bessey OA, Lowry OH, Brock MJ (1946). A method for rapid determination of alkaline phosphatase with five cubic millimeters of serum. J. Bio. Chem. 164: 321-329.
Blaxhall PC, Daisley KW (1973). Routine haematological methods for use with fish blood. J. Fish Biol. 5(6): 771-781.
Buckley JA, Whitemore CM, Matsuda RI (1976). Changes in blood chemistry and blood cell morphology in coho salmon, *Oncorhynchus kisutch* following exposure to sub-lethal levels of total residual chlorine in municipal waste water. J. Fish. Res. Board Canada 33: 776-782.
Cech Jr JJ, Bartholow SD, Young PS, Hopkins TE (1996). Striped bass exercise and handling stress in freshwater: physiological responses to recovery environment. Trans. Am. Fish. Soc. 125(2): 208-320.
Celik ES (2004). Blood chemistry (electrolytes, lipoproteins and enzymes) values of black scorpion fish (*Scorpaena porcus*) in the Dardanelles. Turkey J. Biol. Sci. 4(6): 716-719.
Edquist LE, Madej A, Forsberb M (1992). Biochemical blood parameters in pregnant mink fed pcb and fractions of pcb. Ambio 21(8):577-581.
Fafioye OO, Adebisi AA, Fagade SO (2004). Toxicity of *Parkia biglobosa* and *Raphia vinifera* extracts on *Clarias gariepinus* juveniles. Afr. J. Biotech. 3(11): 627–630.
Gabriel UU, George ADI (2005). Plasma enzymes in *Clarias gariepinus* exposed to chronic levels of round up (glyphosate). Environ. Ecol. 23(2): 271-276.
Gabriel UU, Amakiri NE, Ezeri GNO (2007). Haematology and gill pathology of *Clarias geriepinus* exposed to refined petroleum oil, kerosene under laboratory conditions. J. Anim. Vet. Advances 6(3): 461-465.
Huckabee WE (1961). In : Hawk's Physiological Chemistry, 14th edition (B.L. Oser, Ed.), New Delhi, Tata McGraw-Hill p. 1103.
Jenkins JA (2004). Fish bioindicators of ecosystem condition at the Caleasieu Estuary, Louisiana. National Wetlands Research Center, USGS, Lafayette p. 54.
Luskova V, Svoboda M, Kolarova J (2002). The effects of diazinon on blood plasma biochemistry in carp (*Cyprinus carpio*). ACT VET. BRNO 71: 117-125.
Mason CF (1991). Biology of fresh water fishes. Longman Scientific and Technical, New York, U.S.A. p. 351.
Obomanu FG, Ogbalu OK, Gabriel UU, Fekarurhobo SGB, Abadi SU (2007). Piscicidal effects of *Lepidagathis alopecuroides* on mudskipper, *Periophthalmus papillio* from the Niger Delta. Res. J. Appl. Sci. 2(4): 382-387.
Omoniyi I, Agbon AO, Sodunke SA (2002). Effect of lethal and sublethal concentrations of tobacco (*Nicotiana tobaccum*) leaf dust exract on weight and hematological changes in *Clarias gariepinus* (Burch.). J. Appl. Sci. Environ. Manage. 6(2): 37-41.
Reddy PM, Bashamohideen M (1989). Fenvalerate and cypermethrin induced changes in the haematological parameters of Cyprinus carpio. Acta. Hyrochim. Hydrobiol. 17: 101-107.
Reitman S, Frankel S (1957). Colorimetric assay of alamine and aspartate aminotransferase. Am. J. Clin. Path. 33: 1-13.
Shaffi SA (1979). Effects of starvation on tissue and serum gluconeogenic enzymes, alkaline phosphatase and tissue glycogen in the freshwater catfish *Heteropneustes fossilis* (Bloch). Acta Physiol. Sci. Hung. 53(4): 501-505.
Shakoori AR, Mughal AL, Igbal MJ (1996). Effects of sublethal doses of fenvalerate (a synthetic pyrethroid) administered continuously for four weeks on the blood, liver and muscles of a freshwater fish, *Ctenopharyngodon idella*. Bull. Environ. Contam. Toxicol. 57: 487-494.
Singh D, Singh A (1982). Effects of copper and zinc sulphate on the blood parameters of *Mystus vittatus. Matsya* 8: 1-6.
Singh A, Singh DK, Mishra TN, Agarwal RA (1996). Molluscicides of plant origin. Biol. Agric. Horti. 13: 205-252.
Sampath K, Velamnial S, Kennedy IJ, James R (1993). Haematological changes and their recovery in *Oreochromis massambicus* as a function of exposure period and sublethal levels of Ekalus. Acta Hydrobiol. 35: 73-83.
Smith GL, Hattingh J, Bunger AP (1979). Haematological assessment of anesthesia, MS 222 in natural and neutral form in three fresh water fish species: interspecific differences. J. Fish. Biol. 15: 633-643.
Tiwari S, Singh A (2004). Pscicidal activity of alcoholic extracts of nerium indicum leaf and their biochemical stress response on fish metabolism. Afr. J. Trad. CAM. 1: 15-29.
Watts RD, Watts DC (1974). In: Chemcial zoology (M Florkin and BT Scheer Eds.) Vol VIII. Deuterostomians. Cyclostomers and fishes. Academic Press, New York.
Wells RMG, McIntyre RH, Morgan AK, Devie PS (1986). Physiological stress responses in big gamefish after capture: observation on plasma chemistry and blood factors. Comp. Biochem. Physiol. 84A: 565-571.
Wendelaar Bonga SE (1997). The stress response in fish. Physiol. Rev. 77(3): 591-624.
Wahua TAT (1999). Applied statistics for scientific studies. Afrika Link Books, Aba, Nigeria p. 365.

Thermostable, alkaline tolerant lipase from *Bacillus licheniformis* using peanut oil cake as a substrate

N. Annamalai*, S. Elayaraja, S. Vijayalakshmi and T. Balasubramanian

Marine Sciences, CAS in Marine Biology, Annamalai University, Parangipettai 608502, Tamil Nadu, India.

Thermostable alkaline tolerant lipase producing *Bacillus licheniformis* isolated from marine environment. Peanut oil cake was used as substrate for enzyme production and purified to 3.6 fold with a specific activity of 148.4 U/mg of protein with molecular weight of 35 kDa. The enzyme was 100% stable at 60°C, it retains more than 80% activity after 1 h incubation at 70°C and pH 9-11.5. Metals like Ca^{2+}, Ba^{2+} and Mg^{2+} enhanced whereas Ni^{2+} and Fe^{3+} inhibited the enzyme activity. No effect on enzyme activity found when treated with Tween 80 and sodium deoxycholate, but completely inhibited with SDS.

Key words: *Bacillus licheniformis,* lipase, peanut oil cake, enzyme activity, thermostable, metals.

INTRODUCTION

Lipases (EC 3.1.1.1) and esterases are act on the hydroxyl ester bonds present in triglycerols, liberating fatty acids and glycerols. Esterases act on water soluble substrates with a preference for short fatty acid chains but true lipases work at water-lipid interface and their major substrates are long chain triglycerols (Jaeger et al., 1994). Much attention is paid to lipases of microbial origin, such as *Candida* sp., *Pseudomonas* sp., *Bacillus* sp., and *Rhizopus* sp. for their unsurpassed roles in biotechnology. They function as biocatalyst, capable of catalyzing hydrolytic ester cleavage, transesterification, alcoholysis, acidolysis, and aminolysis and are thus widely used in the food, detergent, chemical and biomedical industries (Pandey et al., 1999).

Lipases have also been used as biocatalyst for the synthesis of chiral compounds, which offer tremendous potential in production of compounds of pharmaceutical interest. Since the last decade, lipases have been perceived by research scientists as one of the most important classes of industrial enzymes (Arbige and Pitcher, 1989) for instance lipases have been used extensively in the dairy industry, for household detergents (Falch, 1991), in the oleochemical industry (McNeil et al., 1991) and to produce structured triacylglycerols (Macrae, 1983).

A relatively smaller number of bacterial lipases have been well studied compared to plant and fungal lipases (Peterson and Drablos, 1994). In recent years there has been a great demand for thermostable enzymes in industrial fields. The importance of thermostable lipases for different applications has been growing rapidly (Shah et al., 2007; Chen et al., 2007). Thermostable enzymes from microbial sources are highly advantageous, for biotechnological applications, since they can be produced at low cost and exhibit improved stability (Handelsman and Shoham, 1994).

Ramakrishnan and Banerjee (1950) investigated that oil seed cakes for their lipolytic activity with a view to get a cheap and active lipase for fat hydrolysis. Among these, peanut oil cake is a cheaper as well as fat rich source. Hence the present study was carried out on production of thermostable lipase using peanut oil cake (*Arachis hypogea*) by alkalophilic thermophilic bacteria, purification and characterization such as effects of temperature, pH and various metal ions, detergents on enzyme activity and stability of lipase.

MATERIALS AND METHODS

Microorganism

Lipase producing bacteria was isolated from marine sediments of parangipettai coast (Lat; 11°46' Long; 79°46'), Tamilnadu, east coast of India using an enrichment procedure. The enriched culture

*Corresponding author. E-mail: annabact@gmail.com.

broth was serially diluted and spreaded on Tween 80 agar plates (pH-9). Inoculated plates were incubated at 55°C for 48 h. The isolated colonies were screened for lipase production on Tributyrine agar plates, colonies with highest clear zone on the plates were selected as potential lipase producing strain and identified through morphological, physiological and biochemical characteristics (Buchanan et al., 1974) and it was confirmed by 16S rDNA sequencing. The DNA was isolated by phenol chloroform method (Marmur, 1961). The primer sequences were chosen from the conserved regions previously reported for the bacterial 16S rDNA gene (Marchesi et al., 1998). Sequencing was done using forward primer (5'-CAGGCCTAACACATGCAAGTC-3') and reverse primer (5'-GGGCGGTGTGTACAAGGC-3'). PCR reactions were performed with the following program for the 16S rDNA gene: 30 cycles consisting of 95°C for 1 min, 55°C for 1 min and 72°C for 1.5 min, followed by a final extension step of 5 min at 72°C. The 16S rDNA sequence was analyzed by an automated DNA sequencer (Megabace, GE). The 16S rDNA gene sequence obtained was compared with other bacterial sequences by using NCBI BLAST for their pair wise identities. Multiple alignments of these sequences were carried out by ClustalX 2.0 version and the phylogenetic trees were constructed in MEGA (4.2 version) using neighbor joining (NJ) algorithms.

Lipase production using peanut oil cake

For the production of lipase, the bacterial cells were inoculated in 1000 ml erlenmeyer flasks containing 300 ml production media containing (g/L) soybean 20, peanut oil cake 20, soluble starch 10, K_2HPO_4 5, Na_2CO_3 10, NaCl 5 and pH of the medium was adjusted to 9.0 using 6 N NaOH (Sodium carbonate was sterilized separately and added to the medium) were incubated at 55°C on shaker incubator (125 rpm). The growth was measured at every 6 h intervals in terms of optical density at 660 nm (Systronics, Visiscan 167) and cell free supernatant was used as crude enzyme.

Lipase assay

Estimation of lipase activity was carried out by titrating the fatty acids liberated from olive oil with an alkali (Kunio et al., 1994). The liberated fatty acids were titrated with 1 mol/L NaOH solution. Simultaneously, a blank titration was also performed. One unit (U) was defined as the amount of lipase capable of releasing 1 μmol of titratable fatty acid per min under the assay conditions used.

Production and partial purification of enzyme lipase

The bacterial cells were removed by centrifugation at 3000 *xg* for 30 min at 4°C and enzyme from the cell free supernatant was precipitated by ammonium sulphate at 80% saturation at 4°C. The precipitate was collected by centrifugation at 3,000 *xg* for 30 min at 4°C and dissolved in 100 mM Tris-HCl buffer of pH 9.0. This aliquot was dialyzed against the same buffer overnight under refrigerated conditions. Dialyzed enzyme solution was loaded onto a DEAE-cellulose column (2.0 x 25 cm) equilibrated with 100 mM Tris-HCl buffer, pH-9.0. The enzyme was eluted with linear gradient of NaCl (0-1 M in 100 mM Tris-HCl buffer) at a flow rate of 25 ml/h. The eluted fractions were pooled together and assayed for enzyme activity. The resultant enzyme solution was loaded onto a Sephadex G-50 gel filtration column (2.5×120 cm), which had been equilibrated with 100 mM Tris-HCl buffer (pH 9.0), then eluted with the same buffer. The peaks exhibiting lipase activity were pooled together and used as a purified enzyme. This partially purified enzyme solution was used for investigating the effects of temperature, pH and various metal ions on enzyme activity.

Protein content in the crude and purified enzyme was determined according to the method of Lowry et al. (1951) using bovine serum albumin (Hi-Media chemical laboratories, India) as a standard.

Molecular weight determination

The molecular weight of the purified protein was estimated using SDS-PAGE (Laemmli, 1970). The electrophoresis was performed in Genei electrophoretic unit and stained with Coomassie Brilliant Blue R-250. Molecular weight was determined by comparing the mobility of standard molecular weight markers for 29, 43, 66, 97 and 205 kDa (Genei, Bangalore, India).

Effect of Temperature, pH on enzyme activity and stability

The effect of temperature on enzyme activity was tested by preincubating the reaction mixture (substrate + crude lipase solution and 1 ml 100 mM Tris HCl buffer, pH 9.0) at temperatures ranging from 30 to 90°C. To check the thermostability, activity of the enzyme incubated at different temperature for 1 h was checked at different time intervals (0 to 60 min). The effect of pH on the enzyme was determined by preincubating with four different buffers (100 mM): citric acid-Na_2HPO_4 (pH 4 to 6), phosphate buffer (pH 6 to 8), boric acid-NaOH buffer (pH 8 to 9.5) and phosphate-NaOH buffer (pH 9.5 to 12), hydroxide chloride buffer (12 to 13) for 2 h and the residual activity were determined by the standard assay.

Effect of EDTA detergent and various metals on lipase activity

Effect of detergents on lipase activity was analyzed by incubating enzyme for 1 h at 45°C in 100 mM Tris HCl buffer (pH 9.0) containing 0.1 or 1% (w/v) detergent. To determine the effects of various metal ions on the lipase activity, enzyme assay was performed for 1 h in the reaction mixture as described previously, with EDTA, SDS and different metal ions such as $MnSo_4$, $FeSo_4$, $CoCl_2$, $MgSO_4$, $CuSO_4$, $CoSO_4$, $MgCl_2$, $CaCl_2$, $FeCl_3$, $BaSO_4$, $HgCl_2$ and $NiSO_4$ at a final concentration of 1 mM.

RESULTS

The thermostable alkaline lipase producing bacteria is a Gram positive, rod shaped, spore forming, motile, positive for citrate utilization, indole production, nitrate reduction, arginine, and acid production from glucose. Based on the morphological, physiological and biochemical characteristics, the isolate was tentatively identified as *Bacillus licheniformis* and it was confirmed by the 16 S rDNA sequencing (Gene bank accession number-HM370569). The phylogenetic tree was constructed and represented in Figure 1.

Lipase production by *Bacillus licheniformis* was highly correlated with the cell growth (Figure 2). Maximum lipase production (730 U/ml) was recorded in the stationary phase. The culture supernatant was used as a starting material for the purification of the lipase. A three-step procedure yielded a 3.6-fold enrichment of the protein with a 19% recovery. The summary of the purification depicted in Table 1 and molecular weight was estimated as 35 kDa (Figure 3).

The lipase from *Bacillus licheniformis* exhibited good

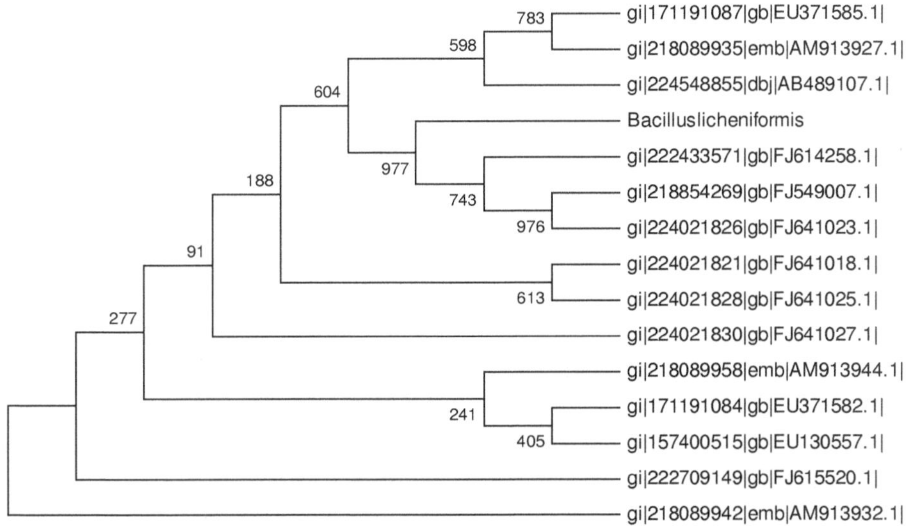

Figure 1. Phylogenetic tree showing the taxonomic position of the *Bacillus licheniformis*.

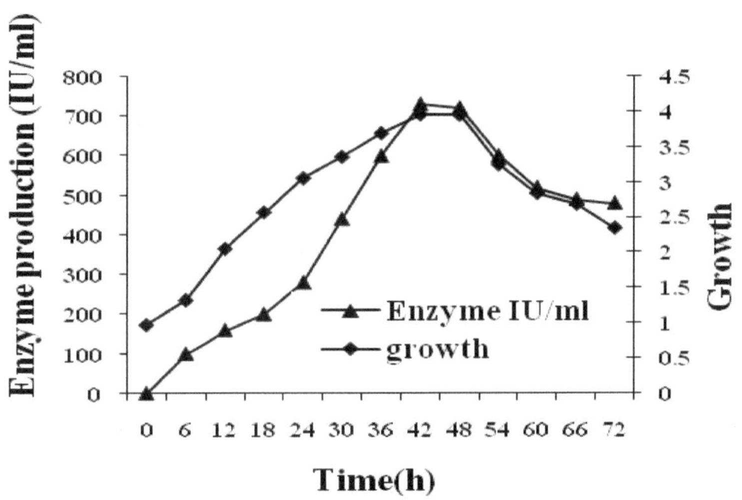

Figure 2. Kinetics of growth and lipase production.

Table 1. Summary of the purification of lipase from *Bacillus licheniformis.*

Fraction (mg)	Total protein (U)	Total activity (U/mg)	Specific activity (Fold)	Purification (%)	Yield
Culture broth	416	17000	40.8	1.0	100
$(NH_4)_2SO_4$ Precipitate	83	6238	75.1	1.8	36.6
DEAE- Cellulose	34	4250	125	3.06	25
Gel filtration	22	3265	148.4	3.6	19

enzyme activity between the temperature of 37 to 60°C and optimum being at 45°C (Figure 4). Regarding stability, enzyme was 100% stable after 1 h incubation at 60°C. It retained 82% original activity when heated at 70°C for 60 min and enzyme was 65% stable up to 30 min at 80°C (Figure 5).

The effect of pH on enzyme activity and stability were investigated by using different buffer solutions. The

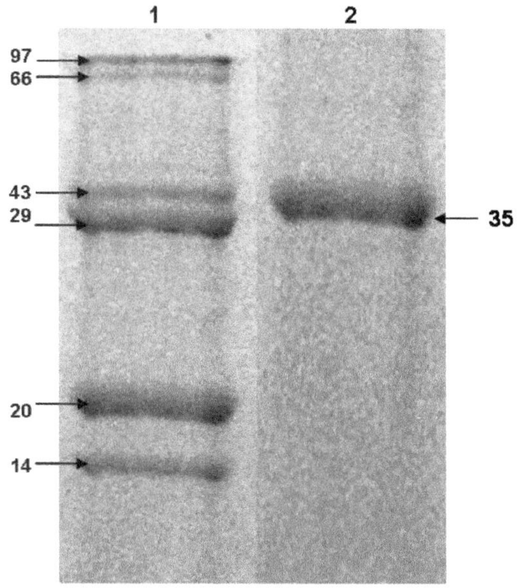

Figure 3. SDS-PAGE of lipase from *Bacillus licheniformis*. Lane. 1. Molecular markers, Lane 2. Purified protein.

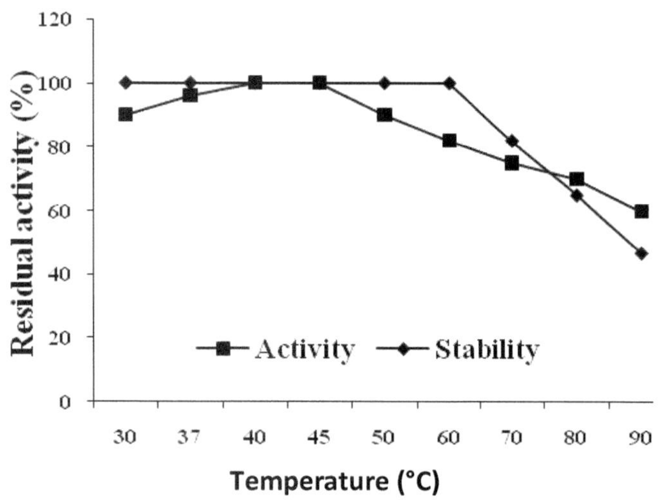

Figure 4. Effect of temperature on the activity and stability of the lipase enzyme. (Activity is expressed in enzyme units and stability is expressed in % of retained activity).

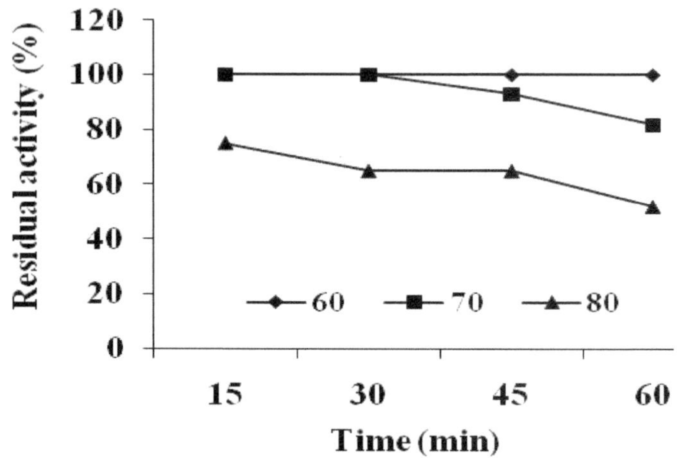

Figure 5. Stability of the lipase at different time in different temperature levels.

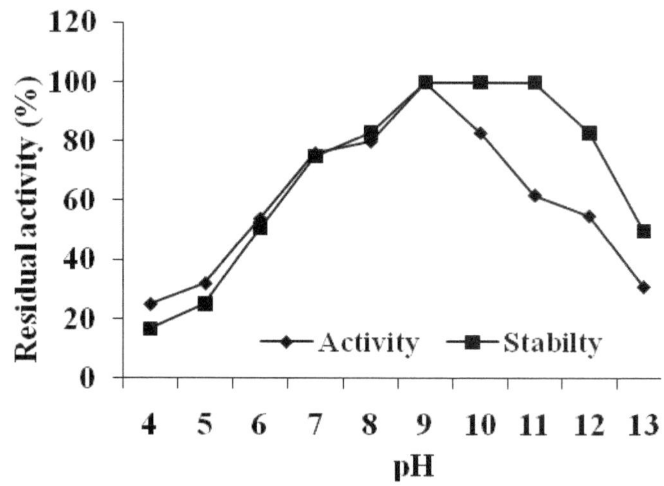

Figure 6. Effect of pH on the activity and stability of the lipase enzyme. (Activity is expressed in enzyme units and stability is expressed in % of retained activity.

enzyme was active between pH range of 6 (53%) and 11 (58%) optimum at pH 9.0 (Figure 6). The enzyme was most stable between at pH 9.0 to 12 (83%).

Among metal ions, Ni^{2+}, Mn^{2+}, Hg^{2+}, Fe^{2+}, Fe^{3+} and Co^{2+} were slightly inhibited enzyme activity whereas Ca^{2+}, Mg^{2+} and Ba^{2+} were enhanced the enzyme activity (Table 2) and no effect was found with Cu^{2+}. Only 25% inhibition was found at 1 mM concentration of SDS and EDTA, but completely inhibited at 10 mM. Tween 80 and sodium deoxycholate had no effect on enzyme activity at 10 mM and it exhibits 110% and 72 activity at 1mM concentration respectively (Table 3). Triton X 100 enhanced (126%) at 10 mM concentration.

DISCUSSION

In the present study, thermostable lipase produced by *Bacillus licheniformis* using peanut oil cake as cheaper source, was purified and characterized. Deive et al. (2003) and Ota et al. (2000) were found trybutyrin induced the lipase production compared with other carbon sources by *Kluyvermyces marxianus* and *Geotrichum* sp. F0401B respectively. In large scale

Table 2. Effect of EDTA and metal ions on the enzyme activity.

Metals	Residual activity (%) (5 mM concentration)
Control	100
$MnSO_4$	75
$FeSO_4$	69
$CoCl_2$	95
$MgSO_4$	150
$CuSO_4$	100
$MgCl_2$	127
$CaCl_2$	225
$FeCl_3$	50
$BaSO_4$	175
EDTA	75
$HgCl_2$	83
$NiSO_4$	43

Table 3. Effect of detergents on stability of the enzyme.

Detergents	Residual activity (%)	
	1 mM	10 mM
Control	100	100
Triton X 100	80	126
Tween 80	110	100
SDS	25	0
Sodium deoxycholate	72	100

production of lipase, the addition of other inorganic carbon sources can be costly and a burden in downstream processing (Seghal et al., 2008). Hence, peanut oil cake is the potential cheaper substrate lipase production.

As described for most extracellular lipases, maximum enzyme production appeared when bacterial cells reached the late logarithmic phase (Jaeger et al., 1994). Purification yielded 3.6-fold with a 19% recovery and molecular weight was estimated as 35 kDa. In general many of the *Bacillus* lipases have low molecular mass about 20 kDa (Sugihara et al., 1991).

The optimum temperature for enzyme activity was 45°C and it is comparatively high with lipase of *Rhodotorula glutinis* (Dimitris et al., 1992) and *Pseudomonas* sp. (Sugihara et al., 1991). Lipase from *Bacillus licheniformis* shown to have good thermostability, it retains 100% activity at 60°C and 65% activity even at 80°C. The thermostability of the lipase was somewhat higher to other reported lipases from *Bacillus* J 33 (Nawani et al., 1998) and *Pseudomonas fluorescens* NS2 W (Kulkarni and Gadre, 2002). Many lipases with moderate thermostability have been found in different strains of fungi *Penicillium wortmanii*, stable at 40°C (Costa and Peralta, 1999) and lipase from *R. glutinis* stable only at 35°C (Papaparaskevas et al., 1992).

The lipase from *Bacillus licheniformis* was 100% stable at pH 9 to 11.5 and 83% of activity retains even at pH 12.0. But the purified lipase from *S. grimisii* (Abdou, 2003), *Pseudomonas* sp. (Fox and Stepaniak, 1983) were stable only between the pH ranges of 8.0 to 9.0.

The activity of the enzyme was inhibited by metal ions such as Mn^{2+}, Hg^{2+}, Fe^{2+}, Fe^{3+} and EDTA and enhanced by Ca^{2+}, Mg^{2+} and Ba^{2+}. Several lipases had enhanced lipolytic activity in the presence of Ca^{2+} and hence it is called as Ca^{2+} dependent metalloenzyme. Similar results have been reported for *Bacillus* sp. Strain 42 (Eltaweel et al., 2005) and *Bjerkandera adusta* R59 lipase (Bancerz and Ginalska, 2007). In contrast, Triton X-100 inhibited activity of the lipase from *Mucor* sp. (Abbas et al., 2002) and *Rhizopus* sp. (Sroka, 1994).

In the present study, the lipase produced by *Bacillus licheniformis* isolated from mangrove sediments was cost effective because of cheaper source as peanut oil cake, stable up to 70°C and pH 12.0. Hence it can be useful for industrial application because many of the processes performed at temperatures around 50°C and alkaline pH.

REFERENCES

Abbas H, Hiol A, Deyris V, Comeau L (2002). Isolation and characterization of an extracellular lipase from *Mucor* sp. strain isolated from palm fruit. Enzyme Microb. Technol., 31: 968-975.

Abdou AM (2003). Purification and partial characterization of psychrotrophic *Serratia marcescens* lipase. J. Dairy Sci., 86: 127-132.

Arbige MV, Pitcher WH (1989). Industrial enzymology: a look towards the future. Trends in Biotechnol., 7: 330-335.

Bancerz R, Ginalska G (2007). A novel thermostable lipase from Basidiomycete *Bjerkandera adusta* R59: Characterisation and esterification studies. J. Ind. Microbiol. Biotechnol., 34: 553-560.

Buchanan RE, Gibbons NE, Cowan ST, Holt TG, Liston J, Murry RG, Niven CF, Ravin AW, Stainer RY (1974). Bergey's manual of determinative bacteriology. Williams and Wilkins co. Baltimore, p. 1246.

Chen S, Qian L, Shi B (2007). Purification and properties of enantioselective lipase froma newly isolated *Bacillus cereus* C71 Process. Biochem., 42: 988-994.

Costa MAP, Peralta RM (1999). Production of lipase by soil fungi and partial characterization of lipase from a selected strain (*Penicillium wortmanii*). J. Basic Microbiol., 39: 110-15.

Deive FJ, Costas M, Longo MA (2003). Production of thermostable extracellular lipase by *Kluyveromyces marxianus*. Biotechnol. Lett., 25: 1403-1406.

Dimitris P, Paul C, Dimitris K, Basil JM (1992). Optimizing production of extracellular lipase from *Rhodotorula glutinis*. Biotechnol. Lett., 14: 397-402.

Eltaweel MA, Rahman RNZRA, Salleh AB, Basri M (2005) An organic solvent stable lipase from *Bacillus* sp. Strain 42. Ann. Microbiol., 55 (3): 187-192.

Falch EA (1991). Industrial enzymes-developments in production and application. Biotechnol. Adv., 9: 643-658.

Fox PF, Stepaniak L (1983). Isolation and some properties of extracellular heat stable lipases from *Pseudomonas fluorescens* strain AFT 36. J. Dairy Res., 50: 7-89.

Handelsman T, Shoham Y (1994). Production and characterization of an extracellular thermostable lipase from a thermophilic *Bacillus* sp. J. Gen. Appl. Microbiol., 40: 435-443.

Jaeger KE, Ransac S, Dijkstra BW, Colson C, Van Huwvel M, Misset O (1994). Bacterial lipases. FEMS Microb. Rev., 15: 29-63.

Kulkarni N, Gadre RV (2002). Production and properties of an alkaline, thermophilic lipase from *Pseudomonas fluorescens* NS2 W. J. Ind. Microbiol. Biotechnol., 28: 344-348.

Kunio O, Yonko Y, Junichi S (1994). Lipase production of *Aspergillus oryzae*. J. Ferment. Bioeng., 77: 490-495.

Laemmli UK (1970). Cleavage of structural proteins during the assembly of the head bacteriophage T4. Nature, 227: 680-685.

Lowry OH, Rosebrough NJ, Farr AL, Randall RJ (1951). Protein measurement with the folin phenol reagent. J. Biol. Chem., 193: 265-275.

Macrae AR (1983). Lipase catalysed interesterification of oils and fats. J. Am. Oil Chem. Soc., 60: 291-294.

Marchesi JT, Sato J, Andrew T, Martin J, Fry S (1998). Hiom: Design and evaluation of useful bacterium specific PCR primers that amplify genes coding for bacterial 16S rRNA. Appl. Environ. Microbiol., 64: 795-799.

Marmur J (1961). A procedure for the isolation of deoxyribonucleic acid from microorganisms. J. Mol. Biol., 3: 208-18.

McNeil GP, Shimizu S, Yamanae T (1991). High-yield enzymatic glycerolysis of fats and oils. J. Amer. Oil. Chem. Soc., 68: 1-5.

Nawani N, Dosanjh NS, Kaur J (1998). A novel thermostable lipase from a thermophilic *Bacillus* sp.: characterization and esterification studies. Biotechnol. Lett., 20: 997-1000.

Ota Y, Saroamoto T, Hasuo M (2000). Tributyrin specially induces a lipase with preference for the Sn-2 position of triglyceride in *Geotrichum* sp. F0401B. Biosci. Biotechnol. Biochem., 64: 2497-2499.

Pandey A, Benjamin S, Soccol CR, Nigam P, Krieger N, Soccols VT (1999). The realm of microbial lipase in biotechnology. Biotechnol. Appl. Biochem., 29: 119-131.

Papaparaskevas D, Christakopoulos P, Kekos D, Macris BJ (1992). Optimizing production of extracellular lipase from *Rhodotorula glutinis*. Biotechnol. Lett., 14: 397-402.

Peterson SB, Drablos F (1994). In Lipases: Their Structure, Biochemistry and Application (Wooley, P. and Peterson, S.B.), Cambridge University Press, UK, pp. 23-48.

Ramakrishnan CV, Banerjee BN (1950). Lipase. Comparative study of lipases from oil-seed cakes. J. Ind. Chem. Soc., 27: 655-56.

Seghal KG, Shanmughapriya S, Jayalakshmi J, Joseph S, Gandhimathi R, Sivaramakrishnan S, Arunkumar M, Thangavelu T, Natarajaseenivasan K (2008). Optimization of extracellular psychrophilic alkaline lipase produced by marine *Pseudomonas* sp. (MSI057). Bioprocess Biosyst. Eng., 31: 483-492.

Shah KR, Patel PM, Bhatt SA (2007). Lipase production by *Bacillus* sp. under different physio-chemical conditions. J. Cell Tissue Res., 7(1): 913-916.

Sroka Z (1994). The activity of lipase from *Rhizopus* sp. in native form and after immobilization on hollow-Wber membranes. J. Memb. Sci., 97: 209-214.

Sugihara A, Tani T, Tominaga Y (1991). Purification and characterization of a novel thermostable lipase from *Bacillus* sp. J. Biochem., 109: 211-216.

Kinetic study of different flavonoids as inhibitors of beta-lactamase enzyme

BOUSSOUALIM, Naoual[1], MEZIANE-CHERIF, Djalal[2] and BAGHIANI, Abderrahmane[1]*

[1]Laboratory of Applied Biochemistry, Department Biochemistry, Faculty of Natural Sciences and Life, University Ferhat Abbas-Setif, Algeria.

[2]Institut Pasteur, Plateforme 6, CNRS URA 2185, 25 rue du Dr Roux, 75724 Paris Cedex 15, France.

The phenotypic study of the clinical strain *Escherichia coli* K18 showed that it is resistant to penicillin and amoxicillin-clavulanate, but susceptible to the combination of piperacillin-tazobactam and the cephalosporins 3rd generation and meropenem. The isolated β-lactamase, using ion-exchange chromatography on resin Mono-Q, gave a yield of 56%, a purification factor equal to 5.9, a molecular weight of approximately 24 KDa and a p*I* of 5.2. Clavulanic acid, tazobactam and sulbactam were competitive inhibitors with K_I and IC_{50} far beyond the values of those of TEM-1. 13 molecules of different flavonoïds, three phenolic acids (cafeic acid, ellagic acid and gallic acid) are tested for their ability to inhibit β-lactamase. The type of inhibition and K_I are determined and the relationship between the structure and activity was established. The results showed that fisetin, flavone, quercetin, catechin and gossypin are non-competitive inhibitors. Myricetin, quercitrin, naringenin, morin, kaempferol and rutin are non-competitive inhibitors. The K_I of the most flavonoïds tested are above 100 µM, which are 200 folds less effective than clavulanic acid (K_I = 0.5 µM). We can conclude that the position of OH in the structures of molecules has a major influence.

Key words: β-lactamase, *Escherichia coli*, β-lactam, clavulanic acid, flavonoids, phenolic acids.

INTRODUCTION

β-Lactamases penicillin amido- β-lactam hydrolase, (E.C.3.5.2.6) are the main factors responsible for bacterial resistance to β-lactam antibiotics. They are divided into four classes A, B, C and D, on the basis of their amino acid sequence homologies. In contrast to the class B β-lactamases, which are metallo-enzymes, all others are active-site serine enzymes (Sykes and Matthew, 1976; Bush et al., 1995). The active-site serine β-lactamases catalyzes the hydrolysis of β-lactam antibiotics according to the acylation deacylation pathway (Matagne and Frère, 1995). Bacterial resistance to amoxicillin-clavulanic has significantly increased in recent years (Cantón, 2008). In *Escherichia coli*, this resistance can be caused by: (1) Hyper production of the plasmid-encoded β-lactamases TEM-1, TEM-2 or SHV-1 (Wu et al., 1995), (2) overproduction of the constitutive AmpC cephalosporinase (Stapleton et al., 1995), (3) production of OXA enzymes (penicillinases poorly inhibited by β-lactamase inhibitors) (Zhou et al., 1994), (4) a deficiency in the OmpF and / or OmpC porins (Reguera et al., 1991) and (5) production of inhibitor-resistant TEM (IRT) β-lactamases (Vedel et al., 1992; Martín et al., 2010). A novel class of TEM-1 derived β-lactamases resistant to β-lactamase inhibitors has emerged in clinical isolates of *E. coli*. These enzymes have been called IRT for Inhibitor-Resistant TEM-derived β-lactamases. On the basis of functional characteristics of β-lactamases, these enzymes are classified in group 2br broad-spectrum β-lactamases with reduced binding of clavulanic acid (Videl et al., 1992).

Medicinal plants reveal an important pharmacological activities used in developing a novel therapeutic agents, extracts from various medicinal plants containing

*Corresponding author. E-mail: baghiani@yahoo.fr.

favonoids. Flavonoids have been reported to possess antimicrobial activity (Tim and Lamb, 2005). Flavonoids possess antibacterial properties mainly against gram-negative bacterial strains. They may be important tools in antimicrobial strategies (Rong-Dih et al., 2005). The purpose of this study was to characterize the phenotypic and biochemical profile of a β-lactamase isolated from a clinical isolate, *E. coli* resistant to clavulanic acid; and the activities of natural flavonoids against a β-lactamase isolated strain of *E. coli* resistant to clavulanic acid were also evaluated.

MATERIALS AND METHODS

Bacterial strain

The bacterial strain used in this work was isolated from the urine of a hospitalized patient at Setif University Hospital in 2002. The strain was identified with an API-20 E gallery (Biomérieux).

Chemicals

Penicillin G, oxacillin, ampicillin, cloxacillin, carbenicillin cafeic acid, ellagic acid, gallic acid, fisetin, flavone, quercetin, catechin, gossypin, myricetin, quercitrin, naringenin, morin, kaempferol, rutin were purchased from Sigma. Piperacillin, clavulanic acid and tazobactam were from Wyeth Laboratory. Cefuroxim and ceftazidim were from Claxo. Cefazolin (Panpharma), cephotaxim and imipenem, sulbactam, aztreonam, E-test bandlettes and nitrocefin were from Merk, Pfizer, Sanofi Wyeth, Biodisk and Oxoid, respectively.

Determination of the minimum inhibitory concentrations (MICs)

MICs were determined by E-test technique by the application of E-test bandlettes (biodisk) on Mueller-Hinton agar (sigma) with a Steers multiple inoculator and an inoculum of 10^6 CFU. All plates were incubated at 37°C for 18 h (CLSI, 2007).

β-lactamase isolation and purification

The isolation of β-lactamase from *E. coli* K18 was carried out according to Hedberg et al. (1995) with slight modification. *E. coli* K18 was grown overnight at 37°C in 1.5 L of brain heart infusion (BHI). Cells were harvested by centrifugation at 18000 × g for 10 min, washed in 20 mM Tris pH 8.4 and re-suspended in the same buffer. Cells were disrupted by ultrasonic treatment and the crude extract obtained was centrifuged at 20000 × g for 30 min. Nucleic acids were precipitated by the addition of 2% streptomycin sulfate and centrifugated at 18000 × g for 30 min. The supernatant was dialyzed overnight against the same buffer. The supernatant was applied to a 1 × 10 cm Q-Sepharose Fast Flow column equilibrated with the dialysis buffer. The column was washed with the same buffer and the enzyme was eluted with a flow rate of 2 ml /min using linear gradient of 0 to 0.5 M NaCl. The β-lactamase activity was determined using the penicillin G as substrate. Protein contents of the supernatant and β-lactamase preparations were estimated by Macart method (Macart and Gerbaut, 1982).

Sodium dodecyl sulfate polyacrylamide gel electrophoresis (SDS-PAGE) was performed on a Bio-Rad Protean II mini-gel apparatus according to Laemmli (1970), using a 5% concentrating gel and 12% separating gels. Samples and standards of known molecular weights (GE Healthcare) were run at 100 (stacking gel) and 200 V (running gel). Gels were stained by Coomassie brilliant blue.

Analytical isoelectric focusing (IEF)

Isoelectric focusing of *E. coli* K18 extracts was performed on 6% polyacrylamide gel containing an ampholine gradient in a pH range of 3 to 10 (biorad cuve) (Righetti, 1983). β-Lactamase bands were revealed by the starch-iodine method in agar gel (Perret, 1954).

Determination of relative molecular mass

The relative molecular mass of the β-lactamase obtained from *E. coli* K18 was estimated by SDS-PAGE 12% analysis (Laemmli, 1970). The β-lactamase activity, being localized by the use of an iodine-starch method in agar gel and the relative rate of correspondent β-lactamse, was determined.

Determination of β-Lactamase activity and kinetic constants

The semi-purified *E. coli* K18 β-lactamase activity was determined spectrophotometrically (chimatzu Instruments) at 25°C in 100 mM phosphate buffer (pH 7.0) in a total volume of 0.5 ml. The following wavelengths were used, according to Rossolini et al. (1999): penicillin G ($\varepsilon_{235\ nm}$ = 560 $M^{-1}.cm^{-1}$), piperacillin ($\varepsilon_{235\ nm}$ = 936 $M^{-1}.cm^{-1}$), oxacillin ($\varepsilon_{260\ nm}$ = 1000 $M^{-1}.cm^{-1}$), cloxacillin ($\varepsilon_{260\ nm}$ = 1000 $M^{-1}.cm^{-1}$), carbenicillin ($\varepsilon_{235\ nm}$ = 780 $M^{-1}.cm^{-1}$), ampicillin ($\varepsilon_{235\ nm}$ = 809 $M^{-1}.cm^{-1}$), cephaloridin ($\varepsilon_{255\ nm}$ = 1000 $M^{-1}.cm^{-1}$), nitrocefin ($\varepsilon_{486\ nm}$ = 20500 $M^{-1}.cm^{-1}$), cefotaxim ($\varepsilon_{260\ nm}$ = 7040 $M^{-1}.cm^{-1}$), cefazolin ($\varepsilon_{320\ nm}$ = 1067 $M^{-1}.cm^{-1}$), cephalothin ($\varepsilon_{265\ nm}$ = 8790 $M^{-1}.cm^{-1}$), cefuroxim ($\varepsilon_{260\ nm}$ = 9320 $M^{-1}.cm^{-1}$), ceftazidim ($\varepsilon_{265\ nm}$ = 10300 $M^{-1}.cm^{-1}$), imipenem ($\varepsilon_{299\ nm}$ = 9000 $M^{-1}.cm^{-1}$) etaztreonam (ε_{320nm} = 700 $M^{-1}.cm^{-1}$). Antibiotic solutions were freshly prepared in 100 mM phosphate buffer (pH 7.0). Kinetic parameters were derived from the initial velocity using SIGMAPLOT 10.0.

Enzyme inhibition was determined using penicillin G as substrate. Inhibitors, at various concentrations (clavulanate (0.01 to 10 μM), sulbactam (5 to 35 μM), tazobactam (0.02 to 10 μM)), were pre-incubated with the enzyme for 10 min at 25°C before addition of the substrate. The inhibitor concentration required to inhibit 50% of enzyme activity (IC50) and KI was determined. Three potential inhibitors; Clavulanic acid, sulbactam, and tazobactam and other flavonoids and phenolic acids were tested.

RESULTS AND DISCUSSION

Determination of the minimum inhibitory concentrations (MICs)

The E-test assay showed that *E. coli* K18 strain was resistant to all penicillins tested with MIC > 256 μg /ml (Table 1). It is also resistant to the combination, amoxicillin-clavulanate (MIC = 48 μg/ml). However, the strain is sensitive to the combination of piperacillin with tazobactam and cephalosporins such as ceftazidim and cefotaxim with a MIC of 0.023 and 0.25 μg/ml, respectively. It is very sensitive to meropenem where MIC = 0.008 μg /ml. The addition of tazobactam reduced substantially to piperacillinc and gave a MIC value of 1.5

Table 1. Minimal inhibitor concentrations (MIC) of *E. coli* K18 determination, using different antibiotics.

ß-lactams	MIC (µg/ml)	
	E. coli ATCC 25922	*E. coli* K18
Ticarcillin	0.75	> 256
Benzylpenicillin	6	> 256
Ampicillin	0.5	> 256
Amoxicillin	1.5	> 256
Amoxicillin-clavulanat	1	48
Piperacillin	0.5	> 256
Piperacillin-tazobactam	0.5	1.5
Ceftazidim	0.094	0.25
Cefotaxim	< 0.016	0.023
Meropenem	0.047	0.008

Table 2. Purification of β-lactamase from the strain *E. coli* K18 using Ion-exchange chromatography on sepharose FF?

Purification step	Protein (mg)	Enzyme activity (µmole/min)	Specific activity (µmole/min/mg)	Purification (fold)	Yield (%)
Crude extract	114	764	6.7	1	100
Ion-exchange chromatography on sepharose FF	10.8	427	39.5	5.9	56

µg/ml. Bonomo et al. (1997) reported that the best bacteriostatic effect against the isolate producing IRT enzymes was the combination of tazobactam-piperacillin. This suggests that the strain of *E. coli* K18 could produce a β-lactamase type IRT (Bonomo et al., 1997; Nicolas-Chanoine, 1997).

Wu et al. (1995) showed that hyperproduction of TEM-1 β-lactamase was defined as a specific activity of over 200 nmol nitrocefin/min/mg protein. The specific activity measured spectrophotometrically at 482 nm using nitrocefin as substrate (final concentration of 20 µM) in our present study gave a crude extract of β-lactamase specific activity of 127.94 ± 11.37 nmol/min/mg protein.

Isolation and purification of β-lactamase

The results of purification are presented in Table 2. The Q-Sepharose Fast Flow anion exchange column was used to remove large amounts of contaminating proteins. The purified β-lactamase gave a purification factor of 6-fold with a yield of approximately 56%. On SDS-PAGE (12%), the semi-purified enzyme showed a major band of around 24 KDa (Figure 1).

Analytical isoelectric focusing and relative molecular mass

The apparent molecular mass of the semi-purified β-lactamase was found to be around 24 KDa using SDS-PAGE (12%), and its p*I* was 5.2 as determined by isoelectric focusing. This gives further evidence that the β-lactamase of the strain of *E. coli* K18 is of the type IRT. These results are in agreement with those reported previously by Vedel et al. (1992), Buch (1995) and Alonso et al. (2002).

Determination of kinetic parameters of the semi-purified β-lactamase

The kinetic parameters for the β-lactamases *E. coli* K18 were determined for 15 β-lactams (Table 3). The semi-purified β-lactamases hydrolyzed all penicillins tested except for cloxacillin. The K_m values of the enzyme for piperacillin and penicillin G were found to be two to four folds higher (Table 4) compared to the TEM-1 described by Canica et al. (1998) and Delaire et al. (1992). Whereas, oxacillin was a poor substrate. From these results, the semi-purified β-lactamase is penicillinase showing the highest activity with penicillin G as substrate ($V_{max} / K_m = 0.91$).

The semi-purified β-lactamases showed no activity against Aztreonam, imipenem, cephalosporins 2nd and 3rd generation (cefuroxim, cefotaxim and ceftazidim). However, the enzyme showed a low activity with the 1st generation cephalosporins as follows: Cephaloridine > cefazolin > cephalothin. The K_m values for cephaloridine and cephalothin were closer to that of TEM-1 (Table 4).

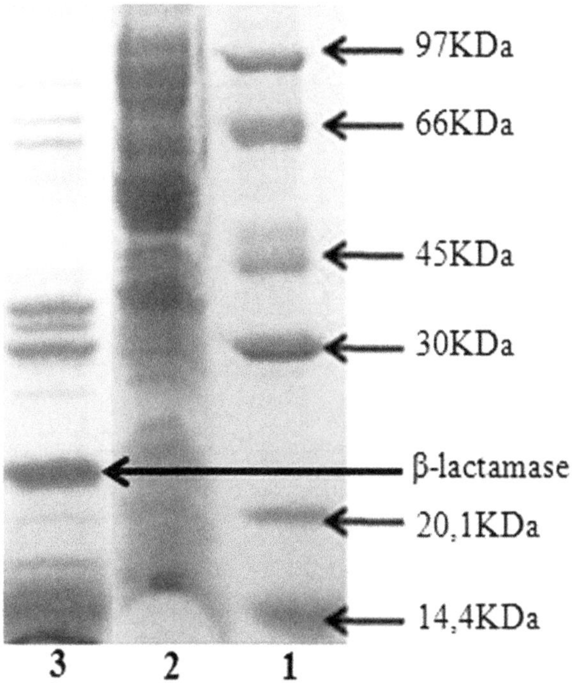

Figure 1. SDS-PAGE (12%), fractions eluted from the ion exchange chromatography on sepharose FF, stained with coomassie brilliant blue. Lines: 1. molecular weight markers (Phophorylase b, albumin, ovalbumin, carbonic anhydrase, trypsin inhibitor, α-Lactalbumin); 2. Crude extract; 3. The semi-purified β-lactamase.

Table 3. Kinetic parameters of semi-purified *E. coli* K18 β-lactamase using different antibiotics as substrate.

Antibiotics	**K_m**	**V_{max}**	**V_{max}/K_m**
Penicillin G	141.7 ± 8.9	129.3 ± 3.6	0.91
Piperacillin	100 ± 9.4	50.3 ± 2	0.50
Carbenicillin	12.6 ± 0.5	39.3 ± 4.6	0.30
Ampicillin	179 ± 43.9	46.3 ± 3.6	0.26
Oxacillin	75.9 ± 10	4.4 ± 0.2	0.06
Cloxacillin	-	-	-
Cephaloridin	833.3 ± 50	84.7 ± 3.5	0.10
Cefazolin	468.7 ± 101.3	53.2 ± 6.8	0.11
Cephalothin	110 ± 6.6	4.9 ± 0.1	0.044
Cefuroxim	-	-	-
Cefotaxim	-	-	-
Ceftazidim	-	-	-
Nitrocefin	70.2 ± 12.9	2.8 ± 0.3	0.039
Aztreonam	-	-	-
Imipenem	-	-	-

(-) Not determined.

The semi-purified enzyme shows less affinity for nitrocefin than TEM-1 (the K_m values of nitrocefin are two folds greater). These results are comparable with those of Bonomo et al. (1997). The most IRT enzymes have lower catalytic efficiency for all substrates than does TEM-1, due to decreased Kcat and high K_m values

Table 4. Semi-purified *E. coli* K18 β-lactamase K_m compared to those of TEM-1 obtained from literatures.

Substrats	K_m (μM)	
	β-lactamase of the strain *E. coli* K18	TEM-1
Penicillin G	141.7	36[a]
Piperacillin	100	67[b]
Carbenicillin	12.6	13[d]
Cephaloridin	833.3	1020[b]
Cephalothin	110	250[d]
Nitrocefin	70.2	32[c]

(a) Delaire et al. (1992), (b) Caniça et al. (1998), Bonomo et al. (1997) and (d) Chaibi et al. (1998).

Table 5. The IC_{50} and the inhibition constant (K_i) of the semi-purified *E. coli* K18 β-lactamase compared to those of TEM-1.

Enzyme	Clavulanate		Tazobactam		Sulbactam	
	IC_{50} (μM)	K_i (μM)	IC_{50} (μM)	K_i (μM)	IC_{50} (μM)	K_i (μM)
The semi-purified β-lactamase	1.05 ± 0.11	0.48 ± 0.08	1.15 ± 0.19	0.09 ± 0.02	20.6 ± 1.3	34.6 ± 8
TEM-1	0.08	0.1	0.1	0.01	6.1	0.9

(Chaibi et al., 1999). Our results are comparable to those reported in the literature (Livermore, 1995; Chaibi et al., 1999), suggesting that the β-lactamase of *E. coli* K18 coli can be an inhibitor-resistant TEM (IRT) β-lactamases.

Effect of clavulanic acid, tazobactam and sulbactam

Measuring the activity of the semi-purified β-lactamase in the presence of penicillin G (200 mM) and clavulanic acid, sulbactam and tazobactam at different concentrations showed that the inhibition of the enzyme activity is dose -dependent. The IC_{50} values are 1.05 ± 0.11, 1.15 ± 0.19 and 20.6 ± 1.3 for clavulanate, tazobactam and Sulbactam, respectively. They are 13, 11 and 3 fold higher than those observed with TEM-1 (Table 5). All the IRTs have IC_{50} and K_i values for ß-lactamase inhibitors higher than those of TEM-1. Bonomo (1999) and Chaibi et al. (1998) have shown that slbactam was the less effective inhibitor against these enzymes with the highest IC_{50} and K_i, whereas, tazobactam was the most active inhibitor with the lowest IC_{50} and K_i. Clavulanic acid and tazobactam have been shown to be more effective β-lactamase inhibitors than sulbactam against IRT.

In the present study the values of IC_{50} and K_i are lower than the mutants IRT-1, IRT-2, IRT-10 and IRT-14, but close to the values of other IRT with a single substitution (IRT-5, IRT-6 and IRT-11) (Yang et al., 1999; Henquell et al., 1995; Stapleton et al., 1995). This suggests that the β-lactamase *E. coli* K18 could have a single substitution at the position 69. According to the different characteristics of IRT, it could be very close to the IRT-11 (TEM-40) (Chaibi et al., 1998). However, all these are only assumptions. It would be very useful to establish a genetic characterization with a determination of the sequence of the gene bla to the origin of the β-lactamase *E. coli* K18.

The inhibitory effect of flavonoids

The objective of this part of study was to evaluate the inhibitory effect of a number of flavonoïds and phenolic acids on the β-lactamase strain of *E. coli* K18 and to establish a structure-activity relationship. In order to establish a relationship between the chemical structures of flavonoïds and their inhibitory activity, the type of inhibition and inhibition constant (K_i) for each molecule were determined. The results are shown in Table 6.

The suicide inhibitors (clavulanic acid, tazobactam and sulbactam) inhibiting β-lactamases in a competitive manner, however the flavonoïds myricetin, quercitrin, naringenin, morin, kaempferol and rutin inhibit β-lactamase of *E. coli* K18 in a non-competitive manner. While fisetine, flavone, quercetin, catechin and gossypin were uncompetitive inhibitor (Table 6).

The position of the hydroxyl (-OH) is probably responsible for such inhibition. The flavonoïds that are uncompetitive have OH in positions C3', C4', C5 and C7, while those are uncompetitive have a hydroxyl in positions C4 ', C5 and C7. The hydroxyl at C3 'seems to be able to modify the interaction of flavonoïd with the β-lactamase and thus the type of inhibition. The presence of a hydroxyl group at C-3 in ring B is associated with a high K_i. This is the case of kaempferol, naringenin and morin. The diosmin has no effect on the enzyme activity. This could be due to the absence of hydroxyl groups on the C5 and C7. Caffeic acid, ellagic acid and gallic acid have no inhibitory effect on β-lactamase, suggesting

Table 6. Inhibition type and inhibition constant (K_I) determination of the semi-purified *E. coli* K18 β-lactamase, using different flavonoids.

Flavonoids	Structure	K_I (μM)	Type of inhibition
Quercitrin	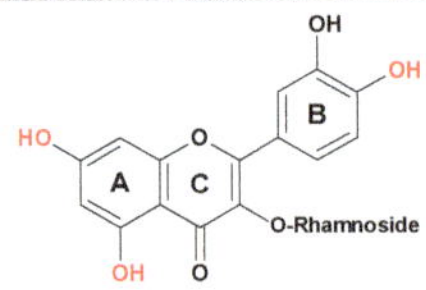	140 ± 10	Non-competetive
Myricetin	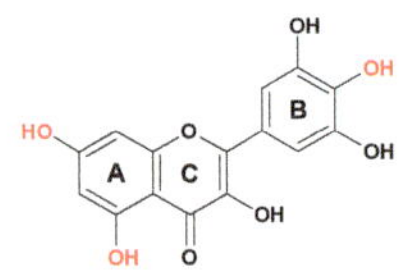	158 ± 10	Non-competetive
Rutin	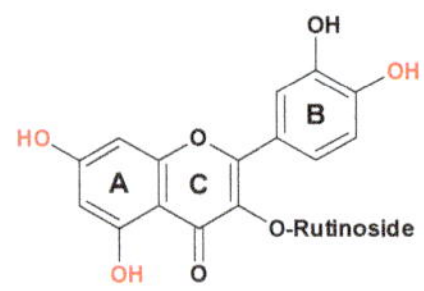	182 ± 16	Non competetive
Kaempferol	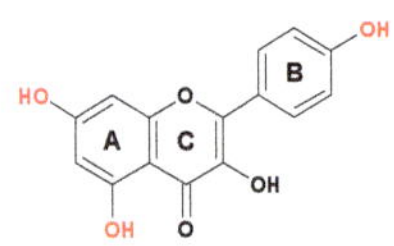	221 ± 10	Non-competetive
Morin	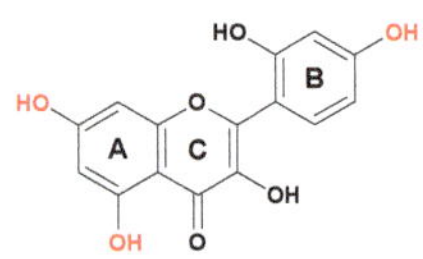	326 ± 13	Non-competetive
Naringenin	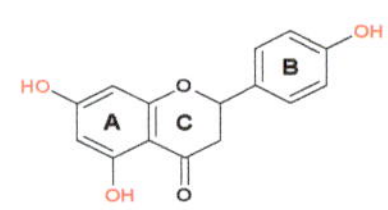	493 ± 40	Non-competetive
Gossypin	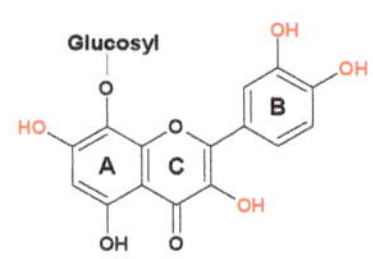	92 ± 5	Uncompetetive
Catechin	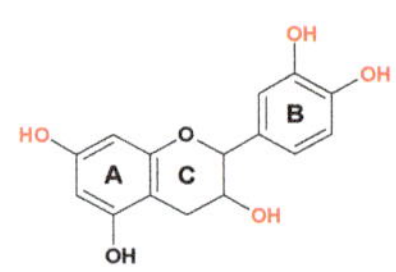	128 ± 11	Uncompetetive
Fisetin	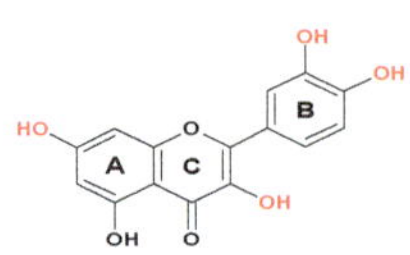	181 ± 16	Uncompetetive
Quercetin	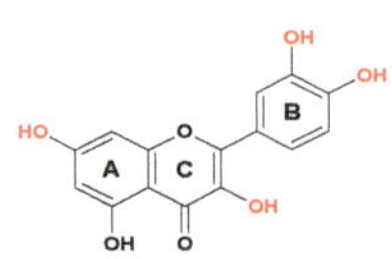	213 ± 20	Uncompetetive
Flavone	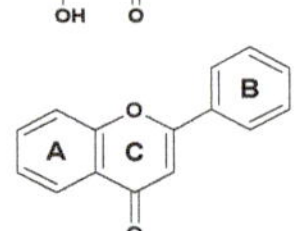	410 ± 37	Uncompetetive

Figure 6. Continued.

Apigenin	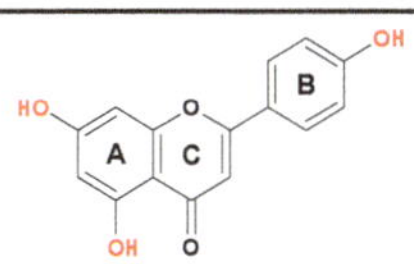	IC_{50} > 500 µM
Diosmin	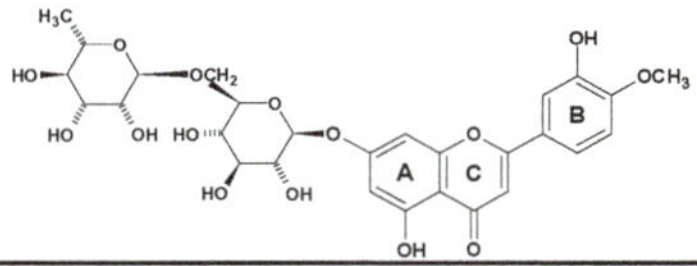	No effect

that the structure C6-C3-C6 is important for this inhibition.

REFERENCES

Alonso R, Gerbaud G, Galimand M, Courvalin P (2002). TEM-103/IRT-28 beta-lactamase, a new TEM variant produced by *Escherichia coli* BM4511. Antimicrob. Agents Chemother., 46(11): 3627-3629.

Bonomo RA, Rudin SA, Shlaes DM (1997). Tazobactam is a potent inactivator of selected inhibitor-resistant class A beta-lactamases. FEMS Microbiol Lett., 148: 59-62.

Bush K, Jacoby GA, Medeiros AA (1995). A functional classification scheme for β-lactamases and its correlation with molecular structure. Antimicrob. Agents Chemother., 39: 1211-1233.

Caniça MM, Caroff N, Barthélémy M, Labia R, Krishnamoorthy R, Paul G, Dupret JM, (1998). Phenotypic study of resistance of β-lactamase-inhibitor-resistant TEM enzymes which differ by naturally occurring variations and site-directed substitution at Asp270. Antimicrob. Agents Chemother., 42: 1323-1328.

Cantón R, Morosini MI, Martin O, De la Maza S, Gomez de la Pedrosa EG (2008). IRT and CMT β-lactamases and inhibitor resistance. Clin. Microbiol. Infect., 14: 53-62.

Chaibi EB, Péduzzi J, Farzaneh S, Barthélémy M, Sirot D, Labia R (1998). Clinical inhibitor-resistant mutants of the beta-lactamase TEM-1 at amino-acid position 69. Kinetic analysis and molecular modelling. Biochim. Biophys. Acta, 1382(1): 38-46.

Chaïbi EB, Sirot D, Paul G, Labia R (1999). Inhibitor-resistant TEM beta-lactamases: phenotypic, genetic and biochemical characteristics. J. Antimicrob. Chemother., 43(4): 447-458.

Delaire M, Labia R, Samama JP, Masson JM (1992). Site-directed mutagenesis at the active site of *Escherichia coli* TEM-1 β-lactamase. Suicide inhibitor-resistant mutants reveal the role of arginine 244 and methionine 69 in catalysis. J. Biol. Chem., 267: 20600-20606.

Hedberg M, Lindqvist L, Bergman T, Nord CE (1995). Purification and characterization of a new β-Lactamase from *Bacteroides uniformis*. Antimicrob. Agents Chemother., 39(7): 1458-1461.

Henquell C, Chanal C, Sirot D, Labia R, Sirot J (1995). Molecular characterization of nine different types of mutants among 107 inhibitor-resistant TEM β-lactamases from clinical isolates of *Escherichia coli*. Antimicrob. Agents Chemother., 39: 427-430.

Laemmli UK (1970). Cleavage of structural proteins during the assembly of the head of bacteriophage T4.Nature, 227: 680-685.

Livermore D M (1995). β-lactamases in laboratory and clinical resistance. Clin. Microbiol. Rev., 8: 557-584.

Macart M, Gerbaut, L (1982). An improvement of the coomassie blue dye binding method allowing an equal sensitivity to various proteins application to cerebrospinal fluid. Clin. Chem. Acta, 122: 93-101.

Martín O, Valverde A, Morosini MI, Rodríguez-Domínguez M, Rodríguez-Baños M, Coque TM, Cantón R, del Campo R (2010). Population analysis and epidemiological features of inhibitor-resistant-TEM-beta-lactamase-producing *Escherichia coli* isolates from both community and hospital settings in Madrid, Spain. J. Clin Microbiol., 48(7): 2368-2372.

Matagne A, Frère JM (1995).Contribution of mutant analysis to the understanding of enzyme catalysis: the case of class A beta-lactamases. Biochim. Biophys. Acta, 1246(2): 109-127.

Nicolas-Chanoine M H (1997). Inhibitor-resistant β-lactamases. J. Antimicrob. Chemother., 40: 1-3.

Perret CJ (1954). Iodometric assay of penicillinase. Nature, 154: 1012-1013.

Reguera JA, Baquero F, Pérez-Diaz JC, Martinez JL (1991). Factors determining resistance to β-lactam combined with β-lactamase inhibitors in *Escherichia coli*. J. Antimicrob. Chemother., 27: 569-575.

Righetti PG (1983). Isoelectric Focusing: Theory. Methodology and Applications. Elsevier Biomedical Press, Amsterdam.

Rong-Dih L, Yi-Ping C, Mei-Hsien L (2005). Antimicrobial Activity of Antibiotics in Combination with Natural Flavonoids against Clinical Extended-Spectrum β-Lactamase (ESBL)-producing *Klebsiella pneumonia*. Phytotherapy Res., 19(7): 612-617.

Rossolini GM, Franceschini N, Lauretti L, Caravelli B, Riccio ML, Galleni M, Frère JM, Amicosante G (1999). Cloning of a Chryseobacterium (Flavobacterium) meningosepticum Chromosomal Gene (blaACME) encoding an extended-spectrum Class A β-lactamase related to the Bacteroides cephalosporinases and the VEB-1 and PER β-lactamases.Antimicrob. Agents Chemother., 43(9): 2193-2199.

Stapleton P, Wu PG, King A, Shannon K, Frenc G, Phillips L (1995). Incidenses and mechanisms of resistance to the combination of amoxicillin and clavulanic acid in *Escherichia coli*. Antimicrob. Agents Chemother. 39: 2478-2483.

Sykes RB, Matthew M (1976).The β-lactamases of gram negative bacteria and their role in resistance to β-lactam antibiotics. J. Antimicrob. Chemother. 2: 115-157.

Tim Cushnie TP, Lamb AJ (2005). Antimicrobial activity of flavonoïds. Int. J. Antimicrob. Ag., 26: 343-356.

Vedel G, Belaaouaj A, Gilly L, Labia R, Philippon A, Névot P, Paul G (1992). Clinical isolates of *Escherichia coli* producing TRI β-lactamases: novel TEM enzymes conferring resistance to β-lactamase inhibitors. J. Antimicrob. Chemother., 30: 449.

Wu P J, Shannon K, Phillips I (1995).Mechanisms of hyperproduction of TEM-1 β-actamase by clinical isolates of *Escherichia coli*. J. Antimicrob. Chemother. 36: 927-939.

Yang Y, Rasmussen BA, Shlaes DM (1999). Class A β-lactamases-enzyme-inhibitor interactions and resistance. Pharmacol. Therapeut., 83: 141-151.

Zhou XY, Bordon F, Sirot D, Kitzis MD, Gutmann L (1994). Emergence of clinical isolates of *Escherichia coli* producing TEM-1 derivates or an OXA-1 β-lactamase conferring resistance to β-lactamase inhibitors. Antimicrob. Agents Chemother., 38: 1085-1089.

Properties of rhodanese from the liver of tilapia, *Oreochromis niloticus,* in Asejire Lake, Nigeria

Atere Tope Gafar[1], Ademakinwa Adedeji Nelson[2] and Agboola Femi Kayode[2]

[1]Department of Biochemistry, College of Health Sciences, Osun State University, Osogbo, Nigeria.
[2]Department of Biochemistry, Obafemi Awolowo University, Ile-Ife 220005, Nigeria.

The study investigates the purification and characterisation of rhodanese from the liver of the tilapia fish (*Oreochromis niloticus*) collected from Asejire Lake in Nigeria. This was with a view to understanding the biochemical basis of the survival of the fish in cyanide polluted water. Rhodanese was isolated and purified from liver tissue homogenate of tilapia using CM-Sephadex ion exchange chromatography and Sephadex G-75 gel filtration. The specific activity of the enzyme was 56.86 U/mg. The K_m values for KCN and $Na_2S_2O_3$ as substrates were 0.1240 ± 0.0021 mM and 0.0516 ± 0.0097 mM, respectively. The apparent molecular weight was estimated by gel filtration on a Sephacyl S-400 column to be 35,460 Da. The optimal activity was found at pH 6.5 and the temperature optimum was 40°C. The rhodanese enzyme showed that the activity of the enzyme was not affected by $MgCl_2$, KCl, NH_4Cl, $MnCl_2$ and $CaCl_2$ while $AlCl_3$, inhibited the enzyme.

Key words: Cyanide, detoxification, tilapia, liver, rhodanese.

INTRODUCTION

Cyanide is a potent health hazard for human and ecosystem (Gupta et al., 2010). The toxic effect of cyanide (CN) is predominantly attributed to the inhibition of cytochrome c oxidase, a terminal oxidase of the mitochondrial respiratory pathway resulting in a condition of histotoxic anoxia.

Tolerance of organisms to cyanide contaminated environment and/or feed are due to a number of cyanide detoxification pathways, among the enzymatic pathways are 3-mercaptopyruvate sulphurtransferase (EC 2.8.1.2) and rhodanese (thiosulphate: cyanide sulphurtransferase, EC 2.8.1.1) which are widely distributed in living organisms. A major mechanism which removes CN^- from the body by enzymatic conversion to the less toxic thiocyanate (SCN) in the presence of a sulfur donor is by rhodanese (Rh) (Way, 1983; Petrikovics et al., 2010).

Rhodaneses are highly conserved and widespread enzymes, currently regarded as one of the most effective mechanism evolved for cyanide detoxification. *In vitro*, rhodanese catalyzes the irreversible transfer of a sulphur atom from a suitable donor (i.e. thiosulphate) to cyanide, leading to formation of less toxic sulphite and thiocyanate (Gupta et al., 2010). The enzyme has been purified and characterised from a number of animal tissues (Lee et al., 1995; Agboola and Okonji, 2004; Akinsiku et al., 2010) some of which are actively exposed to cyanide contaminated environments or food. The specific activity of rhodanese in many animal tissues is present in Table 1

Table 1. Specific activity of rhodanese (U/mg protein) in selected tissues of human and some domestic animals.

Tissue	Specie								
	Sheep[a]	Cattle[a]	Goat[b]	Camel[a]	Horse[a]	Pig[c]	Dog[a]	Chicken[d]	Human[e]
Liver	5.21	4.95	2.93	1.7	1.7	0.56	0.18	0.31	0.15
Rumen				0.28					
Epithelial Layer	16.3	10.3	3.31						
Muscular Layer	0.09	0.05	0.01						
Kidney				0.08	1.4		0.06	0.08	0.96
Cortex	1.72	0.82	0.94			0.4			
Medullar	0.24	0.05	0.25			0.1			
Brain	0.48	0.36	0.47	0.13	0.15	ND	0.32	ND	0.03
Lung	0.24	0.13	0.36	0.24	0.03	0.04	0.02	0.01	0.02
Stomach					0.1		0.07		0.01
Abomasal Fundus	0.09	0.04	0.35	0.07		0.13			
Abomassal Pylorus	0.1	0.04	0.31	0.02		0.006			
Muscular layer of proventriculus								0.09	
Epithelial layer of proventriculus								0.59	
Muscle	0.04	0.02	0.08	0.05	0.15	0.07	0.04	ND	0.01

*ND = Not determined; [a]Aminlari et al., 1989; [b]Nazifi, 2003; [c]Aminlari et al., 2002; [d]Aminlari et al., 1997; [e]Aminlari et al., 2007.

Asejire Lake (07° 21'N 04° 05'E), the source of the tilapia used in this study, is a manmade lake on River Osun, has an impounded area of 2342 ha and located about 30 km east of Ibadan, Southwest Nigeria. River Osun is one of the series of West African rivers which do not drain into Niger system but discharge into coastal lagoons and creeks bordering the Atlantic Ocean (Ayoade et al., 2006).

River Osun constantly received cyanide contaminants through agricultural and industrial activities (Olajire and Imeokparia, 2001) yet the river and lake support aquatic life (Ayoade et al., 2006; Adeyemo and Akomolafe, 2011). Tilapia *Oreochromis niloticus* is the most widely harvest freshwater fish in the Lake.

The survival nature of the fish is suspected to be due to the biotransformation of cyanide to thiocyanate by rhodanese. In early report from our laboratory we described the physicochemical and kinetic characterization of rhodanese from the liver of *Clarias gariepinus* Burchell in Asejire lake (Akinsiku et al., 2010), this paper describes the isolation and characterization of rhodanese from the cytosolic fraction of tilapia liver from the lake.

MATERIALS AND METHODS

CM-Sephadex C-50, Sephadex G-75 and Sephacyl S-400 were obtained from Pharmacia Fine Chemical, Uppsala, Sweden. All other reagents were of analytical grade and were obtained from either Sigma or BDH. The tilapia fishes, *Oreochromis niloticus*, were obtained from Asejire Lake located in the outskirt of Ibadan, Oyo State, Nigeria.

Enzyme and protein assays

The activity was expressed in rhodanese unit (RU). One rhodanese unit was taken as the amount of enzyme, which under the given condition produced an optical density reading of 1.08 at 460 nm per min (Sorbo, 1953). Protein concentrations were determined by the method of Bradford (1976) using bovine serum albumin (BSA) as the standard.

Enzyme isolation and homogenisation

Scalpels were used to open up the abdomen. The liver was quickly removed and stored in the freezer until required. The frozen liver was thawed at room temperature and rinsed with cold saline (0.9% NaCl, pH 7.4) to remove blood and other impurities. Fifty grams (50 g) of the liver (obtained from 15 fishes) was minced with a pair of scissors into smaller pieces and homogenized in two volumes of 0.1 M acetate glycine buffer, pH 7.8 containing 1 mM ε-amino-n-caprioc acid, and 10 mM sodium thiosulphate for 5 min with a Warring Blender. The homogenate was centrifuged for 60 min at 20,000 rpm. The crude enzyme was assayed for protein and enzyme.

C+M-Sephadex ion exchange chromatography

CM-Sephadex C-50 was pretreated according to Agboola and Okonji (2004). The resin was packed into a 2.5 × 40 column and equilibrated with 50 mM citrate buffer, pH 5.0 containing 10 mM $Na_2S_2O_3$. The protein from the preceding step was layered on the column. The column was first washed with 200 ml of the equilibration buffer to wash the unbound protein, followed by elution with a 250 ml linear gradient of 0-0.5 M NaCl in same buffer. Fractions of 4 ml were collected at a rate of 24 ml/h. The protein profile was monitored at 280 nm. The fractions were also assayed for enzyme. The active fractions were pooled and dialysed against 50% glycerol.

Gel filtration on sephadex G-75

Sephadex G-75 gel filtration resin was prepared by swelling 40 g in boiling distilled water for 3 h. The resin was then washed with several changes of 0.1 M phosphate buffer containing 10 mM

Table 2. Purification processes.

Fraction	Total activity (mg)	Total protein (mg)	Specific activity (U/mg protein)	Percentage Yield	Purification fold
Crude extract	3113.43	2700.00	1.15	100	1.00
CM Sephadex ion exchange	2087.60	123.16	16.95	67.00	14.74
Sephadex-G75 Gel filtration	457.01	8.02	56.98	14.68	49.55

Each purification procedure is as described in the materials and methods section. Protein concentration was determined using Bradford and activity was determined by the rate of formation of thiocyanate. 1U of activity is defined as the amount of enzyme, which under the given conditions, produced an absorbance reading of 1.08 at 460 nm per min (Sorbo, 1951).

sodium thiosulphate, pH 7.2 and then packed into 2.5 × 70 cm glass column and equilibrated with the same buffer.

Forty millilitres (40 ml) of the dialyzed protein from the ion-exchange step was layered on the column. Fractions (4 ml) were collected from the column at a rate of 15 ml per hour. Protein profile was monitored at 280 nm and assayed for rhodanese activity. The active fractions were pooled and dialysed against several changes of 50% glycerol for storage.

Polyacrylamide gel electrophoresis

Polyacrylamide gel electrophoresis and subunit molecular weight determination in the absence of sodium dodecyl sulphate (SDS) was carried out according to the procedure of Weber and Osborn (1975) on 7.5% acrylamide gel. Destaining was carried out with a solution containing 25 ml of methanol, 37.5 ml of glacial acetic acid made up to 500 ml with distilled water.

SDS-polyacrylamide gel electrophoresis in 10% acrylamide gel was used for the determination of the subunit molecular weight. Standard proteins that were used for the calibration of the gel were bovine serum albumin (66,000 Da), ovalbumin (45,000 Da), pepsin (34,700 Da), trypsinogen (24,000 Da) and β-lactoglobulin (20,100 Da). The gels were run at 8 mA per gel at room temperature and stopped after 5 h. The staining and destaining were performed as described earlier. R_f values of the standards calculated were then plotted against the logarithms of their molecular weight. The molecular weight of the enzyme preparation was then intrapolated from the curve.

Determination of apparent molecular weight

The native molecular weight was determined on a Sephacyl S-400 column (1.5 × 100 cm) using standard proteins (BSA (67,000; 3 mg/ml), α-chymotrypsinogen (25,000; 3 mg/ml) and ovalbumin (45,000; 3 mg/ml)). 3 ml of each standard protein was layered on the column. The column was eluted with 10 mM phosphate buffer, pH 7.0 at a flow rate of 17 ml/h. The void volume (V_o) of the column was determined by the elution volume of Blue Dextran (2 mg/ml). 4 ml of the pure enzyme preparation was then passed through the same column and the elution volume (V_e) estimated as describe earlier. Partition coefficient (K_{av}) was calculated from V_o and (V_e). A plot of logarithm of the molecular weight of the standard proteins against K_{av} was made, the molecular weights of the enzyme preparation was then intrapolated from the curve.

Determination of kinetic parameters

The kinetic parameters (K_m and V_{max}) of the enzyme were determined by varying concentrations of KCN between 12.5 and 50 mM at fixed concentration of 50 mM $Na_2S_2O_3$. Also, the concentration of $Na_2S_2O_3$ was varied between 12.5 and 50 mM at fixed concentration of 50 mM KCN. The kinetic parameters were estimated from the double reciprocal plots (Lineweaver and Burk, 1934).

Effect of pH on the enzyme activity

The enzyme was assayed using the buffers of different pH values: 50 mM citrate buffer (pH 4.0-6.5), 10 mM phosphate buffer (pH 7.0-8.5), and 50 mM borate buffer (pH 9.0-11). A reaction mixture of 1 ml contained 0.5 ml of the respective buffer, 0.2 ml of 0.25 mM KCN, 0.2 ml of 0.25 mM of $Na_2S_2O_3$ and 0.1 ml of enzyme solution.

Effect of temperature on the enzyme activity

The enzyme was assayed at temperatures between 0 and 70°C. The assay mixture was first incubated at the indicated temperature for 10 min before initiating reaction by the addition of an aliquot of the enzyme which had been equilibrated at the same temperature. The residual enzyme was then assayed.

Determination of heat stability

The heat stability of the enzyme was also determined by incubating 0.6 ml of the enzyme for 1 h at 30, 40, 50, 60 and 70°C, respectively. 0.1 ml was withdrawn at 10 min interval and assayed for residual activity. The activity at 30, 40, 50, 60 and 70°C was expressed as a percentage of activity of the enzyme incubated at 30°C which was the control.

Effect of chloride salts on the enzyme activity

The salts include $MnCl_2$, NH_4Cl, $CaCl_2$, $AlCl_3$ and $MgCl_2$ at the final concentrations of 0.5 mM and 1.0 mM in a typical enzyme assay reaction mixture.

RESULTS

The summary of the result of the purification of rhodanese from the liver of tilapia is summarized in Table 2. The specific activity of the pure enzyme was 56.86 U/mg protein. The elution profile of the enzyme in ion-exchange chromatography on CM-Sephadex C-50 is shown in Figure 1. The peak was obtained in the breakthrough volume without the salt gradient elution. The elution profile on Sephadex G-75 is shown in Figure 2. Only one protein band was observed after gel electrophoresis of the purified rhodanese either in the presence or absence of sodium dodecyl suphate (figures not shown).

The calibration curve on Sephacyl S-400 for the determination of the native molecular weight is shown in Figure 3. The molecular weight of rhodanese from the liver of tilapia after the gel filtration was estimated to be

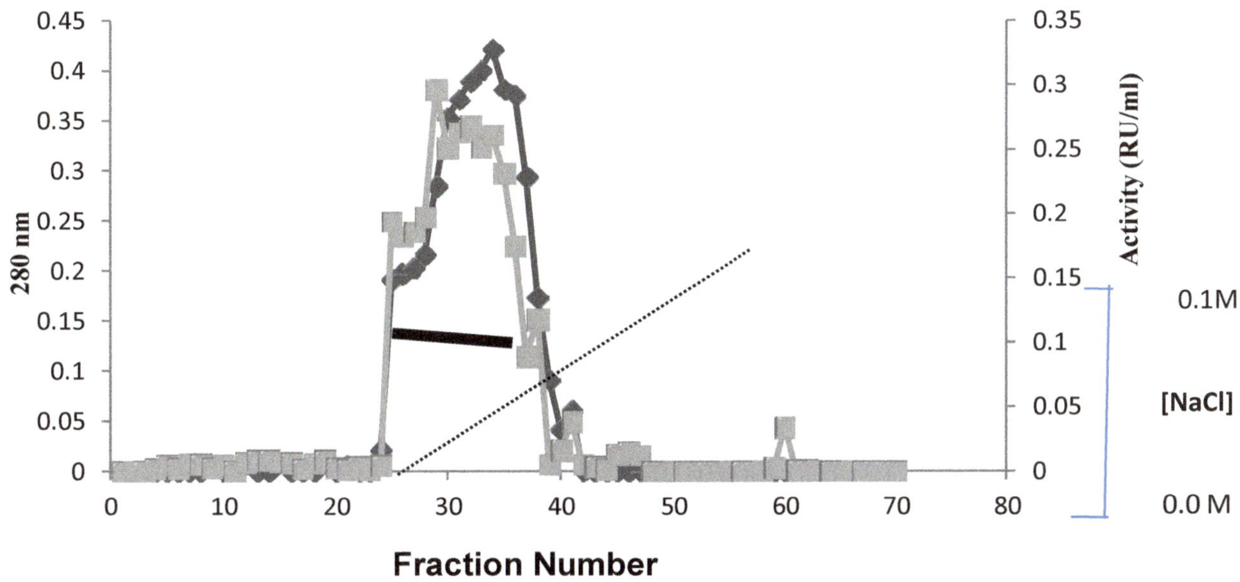

Figure 1. CM-Sephadex C-50 ion exchange chromatography elution profile of tilapia rhodanese. The column was equilibrated with 50 mM citrate buffer, pH 5.0 containing 10 mM Na_2SO_3. The dialysed protein was then layered on the column. The column was washed firstly with 200 ml buffer followed by elution with a 250 ml linear gradient of 0-0.5 mM NaCl in elution buffer. 4 ml fractions were collected at a rate of 24 ml/h. Protein profile was monitored using Bradford at 595 nm. ▬ Pooled Fraction; ········ Linear NaCl Gradient; Enzyme activity; ▢ OD890

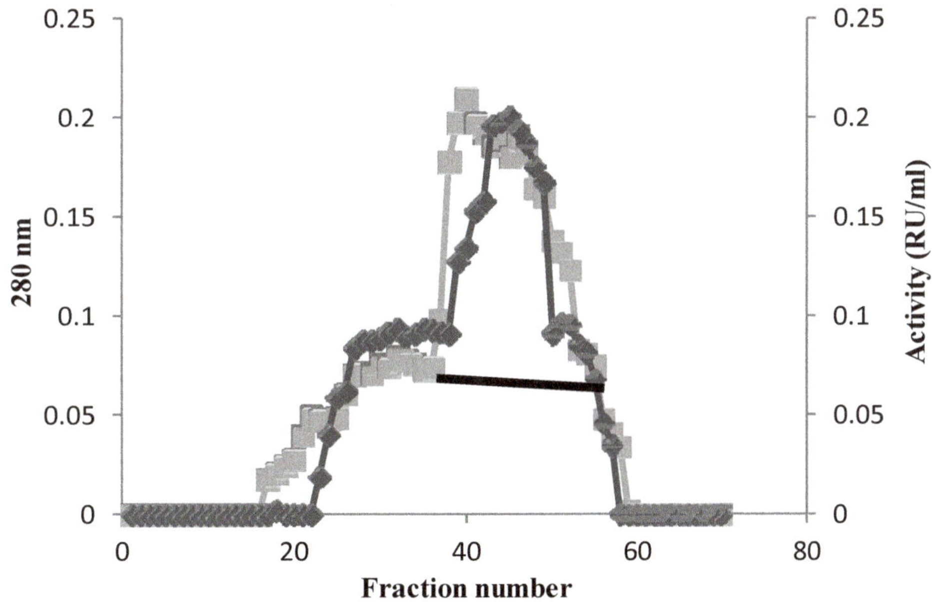

Figure 2. Sephadex G-75 Gel chromatography elution profile of Tilapia rhodanese. 40 ml of the dialysed protein from the ion exchange step was layered on a 1.5 × 40 cm column. 4 ml fractions were collected at a flow rate of 15 ml/h. Protein concentration was determined using Bradford at 595 nm. ▬, Pooled fraction; , Enzyme activity; ◆, OD_{280}

35,460 ± 450 daltons. The calibration curve obtained for polyacrylamide gel electrophoresis on 10% acrylamide gel electrophoresis is shown in Figure 4. The molecular weight was estimated to be 33,700 ± 714 daltons.

The double reciprocal plot of the change in KCN concentrations of fixed $Na_2S_2O_3$ is shown in Figure 5a, while that of varying concentrations of $Na_2S_2O_3$ at fixed KCN concentration is shown in Figure 5b. Table 3 shows the results of the kinetic parameters. The influences of pH on the rate of enzyme activity are shown in Figure 6. There were steady increases in enzyme activities until pH 6.5.

The effect of temperature on the activity of rhodanese

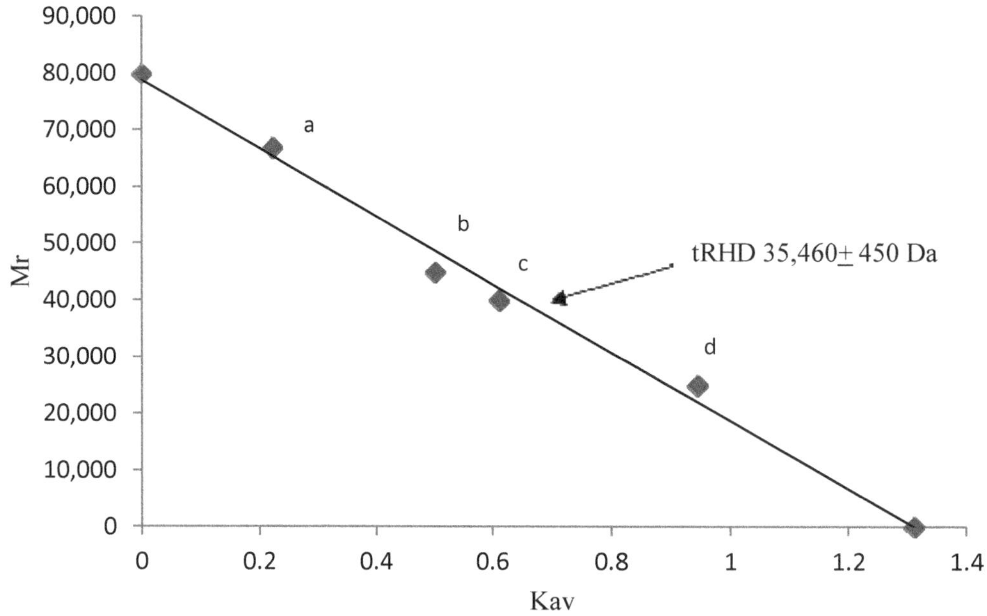

Figure 3. Calibration curve for molecular weight determination by gel filtration on sephacyl S-400. 3 ml of standard proteins was applied to the column (1.5 × 100 cm) and eluted with 10 mM phosphate buffer, pH 7.0 at a flow rate of 17 ml/h. Blue dextran (2 mg/ml) elution was used to determine the void volume. The standard proteins are (**a**) bovine serum albumin (66 kDa), (**b**) ovine albumin (45 kDa), (**c**) peroxidase (40 kDa), (**d**) chymotrypsinogen (25 kDa) and (**e**) β-lactoglobulin (20.1 kDa). The arrow indicates the position of the rhodanese enzyme (tRHD).

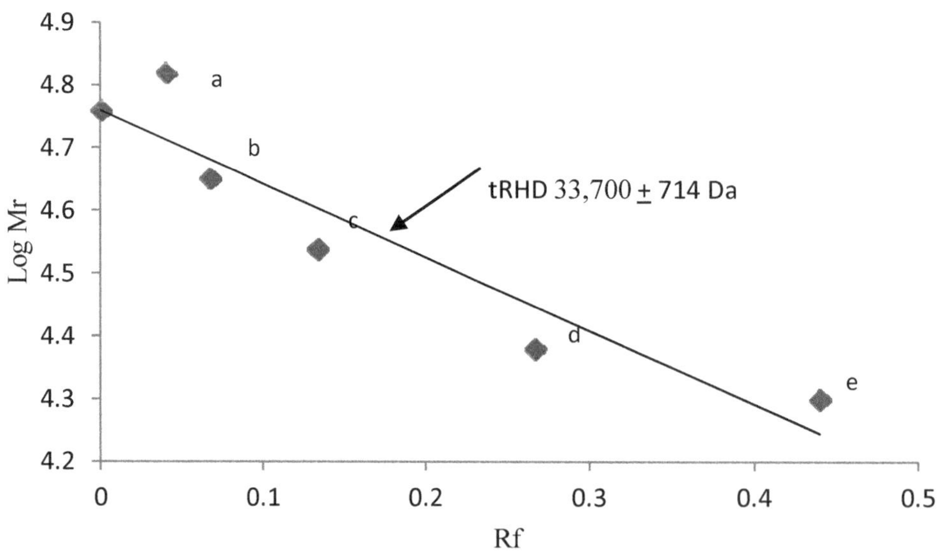

Figure 4. Calibration Curve for sub unit molecular weight determination on SDS-PAGE. The subunit molecular weight was determined by interpolation. The standard proteins are (**a**) bovine serum albumin (66 kDa), (**b**) ovine albumin (45 kDa), (**c**) peroxidase (40 kDa), (**d**) chymotrypsinogen (25 kDa) and (**e**) β-lactoglobulin (20.1 kDa). The arrow indicates the position of the rhodanese enzyme (tRHD).

from the tilapia liver is shown in Figure 7. Figure 8 shows the heat stability of the enzyme from the tilapia liver. The result of the effect of various chloride salts on the activity of catfish liver rhodanese (Table 4) showed that $MgCl_2$, KCl, NH_4Cl, $MnCl_2$ and $CaCl_2$ had little or no effect on the activity of the enzyme. Both concentrations of $AlCl_3$ have

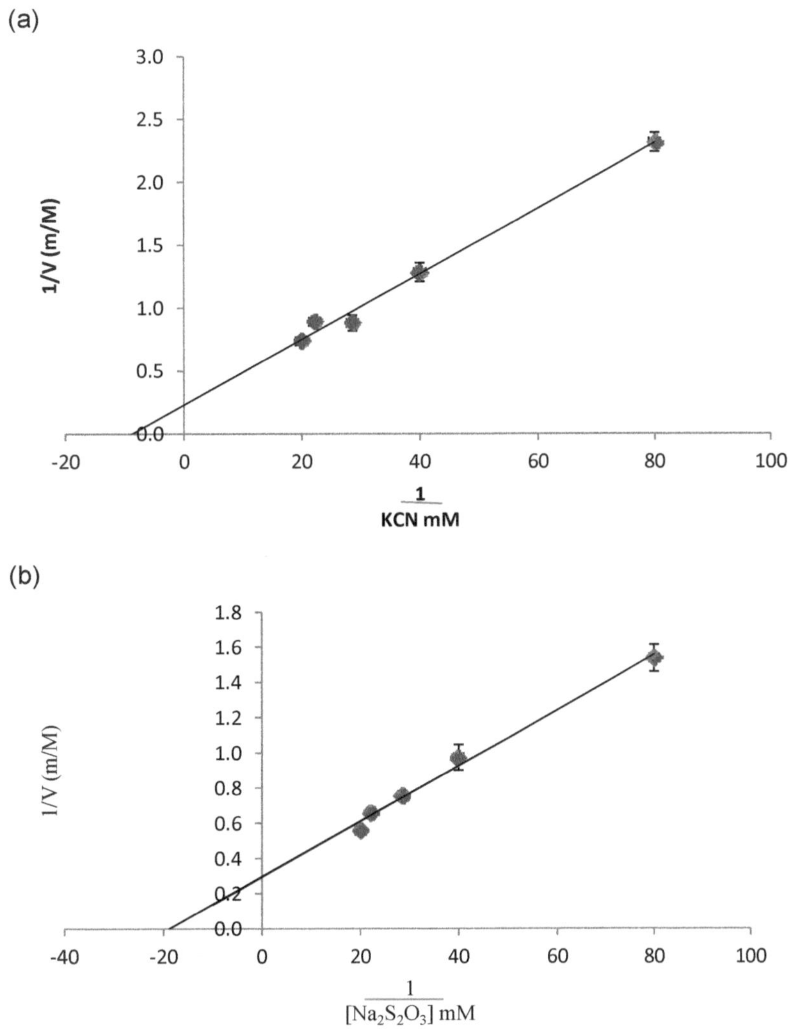

Figure 5. Lineweaver-Burk plot for the determination of K_m and V_{max}. **(a)** The concentration of $Na_2S_2O_3$ was varied between 12.5 mM and 50 mM at fixed concentration of 50 mM KCN while the concentration of KCN was determined at fixed concentrations of 50 mM $Na_2S_2O_3$. **(b)**. The reaction mixture contained 25 mM borate buffer pH 9 and 0.1 ml of enzyme in a total reaction volume of 3.0 ml at room temperature.

Table 3. Kinetics Parameters of KCN and Na_2SO_4 as Substrates of Rhodanese from Tilapia Fish liver.

Parameter	KCN	$Na_2S_2O_3$
K_m (mM)	0.1240 ± 0.0201	0.0516 ± 0.0097
V_{max} (RU/ml)	4.395 ± 0.3800	3.4208 ± 0.3690

The above kinetic parameters K_m and V_{max} were determined from the double reciprocal plots and were a mean of triplicate determinations

the inhibition effect on the enzyme.

DISCUSSION

Tilapia (*Oreochromis niloticus*) is more tolerant than most commonly freshwater fish to high salinity, high water temperature, low dissolved oxygen, and high ammonia concentrations (Popma and Masser, 1999). Their habitants such as lakes, streams, rivers are constantly subject to contaminations from various sources such as effluents from manufacturing companies (Bruton, 1979). Cyanides are present in many industrial wastewaters,

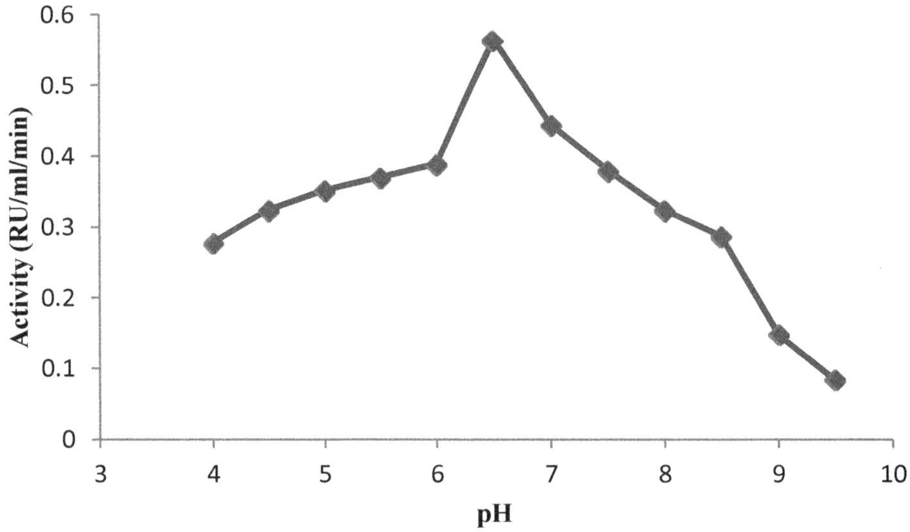

Figure 6. Effect of pH on the tilapia liver rhodanese. The assay mixture contained the appropriate buffers: 50 mM citrate buffer (pH 4.0-6.5), 10 mM phosphate buffer (pH 7.0-8.5), 0.2 ml of 0.25 KCN, 0.2 ml of 0.25 mm of $Na_2S_2O_3$ and 0.1 ml of enzyme inn final reaction volume of 3.0 ml.

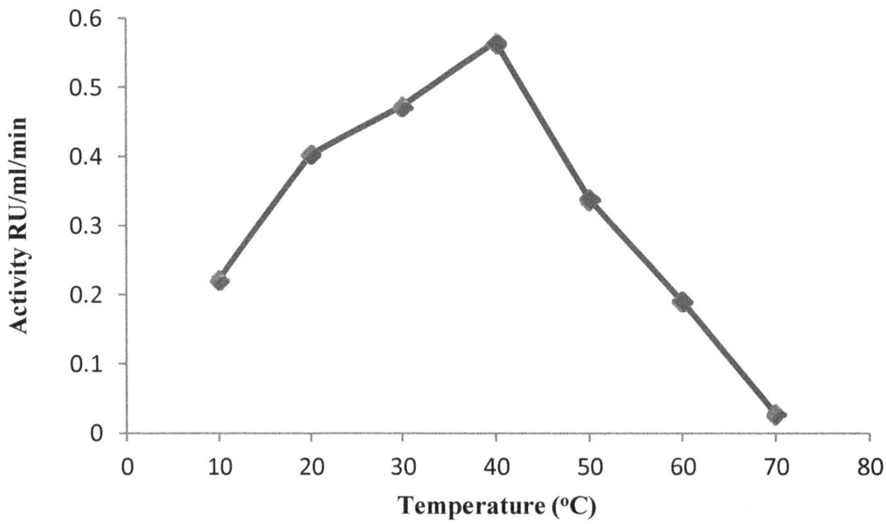

Figure 7. Activity-temperature profile showing optimum temperature of rhodanese. The assay mixture was first incubated at the incubated temperature for 10 min before initiating the reaction by the addition of an aliquot of the enzyme that had been previously equilibrated at same temperature.

especially effluents from iron and steel processing plants, petroleum refineries, and metal-plating plants, and constitute a hazard to aquatic ecosystems in certain waste-receiving waters (Eisler and Wiemeyer, 2004).

Study on feeding behaviour and nutrition requirements of *O. niloticus* by Popma and Masser (1999) shown that Tilapia ingest a wide variety of natural food organisms, including plankton, some aquatic macrophytes, planktonic and benthic aquatic invertebrates, larval fish, detritus, insects, even small fish fry and decomposing organic matter. With heavy supplemental feeding, natural food organisms typically account for 30 to 50% of tilapia growth. Most of these food items contain cyanogenic glycosides which upon hydrolysis release cyanide or hydrocyanic acid (HCN).

The Asejire Lake, the source of tilapia for this study also received cyanide from cassava processing activities around the lake. Rhodanese and 3-mercaptopyruvate sulphur transferase represent the chief enzymes of cyanide detoxification (Nagahara et al., 1999; Aminlari et

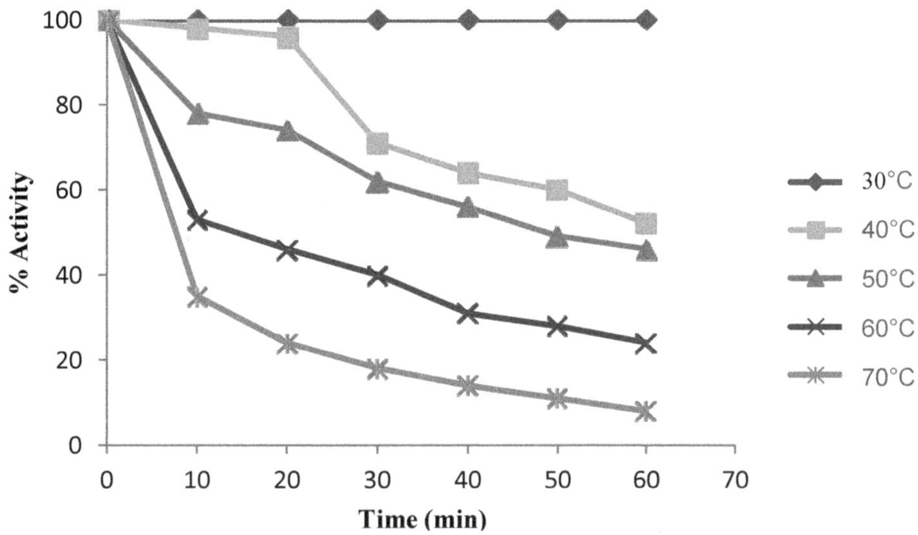

Figure 8. Effect of temperature on heat stability of tilapia rhodanese. 0.6 ml of enzyme was incubated at temperature ranges indicated above and aliquots of 0.1 ml was withdrawn at respective temperatures and assay was carried out for residual enzyme activity in 50 mM borate buffer. Percent residual activity is compared to the activity of the native enzyme at 30°C.

Table 4. Effect of chloride salts on tilapia liver rhodanese.

Parameter	Percentage enzyme residual activity	
Salt	0.5 mM	1.0 mM
None	100	100
$MgCl_2$	92 ± 6.14	90 ± 3.25
KCl	91 ± 2.07	89 ± 3.32
NH_4Cl	84 ± 1.25	79 ± 2.20
$MnCl_2$	83 ± 2.01	86 ± 3.25
$CaCl_2$	74 ±2.98	77 ± 2.82
$AlCl_3$	67 ± 5.45	67 ± 0.19

Enzyme assay were carried out as described in methods with each assay mixture containing the salts at final concentration of 0.5 and 1.0 mM. The values are a mean of at least triplicate determinations

al., 2002).

The purification procedures involved ion-exchange chromatography on CM-Sephadex C-50 and gel filtration on Sephadex G-75. Our initial attempts to use ammonium suphate fractionation resulted into a great lost of enzyme activities hence we omitted the purification process.

The specific activity of rhodanese from the liver of tilapia was found to be 56.98 RU/mg of protein. Sorbo (1953) and Himwich and Saunders (1948) obtained 256 RU/mg and 131 RU/mg, respectively for bovine liver rhodanese. Lee et al. (1995) obtained 1,070 RU/mg of protein for purified mouse liver rhodanese. Agboola and Okonji (2004) obtained 136.6 RU/mg protein for fruit bat liver rhodanese, Akinsiku et al. (2010) obtained 73 and 72 RU/mg for cRHD I and cRHD II, respectively for cat-fish liver rhodanese. Tayefi-Nasrabadi and Rahmani (2012) obtained 0.206 RU/mg protein for rhodanese from liver Rainbow Trout (*Oncorhynchus mykiss*).

The apparent relative molecular weight of 35,460 Da compares very well with that reported for the enzyme from other sources. It is in line with the reports that rhodanese has molecular weight value approximately between 33000 and 37000 daltons (Nagahara et al., 1999). The subunit molecular weight was 33,700±714 daltons. This result suggests that the enzyme is a monomeric protein. While Jarabak and Westley (1974) and Lee et al. (1995) showed that purified mouse and human liver rhodaneses are monomeric proteins, other workers have shown that they are dimers of two identical subunits of molecular weights 18,000-19,000 daltons (Blumenthal and Heinrikson, 1971).

The apparent K_m values, KCN and $Na_2S_2O_3$, were 0.1240 and 0.0516 mM, respectively. These values are lower than those of rainbow trout (KCN = 36.81mM; Na_2SO_4 = 19.84 mM) (Tayefi-Nasrabadi and Rahmani, 2012), african catfish liver (KCN = 25.40 mM; Na_2SO_4 = 18.60 mM) (Akinsiku et al., 2010), fruit bat liver (KCN = 13.36 mM; Na_2SO_4 = 19.15 mM) (Agboola and Okonji, 2004), mouse liver (KCN = 12.50 mM; Na_2SO_4 = 8.30 mM) (Lee et al., 1995), bovine liver (KCN = 19.0 mM; Na_2SO_4 = 6.7 mM) (Sorbo, 1953), and human liver (KCN = 9.50 mM; Na_2SO_4 = 4.50 mM) (Jarabak and Westley, 1974), indicating that the affinity of tilapia enzyme for these substrates is more than that of the other enzymes and that it would catalyze the detoxification reaction with high efficiency. As reported by Agboola and Okonji (2004) less effective enzymatic system may be due to a lower exposure to cyanide and high affinity in some mammalian animals is due to continue exposure through their diet. This high affinity of tilapia rhodanese is supported by feeding and nutrition requirements of the fish that include wide range of natural foods which are cyanogenic (Popma and Masser, 1999). Madalla (2008) reported that (*Oreochromis niloticus* L) survived on diet which containing 40% and 10% of hydrogen cyanide from cassava leaf meal and cassava root meal respectively. It should be noted that cyanide detoxification is a secondary benefit of rhodanese, which has a number of major physiological roles, including the supply of sulphide for the formation of iron-sulphur centres for the electron transport chain (Cerletti, 1986), participation in energy metabolism (Ogata et al., 1989), and function as a thioredoxin oxidase (Nandi et al., 2000).

An optimum temperature of 40°C was obtained for tilapia liver rhodanese. This result is in good agreement with the results reported for rhodaneses from different sources. Sorbo (1953) reported an optimum temperature of 50°C for bovine liver rhodanese. Ezzi et al. (2003) obtained a wide temperature optimum of 35-55°C for rhodanese enzyme in all different *Trichoderma* strains. Agboola and Okonji (2004) reported 35°C for the rhodanese in the cytosolic fraction of fruit bat liver. Also, Akinsiku et al. (2010) reported 40°C for the rhodanese in the liver of catfish (*C. gariepinus*) from Asejire Lake, same source of tilapia for this study. Tayefi-Nasrabadi and Rahmani (2012) reported 25°C for rhodanese from the liver of rainbow trout. Akinsiku et al. (2010) posited that with the level of pollution in this water, there will be various metabolic activities going on in it which likely results in the release of heat and the ambient temperature of 40°C which might have conditioned the enzyme to function at higher temperature. The heat stability experiment showed that the enzyme was stable up to 60°C for about 30 min which means that the enzyme is thermostable.

An optimum pH of 6.5 was obtained for rhodanese from the liver of the tilapia from Asejire Lake. Same pH value was obtained by Akinsiku et al. (2010) for the rhodanese of catfish from the same lake. They posited that the value is as a result of the acidity of the water. Pollutants in water include wide spectrum of chemicals and pathogens which sometimes alters the acidity, conductivity and temperature of the water. The water sample collected from Asejire Lake showed a pH of 5.7, 6.3 and 6.5 at different points of the Lake (Lameed and Obadara, 2006).

Studies on effect of chloride salts on the enzyme showed that the activity of the enzyme was not affected by $MgCl_2$, KCl, NH_4Cl, $MnCl_2$ and $CaCl_2$ while $AlCl_3$ inhibited the enzyme greatly. Metal ions showing inhibitions are those that have strong affinity for cysteinyl and histidyl side chains of proteins (Ulmer and Vallee, 1972). The properties of rhodanese from the liver of tilapia are similar to those from other animal sources and might properly function in the same manner.

Conflict of interests

The author(s) have not declared any conflict of interests.

REFERENCES

Adeyemo OK, Akomolafe OT (2011). Copper level in fish, selected fresh and marine aquatic ecosystems in Nigeria. Afr. J. Biomed. Res. 14:105-111.

Agboola FK, Okonji RE. (2004). Presence of Rhodanese in the Cytosolic Fraction of the Fruit Bat (Eidolon helvum) Liver. J. Biochem. Mol. Biol. 37(3):275-281.

Akinsiku OT, Agboola FK, Kuku A, Afolayan A (2010). Physicochemical and kinetic characteristics of rhodanese from the liver of African catfish Clarias gariepinus Burchell in Asejire lake. Fish Physiol. Biochem. 36:573-586. PMid:19536635

Aminlari M, Li A, Kunanithy V, Seaman CH (2002). Rhodanese distribution in porcine (Sus scrofd) tissues. Comp. Biochem. Physiol. 132B:309-313.

Ayoade AA, Fagade SO, Adebisi AA. (2006). Dynamics of limnological features. Biochem. Sci. 11:369-372.

Blumenthal KM, Heinrikson RL (1971). Structural Studies of Bovine Liver Rhodanese: I: Isolation and Characterization of two Active Forms of the Enzymes. J. Biol. Chem. 246(8):2430-2437. PMid:5553401

Bradford M (1976) A rapid and sensitive method for the quantitation of microgram quantities of protein utilizing the principle of protein-dye binding. Anal. Biochem. 72:248-25.

Bruton, MN (1979). The breeding biology and early development of Clarias gariepinus (Pisces: Clariidae) in Lake Sibaya, South Africa, with a review of breeding in species of the subgenus Clarias. Trans. Zool. Soc. Lond. 35:1-45.

Cerletti P(1986). Seeking a better job for an under-employed enzyme: rhodanese. Trends Biochem. Sci. 11:369-372.

Eisler R, Wiemeyen SN (2004).Cyanide hazards to plants and animals from gold mining and related water issues. Rev. Environ. Contam. Toxicol.183:21-54. PMid:15369321

Ezzi MI, Pascual JA, Gould BJ, Lynch JM (2003). Characterisation of the rhodanese enzyme in Trichoderma spp. Enzyme Microb. Technol. 32(5):629-634.

Gupta N, Balomajumder C, Agarwal VK (2010). Enzymatic mechanism and biochemistry for cyanide degradation: A review. J. Hazard. Mater. 17:1-13. PMid:20004515

Himwich WA, Saunders JB (1948). Enzymic conversion of cyanide to thiocyanate. Am. J. Physiol. 53:348-354.

Jarabak R, Westley J (1974). Steady-state kinetics of 3-mercaptopyruvate sulphurtransferase from bovine kidney. Arch. Biochem. Biophys. 185:458-465.

Lameed GA, Obadara PG (2006). Eco-Development Impact of Coca-Cola Industry on Biodiversity Resources at Asejire Area, Ibadan; Nigeria. J. Fish. Int. 1(2-4):55-62.

Lee CH, Hwang JH, Lee YS, Cho KS (1995). Purification and Characterization of mouse liver rhodanese. J. Biochem. Mol. Biol. 28: 170-176.

Lineweaver H, Burk D (1934). The determination of enzyme dissociation constants. J. Amer. Chem. Soc. 56:658-666.

Madalla N (2008). Novel Feed Ingredients for Nile Tilapia (Oreochromis niloticus L.). PhD Thesis submitted to Institute of Aquaculture, University of Stirlin, Scotland, United Kingdom.

Nagahara N, Ito T, Minam M (1999). Mecaptopyruvate sulphur transferase as a defence against cyanide toxications: Molecular properties and mode of detoxification. Histol. Histopath. 14:1277-1286. PMid: 10506943.

Nandi DL, Horowitz PM, Westley J (2000). Rhodanese as a thioredoxin oxidase. Int. J. Biochem. Cell Biol. 32:465-473.

Nazifi S, Aminlari M, Alaibakhsh MA. (2003). Distribution of rhodanese in tissues of goat (Capra hircus). Comp. Biochem. Physiol. 134B, 515-518.

Ogata K, Xing D, Volini M (1989). Bovine Mitochondrial Rhodanese is a Phosphoprotein. J. Biol. Chem. 246(5):2718-2725.

Olajire AA, Imeokparia FE (2001).Water quality assessment of Osun River: Studies on inorganic nutrients. Envin Monit. Asses. 69:17-28. PMid:11393542.

Petrikovics I, Baskin SI, Beigel KM, Schapiro BJ, Rockwood GA., ManageAB, Budai W, Szilasi M (2010). Nano-intercalated rhodanese in cyanide antagonism. Nanotoxicology 4(2):247-254. PMid:20795898

Popma T, Masser M (1999). Southern Region AquaCulture Publication No. 283, MSU. Mississippi, USA. pp. 1-4.

Sorbo, BH (1951). Rhodanese. Method Enzymol. 2(2):334-337.

Sorbo BH (1953). Crystalline Rhodanese. Enzyme catalyzed reaction. Acta Chem. Scand. 7:1137-1145.

Tayefi-Nasrabadi H, Rahmani R (2012). Partial Purification and Characterization of Rhodanese from Rainbow Trout (Oncorhynchus mykiss) Liver. J. Sci. World 5:1-5. PMid: 22629177 PMCid:PMC3354555

Ulmer DD Vallee BL (1972). Role of metals in sulphurtranferases activity. Ann. Rev. Biochem. 32:86-90.

Way JL (1983). Cyanide antagonism. Fundam. Appl. Toxicol. 3:383-386.

Weber K, Osborn M (1975). Protein and Sodium dodecyl sulphate: Molecular weight determination on polyacrylamide gel and related procedures. In: The Proteins. Neutrath. H. and Hill, R. L., eds., vol. I., Academic Press. New York, pp. 179-223.

Collaborative effects of some anti-diabetic plants on the liver marker enzymes of diabetic rats

GOMETI Ajiroghene S. , OGUGUA Victor N., ODO Christian E., JOSHUA Parker E. and OBI Kelechi N.

Department of Biochemistry, University of Nigeria, Nsukka, Nigeria.

The leaves of *Psidium guajava, Anacardium occidentale, Eucalyptus globulus* and fruits of *Xylopia aethiopica* are used in the management of diabetes mellitus. The phytochemical constituents as well as the acute toxicity of the combined chloroform extracts (*A. occidentale* + *E. globulus* and *P. guajava* + *X. aethiopica*) and their effects (at graded doses of 100 and 250 mg/kg body weight each) on the activities of alanine aminotransferase (ALT), aspartate aminotransferase (AST) and alkaline phosphatase (ALP) were therefore, assayed in diabetic and normal rats using standard methods. The phytochemical analyses of the four extracts showed the presence of flavonoids, terpenoids and fats and oil in all of them. Each of the combined extract was found to be non-toxic even at a dose as high as 5000 mg/kg body weight. The combined extracts at the tested doses significantly (p<0.05) and dose-dependently decreased the activities of ALT, AST and ALP. The effects of the combined extracts (especially 250 mg/kg body weight of *P. guajava* + *X. aethiopica*) were better than that of the standard anti-diabetic drug [glibenclamide (5 mg/kg body weight)]. The results generally indicate that the combined chloroform extracts of the leaves of *A. occidentale, E. globulus, P. guajava* and fruits of *X. aethiopica* might be adopted for the management and/or amelioration of diabetes mellitus and its accompanying complications.

Key words: *Psidium guajava, Anacardium occidentale, Eucalyptus globules*, *Xylopia aethiopica,* chloroform extracts

INTRODUCTION

Diabetes mellitus is a heterogenous metabolic disorder characterized by hyperglycemia arising from defective insulin secretion, resistance to insulin action or both. This disease generally affects organs that participate in carbohydrate metabolism such as the liver, a major site of insulin clearance and glucose homeostasis during the concentration of postprandial fasting blood glucose (Gavin et al., 1997). The toxicity of excessively high concentration of plasma glucose arising from prolonged hyperglycemia or postprandial glucose cause the autoxidation of glucose and non-enzymatic glycosylation of proteins leading to the production of advanced glycosylation end products (AGEs) which through their receptors (RAGEs) inactivate enzymes, altering their structures and functions and resulting in free radical formation (McCarthy et al., 2001).

The free radicals through oxidative stress cause lipid peroxidation resulting in cell injury and decreased concentration of cellular anti-oxidants (Hijora et al., 2000). Malondialdehyde (MDA), a product of lipid peroxidation, if elevated in the blood can lead to activation of leucocyte lysosomal membrane rupture releasing the enzymes into general circulation. These results in the increase of cell and parenchymal tissue damage for example, liver cell parenchymal destructions causing liver enzymes to leak into the blood stream and this serves as an indicator for free radial-mediated destruction of liver parenchymal cells (Pratibha et al., 2004). Plants such as *Anacardium occidentale, Eucalyptus globulus, Psidium guajava,* and *Xylopia aethiopica* have been implicated in the treatment of diabetes mellitus because they possess arrays of phytochemicals with multifarious pharmacological effects (Sagheb et al., 2010; Gupta et al., 2011; Ukwenya et al., 2012). Hence, the present study was targeted at evaluating the combined effects of the chloroform extracts of the leaves of *A. occidentale, E. globulus, P. guajava* and fruits of *X. aethiopica* on liver marker enzymes of diabetic rats.

MATERIALS AND METHODS

The plant samples

The leaves of *A. occidentale*, *E. globulus* and *P. guajava* were collected from the premises of University of Nigeria, Nsukka while the fruits of *X. aethiopica* were purchased from a local market in Delta State. The plant samples were identified by Prof. (Mrs.) May Nwosu of the Department of Botany, University of Nigeria, Nsukka where the voucher specimens were deposited in the herbarium.

Preparation of the extracts

The leaves of *A. occidentale, E. globulus, P. guajava* and the fruits of *X. aethiopica* were air dried to constant weight at room temperature and then reduced to powder. 600 g of each plant material was macerated in 2.7 L of analytical grade chloroform. After 48 h, the resulting extracts were filtered and concentrated with rotary evaporator at reduced pressure and the yield of extracts calculated. A standard weight (8 g) of each of the two proportionally combined extracts was dissolved in 16 ml of 10% dimethyl sulphuroxide (DMSO). The doses of each extracts administered was estimated by the method of Tedong et al. (2007) where volumes given were calculated as follows:

$$V\ (ml) = \frac{D \times P}{C}$$

Where D = Dose used (g/kg body weight of test animals); P = Body weight (kg); C = Concentration (g/ml); V = Volume (ml)

Animals

Thirty-five male Wistar albino rats of weight (180-230 g) and 64 male mice of weight (20-30 g) were used for this study. The University Animal Research Ethical Committee approved the experimental protocol. The animals were housed and maintained at a 12 h light and dark cycle and fed with rat diet *ad libitum*. The mice were used for acute oral toxicity study while the rats were made diabetic by a single dose of 180 mg/kg body weight of alloxan monohydrate intraperitonially and 35 rats selected for the study, 72 h after diabetes has been established. Treatments were for 40 h and administrations of the combined extracts were twice daily. After 40 h, rats were sacrificed and their blood collected for further biochemical analyses.

Chemicals and reagents

Dimethyl sulfoxide [DMSO (Serva Heidelberg, New York)], chloroform (Sigma Aldrich Chemicals, Germany), alloxan monohydrate (Sigma Aldrich Chemicals, Germany), sodium chloride, dilute tetraoxosulphate (vi) acid, 2% (v/v) hydrochloric acid, 1% (w/v) picric acid, methyl orange, Dragendorff's reagent, Mayer's reagent, Wagner's reagent, Fehling's solution, 5% (w/v) ferric chloride solution, aluminium chloride solution, lead subacetate solution, ammonium solution and distilled water were used for the study. The commercial kits were purchases from Randox Laboratories Ltd, Crumlin Co Antrim, UK.

Phytochemical analyses

Qualitative phytochemical analyses were carried out on the extracts of the various plant samples according to the procedures outlined by Harborne (1998) and Trease and Evans (1989).

Acute oral toxicity test (LD_{50})

A lethal dose toxicity study of each of the two proportionally combined extracts was carried out by the method described by Lorke (1983).

Experimental procedures

Alanine and aspartate aminotransferase (ALT and AST) activities were assayed using Randox commercial enzyme kit as described by Reitman and Frankel (1957) and Schmidt and Schmidt (1963). Alkaline phosphatase (ALP) activity was estimated using Randox commercial enzyme kit, based on the methods of Rec (1972) and Englehardt (1970).

Statistical analysis

Data generated from this study were represented as mean ± SEM. Variables were analyzed by one-way Analysis of Variance (ANOVA) and comparison done by multiple comparisons using Duncan test.

RESULTS

Qualitative phytochemical constituents of the chloroform extracts of the leaves of *A. occidentale, E. globulus* and *P. guajava* and the fruits of *X. aethiopica*

The qualitative phytochemical analyses showed the presence of flavonoids, terpenoids and fats and oil in the four extracts (Table 1). Saponins and tannins were present in the extracts of *A. occidentale*, *E. globulus* and *P.* guajava. Glycosides and alkaloids were not detected in the extracts of *A. occidentale* and *P.* guajava respectively. Saponins and tannins were not detected in *X. aethiopica* extract.

The acute toxicity and lethality (LD_{50}) of the combined plant extracts

There was no lethality or any sign of toxicity in the four groups of four mice each that received 10, 100 and 1000 mg/kg body weight of each of *A. occidentale* + *E. globulus* and *P. guajava* + *X. aethiopica* as well as 5 ml/kg body weight of 10% DMSO respectively at the end

Table 1. Qualitative phytochemical constituents of *Anacardium occidentale*, *Eucalyptus globulus*, *Psidium guajava* and *Xylopia aethiopica*.

Phytochemical constituent	*Anacardium occidentale*	*Eucalyptus globulus*	*Psidium guajava*	*Xylopia aethiopica*
Alkaloids	+	+	ND	+
Flavonoids	+	+	+	+
Glycosides	ND	+	+	+
Saponins	+	+	+	ND
Tannins	+	+	+	ND
Terpenoids	+	+	+	+
Fats and oil	+	+	+	+

+ = Present; ND = Not detected.

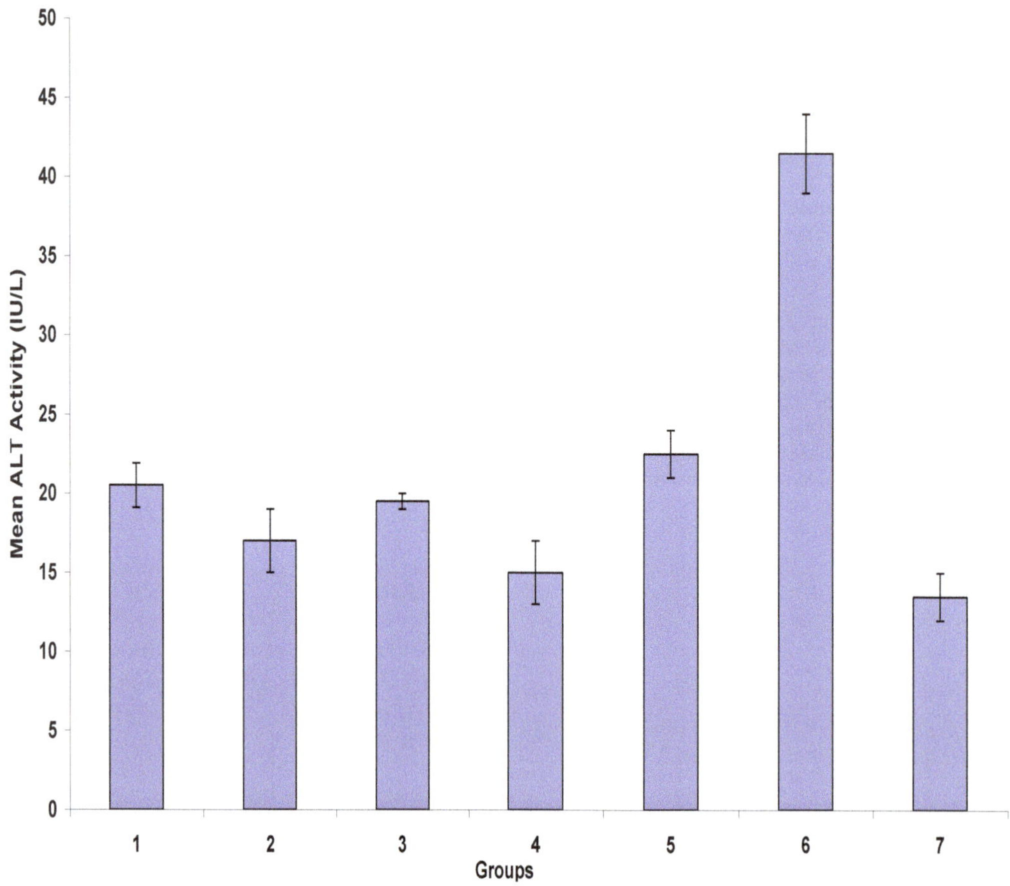

Figure 1. Effects of varying doses of the combined plant extracts on alanine aminotransferase (ALT) activity. Group 1= *A. occidentale* + *E. globulus* (100 mg/kg b.w). Group 2 = *A. occidentale* + *E. globulus* (250 mg/kg b.w). Group 3 = *P. guajava* + *X. aethiopica* (100 mg/kg b.w). Group 4 = *P. guajava* + *X. aethiopica* (250 mg/kg b.w). Group 5 = Glibenclamide (5 mg/kg b.w). Group 6 = Diabetic untreated (5 ml/kg b.w of DMSO). Group 7 = Non diabetic control (5 ml/kg b.w of DMSO).

of the first phase of the study. At the end of the second phase of the study, there was neither death nor obvious sign of toxicity in the groups of mice that received 1900, 2600 and 5000 mg/kg body weight of each of the combined plant extracts.

Effects of varying doses of the combined plant extracts on alanine aminotransferase (ALT) activity

Figure 1 shows that the diabetic untreated group had the highest activity of ALT but administration of the different

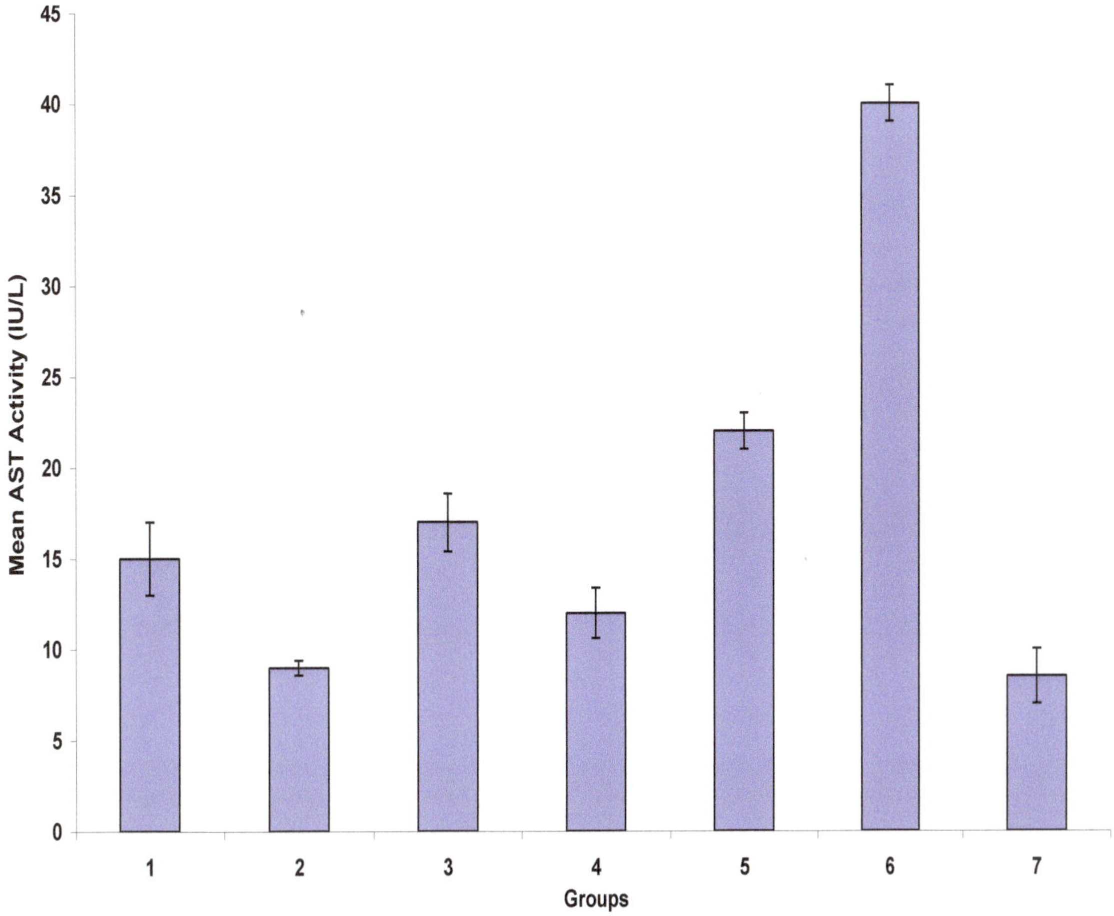

Figure 2. Effects of varying doses of the combined plant extracts on aspartate aminotransferase (AST) activity. Group 1= *A. occidentale* + *E. globulus* (100 mg/kg b.w) Group 2= *A. occidentale* + *E. globulus* (250 mg/kg b.w). Group 3 = *P. guajava* + *X. aethiopica* (100 mg/kg b.w). Group 4 = *P. guajava* + *X. aethiopica* (250 mg/kg b.w). Group 5 = Glibenclamide (5 mg/kg b.w). Group 6 = Diabetic untreated (5 ml/kg b.w of DMSO). Group 7 = Non diabetic control (5 ml/kg b.w of DMSO).

doses of the combined extracts resulted in significant ($p<0.05$) decrease in ALT activity. The 100 and 250 mg/kg body weight of each of *A. occidentale* + *E. globulus* and *P. guajava* + *X. aethiopica* in a similar manner as the standard anti-diabetic drug [glibenclamide (5 mg/kg body weight)] significantly ($p<0.05$) and dose-dependently reduced the activity of ALT when compared with the value obtained for the diabetic untreated group. However, the 250 mg/kg body weight of *P. guajava* + *X. aethiopica* caused the greatest reduction in the activity of ALT.

Effects of varying doses of the combined plant extracts on aspartate aminotransferase (AST) activity

The AST activities of groups 1, 2, 3, 4, 5 and 7 were significantly ($p<0.05$) and dose-dependently lower than that of the diabetic untreated group (group 6). There were no significant ($p>0.05$) differences between the AST acti-vities of groups 2 and 4 and that of the group 7 [non diabetic control group (5 ml/kg body weight of DMSO)] as shown in Figure 2.

Effects of varying doses of the combined plant extracts on alkaline phosphatase (ALP) activity

As shown in Figure 3, the diabetic untreated group had the highest ALP activity but administration of the different doses of the combined extracts caused significant ($p<0.05$) decrease in ALP activity. The 100 and 250 mg/kg body weight of each of *A. occidentale* + *E. globulus* and *P. guajava* + *X. aethiopica* in a similar man-ner as the standard anti-diabetic drug [glibenclamide (5 mg/kg body weight)] significantly ($p<0.05$) and dose-relatedly led to a drop in the activity of ALP when com-pared with the value obtained for the diabetic untreated group. The 250 mg/kg body weight of *P. guajava* + *X. aethiopica* however, depressed the activity of ALP best.

DISCUSSION

Acute toxicity test on the combined extracts (*A. occidentale* + *E. globulus* and *P. guajava* + *X. aethiopica*) using mice showed an LD_{50} value of greater than 5000

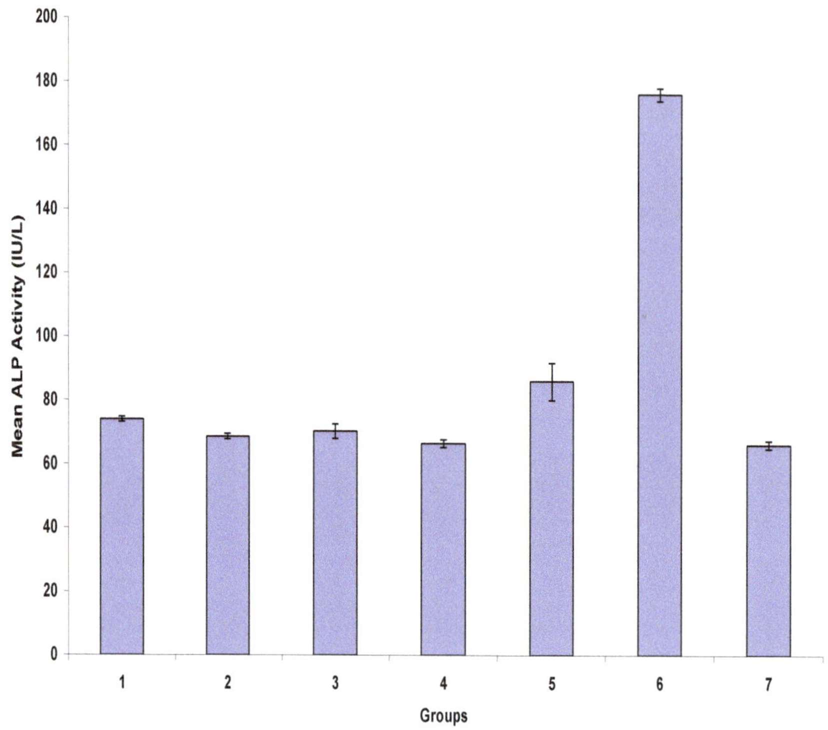

Figure 3. Effects of varying doses of the combined plant extracts on alkaline phosphatase (ALP) activity. Group 1= *A. occidentale* + *E. globulus* (100 mg/kg b.w) Group 2 = *A. occidentale* + *E. globulus* (250 mg/kg b.w). Group 3= *P. guajava* + *X.. aethiopica* (100 mg/kg b.w). Group 4 = *P. guajava* + *X. aethiopica* (250 mg/kg b.w). Group 5 = Glibenclamide (5 mg/kg b.w). Group 6 = Diabetic untreated (5 ml/kg b.w of DMSO). Group 7 = Non diabetic control (5 ml/kg b.w of DMSO).

mg/kg body weight for each combined extract which indicates that the leaves of *A. occidentale, E. globulus, P. guajava* and fruits of *X. aethiopica* have low toxicity.

The 250 mg/kg body weight of the combined extract (*A. occidentale* + *E. globulus*) exerted better effects in the treated rats than all the doses of *P. guajava* + *X. aethiopica* considering all the studied parameters. The remarkable reductions observed in ALT and AST activities could be said to have been caused by the hepatocellular and cardiac protection offered by the combined extracts. Hepatic and cardiac tissues release aspartate and alanine aminotransferases and therefore, the elevation of plasma concentrations of these enzymes are indicators of hepatic and cardiac damage as in the case of complications in diabetes mellitus (Crook, 2006). The observation is in support of the report of Ogbonnia et al. (2010) who studied the effect of a poly-herbal formulation on liver function enzymes in diabetic rats. It was noted that the administration of the poly-herbal formulation (which has *X. aethiopica* as one of its component) to rats led to a pronounced decrease in ALT and AST activities in the treated rats. The implication of this is that the combined extracts did not produce harmful effects on both the cardiac or hepatic tissues of the treated rats while in the diabetic untreated group, there were notable elevations in the activities of these two enzymes, an indication of hepatic and cardiac tissue damage.

The notable reduction in serum ALP activity recorded is suggestive of cellular membrane/hepatocellular membrane protective effects of the combined extracts. ALP functions as a biochemical marker enzyme for maintaining membrane integrity. Increase in its plasma activity indicates peroxidation of cell membrane which occurs during diabetes mellitus (Akanji et al., 1993). Uboh et al. (2010) showed that the aqueous extract of *P. guajava* confers hepatocellular protection in rats. The hepatocellular protection evidenced in the present study might be due to the presence of flavonoids in the plant extracts. Flavonoids have been reported to possess antioxidant activity (Middleton, 1996) and thus, are capable of protecting cell membranes from peroxidative actions of free radicals.

In conclusion, this study shows that the combined chloroform extracts of the leaves of *A. occidentale, E. globulus, P. guajava* and fruits of *X. aethiopica* as employed in the study had better decreasing effects on the activities of liver marker enzymes than glibenclamide (a standard anti-diabetic drug) and therefore, may be adopted for the management and/or amelioration of diabetes mellitus and its accompanying complications.

REFERENCES

Akanji MA, Olagoke OA, Oloyede OB (1993). Effect of chronic consumption of metabisulphite on the integrity of the kidney cellular system. Toxicology 81: 173–179.

Crook MA (2006). Clinical Chemistry and Metabolic Medicine. 7th edn. Hodder Amold, London. p. 246.

Englehardt A (1970). Measurement of alkaline phosphatase. Aerztl Labor. 16: 42-43.

Gavin JR, Alberti KG, Davidson MB, DeFronzo RA, Gabbe SG, Harris MI, Knowler WC, Maclaren NK, Palmer JP, Rizza RA, Stem MP (1997). Report of the expert committee on the diagnosis and classification of diabetes mellitus. Diabetes Care 20: 1183-1197.

Gupta GK, Chahal J, Arora D (2011). *Psidium guajava* Linn: Current research and future prospects. J. Pharm. Res. 4(1): 42-46.

Harborne JB (1998). Phytochemical Methods A Guide to Modern Techniques of Plant Analysis. Chapman and Hall, Thompson Science, London. p. 107.

Hijora E, Nistiar F, Sipulava A (2000). Changes in ascorbic acid and malondialdehyde in rats after exposure to mercury. Bratislava Med. J. 106(8): 248-251.

Lorke D (1983). A new approach to practical acute toxicity testing. Arch. Toxicol. 54: 275-287.

McCarthy AD, Etcheverry SB, Cortizo AM (2001). Effect of advanced glycation end products on the secretion of insulin-like growth factor-1 and its binding proteins: Role in osteoblast development. Acta Diabetol. 38(3): 113-122.

Middleton E (1996). Biological properties of plant flavonoids. 'An overview'. Int. J. Pharma. 34(95): 344-348.

Ogbonnia SO, Mbaka GO, Adekunle AA, Anyika EN, Gbolade OE, Nwakakwa N (2010). Effect of a poly-herbal formulation, Okudiabet on alloxan-induced diabetic rats. Agric. Biol. J. N. Am. 1: 139–145.

Pratibha K, Anand U, Agarwal A (2004). Serum adenosine deamina 8, 5-nucleotidase and malondialdehyde in acute infective hepatitis. Indian J. Clin. Biochem. 19(2): 128-131.

Rec GSCC (1972). Colorimetric method for serum alkaline phosphatase determination. J. Clin. Biochem. 10(20): 182-184.

Reitman S, Frankel S (1957). A colorometric method for the determination of serum glutamic oxaloacetic and glutamic pyruvic transaminase. Am. J. Clin. Pathol. 28: 56-63.

Sagheb HM, Heidari Z, Bokaeian M, Moudi B (2010). Anti-diabetic effects of *Eucalyptus globulus* on pancreatic islets: A stereological study. Folia Morphol. 69(2): 112-118.

Schmidt E, Schmidt FW (1963). Determination of serum glutamic oxaloacetic and glutamic pyruvic transaminase. Enzyme Biol. Clin. 3: 1–5.

Tedong L, Dzeufiet PDD, Dimo T, Asongalem EA, Sokeng SN, Flejou JF, Callard P, Kamtchouing P (2007). Acute and sub-chronic toxicity of *Anacardium occidentale* leaves hexane extract in mice. Afr. J. Tradit. Complement. Altern. Med. 4(2): 140-147.

Trease GE, Evans WC (1989). Pharmacognosy. 9th edn. Macmillan Publishers Limited, London. p. 208.

Uboh EF, Okon EI, Ekong BM (2010). Effects of aqueous extract of *Psidium guajava* leaves in liver enzymes, histological integrity and hematological indices in rats. Gastroenterol. Res. 3(1): 32-38.

Ukwenya VO, Ashaolu JO, Adeyemi DO, Akinonola OA, Caxton-Martins EA (2012). Anti-hyperglycemic activities of methanolic leaf extract of Anacardium occidentale (Linn.) on the pancreas of streptozotocin-induced diabetic rats. J. Cell Anim. Biol. 6(14): 207-212.

Effects of diet-induced hypercholesterolemia on the lipid profile and some enzyme activities in female Wistar rats

Gloria A. Otunola[1]*, Oyelola B. Oloyede[2], Adenike T. Oladiji[2] and Anthony A. Afolayan[3]

[1]Department of Home Economics and Food Science, University of Ilorin, Ilorin, Kwara State, Nigeria.
[2]Department of Biochemistry, University of Ilorin, Ilorin, Kwara State, Nigeria.
[3]Department of Botany, Phytomedicine Research Centre, University of Fort Hare, Alice 5700, South Africa.

The effects of a high dietary soybean oil and cholesterol on serum total cholesterol, low density lipoprotein (LDL-C), high density lipoprotein (HDL-C), and triglycerides were investigated. Total protein, albumin, glucose, gamma glutamyl transferase (GGT), aspartate aminotransferase (AST) and alanine amino transferase (ALT) activities were also investigated in weanling female Wistar rats for eight weeks. Two groups of weanling Wistar rats were use in this study. The first group of rats were fed with a control diet made up of the normal rat chow (C), while the second group was given a hypercholesterolemic diet (HPC) enriched with 25% soybean oil and 1% cholesterol for eight weeks. The dietary intake of the HPC diet significantly increased the level of total cholesterol, LDL-C and triglycerides in the serum of animals fed the (HPC) diet. GGT, AST and ALT activities were also markedly elevated in rats fed with the HPC diet. While total protein and glucose level of the animals fed with the HPC diet was remarkably reduced, there was no significant difference in the HDL-C and albumin contents of both groups. This study established that hypercholesterolemia is induced by high soybean oil/cholesterol diet, despite the fact that soybean oil has high content of poly unsaturated fatty acids. Also, the HPC diet led to weight loss in the rats and injury to both heart and liver of the rats.

Key words: Diet-induced, injury, lipid profile, enzyme activities, hypercholesterolemia, soybean oil, cholesterol.

INTRODUCTION

Hypercholesterolemia is a lipoprotein metabolic disorder characterized by high serum low density lipoprotein and blood cholesterol. It has been reported by Rerkasan et al. (2008) as one of the most important risk factors in the development and progression of atherosclerosis that lead to cardiovascular diseases (CVDs). Hypercholesterolemia poses a major problem to many societies as well as health professionals because of the close correlation between cardiovascular diseases and lipid abnormalities (Matos et al. 2005; Ramachandran et al., 2003). Clinical trials have demonstrated that intensive reduction of plasma LDL-C levels could reverse atherosclerosis and decrease the incidence of CVDs (Brown et al., 1990; Ichihashi et al., 1998).

Dietary factors such as continuous ingestion of high amounts of saturated fats and cholesterol are believed to be directly related to hypercholesterolemia and susceptibility to atherosclerosis (Asashina et al., 2005). Furthermore, dietary trials have revealed that the concentration of serum cholesterol is affected by both the content and source of proteins (Serounge et al., 1995; Forsythe, 1995). Lipid structure, composition, configuration, in addition to excessive fat and cholesterol consumption are also believed to affect the lipid profile in the plasma (Zulet et al., 1999).

Hypercholesterolemic animals are useful models for studies on cholesterol homeostasis, and drug trials to better understand the relationship between disorders in cholesterol metabolism, atherogenesis as well as

*Corresponding author. E-mail: adeglo2004@yahoo.com.

Table 1. Composition of the control and hypercholesterolemic diets fed to the female Wistar rats.

Ingredient (g/kg)	Control diet	Hypercholesterolemic diet
Casein	120.00	120.00
Corn starch	729.60	429.60
Cholesterol	0.00	10.00
Cellulose	10.00	130.00
Soybean oil	80.00	250.00
Mineral/vitamin mix	60.40	60.40

possible treatments for the reduction of circulatory cholesterol levels (Pellizon, 2008; Jang and Wang, 2009).

Inducing hypercholesterolemia in rats is often through a high fat, high cholesterol diet, with the fat source varying from lard to canola, coconut, soybean or palm oil. Commercial rations supplemented with cholesterol have also been used for these investigations (Doucet, 1987).

The present study was aimed at investigating the effect of the inclusion of soybean oil and cholesterol in the diet on hypercholesterolemia in Wistar rats. This was with the view of using our findings as a model for subsequent atherogenic studies.

MATERIALS AND METHODS

Animals and diets

Twelve female weanling rats each weighing approximately 66.5 g were obtained from the Animal House at the Department of Biochemistry, University of Ilorin, Nigeria. The animals were housed in groups of six in a controlled environment with 12 h light and dark cycles.

They were allowed free access to different dietary formulations (Table 1) and water *ad libitum* for eight weeks. Six rats were fed with the control diet (C), while the second group was fed with a HPC diet (using a slight modification of the diet of Matos et al., 2005). The animals were weighed gravimetrically every week throughout the period of the experiment.

Assay kits and reagents

The assay kits for cholesterol, HDL-C, LDL-C, triglycerides, total protein, albumin, glucose, gamma glutamyl transferase, alanine and aspartate amino transferases were obtained from Randox Laboratories Ltd., Ardmore, Co. Antrim, UK. All other reagents used were of analytical grade.

Biochemical analysis

Proximate analysis of both the control and hypercholesterolemic diets were determined using the methods described by AOAC, (2002) for crude protein, crude fat, crude fibre, and ash while carbohydrate was determined by subtracting the sum of the other nutrient parameters from 100.

At the end of the eight weeks, the animals were fasted overnight, sacrificed under ether anaesthesia and blood collected from the jugular vein for analysis. The blood samples were centrifuged at 1282 × g for 5 min and the serum collected was stored in Eppendorf tubes at -20°C before analysis.

The heart, liver and kidney of each animal was removed, weighed and stored at -80°C; abdominal fat was also carefully dissected and weighed. The relative organ to body weight was calculated.

Total cholesterol (TC) , HDL-C and triglycerides were determined in the serum of the rats by adopting the protocol outlined in the manufacturer's assay kit from Randox Laboratories Ltd, Ardmore, Co. Antrim, UK. LDL-C was calculated using the Friedewald formula – LDL-C=TC- [HDL-C + TG/5). Total protein, albumin, glucose and enzyme activities were also determined using assay kits and protocol from the same manufacturer.

Statistical analysis

All data were expressed as mean ± S.D and were statistically analyzed using one way analysis of variance (ANOVA). Means were separated by the Duncan multiple test using SAS (2002). Values were considered significant at $p < 0.05$.

RESULTS AND DISCUSSION

Table 2 shows the proximate analysis of the control and hypercholesterolemic diets. There was no significant difference in the protein content of both diets, while there was a significant difference $P < 0.05$, in the crude fat content of the hypercholesterolemic diet when compared with the control. This was expected since the HPC diet had higher fat contents (25% soybean oil and 1% cholesterol). However, the ash and total carbohydrate content of the HPC was markedly low.

It has been reported that the relative content of saturated and poly unsaturated fatty acids (PUFA) in the diet affects plasma level of cholesterol (Ohtani et al., 1990). Although metabolic studies have shown that consumption of n-6 PUFA lowers circulating cholesterol level (Ramachandran et al., 2003), in this study however, soybean oil, a rich source of PUFA and cholesterol without cholic acid was successfully used to induce hypercholesterolemia.

Figure 1 shows the body weight gain of the rats during the experimental period. There were significant differences in the weight gain pattern of the animals fed with HPC diet compared to those fed with the control diet. While the animals fed on the control diet maintained a

Table 2. Proximate composition of the control and hypercholesterolemic diets fed to the female Wistar rats.

Parameter (%)	Control diet	Hypercholesterolemic diet
Moisture	6.90 ± 0.14[b]	7.15 ± 0.07[a]
Ash	8.24 ± 0.16[a]	5.73 ± 0.04[b]
Crude protein	17.95 ± 0.14[a]	17.74 ± 0.02[a]
Crude fat	14.64 ± 0.04[b]	29.44 ± 0.14[a]
Crude fibre	2.99 ± 0.14[b]	3.25 ± 0.14[a]
Total carbohydrate	49.53 ± 0.22[a]	37.25 ± 0.13[b]

Values are mean ± SD of 3 replicates. [a - b] Test values along the same row carrying different superscripts for each parameter are significantly different ($P < 0.05$).

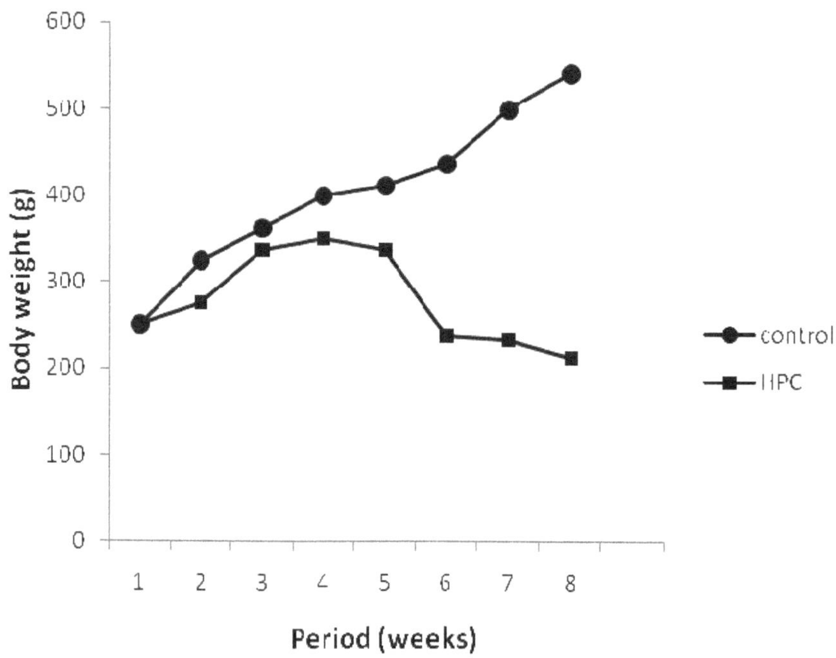

Figure 1. Growth response of female Wistar rats fed with control and hypercholesterolemic (HPC) diet.

consistent weight gain throughout the period, those fed on the HPC only gained weight up till the 4th week, then started a steady decline right till the end of the experiment. The resultant weight loss confirmed by the organ to body weight ratio (Table 3) could be as a result of a reduction in nutrient intake because of the high fat content of the diet which might have impaired the absorption of protein and other nutrients (Zulet and Martinez, 1995; Woo and Henry, 1996; Matos et al., 2005). Records of similar works showed no significant changes in weight gain among the control and hypercholesterolemic animals (Ramachandran et al., 2003), while Harnafi et al. (2009) reported a linear weight increase in all the animals fed with both experimental and control diets. However, Matos et al. (2005), and Hartvigsen et al. (2007) reported similar weight loss as observed in this study with diet-induced hypercholesterolemic rat models. The marked reduction in the abdominal fat observed in the HPC fed rats could be as a result of ketogenic mechanisms (Zulet et al., 1999).

Table 4 shows that there was a significant elevation of serum total cholesterol (TC), low density lipoprotein (LDL-C) and triglycerides in the HPC fed rats, while there was no such difference in the HDL-C of both groups. Increase in LDL-C has been pointed out as one of the risk factors for the development of atherosclerosis and related cardiovascular diseases (Getz and Reardon, 2006). Hence, the result obtained in this study indicated that the dietetic model is suitable for studies in atherogenesis. High serum triglyceride levels has also been reported to

Table 3. Effect of hypercholesterolemic diet on organ to body weight of female Wistar rats.

Organ (g)	Control rats	Hypercholesterolemic rats
Body weight	177.84 ± 4.01^a	112.50 ± 12.50^b
Liver	10.00 ± 1.41^a	7.18 ± 0.25^b
Heart	3.13 ± 0.18^a	2.19 ± 0.09^b
Abdominal fat	5.47 ± 0.69^a	3.40 ± 0.08^b
Kidney	3.00 ± 0.00^a	3.00 ± 0.00^a

Values are mean ± SD of 3 replicates. $^{a-b}$ Test values along the same row carrying different superscripts for each parameter are significantly different ($P < 0.05$).

Table 4. Effect of hypercholesterolemic diet on serum lipid profile of female Wistar rats.

Parameters (mmol/l)	Control rats	Hypercholesterolemic rats
Total cholesterol	2.50 ± 0.57^b	5.10 ± 0.28^a
High density lipoprotein	0.49 ± 0.21^a	0.49 ± 0.28^a
Low density lipoprotein	2.00 ± 0.35^b	4.50 ± 0.28^a
Triglycerides	1.30 ± 0.28^b	3.50 ± 0.28^a

Values are mean ± SD of 3 replicates. $^{a-b}$ Test values along the same row carrying different superscripts for each parameter are significantly different ($P < 0.05$).

Table 5. Effect of Hypercholesterolemic diet on liver and heart functions of female Wistar rats

Parameter	Control rats	Hypercholesterolemic rats
Total protein (g/l)	71.50 ± 1.41^a	61.00 ± 1.41^b
Albumin (g/l)	41.50 ± 1.41^a	41.00 ± 1.41^a
Glucose (mmol/l)	6.98 ± 0.39^a	3.85 ± 0.21^b
Gamma glutamyl transferase (U/l)	17.50 ± 0.00^b	37.75 ± 0.18^a
Aspartate amino transferase (U/l)	41.25 ± 0.35^b	47.25 ± 0.35^a
Alanine amino transferase (U/l)	8.50 ± 0.35^b	27.50 ± 0.71^a

Values are mean ± SD of 3 replicates. $^{a-b}$ Test values along the same row carrying different superscripts for each parameter are significantly different ($P < 0.05$).

be an important risk factor as it influences lipid deposition and clotting mechanisms (Harnafi et al., 2009). Many studies have reported high dietary fat and cholesterol induce hypercholesterolemia in animal models (Cherng and Shih, 2005; Martinello et al., 2006). Similar results were also observed with the HPC fed animals having elevated lipid status than the control. This was an indication that hypercholesterolemia was successfully established in this study.

The data in Table 5 indicates that there was no significant difference $P < 0.05$ in the albumin level of both the animals fed on the control diet and those fed on the hypercholesterolemic diet. Low serum protein was observed in the HPC fed rats and could be due to poor feed intake and utilization. The high fat diet, because of the energy density, might have led to early satiety, reduced feed intake and impaired absorption of protein and other nutrients (Rolls, 2000). The major reason for the marked decrease ($P < 0.05$) in serum glucose in the HPC fed rats was uncertain but it could be that the high serum cholesterol increases the level of glucagon-like peptide-1 which enhances insulin secretion from the pancreatic beta-cells leading to hypoglycemia (Xu et al., 1999; Prasad, 2008). It could also be as a result of reduced supply of glucose through the diet and subsequent demand for pentose and glycogen formation. Reports have indicated that saturated fats and polyunsaturated fats may work together to regulate the expression of several enzymes involved in carbohydrate metabolism (Jump et al., 1994).

The enzyme activities shown in Table 5 indicated that all three enzymes – gamma glutamyl transferase (GGT), aspartate aminotransaminase (AST) and alanine amino transaminase (ALT) were elevated in the serum of the HPC fed rats. This could be as a result of leakage of the enzymes into the serum as a result of damage to the integrity of the heart and liver. Elevated serum activity of these enzymes has been reported to be indicators of calculated risk of cardiovascular disease. According to Pincus and Schaffner (1996), AST and ALT are released into serum when there is severe hepatocellular injury. Elevated serum ALT levels in the absence of viral hepatitis and alcoholism has been reported to lead to a higher risk of cardiovascular disease with the risk greater in women (Ioannou, 2006). The heart also has a high AST content which becomes elevated in myocardial infarction. These reports agree with the present study which showed hepatic injury and cardiovascular distress in the rats fed with HPC.

GGT has been reported to be a very strong risk factor, taking third place, for all forms of heart diseases and a possible indicator for early development of atherosclerosis. Ruttmann et al. (2005), reported a correlation between GGT and cardiovascular mortality indicating that the higher the elevation of GGT, the greater the risk of death.

Conclusion

This study has revealed that hypercholesterolemia could be induced using soybean oil as a source of fat despite its high content of both mono and polyunsaturated fatty acids; that hypercholesterolemia led to reduced serum level of glucose, had no significant effect on serum albumin, HDL-C and led to weight loss and malnutrition in the rat models. Also, the diet led to impaired hepatic and cardiovascular distress in the rats fed with the hypercholesterolemic diet.

REFERENCES

AOAC (2002). Official Methods of Analysis. Association of Analytical Chemists. 17th edn, Washington DC.

Asashina M, Sato M, Imaizumi K (2005). Genetic analysis of diet induced hypercholesterolemia in exogenously hypercholesterolemic (ExHC) rats. J. Lipid. Res., 46: 2289-2294.doi:10.1194/jlr.M500257-JLR200.

Brown AJ, Jessup W (1999). Oxysterols and atherosclerosis. Atherosclerosis. 142: 1-28.

Cherng JY, Shih MF (2005). Preventing dyslipidemia by chorella pyreniidosa in rats and hamsters after chronic high fat diet treatment. Life Sci., 76: 3001-13.

Doucet C, Flament C, Sautier C, Lemonnies D (1987). Effect of an hypercholesterolemic diet on the level of several serum lipids and apolipoproteins in nine rat strains. Reprod. Nutr. Dev. 27: 897-906.

Forsythe WA (1995). Soy protein, thyroid regulation and cholesterol metabolism. J. Nutr., 125: 619-623.

Getz GS, Reardon CA (2006). Diet and murine atherosclerosis. Arteriosclerosis, Thrombosis Vascular Biol., 26: 242-249 doi:10.1161/01.ATV.0000201071.49029.17.

Harnafi H, Aziz M, Amrani S (2009). Sweet Basil (*Ocimum basilicum* L.) improves lipid metabolism in hypercholesterolemic rats. E- SPEN, the European e-J. Clinical Nutr. Metabolism 4:e181-e186 doi:10.1016/j.eclnm.2009.05.011.

Hartvigsen K, Binde CJ, Hansen LF, Rafia A, Juliano J, Horko S, Steinberg D, Palinski W, Witztum JL, Li AC (2007). A diet-induced hypercholesterolemic murine model to study atherogenesis without obesity and metabolic syndrome. Arterioscler Thromb Vasc. Biol., 27: 878-885 doi:10.1161/01.ATV.0000258790.3581002.

Ichihashi T, Izawa M, Miyata K, Mizui T, Hirano K, Takagishi Y (1998). Mechanisms of hypercholesterolemic action of S-89211 in rats: S-8921 inhibits ileal bile acid absorption. J. Pharmacol. Exper. Therapeutics, (JPET) 284(1): 43-50.

Ioannou GN, Weiss NS, Boyko EJ, Mozaffarian D, Lee SP (2006). Elevated serum Alanine aminotransferase activity and calculated risk of coronary Heart Disease in the United States. Hepathology, 43(5): 1145-1151.

Jang MJ, Wang MH (2009). Effect of fermented soybean-derived chung koo kjang on diet-induced hyperlipidemia in Bio FIB hamsters. Food Biotechnol., 23(1): 74-82.doi:10.1080/08905430802672020.

Jump DB, Clarke SD, Thelen A, LiiMatta M (1994). Coordinate regulation of glycolitic and lipogenic gene expression by polyunsaturated fatty acids. J. Lipid Res., 35: 1076-1084.

Martinello F, Soares SM, Franco JJ, Santos AC, Sugohara A, Garcia SB, Curtis, Uyemura SA (2006). Hypolipemic and antioxidant activities from Tamarindus indica L; pulp fruit extract in hypercholesterolemic hamsters. Food Chem. Toxicol.. 44(6): 810-818.

Matos SL, Paula H, Pedrosa ML, Santos RC, Oliveira EL, Chianca Jr DA, Silva ME (2005). Dietary models for inducing hypercholesterolemia in rats. Brazilian Archives Biol. Technol., 48(2): 203-209.

Ohtani H, Hayashi K, Hirata Y, Dojo S, Nakashima K, Nishio E, Kurushima H, Saeki M, Kajiyama G (1990). Effects of dietary cholesterol and fatty acids on plasma cholesterol level and hepatic lipoprotein metabolism. J. Lipid Res., 31: 1413-1422.

Pellizon MA (2008). Diet-induced atherosclerosis/hypercholesterolemia in rodent models. Brief scientific literature review. Retrieved from www.researchdiets.com/pdf/product/atherosclerosis_3%page.pdf.

Pincus MR, Schaffner JR (1996). Assessment of liver function in clinical diagnosis and management by laboratory methods. Philadelphia: W.B. Saunders Company.

Prasad K (2008). Serum biochemical changes in rabbits on a regular diet with and without flax lignin complex following a high-cholesterol diet. Int. J. Angiol., 17(1): 27-32.

Ramachandran HD, Narasimhamurthy K, Raina PL (2003). Modulation of cholesterol induced hypercholesterolemia through dietary factors in Indian desert gerbils (Meriones hurricinae). Nutr. Res., 23: 245-256.

Rerkasem K, Gallagher PJ, Grimble RF, Calder PC, Shearman CP (2008). Managing hypercholesterolemia and its correlation with carotid plaque morphology in patients undergoing carotid endoterectomy (A Review). Vascular Health Risk Manage., 4(6):1259-1264.

Rolls BJ (2000). The role of energy density in the over consumption of fat. J. Nutr., 130: 268S-271S.

Ruttmann E, Brant LJ, Concin H, Diem G, Rapp K, Ulmer H (2005). Gamma glutamyl transferase as a risk factor for cardiovascular mortality: An epidemiologic investigation in a Cohort of 163,944 Austrian adults. Circulation, 112: 2130-2137.

SAS (2002). SAS Systems for windows 9.0, SAS INSTITUTE INC. CARY, NC, USA.

Serounge C, Felgines C, Ferezon J, Hajri T, Bertin C, Mazur A (1995). Hypercholesterolemia induced by cholesterol or cysteine-enriched diets is characterized by different plasma lipoprotein and apoprotein concentrations in rats. J. Nutr., 125: 619-623.

Woo J, Henry JB (1996). Metabolic intermediates and inorganic ions in clinical diagnosis and management by laboratory animals.Philadelphia: W.B Saunders Company, pp. 162-193.

Xu G, Stoffers DA, Habener JF, Bunner-Weir S (1999). Exendin-4 stimulates both beta-cell replication and neogenesis, resulting in increased beta-cell mass and improved glucose tolerance in diabetic rats. Diabetes, 48: 2270-2276.

Zulet MA, Barber A, Garcin H, Higueret P, Martinez JA (1999). Alterations in carbohydrate and lipid metabolism induced by a diet-

rich in coconut oil and cholesterol in a rat model. J. AM. Coll. Nutr., 18(1): 36-42.
Zulet MA, Martinez JA (1995). Corrective role of chickpea intake on a dietary-induced model of hypercholesterolemia. Plant Foods Human Nutr., 48: 269-277.

32

Effect of oestrous cycle on serum electrolytes and liver enzymes in Red Sokoto goats

Yaqub, L. S.[1], Kawu, M. U.[1], Ayo, J. O.[1], Ambali, S. F.[2], Habibu, B.[1]

[1]Department of Veterinary Physiology, Ahmadu Bello University, Zaria, Nigeria.
[2]Department of Veterinary Pharmacology and Toxicology, Ahmadu Bello University, Zaria, Nigeria.

This study was carried out to determine changes in serum electrolytes and liver enzymes during the oestrous cycle in Red Sokoto goats. Eleven (11) apparently healthy Red Sokoto goats were synchronized with a single injection of 7.5 mg of $PGF_{2\alpha}$. The goats were bled via jugular venipuncture in oestrus or late oestrus, metoestrus/early dioestrus, mid-dioestrus and late dioestrus/proestrus. Mean serum sodium, potassium and chloride levels non-significantly fluctuated during the oestrous cycle. Calcium levels in oestrus or late oestrus were higher ($P < 0.05$) than in late diestrus/pro-oestrus. There was a positive correlation between oestradiol and calcium concentrations ($r = 0.771$; $P < 0.05$). Aspartate aminotransferase (AST) and alanine aminotransferase (ALT) concentrations non-significantly fluctuated during the oestrous cycle. Progesterone and AST concentrations were correlated ($r = 0.925$; $P < 0.05$) during the oestrous cycle. In conclusion, some serum electrolytes and liver enzymes fluctuated and correlated with ovarian steroids in different phases of the oestrous cycle. The information generated may be useful as physiologic reference values for reproductive herd health management.

Key words: Red Sokoto goats, oestrous cycle, electrolytes, liver enzymes.

INTRODUCTION

Blood biochemical values is an important tool for assessment of the health status of animals and this has been shown to vary even in healthy animals due to differences such as: sex, season and oestrous cycle phase (Tamukai et al., 2011; Yaqub et al., 2011). Reproductive status and sex variations in blood electrolytes and liver enzymes have been reported in several studies (Tamukai et al., 2011; Stojevic et al., 2005). These electrolytes are very important in homeo-stasis, nerve impulse transmission, muscle contraction, ovarian steroidogenesis and the process of ovulation (Peracchia 1978; Carnegie and Tsang, 1984).

The involvement of oestradiol in the regulation of fluid and electrolytes balance has long been recognized (Khan, 1993). Oestradiol influences salt retention in systemic circulation and alter ion permeability in various epithelial cells (Zeitlin et al., 1989; Sweezey et al., 1996). The oestrous cycle in domestic goats is characterized by cyclic fluctuations in major ovarian steroids. During the follicular phase of oestrous cycle, oestradiol concentration increases and then declines to nadir during the luteal phase (Gaafar et al., 2005). There is paucity of information on changes in serum biochemical parameters during the oestrous cycle in Red Sokoto goats.

Therefore, this study was designed to evaluate changes in serum electrolytes and liver enzyme concentrations during the oestrous cycle in Red Sokoto goats. The data generated from this study may serve as a veritable input for reproductive herd health programme in goats.

MATERIALS AND METHODS

Location of the experiment

The experiment was performed at Small Ruminant Research Programme of the National Animal Production Research Institute (NAPRI), Ahmadu Bello University, Shika, Zaria, Nigeria located on latitude 110 12' N, longitude 70 33' E, and altitude 610 m. The experiment was carried out in the rainy, hot-humid months of June - August, 2009.

Experimental animals

Eleven (11) cycling and apparently healthy Red Sokoto does, aged between 1.5 to 3 years and with body weight ranging between 15 and 25 kg were obtained from local livestock market located at a distance of approximately 45 km from the location of experiment. They were housed in semi-opened concrete floored pens and were preconditioned for two weeks. During preconditioning, the animals were prophylactically treated with albendazole and long-acting oxytetracycline at dose rates of 7.5 and 20 mg/kg, respectively. Blood and faecal samples were collected for haemoparasitic and helminthic screening, and only clinically healthy animals were used. The animals were confined and fed with *Digitaria smutsi* hay as basal diet and supplemented with concentrate ration of ground maize (12%), cotton seed cake (24%), wheat offal (62%), bone meal (1.5%) and salt (0.5%) at 300 g/head/ day. The animals were provided with water and salt lick *ad libtum*.

Oestrus synchronisation

Experimental animals were synchronised with $PGF_2\alpha$ (Lutalyse® Pharmacia, South Africa) at a dose rate of 7.5 mg/animal intra-muscularly. Each animal was weighed prior to commencement of treatment for oestrus synchronisation.

Oestrus detection

An apronised teaser buck was used for detecting oestrus twice daily at 08:00 and 16:00 h. Heat detection was carried out daily between the first day of the second oestrus (d 0) and the first day of the third oestrus following treatment with $PGF_{2\alpha}$.

Blood sampling

Blood sample (5 ml) was collected from each animal by jugular venipuncture into anticoagulant free sample bottles and allowed to clot. The serum samples were then centrifuged at 2000 x *g* for 10 minand serum harvested. Sampling was done on day 0 or early oestral phase (EE), day 1-2 or late oestral phase (LE), day 7-10 or metoestrus/early dioestral phase (M/ED), day 11-15 or mid-dioestral phase (MD) and day 16-22 or late dioestral/proestral phase (LD/PE).

Determination of serum oestradiol and progesterone

The serum level of oestradiol-17β and progesterone values were determined using commercial Enzyme-Linked Immunosorbent assay (ELISA) kits (CLINITECH®, Canada). The sensitivity of the assay for oestradiol and progesterone were 10 ρ/ml and 0.05 ng/ml, respectively.

Determination of serum electrolytes

The serum electrolytes, Na^{+}, Cl^{-}, K^{+} and Ca^{2+} were determined using flame photometry method (Dacie and Lewis, 1991).

Determination of serum liver enzymes

Serum liver enzymes were assayed using Bayer Express Clinical Chemistry Autoanalyzer (Bayer®, Germany). The serum enzymes assayed were alanine aminotransferase (ALT) and aspartate aminotransferase (AST). The assay was carried out as described in Bayer Express Plus Clinical Chemistry Autoanalyzer Manual (2000).

Data analysis

Values obtained were expressed as mean (± SEM) and subjected to one-way analysis of variance (ANOVA), followed by Tukey's post-hoc test and Pearson's correlation analysis. The statistical package used was GraphPad Prism version 4.0 for windows (2003) from GraphPad software, San Diego, California, USA (WWW. Graphpad.com). Values of $P < 0.05$ were considered significant.

RESULTS

Changes in serum electrolyte levels during the oestrous cycle

Mean sodium, potassium, calcium and chloride levels during the oestrous cycle were 139.4 ± 0.5 mmol/l; 4.2 ± 0.1; 2.5 ± 0.2 mmol/l and 97.4 ± 0.4 mmol/l, respectively (Table 1). Sodium, potassium and chloride levels were highest at days 11-15 (MD) of the oestrous cycle (140.2 ± 1.1, 4.5 ± 0.1 and 98.2 ± 0.6 mmol/l, respectively). The decreased sodium and potassium levels were observed at days 1-2 (LE) (138.7 ± 1.1 l; 40.0 ± 0.2 mmol/l, respectively), while the serum chloride level was decreased at day 7-10 (LD/PE) (97.0 ± 0.8 mmol/l). No changes were found in serum calcium level at day 0 (EE) and day 1-2 (LE), while significantly decreased calcium levels were observed at days 7-10 (M/ED), days 11-15 and days 16-22 (LD/PE) of the oestrous cycle.

There was a significant positive correlation between sodium and potassium concentrations during the oestrous cycle ($r = 0.736$; $P < 0.05$). Similarly, serum calcium and oestradiol concentrations were significantly and positively correlated during the cycle ($r = 0.771$; $P < 0.05$) (Table 2).

Changes in serum liver enzyme levels during the oestrous cycle

Mean serum AST concentration during the oestrous cycle was 23.6 ± 1.4 IU/l. Serum AST concentration was highest at day 0 (EE) and days 16-22 (LE/PE) and, lowest at days 11-15 (MD) phases of the oestrous cycle (24.4 ± 1.4 IU/l vs 21.8 ± 1.3 IU/L; $P > 0.05$) (Table 3). There was a

Table 1. Mean (± SEM) serum electrolytes during the oestrous cycle of Red Sokoto goats (n = 11).

Day/cycle phase	Sodium (mmol/l)	Chloride (mmol/l)	Potassium (mmol/l)	Calcium (mmol/l)
Day 0 (EE)	139.8 ± 0.7 [a] (136 - 144)	98.1 ± 0.6 [a] (94 - 100)	4.1 ± 0.2 [a] (3.0 - 4.7)	2.6 ± 0.1 [b] (2.3 - 2.9)
Day 1 - 2 (LE)	139.1 ± 1.1 [a] (133 - 146)	97.5 ± 1.1 [a] (90 - 100)	4.0 ± 0.2 [a] (3.6 - 5.3)	2.6 ± 0.1 [c] (2.3 - 2.7)
Day 7 - 10 (M/ED)	138.7 ± 1.1 [a] (132 - 144)	96.2 ± 1.3 [a] (90 - 102)	4.0 ± 0.2 [a] (2.8 - 4.6)	2.5 ± 0.04 [a] (2.3 - 2.7)
Day 11 - 15 (MD)	140.2 ± 1.1 [a] (136 - 146)	98.2 ± 0.6 [a] (96 - 100)	4.5 ± 0.1 [a] (3.5 - 5.0)	2.5 ± 0.1 [d] (2.2 - 2.6)
Day 16-22 (LD/P/E)	139.2 ± 1.6 [a] (133 - 151)	97.0 ± 0.8 [a] (90 - 102)	4.3 ± 0.2 [a] (3.0 - 5.5)	2.4 ± 0.02 [bc] (2.3 - 3.5)
Mean (± S E M)	139.4 ± 0.5 (132 - 151)	97.4 ± 0.4 (90 - 102)	4.2 ± 0.1 (2.8 - 5.5)	2.5 ± 0.02 (2.2 - 2.9)

[a,b,c,d] = values with different superscripts within column are statistically significant (b = P < 0.01; c = P <.0.05); n = number of animals sample; values in parenthesis are the range; EE = Early Oestral Phase; LE = Late Oestral Phase; M/ED = Meteostrus/Early Dioestral Phase; MD = Mid-dioestral Phase and LD/PE = Late Dioestrus/Pro-oestrus.

Table 2. Relationships between progesterone, oestradiol and serum electrolytes during the oestrous cycle in Red Sokoto goats (n = 11).

Correlated variable	Pearson's correlation (r)	P - Value
Progesterone and chloride	0.2930	P > 0.05
Progesterone and calcium	0.2174	P > 0.05
Progesterone and potassium	-0.2237	P > 0.05
Progesterone and sodium	0.000	P > 0.05
Oestradiol and chloride	0.2693	P > 0.05
Oestradiol and calcium	0.7707	P < 0.05
Oestradiol and potassium	0.2756	P > 0.05
Oestradiol and sodium	0.2149	P > 0.05
Chloride and sodium	0.9178	P < 0.05
Sodium and potassium	0.7355	P < 0.05

positive correlation between serum AST and progesterone levels during the cycle ($r = 0.924$; $P < 0.05$) (Table 4).

Mean serum ALT concentration was 24.7 ± 0.6 mmol/l. Serum ALT was concentration was highest at d 1-2 (LE) and lowest at d 7-10 (M/ED) phases (27.3 ± 2.0 vs 22.6 ± 0.8; $P > 0.05$) .

DISCUSSION

Serum calcium level obtained in the present study lies within the range reported in Saanen goats (Temizel, et al., 2009), West African Dwarf goats (Daramola et al., 2005) and in the equine (Ali et al., 2004). Peak serum calcium concentration observed at day 0 (EE) and day 1-2 (LE) may be due in part to high serum concentration of oestradiol recorded during these phases. This finding agrees with the observations of Ali et al. (2004), who reported high blood calcium concentration in mares during the oestral phase of the oestrous cycle. High oestradiol levels during oestral phase causes increased intestinal absorption of calcium (Brommage et al., 1993). In addition, the ability of oestradiol to retain salt and to alter ion transport in various other epithelial cells could be partly responsible for high serum calcium observed during the oestral phase. Furthermore, increased muscular activities during the oestral phase as a result of psychic manifestation of heat may be responsible for the increase in serum calcium level in extracellular fluid during the follicular phase of the oestrous cycle. This increase in serum calcium level in oestral phase may be necessary to support the increased neuromuscular activity, and ovarian hormone synthesis and release associated with this phase of the

Table 3. Mean (± S.E.M) serum aspartate aminotransferase and alamine aminotransferase during the oestrous cycle in Red Sokoto goats (n = 11).

Day/cycle phase	Aspartate aminotransferase (IU/l)	Alanine aminotransferase (IU/l)
Day 0 (EE)	24.4 ± 1.4 (18.0 - 31.0)	24.5 ± 1.4 (17.0 - 31.0)
Day 1 - 2 (LE)	24.3 ± 1.0 (18.0 - 28.0)	27.3 ± 2.0 (18.0 - 39.0)
Day 7 - 10 (M/ED)	22.8 ± 1.1 (18.0 - 29.0)	22.6 ± 0.8 (18.0 - 26.0)
Day 11 - 15 (MD)	21.8 ± 1.3 (16.0 - 30.0)	26.4 ± 1.1 (22.0 - 33.0)
Day 16 - 22 (LE/PE)	24.4 ± 1.4 (17.0 - 33.0)	22.9 ± 0.8 (18.0 - 26.0)
Overall Mean (± S.E.M)	23.6 ± 0.6 (16.0 - 33.0)	24.7 ± 0.6 (17.0 - 39.0)

n= Number of animals sampled values in parenthesis are the range; EE = Early Oestral Phase; LE = Late Oestral Phase; M/ED = Meteostrus/Early Dioestral Phase; MD = Mid-dioestral Phase and LD/PE = Late Dioestrus/Pro-oestrus.

Table 4. Relationships between serum oestradiol, progesterone, aspartate aminotransferase and alamine aminotransferase during the oestrous cycle in Red Sokoto goats.

Correlated variable	Pearson's correlation Coefficient (r)	P - Value
Progesterone and aspartate aminotransferase	0.9249	$P < 0.05$
Progesterone and alanine aminotransferase	0.1602	$P > 0.05$
Oestradiol and aspartate aminotransferase	-0.4427	$P > 0.05$
Oestradiol and Alanine aminotransferase	0.4446	$P > 0.05$
Aspartate aminotransferase and Alanine aminotransferase	0.1162	$P > 0.05$

oestrous cycle. Significant positive correlation between oestradiol and serum calcium concentrations suggests an enhancing effect of oestradiol on serum calcium homeostasis during the oestrous cycle. Similarly, increased serum calcium concentrations had been reported by Ayo et al. (2009) in Red Sokoto goats as a result of high muscular activities following road transportation stress.

The gradual but consistent decline in serum calcium recorded from M/ED phase to the end of the cycle (LD/PE) may be due to decline in peripheral oestradiol concentration and, thus, decreased intestinal calcium absorption during this period. Also, the decline in serum calcium may be due to a decrease in neuromuscular excitability during the dioestral phase of the oestrous cycle. Brommage et al. (1993) reported a decrease in intestinal calcium absorption during the diestral phase of the oestrous cycle in rats.

Serum sodium, potassium and chloride concentrations were within the range reported in goats (Ayo et al., 2009). The non-significant fluctuations in potassium, sodium and chloride concentrations during the oestrous cycle and the non-significant correlation with oestradiol and progesterone concentrations suggest that neither the cycle phase nor fluctuations in ovarian steroid levels had any significant effect on these serum electrolyte levels during the oestrous cycle in goats.

Aminotransferases act as catalyst during amino acid and carbohydrate metabolism. Changes in their activity in the blood can be a consequence of their increased activity in the cells or cell structure damage. The mean serum value of AST recorded in this study is similar to that obtained in the WAD goat (Daramola et al., 2005), but lower than the value reported for Sahel goats (Waziri et al., 2010), West African Dwarf sheep (Oduye and Adadevoh, 1976) and camels (Ayoub et al., 2003). This difference may be attributed to breed or species differences, geographical location and method used in analyzing the serum samples (Tibbo et al., 2004). The highest AST concentration in this study was observed at EE, LE and LD/PE phases of the oestrous cycle. This finding agrees with the result obtained by Marai et al. (2006), who recorded higher AST activity in oestrus than in pro-oestrus phase in ewes. The highest AST activity recorded during the day 0 (EE) and day 1-2 (LE) phases may be due to the effect of ovarian steroids in increasing permeability of hepatocellular membrane to this enzyme. Furthermore, since AST is also a muscle enzyme, the relatively high AST activity recorded during the oestral phase in this study may have been caused by the heightened neuromuscular excitability, associated with this phase of the oestrous cycle. On the other hand, the positive correlation between serum AST and progesterone concentration recorded in this study indicates a stimulatory effect of progesterone on liver synthesis and secretion of AST; especially during the day 0 (EE), day 1-2 (LE) and days 16 to 22 (LD/PE) phases of the oestrous cycle, when progesterone concentrations were highest. This may also imply that progesterone activity affects serum AST activity during the oestrous cycle in goats.

The serum activity of ALT recorded in this study was slightly higher than the range reported for West African Dwarf goats (Daramola et al., 2005), Arsi-Bale goats, Long-eared Somalia goats, central Highland goats (Tibbo et al., 2008) and sheep (Oduye and Adadevoh, 1976), but lower than the value reported in rabbits (Mizoguchi et al., 2010). Changes in serum ALT activity observed in this study during the oestrous cycle did not follow a consistent pattern. This may imply that oestrous cycle phase and changes in ovarian steroid concentrations may not have a significant influence on serum ALT activities during the oestrous cycle in goats.

Conclusion

Based on the findings of this study, it is concluded that some biochemical parameters fluctuated and correlated with ovarian steroids during the different phases of oestrous cycle. The results obtained may serve as physiologic reference values for reproductive herd health programme.

REFERENCES

Ali F, Lodhi LA, Qureshi ZI, Samad HA, Shahid RU (2004). Some serum biochemical constituents of mares during different phases of reproductive cycle. Pak. Vet. J. 24 (3): 147-152.

Ayo JO, Minka NS, Sackey AKB, Adelaiye AB (2009). Responses of serum electrolytes of goats to twelve hours of road transportation during the hot-dry season in Nigeria, and the effect of pretreatment with ascorbic acid. Onderstepoort J. Vet. Res. 76: 409-418.

Ayoub MA, El-Khouly AA, Mohammed TM (2003). Some haematological and biochemical parameters and steroid hormone levels in one-humped camel during the different physiological conditions. Emir J. Agric. 15(1): 44-55.

Brommage R, Binacua C, Carrie AL (1993). Ovulation associated increase in intestinal calcium absorption during the rat estrous cycle is blunted by ovariectomy. Biol. Rep. 49 (3):544-548.

Carnegie JA, Tsang BK (1984). The calcium calmodulin system; participation in the regulation of steroidogenesis activity at different stages of granulose cell differentiation. Biol. Rep. 30: 515-522.

Dacie JV, Lewis SM (1991). Practical Haematology, 7th Edition, Chuchill Livingstone, London. pp. 659-661.

Daramola JO, Adeloye AA, Fatoba TA, Soladoye AO (2005). Haematological and biochemical parameters of West African Dwarf goats. Livestock Research for Rural Development, 17, Art. 95.

Gaafar KMG, Mohamed KG, Doaa FT (2005). The hormonal profile during the oestrous cycle and gestation in Damascus goats. Small Rum. Res. 57: 85 - 93.

Khan MA (1993). Effects of oestrogens on the fuid balance in ovariectomized rats. Phd Thesis, Islamia University, Bahawalpur.

Marai IFM, El-Darawny El, Abou-Fandoud Abdel-Hafez MAM (2006). Serum blood components during pre-oestrus, oestrous and pregnancy phases in Egyptian Suffolk ewes as affected by heat stress, under the condition of Egypt. Egypt. J. Sheep Goat Desert Anim. Sci. 1(1): 47-62.

Mizoguchi Y, Matsuoka T, Mizuguchi H, Endoh T, Kamata R, Fukuda K, Tsutomu I, Asano Y (2010). Changes in blood parameters in New Zealand White rabbits during pregnancy. Lab. Anim. 44: 33-39.

Oduye OO, Adadevoh BK (1976). Biochemical values in apparently normal Nigerian sheep. J. Nig. Med. Assoc. 5: 59-67.

Peracchia C (1978). Calcium effect on gap junction structure and cell coupling. Nat. Land. 271: 669-671.

Stojevic Z, Pirsljin J, Milinkovic-Tur S, Zdelar-Tik M, Ljubic BB (2005). Activities of AST, ALT and GGT in clinically healthy dairy cows during lactation and in dry period. Vet. Arhiv. 75(1): 67-73.

Sweezey NB, Gauthier C, Gagnon S, Ferretti E, Kopelman H (1996). Progesterone and oestradiol inhibit CFTR- mediated ion transport by pancreatic epithelial cells. Am. J. Physiol. 271: G747-754.

Tamukai K, Takami Y, Akabawe Y, Kamazaroa Y, Une Y (2011). Plasma biochemical reference values in clinically healthy captive bearded dragons (Pogona Vitticeps) and the effects of sex and season. Vet. Clin. Pathol. 368-373.

Temizel EM, Sentuok S, Kasap S (2009). Clinical, haematological and biochemical findings in Saanen goat kids with naturally occurring heat stroke. Tierazetl. Prax. 37(G): 236 - 241.

Tibbo M, Jibril T, Woldemeskel M, Dawo F, Aragaw K, Rege JEO (2004). Factors affecting haematological profiles in three Ethiopian indigenous goat breeds. Int. J. Appl. Res. Vet. Med. 2(4): 297 - 309.

Tibbo M, Jibril Y, Woldemeskel M, Dawo F, Argaw K, Rege JEO (2008). Serum enzyme levels and influencing factors in three indigenous Ethiopian goats breeds. Trop. Anim. Health Prod. 40(8): 657-666.

Waziri MA, Ribadu AY, Sivachevlan N (2010). Changes in serum proteins, haematological and some serum biochemical profiles in gestation period in the Sahel goats. Vet. Arhiv. 80(2): 215-224.

Yaqub LS, Ayo JO, Rekwot PI, Oyeanusi BI, Kawu MU, Ambali SF, Shittu M, Abdullahi A (2011). Changes in serum proteins and urea during the oestrous cycle in Red Sokoto goats. Adv. Appl. Sci. Res. 2(6): 197-205.

Zeitlin PL, Wagner M, Markakis D, Loughlin GM, Guggino WB (1989). Steroid hormones: modulators of Na^+ absorption and Cl^- secretion in cultured tracheal epithelia. Proc. Natl Acad. Sci. USA 86: 2502-2505.

Optimization of metalloprotease production by *Eupenicillium javanicum* in both solid state and submerged bioprocesses

Youssef Ali Abou Hamin Neto[1], Cristina Maria de Souza Motta[2] and Hamilton Cabral[1]

[1]Enzyme Technology Laboratory, Department of Pharmaceutical Sciences, School of Pharmaceutical Sciences, Ribeirão Preto-USP, Ribeirão Preto, Brazil.
[2]Micology Department, Federal University of Pernambuco, Recife, Brazil.

Proteases are enzymes that have great importance in industries and can be produced through bioprocess by microorganisms. Changes in parameters can influence the production rate of these enzymes. The current study evaluated the influence of these parameters on the production of proteases by the fungus *Eupenicillium javanicum* under both solid state bioprocess (SSB) and submerged bioprocess (SmB), and determined the partial biochemical characterization of the crude enzymatic extract. Our studies have indicated that the best conditions for the production of proteases under SSB were a 10% albumin, 90% wheat bran incubated at 30°C. Under SmB the optimal conditions were medium supplemented with 0.25% casein, medium pH of 5.0 and incubation at 30°C. The biochemical characterization showed that the enzyme is a metalloprotease that presented collagenolytic activity, with optimal pH and temperature of 5.5 and 60°C, respectively.

Key words: Metalloprotease, *Eupenicillium javanicum*, bioprocesses, collagenolytic activity, biochemical characterization.

INTRODUCTION

Since the middle of the last century, the interest in fungi to produce commercially important products such as enzymes has increased (Papagianni, 2004). Proteases are proteolytic enzymes that catalyze the hydrolysis of protein peptide bonds and occupy an important position due to its use in various research fields and its commercial applicability in several areas (Rao et al., 1998).

One important kind of proteases is the collagenases that are used to remove stains and keloids, to treat burns and ulcers, and used in scientific studies (Sukhosyrova et al., 2003). Collagen has a rigid structure and only a few proteases can cleave (Watanabe, 2004).

The action of collagenases in the collagen fibers generates peptides that have diverse biological activities of interest to industry, leading to the establishment of a wide variety of applications such as for immunotherapeutic agents and in cosmetic development (Tsuruoka et al., 2003).

Mandl et al. (1953) identified collagenases produced by *Clostridium hystolyticum* and this microbial collagenase is the most known. Moreover, the production of microbial proteases with collagenolytic activities by *Streptomyces* sp. (Zhao et al., 2012), *Aspergillus fumigatus* (Monod et al., 1993) and *Aspergillus flavus* (Sukhosyrova et al., 2003) were described.

These studies show the importance of proteases produced by fungi with potential applicability in industry.

Eupenicillium javanicum is a filamentous fungus which has shown potential for the production of amylases and proteases (Tanaka et al., 1986). However, there are no reports on the literature on the production of proteases with collagenolytic activity by the fungus *E. javanicum*. This can provide the production of enzymes with different characteristics.

The aim of this paper is to optimize the production of proteases produced by *E. javanicum* in both submerged and solid state bioprocesses that showed collagenolytic potential.

MATERIALS AND METHODS

Isolation, identification and maintenance of the organism

The fungus *E. javanicum* was isolated from silage and identified by research group of Dr. Cristina Maria de Souza Motta (Federal University of Pernambuco). This fungus belongs to a collection of microorganisms in the Enzyme Technology Laboratory under the responsibility of Dr. Hamilton Cabral (School of Pharmaceutical Sciences of Ribeirão Preto-University of São Paulo). The fungus was kept in Sabouraud medium at 4°C, up to one month.

E. javanicum inoculum preparation

The inoculum was obtained by peaking of the fungus in 250 mL Erlenmeyer flasks with Sabouraud medium and maintained for 7 days at 30°C, and the mycelium of the fungus was scraped from the surface of the culture medium in the presence of sterile distilled water for submerged bioprocess or in the presence of saline solution (0.1% (w/v) $(NH_4)_2SO_4$, 0.1% (w/v) NH_4NO_3 and 0.1% (w/v) $MgSO_4x7H_2O$) for solid bioprocess.

Submerged bioprocess (SmB)

In 250 mL Erlenmeyer flasks, 50 mL of liquid medium were added, its composition was 0.7% (w/v) KH_2PO_4, 0.2% (w/v) K_2HPO_4, 0.01% (w/v) $MgSO_4.7H_2O$, 0.01% (w/v) $CaCl_2.2H_2O$, 0.5% (w/v) NaCl, 0.1% (w/v) yeast extract, 0.5% (w/v) casein and 0.5% (w/v) peptone (Tran and Nagano, 2002), with modifications. After the preparation, the medium pH was adjusted to 6.0 and autoclaved at 121°C for 15 min. This medium was called standard liquid medium.

One milliliter of a mycelium suspension was inoculated in the Erlenmeyer flasks containing the standard liquid medium, and this solution was incubated for a period of 24 to 168 h at 30°C (except for the temperature variation study) in a shaker with a rotation of 120 rpm; at each 24 h time point, a flask of each experiment variable was removed. After each bioprocess, the material was filtered with Whatman No 1 filter paper. The filtrate was denominated as crude enzyme extract (CEE), and triplicate enzyme assays were performed for each bioprocess study.

Parameters variation in the SmB

Studies were performed to analyze the effects of bioprocess time, the organic nitrogen (casein) and carbon sources, source concentration, pH of culture medium and incubation temperature.

To evaluate of the effect of the nitrogen source on protease production, the standard liquid medium was supplemented with casein in the following concentrations: 0.25% (w/v), 0.5% (w/v) or 1.0% (w/v). To analyze the effect of the carbon source, the standard medium was supplemented with several carbon sources (glucose, fructose or sucrose) at several percentages (0.1% (w/v), 0.5% (w/v) or 1% (w/v)). To analyze the effect of pH, the pH of the standard liquid medium was adjusted to 5.0, 6.0, 7.0 or 8.0 with NaOH or HCl solutions.

To analyze the effect of temperature, the liquid medium was prepared with the best concentration of casein and pH medium and these culture media were incubated at temperatures of 30, 35, 40 and 45°C.

To determine the optimum time of bioprocess, the standard liquid medium was used with the best conditions of casein concentration, initial pH and temperature.

Solid state bioprocess (SSB)

Protease production by fungus *Eupenicillium javanicum* was evaluated in 250 mL Erlenmeyer flasks containing 5 g of solid medium added with 9.0 mL of saline solution, and then the media were sterilized by autoclaving at 121°C for 40 min. To solid medium was added one milliliter of the inoculum in saline solution prepared with 0.1% (w/v) of $(NH_4)_2SO_4$, NH_4NO_3 and $MgSO_4x7H_2O$ (Merheb et al., 2007) and incubated in a chamber at 30°C. Every 24 h (over a total of 168 h), a flask of each experiment was removed, and 40 mL of distilled water at 4°C was added to each flask for enzymes solubilization. The latter process was aided by maceration with a plastic rod, and then the flasks were agitated in a shaker at 200 rpm for 30 min at 4°C. The material was filtered and centrifuged at 5000 x*g* for 20 min at 4°C. The supernatant thus obtained was called crude enzyme extract (CEE), and it was assessed for proteolytic activity. Initially, to determine the standard solid medium the following agro-industrial residues were evaluated: wheat bran (WB) and cottonseed meal (CM) in the proportions WB 100%, CM 100% or WB 50% + CM 50%. The WB 100% was assumed as standard solid medium because it obtained the best protease production.

Parameters variation in the SSB

Using this standard medium, organic nitrogen source (casein and albumin), temperature and bioprocess time were evaluated for their influence on the production of proteases during solid state bioprocess. The effect of exogenous nitrogen sources was shown using casein and albumin, in concentrations of 5, 10 and 20% (w/w) in relation to WB and the final mass of the solids was equal to 5 g. The influence of temperature was evaluated using the medium that had the best enzymatic activity when incubated at 30, 35, 40 and 45°C.

The optimum time of bioprocess was determined with the solid medium (100%WB) under the best conditions of albumin as nitrogen source concentration and temperature.

Evaluation of proteolytic activity with a casein substrate

The protease production profile was evaluated in the SmB or SSB with casein substrate using the protocol described by others (Sarath et al., 1996), with some modifications. We used 1 mL of 1% (w/v) casein prepared in 50 mM monobasic sodium phosphate buffer pH 6.5, 100 µl of 50 mM monobasic sodium phosphate buffer pH 6.5 and 100 µl of the solution containing the enzyme (CEE). The reaction mixture was incubated for 60 min at 40°C. Upon completing the reaction time, 600 µl of 10% (w/v) trichloroacetic acid (TCA) was added to stop the reaction. The blank tube was prepared with denatured enzyme and in same conditions of reaction of test tube. After the enzymatic reaction, test tubes and blanks were centrifuged at 10000 x *g* at 30°C for 10 min. The supernatants of the test tubes were measured against their blanks in cuvettes at

280 nm in a Genesys 10S spectrophotometer (Thermo).

One unit of activity was defined as the amount of the enzyme required to cause an increase of 0.001 $A_{280\ nm}$ within the reaction conditions using casein as a substrate (Gupta et al., 2002).

Fractionation and concentration by ethanol

The partial characterization assays was performed with crude enzymatic extracts of the SmB and SSB concentrated and partially purified by precipitation with ethanol, in a ratio of 1 (enzymatic extract):2 (ethanol). The material was precipitated overnight at -20°C. The precipitate was centrifuged at 5000 *xg* for 20 min at 4°C for sedimentation and then the supernatant was discarded. The precipitate was stored at -20°C. The precipitate was dissolved in distilled water at 4°C for biochemical characterization and was called the partially pure extract (PPE).

Evaluation of proteolytic activity with the azocasein substrate

We added 100 µl of PPE, 100 µl of 50 mM adequate buffer and 200 µl of 1% (w/v) azocasein prepared in the same buffer. The reaction mixture was incubated for 60 min for SmB and 15 min for SSB at 40°C, and the reaction was stopped with 800 µl of 10% (w/v) TCA. The blank tube was prepared with denatured enzyme and in same conditions of reaction of test tube.

Test and blanks tubes were centrifuged at 10000 x*g* at 30°C for 15 min. A volume of 466.5 µl of the supernatant was separated and added 400 µl of 1 M NaOH, and this mixture was measured against the respective control using cuvettes at 440 nm in a Genesys 10S spectrophotometer (Thermo) (Ducros et al., 2009). All the enzymatic reaction were performed in triplicate.

One unit of activity was defined as the amount of the enzyme required to cause an increase of 0.001 $A_{440\ nm}$ within the reaction conditions using azocasein as a substrate (Morita et al., 1998).

Characterization of the partially pure extract: Effect of pH, temperature and stability

The optimum pH was determined by evaluating the activity of the PPE at various pH levels, from 4.5 to 10.5 with increments of 0.5 pH units. The buffers used were acetate (pH 4.5 - 5.0), MES (pH 5.5 - 6.5), HEPES (pH 7.0 - 8.0), BICINE (pH 8.5 - 9.0) and CAPS (pH 9.5 - 10.5), all at 50 mM. The enzymatic reaction was performed with different pH buffers at 40°C. The pH condition that proportioned the best enzymatic activity was called optimum pH.

The optimum temperature was determined by assessing various temperatures that ranged from 30 to 80°C, with 5°C increments. The enzymatic reaction was performed with optimum pH at different temperatures. The temperature condition that proportioned the best enzymatic activity was called optimum temperature.

The pH stability was assessed by incubating PPE for 60 min at 25°C at various pH levels over a range of values (4.0 to 10.5) with increments of 0.5 pH units. After the previous exposition to different pH the enzymatic reaction was performed with optimum pH buffer at 40°C.

The thermal stability was evaluated by incubating the PPE at temperatures of 30, 35, 40, 45, 50, 55 or 60°C for 5, 15, 30 or 60 min. The enzymatic reaction was performed with optimum pH and temperatures.

All the enzymatic reactions were performed in triplicate as described in evaluation of proteolytic activity with the azocasein substrate.

Characterization of the partially pure extract: Effect of ions and inhibitors on the enzymatic activity

The effect of ions and inhibitors on the enzymatic activity in each reaction was determined, with the addition of ion and inhibitor solutions in tubes containing the enzymatic extract (SmB or SSB) to obtain a final concentration of 10 mM. The ions tested were NaCl, $ZnSO_4$, $CoCl_2$, $CuCl_2$, $CaCl_2$, $MgCl_2$, $BaCl_2$ and $AlCl_3$ all at a concentration of 100 mM (stock). The inhibitors tested were phenylmethylsulfonyl fluoride (PMSF), ethylenediaminetetraacetic acid (EDTA) and iodoacetic acid (IAA), all at a concentration of 100 mM (stock), according to the protocol previously described (Dunn, 1989), with modifications. The tubes containing ions or inhibitors were previously incubated for 5 min at 4°C. The control tube was prepared without addition of ions or inhibitors and proteolytic relative activity was assuming as 100%. The enzymatic reactions were performed in triplicate as described in evaluation of proteolytic activity with the azocasein substrate.

Evaluation of collagenolytic activity with substrate azocoll

One milliliter of PPE was added to a suspension of 1.5 g/l of azocoll in 1 mL of MES buffer (50 mM, pH 5.5) and incubated in a shaking water bath at 40°C for 4.5 h. Samples from the reaction mixtures were transferred to tubes and centrifuged at 10000 x*g* for 10 min at 25°C. The supernatant was measured against its blank at 550 nm in a Genesys 10S spectrophotometer (Thermo). Blanks were prepared in a similar manner as the reaction; however, prior to mixing, the PPE was boiled for 10 min as previously described (Foroughi et al., 2006), with modifications.

RESULTS

Protease production profile by SmB using casein as nitrogen source

The analysis of nitrogen source effects in SmB was performed using casein. The fungus *E. javanicum* showed that the presence of 0.25% (w/v) of casein was adequate to promote protease synthesis and in this condition the peak production was 32.8 U/mL at 144 h. The increase in the casein concentration repressed protease secretion and detection (Figure 1a).

Protease production profile by SmB using different carbon sources

The carbon source effects in SmB were evaluated with glucose, fructose and sucrose. The results shown in Table 1 represent the peak production obtained with each carbon source and their respective concentrations. In the assessment of carbon source supplementation, we observed that the protease production by *E. javanicum* increased 37% only when the medium was supplemented with 0.5% (w/v) glucose at 144 h (Table 1). This higher level of protease production, as compared to medium without supplementation, suggests a dependence on the glucose for the production of proteases.

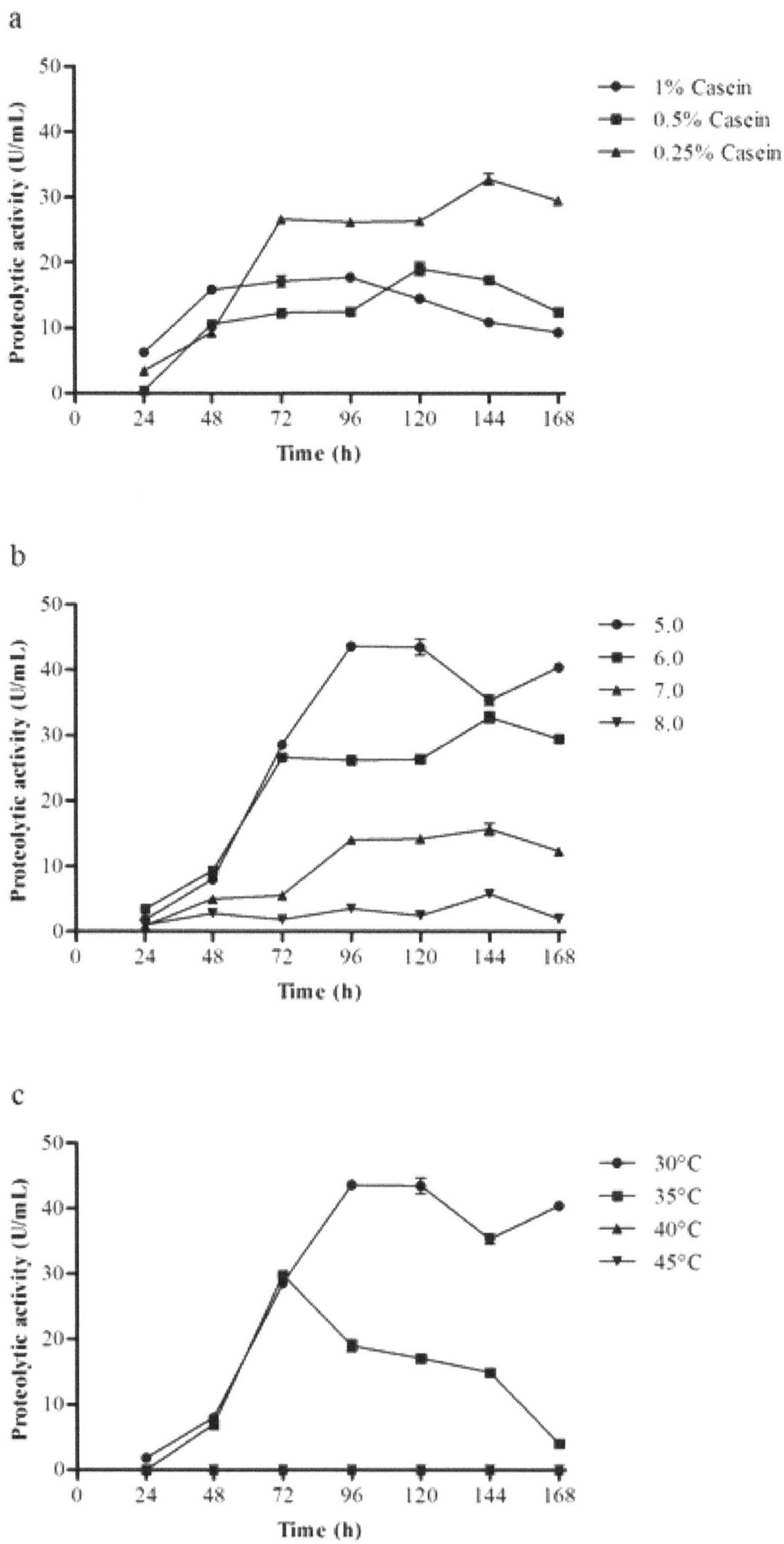

Figure 1. Effect of parameters variation on protease synthesis by the fungus *E. javanicum* under SmB. (a) Casein concentration at 30°C and 120 rpm. (b) Initial medium pH at 30°C and 120 rpm. (c) Temperature variations pH 5.0 and 120 rpm. Values shown are the averages (mean ± SD) of the triplicate samples from one representative experiment. SmB: submerged bioprocess.

Table 1. The effect of different concentrations of carbon source on protease synthesis by *E. javanicum* under submerged bioprocess at 30°C and 120 rpm.

Carbon source	Carbon source concentration (%)	Proteolytic activity (U/mL)
Control	-	19.05 ± 1.00
Fructose	0.1	15.86 ± 0.97
Fructose	0.5	18.35 ± 1.27
Fructose	1.0	18.35 ± 1.36
Glucose	0.1	16.39 ± 1.76
Glucose	0.5	26.28 ± 1.84
Glucose	1.0	16.18 ± 1.00
Sucrose	0.1	18.25 ± 0.88
Sucrose	0.5	10.32 ± 0.28
Sucrose	1.0	12.00 ± 0.18

The proteolytic activity was determined using casein 1% as substrate. The control sample represents medium without supplementation carbon sources. The values shown are the averages (mean ± SD) of the triplicate samples from peak production of each experiment.

Protease production profile by SmB in different initial medium pH

The effect of pH on protease production is related to the amount of available protons in the bioprocess medium that influence the transport of nutrients and enzyme reactions. In the assessment of the effect of pH on protease synthesis by the fungus *E. javanicum* in SmB, the highest production was observed at pH 5.0 (Figure 1b). The increasing pH levels proportioned the decrease in protease detection.

Protease production profile by SmB in different temperature of incubation

The temperature can influence thermal stability and fungal metabolism of the enzyme produced. In the analysis of the temperature effects, the best temperature for protease production was 30°C, since the fermentation process at 40 and 45°C was unsuccessful (Figure 1c). The temperature assessment revealed that an increase in temperature decreased protease secretion by *E. javanicum* in SmB.

The choice of wheat bran as agro-industrial residue to protease production by SSB

Agro-industrial residues are interesting resources to enzymes production in SSB once these substrates are inexpensive. Protease synthesis by the fungus *E. javanicum* in SSB was analyzed using the agro-industrial residues wheat bran (WB) and cottonseed meal (CM) and different proportions (100% WB, 100% CM and 50% WB + 50%CM). The results showed higher levels of protease synthesis when 100%WB was used as the solid medium in SSB (Figure 2a). The condition 100% WB was assumed as standard solid medium.

Protease production profile by SSB using different nitrogen source

After determining the best agro-industrial residue, WB was supplemented with a nitrogen source trying to improve protease synthesis. The additional nitrogen sources were casein and albumin. The analysis of casein influence on protease synthesis in SSB by *E. javanicum* showed that 10% casein concentration proportioned similar profile production as those without a nitrogen source supplementation, 94 and 99 U/mL at 96 h, respectively (Figure 2b). The condition 80% WB+20% casein decreased protease synthesis. In the analysis of albumin influence on protease production in SSB by *E. javanicum* showed that high albumin concentrations improved production. In the conditions with 10 (w/w) and 20% (w/w) albumin increased protease secretion about 10% as compared to 100% WB, and the peak production was 110 U/mL at 72 h (Figure 3a).

The different nitrogen source (casein and albumin) proportioned different effects on proteases production. The casein supplementation had no effect, although 90% WB + 10% albumin had proportional increase in production and was selected further studies.

Protease production profile by SSB in different temperature of incubation

The temperature influences the fungal metabolism and other parameters of the culture. The optimal temperature for protease synthesis was 30°C, however the fermentation process at 40 and 45°C was unsuccessful different

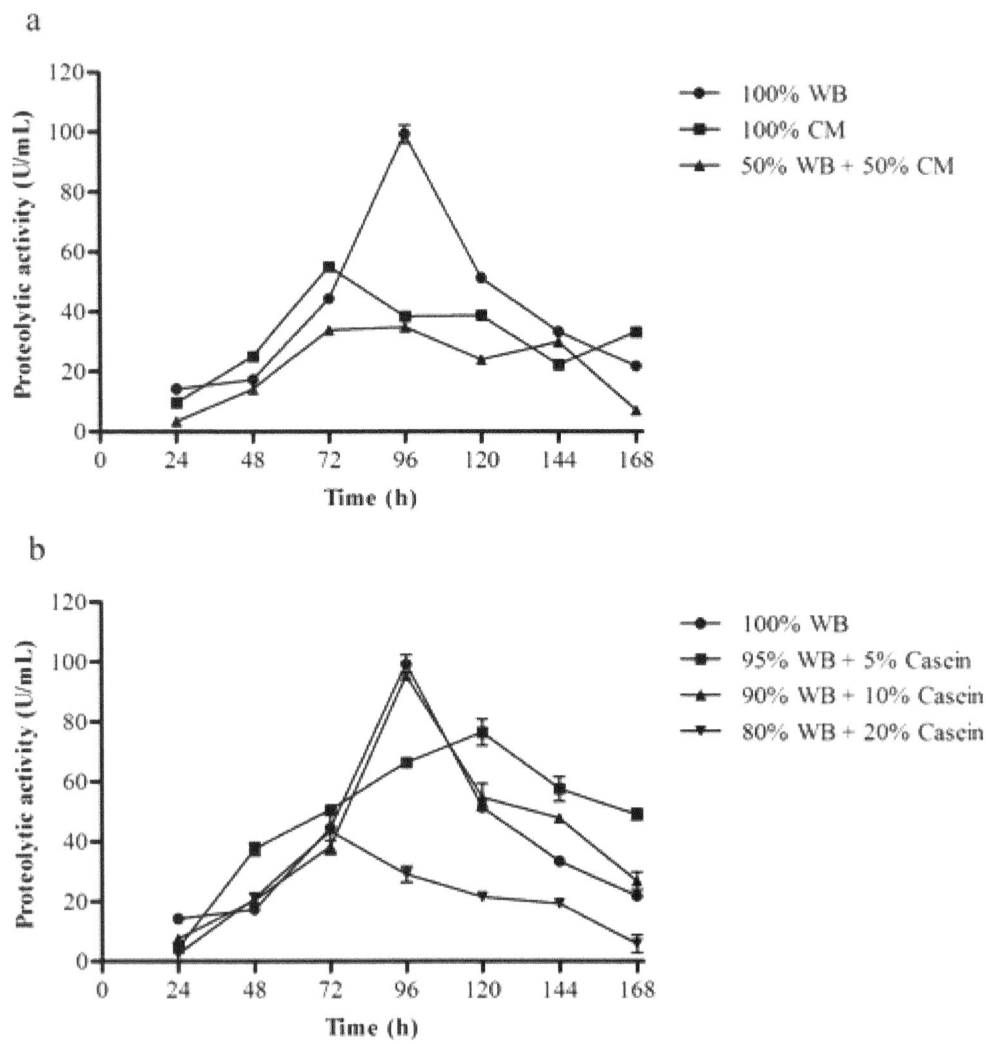

Figure 2. Effect of parameters variation on protease synthesis by the fungus *E. javanicum* under SSB. (a) Effect of agro-industrial residues at 30°C. (b) Nitrogen sources: casein concentration at 30°C. Values shown are the averages (mean ± SD) of the triplicate samples from one representative experiment. SSB: solid state bioprocess, WB: wheat bran, CM: cottonseed meal.

temperatures on production showed that increasing the bioprocess temperature decreased protease synthesis by the fungus *E. javanicum* in SSB.

Optimal conditions to activity of the protease produced by *E. javanicum* in SmB and SSB

Initially, we determined the pH in the one that obtained the best levels of protease activity. In both bioprocesses, the optimum pH was 5.5 (Figure 4a and b), however the protease produced in SmB showed other smaller peaks (at pH 8.0 and 10.0) (Figure 4a), indicating the possible presence of two other proteases produced by SmB. After determining the optimum pH, the enzymatic extract was submitted to various temperatures and the effect of this parameter was evaluated based on proteolytic activity. The results demonstrated that the optimal temperature was 60°C in both bioprocesses (Figure 5a and b).

Ions effects and subclass determination of proteases produced in SmB and SSB using inhibitors assay

The inhibitors assay is used to determine the subclass of proteases that were produced, according to the mechanism of catalytic action. This enzymatic test revealed that enzymes produced under both fermentative processes are inhibited by EDTA (Figure 6a), suggesting that the proteases of both bioprocesses belong to the metalloprotease subclass.

The ions can influence the proteolytic activity by interaction with amino acids residues in protease structure. In SmB the presence of $CoCl_2$ and $ZnSO_4$ increased 87 and

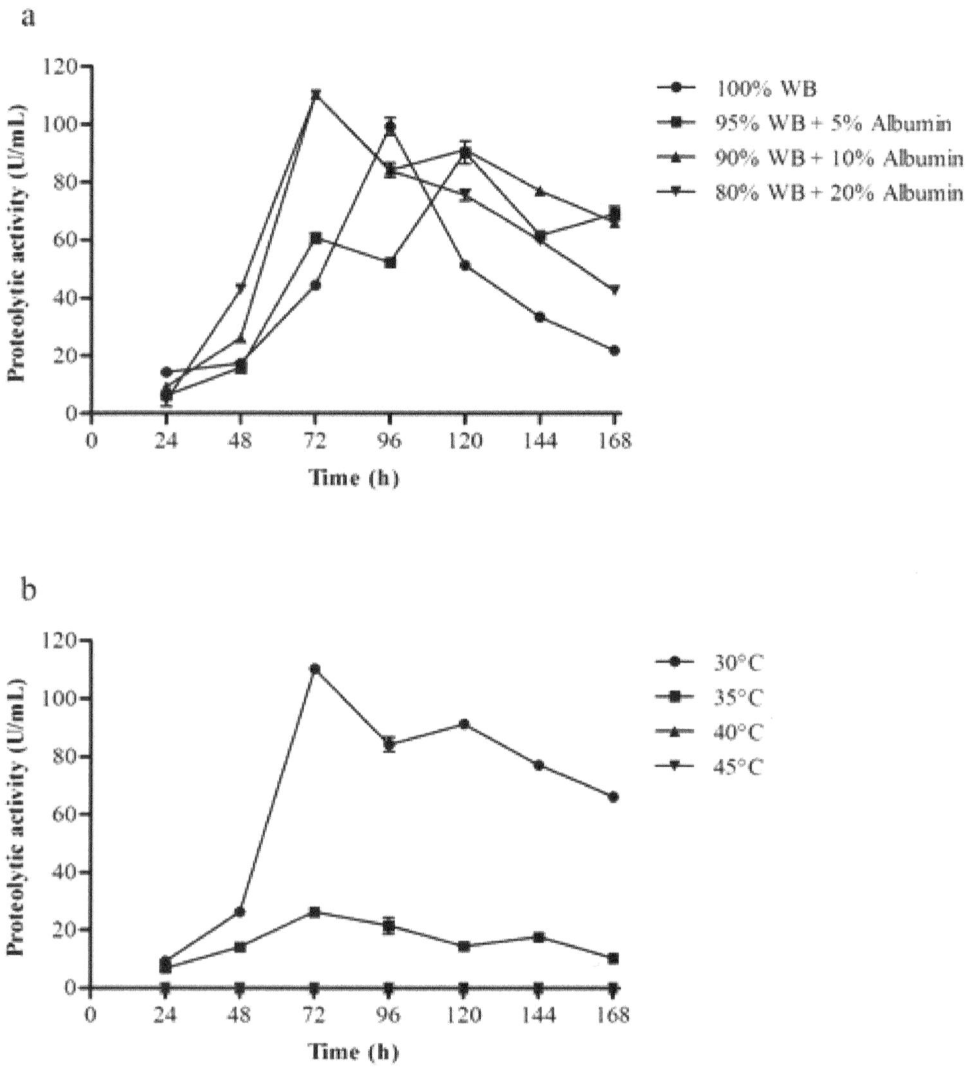

Figure 3. Effect of parameters variation on protease synthesis by the fungus *E. javanicum* under SSB. (a) Nitrogen sources: albumin concentration at 30°C. (b) Temperature variations (10% albumin+90%WB). Values shown are the averages (mean ± SD) of the triplicate samples from one representative experiment. SSB: solid state bioprocess, WB: wheat bran.

12% of the enzyme activity of proteases, respectively. The influence of ions on protease produced in SSB was more evident and different divalent metallic cations presented positive effects such as Co^{+2}, Mg^{+2}, Zn^{+2}, Ca^{+2} and Ba^{+2} or trivalent metallic ions Al^{3+}, however Cu^{+2} demonstrated negative effect. The highest activities in SSB were obtained in the presence of $CoCl_2$ and $ZnSO_4$ that show an increase of 40 and 46% as compared to control sample (Figure 6b). The enzyme assay with the addition of various ions showed that ions can have positive or negative effects on proteolysis.

pH and temperature stability of the proteases produced by *E. javanicum* in SmB and SSB

The recognition of pH and thermal stabilities can guide the potential application of the enzymes. The exposition to various pH levels revealed that the proteases produced in SmB and SSB provides greater stability at acidic pH levels (Figure 7a and b). In the Figure 7a it is observed that the residual activities in all pH levels remained above 80%, although had shown a slight drop when the protease produced in SmB was exposed to a pH level above 4.5. Similarly, in the Figure 7b, the residual activities in all pH levels remained above 70% and the exposition to pH levels above 5.5 for protease produced under SSB also had showed a slight drop. The pH stability results indicate that proteases can be used or stored at several pH levels.

The exposition to different temperatures and times provides the recognition of temperature stability of the proteases produced in SmB and SSB. The proteases produced under SmB retained 40% of their activity when exposed at 55°C for 30 min and 35% when exposed to

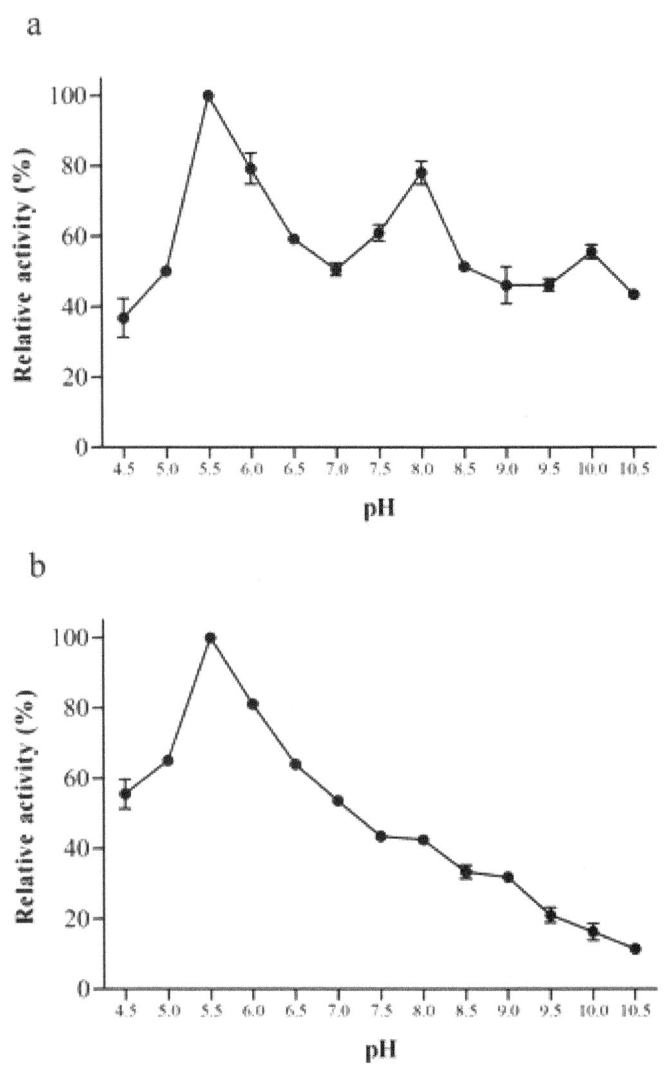

Figure 4. Biochemical characterization of enzyme extract precipitated. Effect of pH on the proteolytic activity of proteases at 40°C: (a) SmB and (b) SSB. Values shown are the averages (mean ± SD) of the triplicate samples from one representative experiment. SSB: solid state bioprocess, SmB: submerged bioprocess.

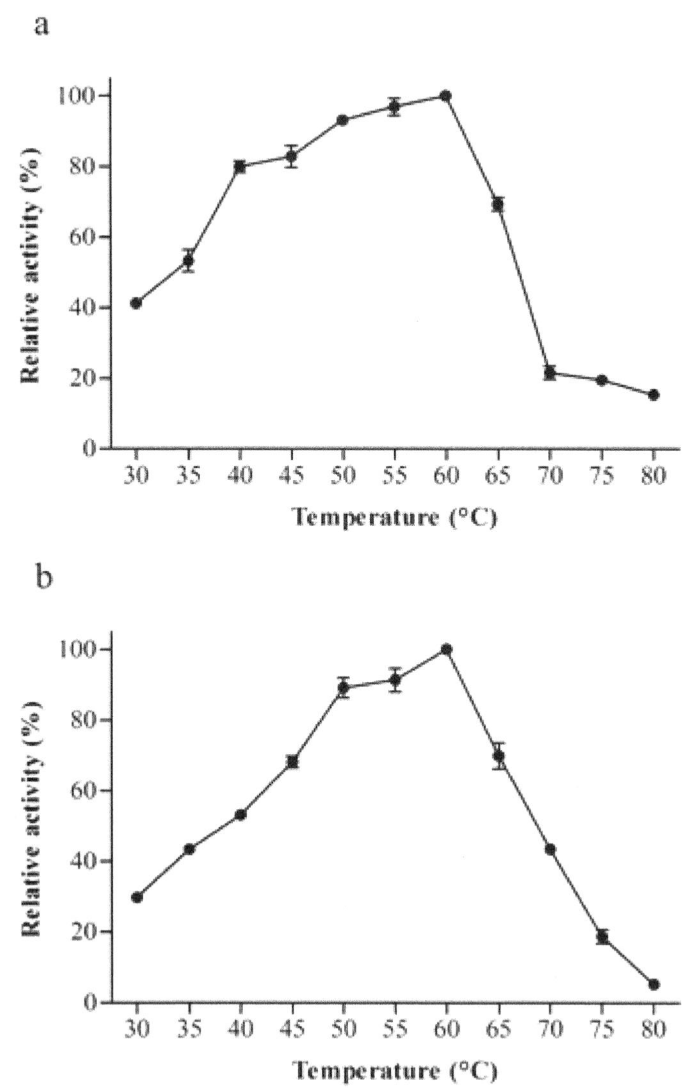

Figure 5. Biochemical characterization of enzyme extract precipitated. Effect of temperature on proteolytic activity of proteases at pH 5.5: (a) SmB and (b) SSB. Values shown are the averages (mean ± SD) of the triplicate samples from one representative experiment. SSB: solid state bioprocess, SmB: submerged bioprocess.

60°C for 15 min (Figure 8a). The proteases produced under SSB maintained 40 and 35% of their activity when exposed at 55°C for 30 and 60 min, respectively (Figure 8b). The activity of proteases produced under both bioprocesses maintained about 50% of initial activity for 60 min of exposition at 50°C.

Collagenolitic potential of proteases produced in SmB and SSB

The potential of proteases from both SmB and SSB to collagen degradation was shown using Azocoll substrate. The results were observed by hydrolysis of Azocoll, SSB and SmB obtained with $A_{550} = 0.095$ and $A_{550} = 0.139$ abs, respectively. The protease produced by SmB demonstrated more activity than SSB.

DISCUSSION

Here we found that *Eupenicillium javanicum* was able to produce proteases in both solid state and submerged bioprocesses, although the SSB protease production was twice higher than SmB and this production can be optimized by parameters studies.

Furthermore, we showed that the proteases produced had collagenolytic potential. The optimal rate of metabolite production of a fungus is obtained through detailed knowledge of its growth and physiological conditions

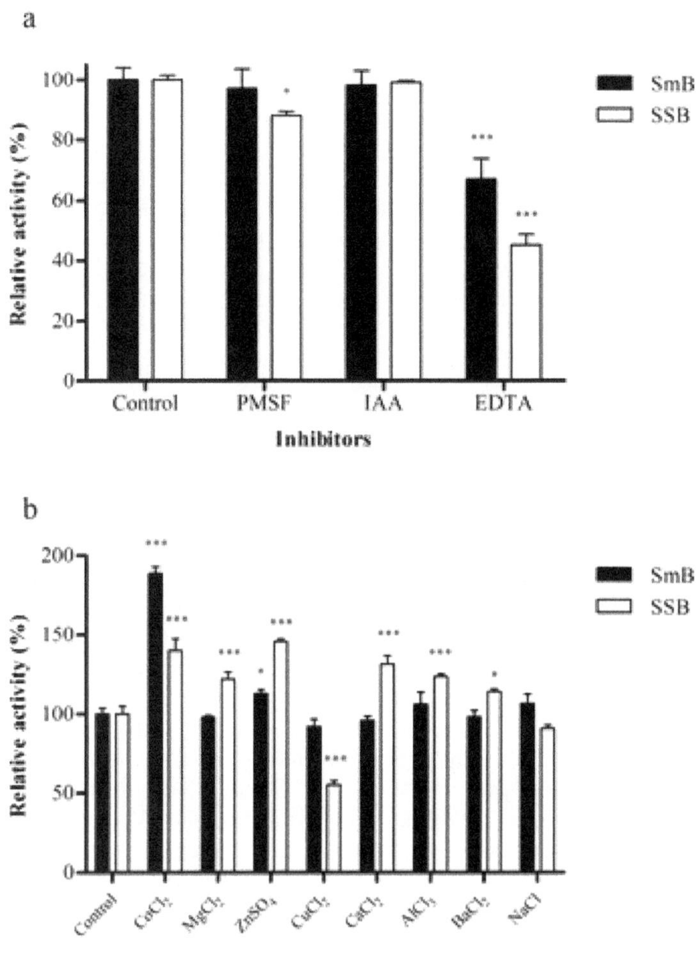

Figure 6. Effect of the addition of metallic ions and inhibitors on proteolytic activity of proteases produced by the fungus *E. javanicum* under SmB and SSB. (a) Metallic ions at 40°C and pH 5.5. (b) Inhibitors at 40°C and pH 5.5. Values shown are the averages (mean ± SD) of the triplicate samples from one representative experiment. Significance level: *p<0.05, ***p<0.001. SSB: solid state bioprocess, SmB: submerged bioprocess, PMSF: phenylmethylsulfonyl fluoride, IAA: iodoacetic acid, EDTA: ethylenediaminetetraacetic acid.

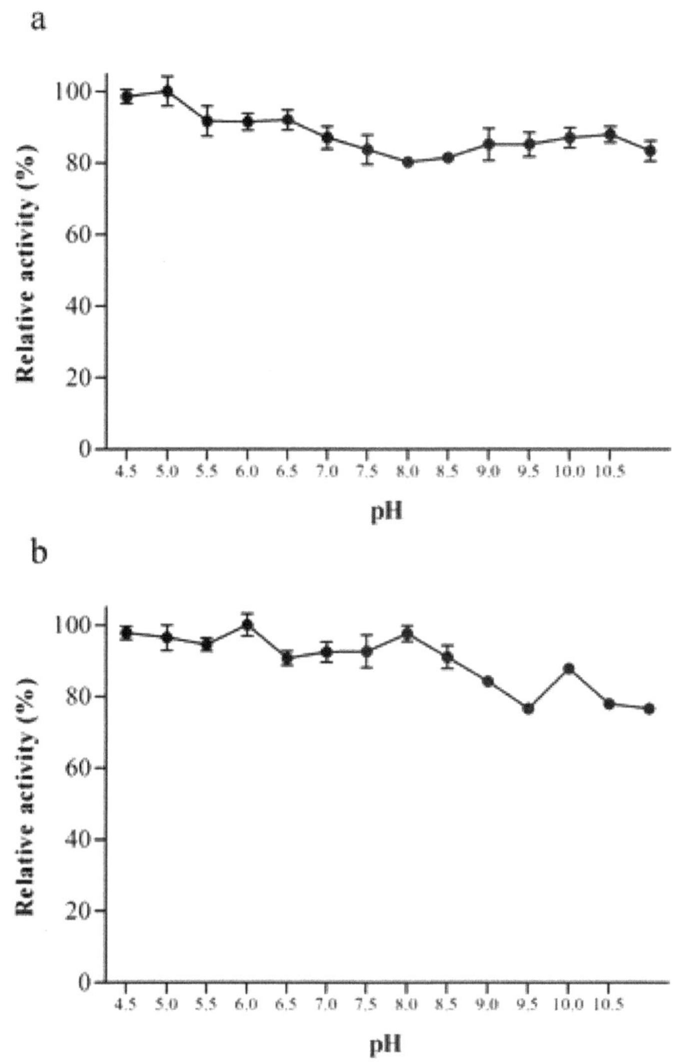

Figure 7. Stability of the proteases produced by the fungus *E. javanicum*. Exposition to various pH levels at 40°C: (a) SmB and (b) SSB. Values shown are the averages (mean ± SD) of the triplicate samples from one representative experiment. SSB: solid state bioprocess, SmB: submerged bioprocess.

(Papagianni, 2004).

The production of proteases can be induced with or without the addition of supplementary protein sources. In analysis of protease produced by SmB, the fungus *Eupenicillium javanicum* showed a repression of protease synthesis with casein concentration above 0.25%. However, other study demonstrated that the addition of 0.5% casein has been shown to increase protease production by the fungus *Mucor mucedo* DSM 809 (Yegin et al., 2010).

We showed that 0.5% glucose increased 37% of protease production by *E. javanicum* in SmB demonstrating the importance of this carbon source. However, the addition of a carbon source can cause catabolic repression, as was observed in the presence of sucrose, because increasing the percentage of sucrose decreased the production of proteases. Similarly, it has been shown that greater acid protease production by *Aspergillus awamori* MTCC 548 was observed when the medium was supplemented with glucose (Sinha and Sinha, 2009), but this situation was not observed in the protease production by the fungus *Botrytis cinerea*, once the addition of glucose to submerged media promotes a decrease in protease production (Abidi et al., 2008). Sinha and Sinha (2009) also showed that higher glucose concentrations decreased the protease production and other carbon sources did not increase production.

The study of effect of pH is important because the pH may influence any activities, yet this topic is rarely addressed (Papagianni, 2004). *E. javanicum* responded to an increase in pH levels with lower protease production, indicating that the absence of protons in the bioprocess

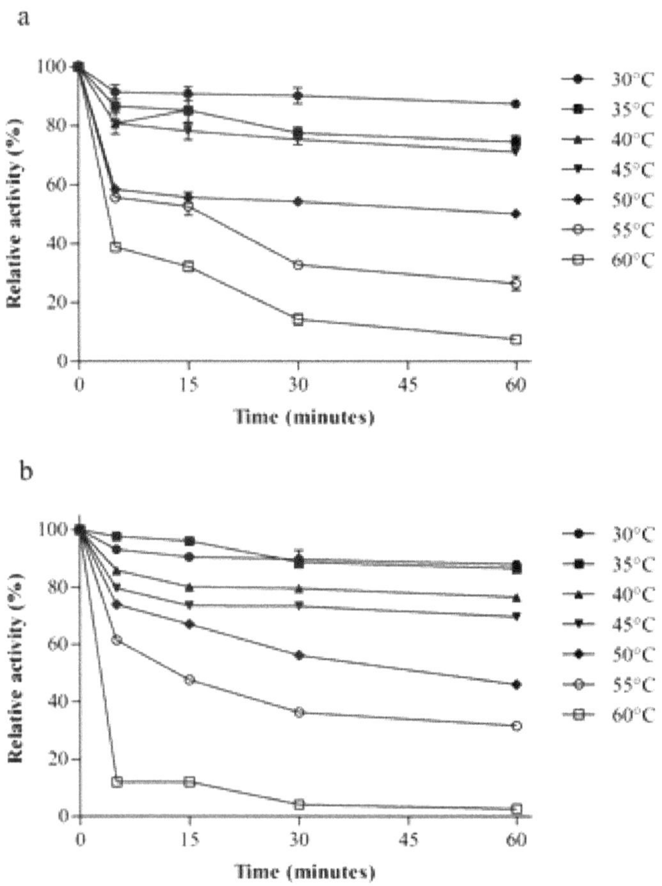

Figure 8. Stability of the proteases produced by the fungus *E. javanicum*. Exposition to various temperatures and times at pH 5.5: (a) SmB and (b) SSB. Values shown are the averages (mean ± SD) of the triplicate samples from one representative experiment. SSB: solid state bioprocess, SmB: submerged bioprocess.

media decreases the protease production. Likewise, the fungus *Aspergillus oryzea* has shown a lower production of proteases with increasing pH (Sandhya et al., 2005), confirming that pH has large effect on the production of proteases. The assay of the temperature effect on the protease production revealed that *E. javanicum* did not respond positively to increases in temperature. The results found during the optimization of protease production by *Bacillus* sp. I-312 also showed that increasing incubation temperature can cause a reduction in the protease production (Joo and Chang, 2005) and indicated the importance of temperature control in bioprocesses.

The choice of agro-industrial residues in bioprocesses is very important. In SSB, the analysis of agro-industrial residues wheat bran (WB) and cottonseed meal (CM) showed better protease production when 100% of WB is used as a bioprocess medium. Sandhya et al. (2005) related that although many agro-industrial residues have been tested, the wheat bran showed the best results.

Other authors suggest that residues with higher amounts of protein in their composition (Madruga and Camara, 2000) as well as the porosity of the residue facilitate the mycelial dispersion under SSB (Sandhya et al., 2005; Pandey, 2004) and can improve the protease production.

Supplementation of the medium with a protein source can positively or negatively influence the production of proteases. In SSB, the addition of 10 and 20% of albumin promoted an improvement in the production of proteases, but the addition of 10% of casein showed no effect when compared with medium without additional nitrogen source. Casein concentrations above 10% demonstrated negative effects on protease synthesis. These data show the importance of choosing the ideal protein source.

The effect of temperature in solid state bioprocess process was similar to submerged bioprocess process, the temperature increase decreased proteases production, and at temperatures above 35°C, there was no apparent fungus growth. According to other authors, high temperatures can negatively influence the metabolic activities of microorganisms and can inhibit the growth of fungus, in general, fungal proteases are thermolabile and show lower activities at higher temperatures (Haq et al., 2006).

The protease production under SSB (90% wheat bran + 10% albumin) at 30°C was twice higher than SmB (glucose) at 30°C; however this difference can be more expressive. Zanphorlin et al. (2010) demonstrated that the pro-duction of proteases by thermophilic *Myceliophtora* sp. by solid state using wheat bran and casein was 4.5 times higher than submerged fermentation (Zanphorlin et al., 2010). Similar result was demonstrated by Silva et al. (2013) (accepted for publication), in this study the fungus *A. fumigatus* presented protease production 30 times higher under solid state fermentation than submerged fermentation.

In the current study, we determined the effect of pH and temperature on activity and stability, and the effect of metal ions and inhibitors. The reaction parameters are essential for the catalytic production of proteases from two bioprocesses. The effect of pH levels on activity show that pH 5.5 provide highest activities for proteases produced under both bioprocesses, suggesting that these has acidic characteristics. Additionally, in the SmB process, there are peaks at pH 8.0 and 10.0, which indicate the potential presence of another protease with alkaline characteristics. The effect of temperature was similar for proteases produced under both bioprocesses.

The results obtained in SSB and SmB using *E. javanicum* suggest that proteases produced have optimal temperature of 60°C and at temperatures above 60°C, the activity decreased. In solid state bioprocess, *Rhizopus oryzae* has been shown to produce acidic proteases with same optimal temperature and pH (Kumar et al., 2005), and similar results also have been seen with the fungus *Thermoascus aurantiancus* (Merheb et al., 2007). The optimal conditions of pH and temperature were similar to proteases produced in SmB and SSB.

The EDTA provides an inhibition of approximately 33

and 55% on proteolytic activity of proteases produced by *E. javanicum* in SmB and SSB, respectively. These results suggest metalloproteases. Despite this enzyme subclass shows optimal pH at near neutrality, some acidic metalloproteases have been described in other studies, including the fungi *T. aurantiancus* (Merheb-Dini et al., 2009) and *Streptomyces septatus* TH-2 (Hatanaka et al., 2005).

Metal ions can modulate the proteolytic activity positively or negatively. The addition of 10 mM cobalt chloride or zinc sulfate increased the enzyme activity of proteases produced under both bioprocesses. According to other authors, metalloproteases need divalent ions to improve their activity levels (Rao et al., 1998). However, other study related the opposite comportment, 1 mM of divalent metallic ions cobalt and zinc were responsible for 100% inhibition of a metalloprotease with collagenolytic activity secreted by *A. fumigatus*. Furthermore, similar to our results in SSB, the presence of copper chloride resulted in a loss of metalloprotease activity (Monod et al., 1993).

The protease produced under SmB and SSB presented greater stability at acidic pH levels which is optimal pH. A pH closer to the alkaline level tends to decrease the stability, and this trend was greater for proteases produced under SSB than under SmB. Similarly, the acidic proteases produced by *A. niger* I1 had greater stability near the optimal pH level, and stability decreased when the pH was shifted away from the optimal level (Siala et al., 2009); however, this drop was larger than that presented in this study.

The protease stability results in relation to temperature as a function of time revealed that the enzymes produced under SSB, when exposed to 55°C for 30 min, retained 40% of their activity. The enzymes produced under SmB were more stable with temperature changes, maintaining approximately 40% of their activity after 15 min of exposure to 60°C. Here, the proteases produced in SmB and SSB maintained about 60 and 50% activity after 30 and 60 min of exposition at 50°C, respectively. Shivakumar (2012) considered the thermostability at 50°C for 30 min of the protease produced by *Aspergillus* sp. with potential for food industries. In another study, the acidic proteases produced by *A. niger* I1 maintained approximately 30% of their activity following exposure to 50°C for 15 min (Siala et al., 2009).

In this study, the metalloproteases produced in SmB and SSB by *E. javanicum* were able to degrade Azocoll substrate. Others authors related collagenolytic metalloprotease with similar characteristics secreted by *A. fumigatus* (Monod et al., 1993). Another study measured the activity of several commercially available enzymes, after 24 h of reaction with an azocoll substrate finding values that ranged from 0.3 to 0.92 abs (Foroughi et al., 2006).

In the current study, after 4.5 h, we found value of 0.139 and 0.095 from metalloproteases produced by *E. javanicum* under SmB and SSB, respectively. The exposure time of the enzyme with a substrate was approximately 5.5 times lower than the previously mentioned study, showing the collagenolytic potential of these specific metalloprotease.

We found that the metalloprotease profile production can be explored to obtain proteases with different applications. Here, the metalloproteases produced by fungus *E. javanicum* showed potential to collagenolytic activity for proteases produced under both submerged and solid state bioprocess, and this type of enzyme can have potential uses in industry.

ACKNOWLEDGEMENTS

This work received financial support from School of Pharmaceutical Sciences of Ribeirão Preto (University of São Paulo – USP), Coordenação de Aperfeiçoamento de Pessoal de Nível Superior (CAPES), Fundação de Amparo à Pesquisa do Estado de São Paulo (2011/06986-0) and Conselho Nacional de Desenvolvimento Científico e Tecnológico (PQ-2). We thank Nathalia Gonsales da Rosa for technical assistance.

REFERENCES

Abidi F, Liman F, Nejib MM (2008). Production of alkaline proteases by *Botrytis cinerea* using economic raw materials: Assay as biodetergent. Process Biochem. 43:1202-1208.

Ducros E, Ferrari M, Pellegrino M, Raspanti C, Bogni C (2009). Effect of aeration and agitation on the protease production by *Staphylococcus aureus* mutant RC128 in a stirred tank bioreactor. Bioprocess Biosyst. Eng. 32:143-148.

Dunn BM (1989). Determination of protease mechanism. In: Beynon, RJ, Bond, JS. Proteolytic Enzymes: a practical approach. Great Britain: IRL Press. 57:81.

Foroughi F, Keshavarz T, Evans CS (2006). Specificities of proteases for use in leather manufacture. J. Chem. Technol. Biotechnol. 81:257-261.

Gupta R, Beg QK, Khan S, Chauhan B (2002). An overview on fermentation, downstream processing and properties of microbial alkaline proteases. Appl. Microbiol. Biotechnol. 60:381-395.

Haq IU, Mukhtar H, Umber H (2006). Production of protease by *Penicillium chrysogenum* through optimization of environmental conditions. J. Agric. Soc. Sci. 2:23-25.

Hatanaka T, Uesugi JAY, Iwabuchi M (2005). Purification, characterization cloning, and sequencing of metalloendoprotease from *Streptomyces septatus* TH-2. Arch. Biochem. Biophys. 434:289–298.

Joo HS, Chang CS (2005). Production of protease from a new alkaliphilic *Bacillus* sp. I-312 grown on soybean meal: optimization and some properties. Process Biochem. 40:1263–1270.

Kumar S, Sharma NS, Saharan MR, Singh R (2005). Extracellular acid protease from *Rhizopus oryzae*: purification and characterization. Process Biochem. 40:1701-1705.

Madruga S, Camara S (2000). The chemical composition of multimistura as a food supplement. Food Chem. 68:41-44.

Mandl I, MacLennan J, Howes E, DeBellis R, Sohler A (1953). Isolation and Characterization of Proteinase and Collagenase from Cl. histolyticum. J. Clin. Invest. 32:1323.

Merheb CW, Cabral H, Gomes E, da Silva R (2007). Partial characterization of protease from a thermophilic fungus, *Thermoascus aurantiacus*, and its hydrolytic activity on bovine casein. Food Chem. 104:127-131.

Merheb-Dini C, Cabral C, Leite RSR, Zanphorlin LM, Okamoto DN, Rodriguez GOB, Juliano L, Arantes EC, Gomes E, da Silva R (2009). Biochemical and Functional Characterization of a Metalloprotease from the Thermophilic Fungus *Thermoascus aurantiacus*. J. Agric. Food Chem. 57:9210 – 9217.

Monod M, Paris S, Sanglard D, Jaton-Ogay K, Bille J, Latgé JP (1993). Isolation and characterization of a secreted metalloprotease of *Aspergillus fumigatus*. Infect. Immun. 61:4099-4104.

Morita Y, Hasan Q, Sakagushi T, Murakami Y, Yokoyama K, Tamiya E (1998). Properties of a cold-active protease from psychrotrophic *Flavobacterium balustinum* P104. Appl. Microbiol. Biotechnol. 50:669-675.

Pandey A (2004). Solid-state fermentation. Biochem. Eng. J. 13:81–84.

Papagianni M (2004). Fungal morphology and metabolite production in submerged mycelial processes. Biotechnol. Adv. 22:189-259.

Rao MB, Tanksale AM, Ghatge MS, Deshpande VV (1998). Molecular and biotechnological aspects of microbial proteases. Microbiol. Mol. Biol. R. 62:597-635.

Sandhya C, Sumantha A, Szakacs G, Pandey A (2005). Comparative evaluation of neutral protease production by *Aspergillus oryzae* in submerged and solid-state fermentation. Process Biochem. 40:2689-2694.

Sarath G, de la Motte RS, Wagner FW (1996). Protease assay methods. In: Beynon RJ, Bond JS (ed) Proteolytic enzymes: a practical approach. New York: Oxford University. 3:25-55.

Shivakumar S (2012). Production and characterization of an acid Protease from a local *Aspergillus* sp. by solid substrate fermentation. Arch. Appl. Sci. Res. 4:188-199.

Siala R, Sellami-Kamoun A, Hajji M, Abid I, Gharsallah N, Nasri M (2009). Extracellular acid protease from *Aspergillus niger* I1: purification and characterization. Afr. J. Biotechnol. 8:4582-4589.

Sinha S, Sinha S (2009). Studies on the production of acid protease by submerged fermentation. Int. J. Food Eng. DOI: 10.2202/1556-3758.1338.

Sukhosyrova EA, Nikitina ZK, Yakovleva MB, Veshchikova EV, Bykov VA (2003). Characteristics of collagenolytic enzymes secreted by deuteromycete fungi *Aspergillus flavus*. B. Exp. Biol. Med. 135:447-451.

Tanaka M, Kawaide A, Matsuno R (1986). Cultivation of microorganisms in an air–solid fluidized bed fermentor with agitators. Biotechnol. Bioeng. 28:1294-1301.

Tran LH, Nagano H (2002). Isolation and characteristics of *bacillus subtilis* CN2 and its colagenase production. J. Food. Sci. 67:1184-1187.

Tsuruoka N, Nakayama T, Ashida M, Hemmi H, Nakao M, Minakata H, Oyama H, Oda K, Nishino, T (2003). Collagenolytic Serine-carboxyl proteinase from *Alicylobacillus sendaiensis* strain NTAP-1: Purification, characterization, gene cloning, and heterologous expression. Appl. Environ. Microbiol. 69:162-169.

Watanabe K (2004). Collagenolytic proteases from bacteria. Appl. Microbiol. Biotechnol. 63:520-526.

Yegin S, Fernández-Lahore M, Guvenc U, Goksungur Y (2010). Production of extracellular aspartic protease in submerged fermentation with *Mucor mucedo* DSM 809. Afr. J. Biotechnol. 9:6380-6386.

Zanphorlin LM, Facchini FDA, Vasconcelos F, Bonugli-Santos RC, Rodrigues A, Sette, LD, Gomes E, Bonilla-Rodriguez GO (2010). Production, partial characterization, and immobilization in alginate beads of an alkaline protease from a new thermophilic fungus *Myceliophthora sp*. in biotechnological processes. In this work, we describe the characterization of a protease produced in solidstate (SSF) and submerged (SmF) fermentations by a newly isolated. J. Microbiol. 48:331-336.

Zhao H, Mitsuiki S, Takasugi M, Sakai M, Goto M, Kanouchi H, Oka T (2012). Decomposition of insoluble and hard-to-degrade animal proteins by enzyme E77 and its potential applications. Appl. Biochem. Biotechnol. 166:1758-1768.

Optimization of culture parameters to enhance production of amylase and protease from *Aspergillus awamori* in a single fermentation

Sangeeta Negi[1]* and Rintu Banerjee[2]

[1]Biotechnology Section, Applied Mechanics Department, Motilal Nehru National Institute of Technology, 221 004 Teliarganj, Allahabad, India.
[2]Agriculture and Food Engineering Department, Microbial Biotechnology, 721302 Kharagpur, India.

Concomitant production of amylase and protease by *Aspergillus awamori* nakazawa MTCC 6652 was enhanced in a single fermentation by media engineering and optimization of other important parameters. Wheat bran was considered as a suitable substrate for production of amylase and protease by *A. awamori* in a single fermentation as none of the additional substrate such as powder of peanut, corn, soybean seeds and sunflower seeds was effective in enhancing the concurrent production of enzymes further. Optimum amylase yield of 4528.4 ± 121U/gds and protease yield of 250.4 ± 10 U/gds was achieved with wheat bran to Czapek-dox ratio of 1:1.5 (w/v), 96 h incubation, 35°C temperature, pH 5.5 and 85% relative humidity. Media engineered with 1% casein and 1% starch solution increased yield of amylase and protease by 2.07 fold (9386.5 ± 101 U/gds) and 3.73 fold (934.8 ± 67 U/gds), respectively, therefore, considered as the most suitable media for concomitant production of amylase and protease by *A. awamori* in a single fermentation.

Key words: Amylase, protease, solid state fermentation, optimization.

INTRODUCTION

Amylolytic enzymes (alpha-amylase E.C. 3.2.1.1, beta-amylase 3.2.1.2 and glucoamylase E.C. 3.2.1.3) and proteases (acid and alkaline proteases) are among the most important industrial enzymes because of their industrial applications in the field of brewing, textile, paper, sugar, distilling, food processing, pharmaceuticals, tannery, waste bio-degradation, detergent etc. (Pandey et al., 2000), concomitant production of these enzymes in a single fermentation is the current need of the enzyme industry.

Selection of a suitable strain capable of producing both these enzymes concomitantly with commercially acceptable yield was a crucial step. After detail study of the relevant literature decision was taken to explore *Aspergillus awamori* for multienzymes production. *A. awamori* has been reported to be very effective for production of amylases and proteases, although not much has been reported on concomitant production of multi-enzymes, Pandey et al. (2000), Peixoto et al. (2003), Sumantha et al. (2006) and Yamamoto et al. (2005).

Conventionally, multienzymes are produced by mixing various enzymes directly through genetically engineered microbes endowed with multi-functions (Zaghloul et al., 2000) and by mixed culture consisting of different well-designed microbes, Zhang et al. (2004). These methods have their own disadvantages; hence, appropriate bioprocesses with a well-established bioengineering were needed to be developed for efficient and simultaneous production of amylase and protease by a single microbial culture in a single fermentation.

Present investigation focused on two important aspects; optimization of parameters and effects of various additives, for achieving optimum production of amylase and protease in a single fermentation by *A.awamori*. Wheat bran was selected as substrate because it has been reported as the most common and effective substrate,

*Corresponding author. Email: sn5@mnnit.ac.in.

Pandey et al. (1999). Optimum utilization of environmental conditions, such as incubation period, temperature, pH, relative humidity etc. was carried out for achieving high yield. Additives play an important role in fermentation processes; hence, investigation was also carried out on the effects of various additives such as nitrogen sources, carbon sources, metal ions, surfactants and hormones on the production of amylase and protease in a single fermentation.

MATERIALS AND METHODS

Chemicals

The chemicals used during this work were of analytical grade.

Microorganism

A. awamori nakazawa MTCC 6652 used for the present investigation was maintained on 2% malt-extract agar slants and stored at 4°C. A spore suspension was made by adding 25 ml of sterile distilled water to a 5 days old slant and scraped aseptically with inoculating loop. A suspension, having spore concentration of about 1.3 x 10^7 cells ml^{-1}, was used as inoculum for the subsequent fermentation.

Selection of substrate

Wheat bran was used as suitable substrate for the present investigation because its composition (15.5% protein, 64.51% carbohydrates, 4.2% fat and other nutrients also in desirable amount) was found appropriate for the production of desired enzymes.

Substrates manoeuvring

In order to enhance the simultaneous production of amylase and protease other substrates such as peanut, maize, soybean and sunflower seeds powder (grinded in mixer-grinder for 5 min) were supplemented with wheat bran in 1% (w/w) to enrich the media. Fermentation was carried out for 5 days at 35°C and other fermentation conditions were kept constant. After 120 h of incubation, fermented broth was filtered and centrifuged. Effects of various combinations of substrate with wheat bran as base were studied in triplicates and combination giving optimum concomitant yields of enzymes was selected as substrate for subsequent experiments.

Fermentation process

Fermentation was carried out in a Erlenmeyer flask (500 ml) containing 10.0 g of wheat bran soaked in 15 ml of salt solution (Czapek-dox: $NaNO_3$ 2.5 g/L, KH_2PO_4 1.0 g/L, $MgSO_4.7H_2O$ 0.5 g/L, KCl 0.5 g/L), shaked till entire substrate got moistened with Czapek-dox solution and then autoclaved. This mixture was inoculated with 2 ml of spores suspension aseptically and incubated for 4 days at 35°C and 85% relative humidity. Fermentation was conducted under various experimental conditions.

Extraction process

Fermented mass was soaked for 2 hrs in distilled water (in a ratio of 1:4) and squeezed with cheese cloth. The extract was centrifuged at 10,000 rpm for 15 min to remove the spores and other insoluble materials.

Determination of amylase activity

Amylase activity was measured following the method described by Bernfeld (1955). A reaction mixture containing 0.5 ml of 1% soluble starch solution prepared in 0.2 M acetate buffer and 0.5 ml of diluted enzyme solution was incubated at 50°C. After 10 min of incubation the reaction was terminated by adding 1.0 ml of DNS solution (1 g of DNS dissolved in 20 ml of 2 M NaOH, to which 30 g of sodium potassium tartarate and water were added to make it 100 ml). Reaction mixtures were boiled for 15 min and after cooling 18 ml water was added. Absorbance was measured at 540 nm. One unit of enzyme activity was defined as the amount of enzyme that releases 1 micromole of glucose as reducing sugar in 1 min under the assay conditions.

Determination of protease activity

Protease activity was determined by caseinolytic method described by Walter (1984). 0.05 ml of enzyme solution was added to 2 ml of casein solution with 0.65 ml of 0.2 M glycine NaOH buffer and incubated at 37°C for 20 min. Reaction was terminated by adding 0.2 ml 1 N HCl and unhydrolyzed casein was precipitated with 5 ml of 5% TCA solution. Thereafter, clear solution was separated by centrifugation at 9780 xg for 10 min and peptide fragment were measured. One unit of enzyme activity was defined as the amount of enzyme that liberates peptide fragments equivalent to 1 mg of Bovine Serum Albumin (BSA) under the assay conditions.

Optimization of various parameters

Optimization of various parameters was carried out with "One at a time" strategy keeping all other variables constant except one. The variable under study was varied over a desired range in triplicates and optimal level was used for further investigation.

Preparation of engineered media

A mixture of starch and casein 1% (w/v) was prepared in Czapek-Dox in media and pure culture of *A.awamori* was inoculated to it. After 4 days of incubation at 35°C a clear film of spores could be seen over liquid media. This broth was vortexed and mycelia were removed to make a clear spore suspension.

Effects of additives

Different additives, such as carbon sources (dextrose, starch, sago, arrowroot, galactose and lactose in 1% w/w ratio), organic nitrogen sources (urea, yeast extract, casein and peptone in 1% w/w ratio), inorganic nitrogen sources (NH_4NO_3, $(NH_4)2SO_4$, NH_4Cl, KNO_3 and $NaNO_3$ in 1% w/w ratio), metal ions (0.05 mg/g $CaCl_2$, $MgCl_2$, $SnCl_2$, $FeCl_3$, $CoCl_2$ and $HgCl_2$), surfactants (0.05 mg/g SLS, Triton X-100, Tween 80 and 40) and hormones (0.0025 mg/g IAA, IBA, 2,4D and NAA) were added to fermentation media and their effects on enzymes production were studied by carrying out fermentation process in triplicates.

RESULTS

Effects of additional substrate

Additional substrates, such as powder of peanut, corn,

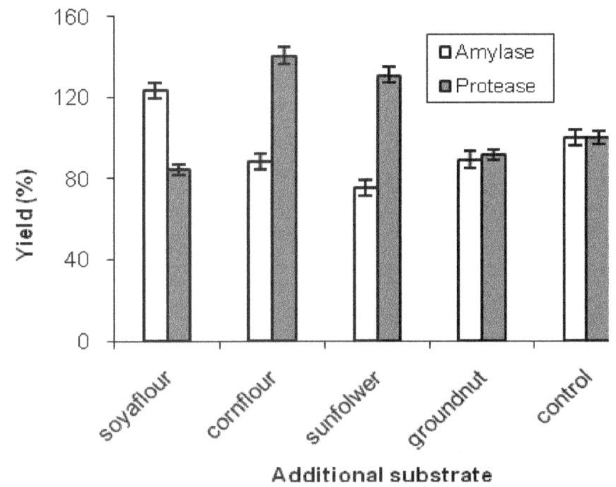

Figure 1. Effect of additional substrate on the production of amylase and protease.

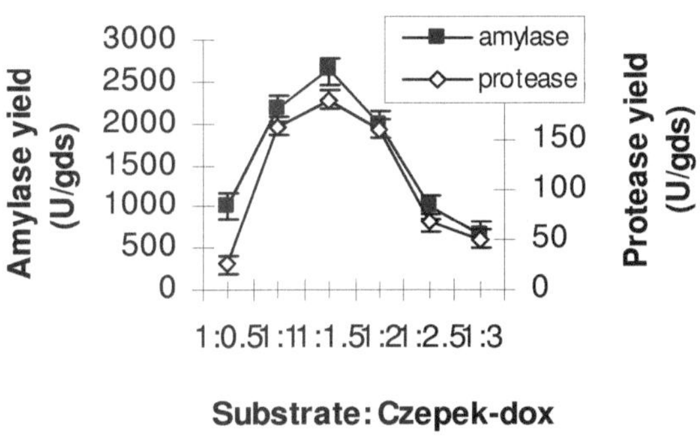

Figure 2. Effect of substrate to Czapek-dox ratio on the production of amylase and protease.

Table 1. Effect of incubation period on the production of amylase and protease.

Incubation period (h)	Amylase (U/gds)	protease (U/gds)
24	946.5 ± 38	21.5 ± 5
48	1797.3 ± 56	77.5 ± 7
72	3060.5 ± 77	146.5 ± 14
96	706.2 ± 87	256.4 ± 25
120	3529.5 ± 85	264.5 ± 32
144	2022.5 ± 75	154.5 ± 23

soybean seeds and sunflower seeds were supplemented with wheat bran in 1% (w/w) to enrich the media. Addition of none of the additional substrates was found effective in enhancing the yield of amylase as well as protease simultaneously (Figure 1). Soy flour increased amylase yield (1.23 fold) only, whereas corn (1.4 fold) and sunflower (1.3 fold) increased protease yield but reduced amylase yield.

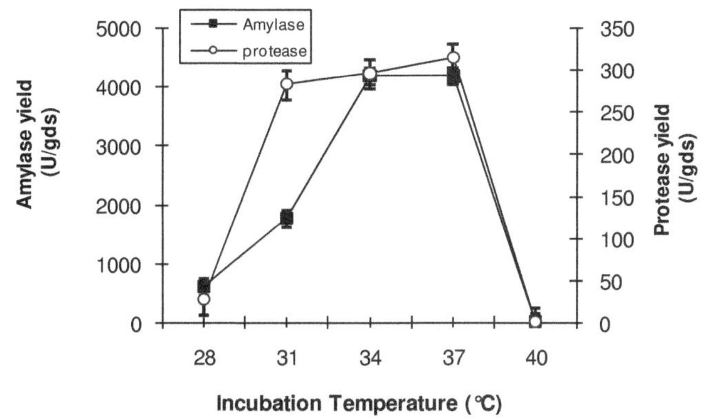

Figure 3. Effect of incubation temperature on the production of amylase and protease.

Optimization of solid to liquid ratio

In order to optimize the liquid-solid proportion, volume of Czapek-dox media was varied from 5 to 40 ml and substrate amount was kept constant at 10 g. As shown in Figure 2, substrate to liquid ratio of 1:1.5 (w/v) gave optimum yield of amylase and protease concomitantly.

Optimization of incubation period

The incubation period was varied from 24 to 144 h. The results in Table 1 showed that production of amylase increased with the increase in incubation period up to 96 h and beyond this there was no remarkable change in the yield of amylase. Yield of protease was also high at 96 h incubation. Its maximum value was at 120 h. Therefore, 96 h incubation was most suitable for optimum concomitant production of both enzymes.

Optimization of incubation temperature

Incubation temperature was varied from 28 to 40°C, keeping all other parameters constant. As shown in Figure 3, yields of amylase and protease were highest at 35 and 37°C, respectively. However, protease yield was also high at 35°C, so further fermentation was carried out at 35°C.

Selection of pH of media

The results in Table 2 showed that the yield of extracellular amylase and protease had been sensitive to initial pH of the medium. Amylase and protease productions were maximum at pH 5.5 and 4.5, respectively. However, further experiments were performed at pH 5.5 as yields of both enzymes were high at this level.

Table 2. Effect of initial pH of the media on amylase and protease activities.

Initial pH	Amylase activity (U/ml)	Protease activity (U/ml)
3.5	3102.9 ± 85	240.6 ± 21
4.5	4128.3 ± 101	309.2 ± 25
5.5	4528.4 ± 121	250.4 ± 10
6.5	4028.4 ± 96	201.5 ± 27
7.5	3521.5 ± 56	185.4 ± 25
8.5	2851.3 ± 45	108.5 ± 32
9.5	1493.2 ± 37	50.5 ± 18

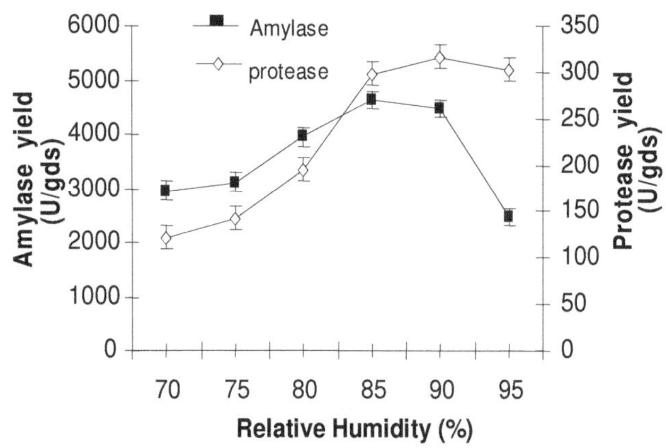

Figure 4. Effect of relative humidity on the production of amylase and protease.

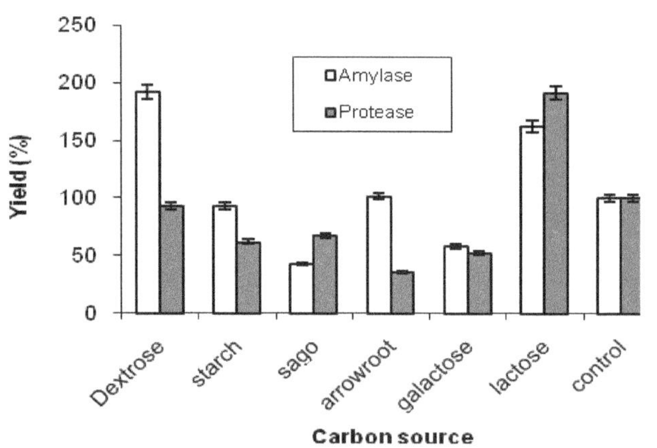

Figure 5. Effect of carbon sources on the production of amylase and protease.

Optimization of humidity

Relative humidity was varied from 70 to 95% keeping other parameters constant. Maximum production of amylase and protease was achieved at 85 and 90%

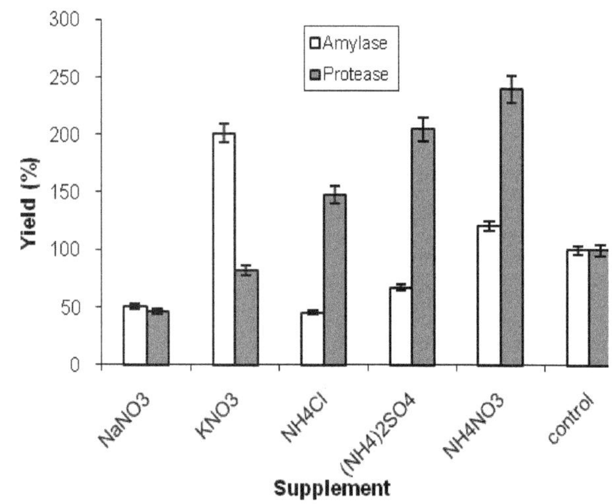

Figure 6. Effect of inorganic nitrogen sources on the production of amylase and protease.

relative humidity, respectively (Figure 4). Further investigation was carried out at 85% relative humidity as yields of both enzymes were high and stable at that level.

Effect of additional carbon supplements

Only lactose increased production of amylase as well as protease by 1.62 and 1.91 fold, respectively (Figure 5). All other supplements either reduced or had no impact on yields of enzymes, except dextrose which increased amylase yield by 1.92 fold but it reduced protease activity.

Effect of additional nitrogen sources

Mycelia growth had been found very high in presence of inorganic nitrogen sources. Only NH_4NO_3 acted as enhancer for production of amylase and protease by 1.21 and 2.1 fold, respectively. Among other inorganic sources $(NH_4)_2SO_4$ and NH_4Cl increased protease production by 2.05 and 1.48 fold, respectively, but both led to reduction in amylase production (Figure 6). Other inorganic supplements had negative impact on production of both the enzymes.

Among organic supplements only peptone increased amylase and protease yield by 2.01 and 1.35 fold, respectively (Figure 7).

Effect of metal ions

Among metal ions only $HgCl_2$ stimulated the production of amylase as well as protease, by 2.44 and 2.62 folds, respectively (Figure 8). $FeCl_3$ adversely affected production

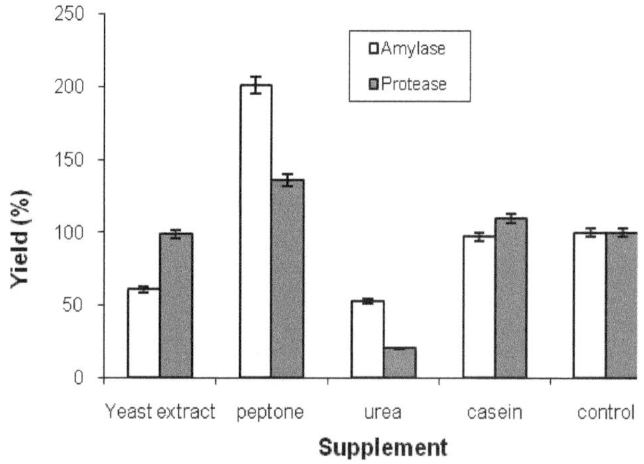

Figure 7. Effect of organic supplements on the production of amylase and protease.

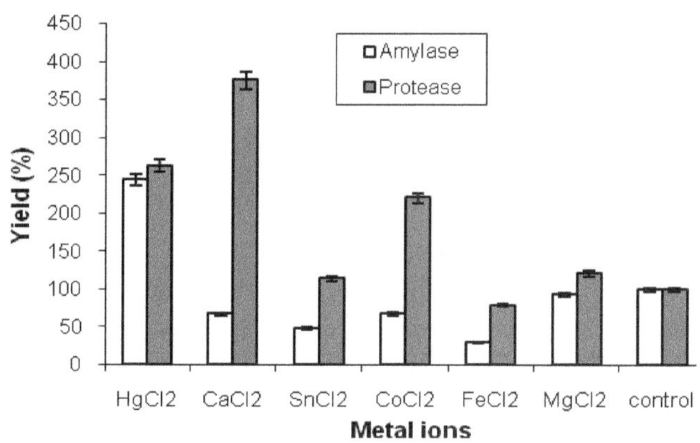

Figure 8. Effect of metal ions on the production of amylase and protease.

production of both the enzymes. All other metal ions enhanced protease production but remarkably reduced amylase production.

Effect of surfactants

Triton-X100 acted as stimulator for production of both amylase and protease, albeit marginally (Figure 9). Sodium laurate sulphate (SLS) increased amylase production by 1.28 fold but reduced protease production. Other non ionic surfactants showed negative influence on production of both the enzymes.

Effect of hormones

Indol acetic acid (IAA) and Indol-3-butyric acid (IBA) stimulated production of both the enzymes. Napthelene

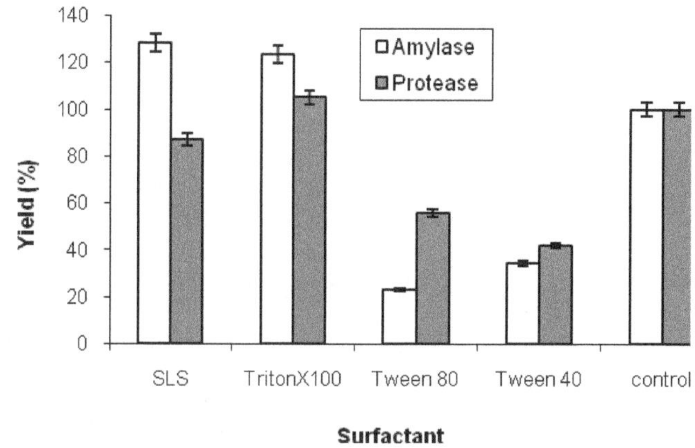

Figure 9. Effect of surfactants on the production of amylase and protease.

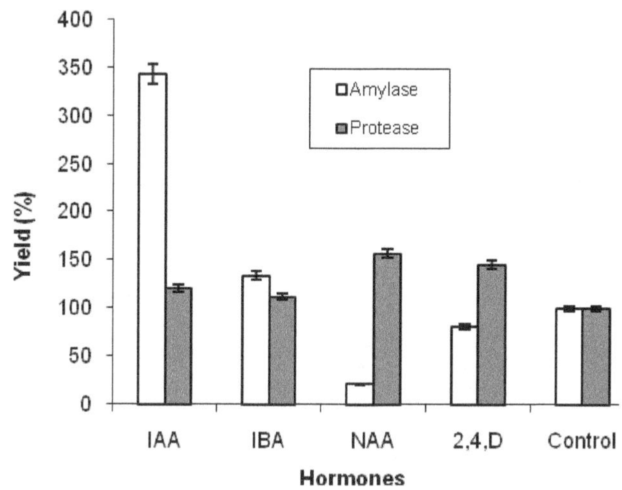

Figure 10. Effect of hormones on the production of amylase and protease.

acidic acid (NAA) and 2, 4, D enhanced protease production but reduced amylase production (Figure 10). IAA was more effective than IBA in enhancing production of amylase and protease.

Multienzymes production using engineered media

Culture was induced by cultivating it on different concentrations of casein and starch solution to get high yield. With starch alone, sporulation started within 8 to 10 days of incubation and high mycelia growth was also observed. Very little growth was observed with casein solution only. However, adequate sporulation was observed within 4 to 5 days with a mix of starch and casein (1:1 ratio) (Figure 11). With this engineered media amylase and protease yields had increased by 2.1 fold

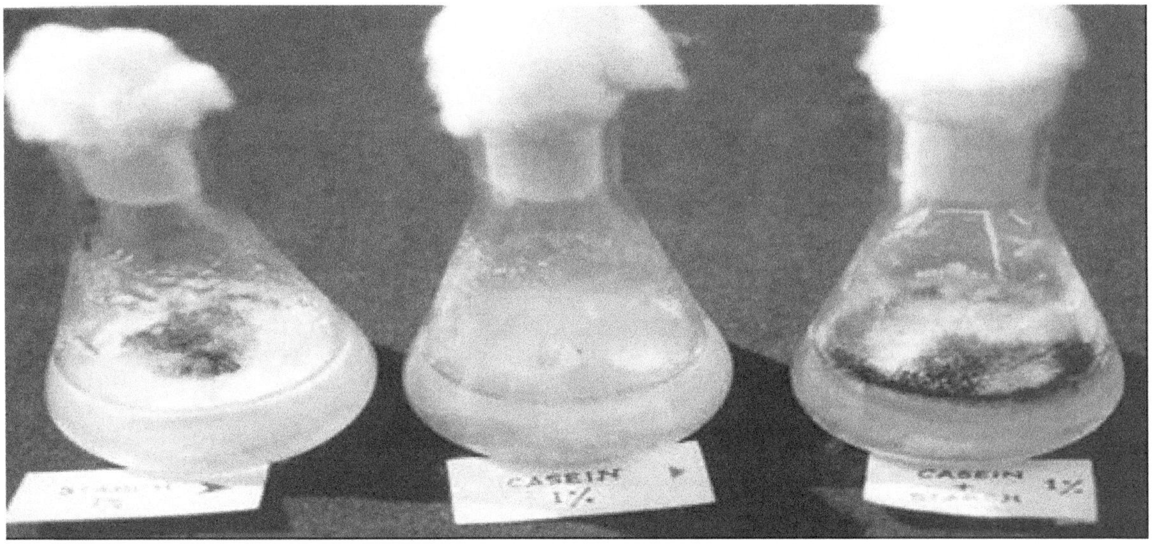

Figure 11. Effect of media engineering on the production of amylase and protease. (1% starch) (1% casein) (1% starch + 1% casein).

Table 3. Comparative analysis of yield of amylase and protease in different media.

Media	Amylase (U/gds)	Protease (U/gds)	Yield (Fold) Amylase	Yield (Fold) Protease
Basal media (PM)	4528.4 ± 121	250.40 ± 10	1.0	1.0
PM + additives	8940.5 ± 198	829.0 ± 75	1.98	3.3
Engineered media (EM)	9420.6 ± 89	903.2 ± 55	2.1	3.61
EM + additives	9386.5 ± 101	934.8 ± 67	2.07	3.73

(9420.6 ± 89 U/gds) and 3.61 fold (903.2 ± 55U/gds), respectively, in comparison to basal media (Table 3).

Comparative study of yields of amylase and protease in different media

As shown in Table 3, basal media alone gave amylase and protease yield of 4528.4 ± 121U/gds and 250.4 ± 10 U/gds, respectively. Amylase and protease yields increased by 1.98 fold (8940.5 ± 198 U/gds) and 3.3 fold (829.0 ± 75 U/gds), respectively, with suitable additives in primary media. Yields of amylase and protease increased by 2.1 and 3.61 fold, respectively, with engineered media and by 2.07 and 3.73 fold, respectively, with additives in engineered media.

DISCUSSION

In present work *A. awamori* nakazawa was found as an effective producer of glucoamylases and proteases in solid state fermentation with wheat bran as substrate. *A.awamori* has been reported as an efficient producer of glucoamylase as well as acid proteases by Yamamoto et al. (2005). Aalbæk et al. (2002) reported production of multiple forms of glucoamylase (GA) and secretion of acid protease by *Aspergillus niger* in pH-controlled batch fermenters and chemostats. Al-Turki et al. (2008) reported improvement of glucoamylase production by *A.awamori* using mutagenesis. Ellaiah et al. (2002) and Yamamoto et al. (2005) have reported wheat bran as the best substrate among various agro-residues for production of glucoamylase by the *Aspergillus* sp. A3. Kaur et al. (2003) also reported production of α-amylase by *A. niger* using wheat bran in submerged and solid state fermentation. Production of glucoamylase by *A.awamori* in Solid-State Fermentation was also reported by Negi and Banerjee (2009).

In present study additional supplements to wheat bran, such as powder of peanut, corn, soybean seeds and sunflower seeds were not found necessary for enhancing concomitant production of amylase and protease because wheat bran with a composition: 15.5% protein; 64.51% carbohydrates; 4.2% fat and other nutrients also in desirable amount, probably provides most of the essential nutrients. Uyar and Baysal (2004) optimized production of alkaline protease employing *Bacillus* sp.

under solid state fermentation (SSF) and also reported highest enzyme production using wheat bran. Akpan and Adelaja (2004) reported significant increase in stability of amylase from *Aspergillus oryzae* using rice bran supplemented with soya bean flour (SBF) and cassava starch (3:1).

In solid state fermentation microbial growth and product formation occur at or near the surface of the solid substrate particle having low moisture contents, hence, it is crucial to provide optimum water content and control the water activity of the fermenting substrate Selvakumar and Pandey (1999). In present study (Figure 2), amylase and protease yields were optimum at substrate to Czapek-dox ratio of 1:1.5 (w/v), which indicates very little requirement of water for extracellular production of these enzymes.

Significant activities of both enzymes were noticed within 48 h of incubation while 96 h incubation had been found sufficient to get optimum yield of these enzymes. This is venerable and remarkable finding for enzymes industry, as not many species have been reported to perform with such efficiency in concomitant production of two enzymes at a time.

A.awamori was very efficient in wide range of temperature and pH. Optimum temperature and pH were 35°C and 5.5, respectively. Narayana and Vijaylakshmi (2008) has also reported optimum temperature and pH as 30°C and 6.5, respectively, for production of α-amylase from *Streptomyces albidoflavus*. Kaur et al. (2003) achieved optimum production of α-amylase during solid state fermentation (SSF) at pH 5.0 and 30°C. Yang and Lin (1998) have also reported maximum production of acid protease from *A. niger* at pH 4.0 and 30°C. Nahas and Waldemarin (2002) explored *Aspergillus ochraceus* for amylase production at pH 5.5 and 30°C.

Investigation on the impact of carbon and nitrogen supplements had revealed that not all carbon and nitrogen sources would act as enhancer for simultaneous production of these enzymes in a single fermentation system. Unlike in single enzyme production role of supplements become very critical in multi-enzyme production as not many supplements enhance simultaneous production of all enzymes in a single bioreactor. Out of all the carbon sources investigated under this study, only lactose resulted in remarkable increase in yields of amylase as well as protease. Similarly, only peptone had been found as the most promising organic nitrogen source. Similar observations have also been made by Pandey et al. (2000), Cheng et al. (1995) in the case of protease and amylase production by different microbial species.

In many fermentation processes presence of metal ions may be inescapable and certain metal ions have been reported as enhancer of enzyme production. In present study only $HgCl_2$ was found enhancing the production of both the enzymes simultaneously. $CaCl_2$ appeared to act as a stabilizer for protease; which has also been reported earlier by Tunga et al. (2003), but greatly reduced activity of amylase. Among surfactants only Triton X-100 accelerated production of both enzymes, probably, because it enhances cell permeability and facilitates extra cellular enzyme secretion.

On comparative study of suitability of different media for concurrent production of amylase and protease in a single fermentation, engineered media with 1% casein and 1% starch was found as the most suitable media. Darani1 et al. (2008) also achieved optimum production of alkaline protease by *Bacillus* sp. 2 - 5 on date waste by addition of starch concentration of 5% (w/v or 0.5 g/L).

Conclusion

Concomitant production of amylase and protease by *A.awamori* in a single fermentation can be achieved by media engineered with 1% casein and 1% starch and maintaining various parameters at optimal levels. It is cost effective, convenient, less tedious and easier to scale up than the conventional method of using blend of enzymes from different fermentations. Production of amylase and protease in single fermentation can be particularly effective and useful for industries where both these enzymes are used together; such as food, detergent, pharmaceutical, oil extraction etc.

REFERENCES

Aalbæk T, Reeslev M, Jensen B, Eriksen SH (2002). Acid protease and formation of multiple forms of glucoamylase in batch and continuous cultures of *Aspergillus niger*. Enzy. Microbial Technol. 30(3): 410-415.

Akpan I, Adelaja FA (2004). Production and stabilization of amylase preparations from rice bran solid medium. World J. Microbiol. biotechnol. 20: 47-50.

Al-Turki AI, Khattab AA, Ihab AM (2008). Improvement of glucoamylase production by *A. awamori* using microbial biotechnology techniques. Biotechnol. 7(3): 456-462.

Bernfeld P (1955). Amylase α and β, In Methods in Enzymology (Colowick SP, Kaplan NO, ed.), Academic Press Inc, New York 1: 149-158.

Cheng SW, Hu HM, Shen SW, Takagi H, Asano M, Tsai YC (1995). Production and characterization of keratinase of a feather degrading *Bacillus licheniformis*. Biosci. Biotech. Biochem. 59: 2239-2243.

Darani1 KK, Falahatpishe HR, Jalali M (2008). Alkaline protease production on date waste by an alkalophilic *Bacillus* sp. 2-5 isolated from soil. African J. Biotechnol. 7(10): 1536-1542.

Ellaiah P, Adinarayana K, Bhavani Y, Padmaja P, Srinivasulu B (2002). Optimization of process parameters for glucoamylase production under solid state fermentation by a newly isolated *Aspergillus* species. Process Biochem. 38: 615-620.

Kaur P, Grewal HS, Kocher GS (2003). Production of α-amylase by *Aspergillus niger* using wheat bran in submerged and solid state fermentations. Indian J. Microbiol. 43:143-145.

Nahas E, Waldemarin MM (2002). Control of amylase production and growth characterizations of *Aspergillus ochraceus*. Review of Latino American Microbiol. 44(1): 5-10.

Narayana KJP, Vijayalakshmi (2008). Production of extracellular α-amylase by *Sreptomyces albidoflavus*. Asian J. Biochem. 3(3): 194-197.

Negi S, Banerjee R (2009). Optimization of extraction and purification of glucoamylase produced by *A. awamori* in solid state fermentation.

Biotechnol. Bioprocess Eng. 14: 60-66.

Pandey A, Nigam P, Soccol CR, Soccol VT, Singh D, Mohan R (2000). Advances in microbial amylases. Biotechnol. Appl. Biochem. 31: 135-152.

Pandey A, Selvakumar P, Soccol CR, Nigam P (1999). Solid state fermentation for the production of industrial enzymes. Curr. Sci. 77: 149-162.

Peixoto SC, Jorge JA, Terenzi HF, DeLourdes M, Polizeli TM (2003). *Rhizopus icrosporus* var. rhizopodiformis: A thermotolerant fungus with potential for production of thermostable amylases. Int. Microbiol. 6: 269-273.

Selvakumar P, Pandey A (1999). Solid state fermentation for the synthesis of inulinase from *Staphylococcus* sp and *Kluyveromyces marxians*. Process Biochem. 34: 851-855.

Sumantha A, Larroche C, Pandey (2006). A Microbiology and Industrial Biotechnology of Food-Grade Proteases. Food Technol. Biotechnol. 44(2): 211-220.

Tunga R, Shrivastava B, Banerjee R (2003). Purification and characterization of a protease from solid state cultures of *Aspergillus parasiticus*. Process Biochem. 38(11): 1553-1558.

Uyar F, Baysal Z (2004). Production and optimization of process parameters for alkaline protease production by a newly isolated *Bacillus* sp. under solid state fermentation. Process Biochem. 29: 1893-1898.

Walter HE (1984). Proteinases: Methods with hemoglobin, casein and azocoll as substrates in: Methods of Enzymatic Analysis, Vol. 5, Bergmeyer HU, Bermeyer J (Eds.), John Wiley and Sons Ltd, New York, USA pp. 270-274.

Yamamoto M, Saleh F, Ohtsuka A, Kunioki (2005). H New fermentation technique to process fish waste. Animal Sci. J. 76: 245.

Yang FC, Lin IH (1998). Production of acid protease using thin stillage from a rice-spirit distillery by *Aspergillus niger*. Enz. Microbial Technol. 23:397-402.

Zaghloul TI, Wahab AE, Mostafa MH (2000). Enhanced alkaline protease production in addition to α-amylase via constructing a *Bacillus subtilis* strain. Appl. Biochem. Biotechnol. 84-86: 319-27.

Zhang C, Xing XH, Liu MS (2004). Production of multienzymes consisting of alkaline amylase and cellulase by mixed alkalophilic culture and their potential use in the saccharification of sweet potato. Biochem. Eng. J. 19: 181-187.

Effect of *egusi* melon oil on lecithin: Cholesterol acyltransferase activity in rats fed a cholesterol diet

G. C. Ojieh[1], G. O. Eidangbe[1] and O. M. Oluba[2*]

[1]Department of Medical Biochemistry, College of Medicine, Ambrose Alli University, Ekpoma, Nigeria.
[2]Department of Biochemistry, University of Benin, Benin-City, Nigeria.

The effect of feeding *egusi* melon oil as a supplement to a cholesterol-based diet on serum lecithin: cholesterol acyltransferase activity of rats was evaluated. The rats were divided into two groups designated: control and test respectively. Rats in the test group were fed 5% cholesterol diet supplemented with 5% *egusi* melon oil while the control rats received 5% cholesterol diet without *egusi* melon oil. After 6 weeks of diet feeding, the enzyme activity was decreased significantly ($p < 0.05$) in the *egusi* melon oil-fed rats compared with the control. The test group also showed relative significant decreases in the serum levels cholesteryl ester and lysolecithin ($p < 0.05$) and increase levels of lecithin ($p < 0.05$). Significant decreases ($p < 0.05$) were also observed in serum total and free cholesterol in the *egusi* melon oil treated group compared to the control group. The implications of these results are discussed with respect to hypercholesterolemia.

Key words: *Egusi* melon oil, cholesterol diet, cholesteryl ester, LCAT, supplementation.

INTRODUCTION

Hypercholesterolemia has become a worldwide epidemic and its prevalence continues to increase at a rapid rate in various populations and across all age groups (Beaglehole et al., 1988). Hypercholesterolemia poses a major public health challenge since it is a well recognized independent predictor of premature mortality (Stamler et al., 1986). Moreover, it often coexists with other cardiovascular risk factors namely, hypertension and diabetes, which further add to the burden of cardiovascular disease. The dramatic increase in the occurrence of hypercholesterolemia over the past several decades is attributed in part to changes in dietary and lifestyle habits, such as rapidly changing diets, increased availability of high fat foods and reduced physical activity of people in both developed and developing countries (Hetzel et al.,1989).

Preventive or therapeutic strategies to control hypercholesterolemia have focused on the manipulation of the amount and nature of dietary fat intakes. In recent years, increased attention has shifted towards the role of nutritional supplement in the management of hypercholesterolemia. *Egusi* melon oil extracted from the seeds of *Citrullus lanatus* that originated in the Western region of Africa is an excellent source of essential fatty acids. The yellow coloured oil which is liquid at room temperature has been reported to contain about 56.9% linoleic acid (Oluba et al., 2008). *Egusi* melon has been shown to lower serum cholesterol concentration in rats when fed as a supplement to a cholesterol-based diet Oluba et al. (2008).

*Corresponding author. E-mail: olubamike2000@yahoo.co.uk.

Lecithin

Cholesterol acyltransferase (LCAT) (EC 2.3.1.43) is the enzyme responsible for the esterification of plasma cholesterol. In humans almost all plasma cholesterole esters is formed by the activity of this enzyme (Glomset, 1968). It has been suggested that the primary function of this enzyme is related to the maintenance of plasma lipoprotein structure during metabolism (Schumaker and Adams, 1970). A role for the enzyme in the transport of cholesterol from the peripheral tissues to liver (Glomset, 1968) and in maintaining the integrity of the plasma membrane has also been proposed (Glomset and Norum, 1973).

Table 1. Diet composition (% by weight).

Feed composition	Control	Test
Maize corn	65	60
Fish meal	10	10
Groundnut cake	20	20
Cholesterol	5	5
Egusi melon oil	-	5
Total	100	100

Experiments have shown that the modification of serum lipoproteins by LCAT (Glomset, 1976) alters rates of cholesterol flux between cells and medium and results in a net loss of cholesterol from the cells accompanied by cellular cholesterol synthesis (Relimpio and Iriarte, 1981). Since most extra hepatic tissues cannot catabolize cholesterol (Pometta et al., 1986), the mechanism of egress of cholesterol from cells and the involvement of LCAT in this reaction is of utmost importance because most of the cholesterol that accumulates in the aorta during atherosclerotic cardiovascular disease is in the esterified form. The reported response of LCAT to nutritional factors (Fielding and Fielding, 1980) and the earlier report of lowering of LCAT activity by plant protein (Forsythe et al., 1980) prompted us to investigate the effect of *egusi* melon oil supplementation to a cholesterol diet on serum LCAT activity in rats.

MATERIALS AND METHODS

Chemicals

All chemicals used were of analytical grade and were products of BDH Chemicals, Poole, England unless otherwise stated.

Collection and preparation of *egusi* melon seeds sample

Egusi melon seeds used for this study were obtained from a local market in Iwaro-Oka Akoko, Ondo State, Nigeria and were identified as *C. lanatus* (*egusi* melon) by a taxonomist in the Department of Crop Science, Faculty of Agriculture, University of Benin, Nigeria. The seeds were screened to remove bad ones, shelled manually and further screened. The seeds were then dried to constant weight in an oven at 70°C, ground using mechanical grinder, put in air-tight containers and stored in desiccators for further analysis, some of the seeds was subsequently deposited at the herbarium of the faculty.

Oil extraction

Oil from the seeds of *egusi* melon was extracted by continuous extraction in Soxhlet apparatus (Cehmglass) for 8 h using petroleum ether (60 - 80°C boiling range) as solvent according to the method of AOAC (1980). At the end of the extraction the extraction solvent was evaporated in a rotary evaporator (Cehmglass). The extracted oil was used for feed formulation and the remaining stored in light proof, airtight and moisture proof container at -4°C.

Animals and diets

Male Wistar rats (n = 28) eight weeks old and having a mean weight of 120.6 ± 11.2 g were obtained from the Nigerian Institute of Medical Research, Lagos (Nigeria) and housed individually in stainless steel cages with raised wire flour in an environment of 28 - 30°C, 50 - 60% relative humidity and a 12 h light-dark cycle. The animals were fed commercial rat pellets (Guinea feeds, Nigeria) and tap water *ad libitum* and were treated according to the Nigerian guidelines for the care and use of laboratory animals. The experiments were conducted according to the ethical norms approved by the Government of Nigeria and Institutional Animal Ethics Committee Guidelines.

The rats were acclimatized to the facility for 5 days before the start of the experiments. Rats were the assigned to two groups of fourteen animals each designated: control and test respectively and placed on their respective diet for a period of 6 weeks. Composition of each diet is presented in Table 1. Before the start of the diet treatment, the rats were fasted overnight but allowed water ad libitum. Rats have free access to their diet and were weighed weekly.

Serum preparation

At weekly intervals two rats from each group was sacrificed and about 2 ml blood collected from each animal by cardiac puncture and pooled together into plain tubes. The blood was allowed to clot at room temperature for 1 h and then centrifuged at 3000 rpm for 10 min the serum was separated into plain tubes and frozen at -20°C and analyzed.

Assays

Determination of total and free cholesterol concentrations in the serum was by the method of Searcy and Bergquist (1960), while the esterified cholesterol concentration was calculated as the difference between total and free cholesterol values. Serum lecithin and lysolecithin concentrations were determined by the method of Stewart (1980) after thin layer chromatography as suggested by Bowyer and King (1977). LCAT activity was determined by the method of Varma and Soloff (1976). The enzyme activity was expressed as mg cholesterol ester formed per 100 ml serum.

Statistical analysis

The data are presented as mean ± SEM. Statistical analysis was by one way analysis of variance and Duncan Multiple Range Test (DMRT) using SPSS 11.0.

RESULTS

The average daily feed intake for the control and test groups is 23.3 ± 1.3 and 23.0 ± 2.1 g respectively. Both the control and test rats showed increase in body weight with a mean weekly increase in weight of 14.9 ± 1.0 and 18.7 ± 1.4 g respectively.

Serum lipids

Table 2 represents the changes observed in serum

Table 2. Changes in serum cholesterol concentration (mg/ 100 ml serum).

Time on diet (weeks)	Total cholesterol		Cholesterol ester		Free cholesterol	
	Group		Group		Group	
	Control	Test	Control	Test	Control	Test
0	53.4±1.5	56.0±3.0	41.1±1.0	42.0±2.1	12.3±1.0	14.0 ± 1.0
1	61.0±2.0	62.5±1.0	45.8±2.2	45.8±1.1	15.2±1.0	13.7 ±2.0
2	69.9±1.2	69.1±2.5	51.7±3.1	51.5±2.0	18.2±2.0	17.6 ±1.0
3	105.3±3.0	82.5±3.0*	80.0±3.2	62.7±1.0*	25.3±2.2	19.8 ±2.1*
4	156.0±5.5	110.0±6.3*	117.4±4.6	81.4±1.0*	38.6±1.5	28.6 ±1.0*
5	170.7±2.0	123.6±3.3*	131.4±6.0	92.2±1.3*	39.3±3.1	31.4 ±1.0*
6	173.2±7.8	125.2±5.2*	128.2±3.8	93.9±2.0*	45.0±1.6	31.3 ±1.2*

Results are mean ± SEM of triplicate determinations. * Significantly different from the control.

Table 3. Changes in lecithin and lysolecithin concentration (mg/ 100 ml serum).

Time on diet (weeks)	Lecithin		Lysolecithin	
	Group		Group	
	Control	Test	Control	Test
0	17.5±1.0	16.9±1.2	3.3±0.3	3.2±0.1
1	19.1±1.2	20.2±1.0	7.8±1.0	7.8±0.5
2	23.3±2.0	23.6±2.3	15.3±1.0	17.2±1.0
3	26.3±2.0	32.2±1.5*	35.2±1.2	45.5±1.2*
4	30.0±3.1	37.2±2.5*	43.5±1.5	53.2±1.0*
5	39.1±3.0	48.7±2.2*	52.1±2.0	73.5±2.1*
6	43.5±1.7	53.5±3.0*	61.8±3.0	87.3±1.5*

Results are mean ± SEM of triplicate determinations. * Significantly different from the control.

cholesterol concentration which indicate no significant change ($p > 0.05$) between the control and test groups in the first 2 weeks of the feeding trial. However, starting from the third week a progressive decrease in serum total, free and esterified cholesterol fractions was observed in the *egusi* melon oil fed rats compared to control.

Table 3 shows serum lecithin and lysolecithin values. Serum lecithin and lysolecithin concentrations were significantly higher ($p < 0.05$) in the test rats compared to control rats after 2 weeks of *egusi* melon treatment.

Table 4. Changes in serum LCAT activity (mg cholesterol ester/100 ml serum).

Time on diet (weeks)	Control	Test
0	5.9 ± 1.0	5.6 ± 0.7
1	7.7 ± 1.1	7.8 ± 3.0
2	12. 1± 2.2	11.0 ± 2.0
3	26.5 ± 3.0	24.5 ± 2.3
4	34.5 ± 2.5	24.0 ± 1.5*
5	48.0 ± 3.0	25.9 ± 2.0*
6	54.2 ± 2.0	32.7 ± 2.0*

Serum LCAT activity

The esterifying activity of serum LCAT expressed as mg cholesterol formed per 100 ml serum is as shown in Table 4. The results show that exogenous cholesterol increases the esterification rate of LCAT as observed in the control group. On the contrary *egusi* melon oil suppresses this expected increase in the enzyme activity following cholesterol feeding as seen in the *egusi* melon fed (test) group. It is important to note that effect of *egusi* melon supplementation became significant after the third week of feeding.

DISCUSSION

Findings in the past several years indicate an important role of dietary fats in influencing the fatty acid profile of serum lipids, including phospholipids which are substrates of lecithin: cholesterol acyltransferase (LCAT), an important enzyme in lipoprotein metabolism. Although LCAT esterifies serum cholesterol solely at the interface of HDL and VLDL, the cholesteryl esters thus produced accumulate in all other lipoproteins (Zilversmith et al., 1975). Studies have shown that there is a correlation between increases in serum esterified cholesterol and susceptibility to coronary heart disease (Maroko et al., 1979). Although elevation of serum or tissue cholesterol after excessive cholesterol intake depends on the animal species, cholesterol feeding has been demonstrated to increase serum cholesterol in rats (Onyeneke et al., 2007).

The results obtained in this study show that feeding *egusi* melon oil as a supplement to a cholesterol enriched diet lowers serum total, free and esterified cholesterol levels as well as serum levels of lecithin and lysolecithin. These observations are in accord with the report of Oluba et al. (2008) on the effect of *egusi* melon oil on serum lipids in cholesterol-fed rats. The decrease in serum esterified cholesterol concentration observed in this study reflects an inhibition in the esterifying activity of LCAT, which is responsible for the regulation of serum cholesteryl ester level (Glomst, 1962; Nordy and Norum, 1975). Since serum lecithin is required as the acyl donor for the transesterification reaction of LCAT, a decrease in the activity level of the enzyme would result in lowering of serum lecithin level, as is observed in the *egusi* melon oil fed rats. The increase in lysolecithin fraction did not correspond to the decrease in the lecithin level, suggesting that the enzyme does not derive its entire acyl moiety from lecithin. The possibility exists that LCAT could utilize other fatty acids within the lipoprotein particles for the esterification process, although this remains to be established.

The decrease in serum LCAT activity leading to a decrease in serum cholesteryl ester fraction observed in this work is of interest, since the offending lipid in atherogenesis is the cholesteryl ester fraction. An increase in cholesteryl ester concentration above the serum threshold level could possibly initiate atherogenesis. It then follows that the hypercholesterolemia that developed when rats are fed a high cholesterol diet is due to stimulated activity of LCAT with the resultant production of excess cholesteryl ester. This is because if the cholesteryl esters produced are not effectively catabolized due to its relatively high concentration, there would be consequent deposition of the excess in the peripheral and vascular tissues thus resulting in atherosclerosis.

The inhibitory role of *egusi* melon oil on LCAT activity with the result that less amount of cholesteryl ester is produced could be beneficial in reducing the incident rate of atherosclerosis. *Egusi* melon oil has been reported to contain nutritionally good amount of linoleic acid and other essential fatty acids which have been reported to have protective effect against coronary heart disease (Baylin et al., 2003; Wijendran and Hayes, 2004). Thus, it is evident from this study that regulation of the activity of LCAT (especially by nutritional methods) could be a new target for therapy to prevent atherosclerosis.

REFERENCES

AOAC (1980). Official methods of analysis of the Association of Official Analytical Chemist. 13th edn, William Horwitz, Washington DC, USA pp. 56-132.

Baylin A, Kabagambe EK, Acherio A, Spiegelman D, Campus H (2003). Adipose tissue alpha linolenic acid and nonfatal acute myocardial infarction in Costa Rica. Circulation, 107: 1586-1591.

Beaglehole R, Bonita R, Stewart A (1988). Cardiovascular disease mortality trends in the Western Pacific, 1968-1984. N. Z. Med. J., 101: 441-443.

Bowyer DE, King JP (1977). Methods for the rapid separation and estimation of the major lipids of arteries and other tissues by thin-layer chromatography of small plates followed by micro-chemical assays. J. Chromatogr., 143: 473-490.

Fielding CJ, Fielding PE (1980). Characteristic of triacylglycerol and partial acylglycerol hydrolysis by human plasma lipoprotein lipase. Biochim. Biophys. Acta, 620: 440-446.

Forsythe WA, Miller ER, Hill GM, Romsos DR, Simpson RC (1980). Effect of dietary protein and fat sources on plasma cholesterol parameters, LCAT activity and amino acid levels and on tissue lipid content of growing pigs. J. Nutr., 110: 2467-2479.

Glomset JA (1962). The plasma cholesterol esterification reaction: Plasma fatty acid transferase. Biochim. Biophys. Acta, 1962; 65: 128-132.

Glomset JA (1976). Lecithin: cholesterol acyltransferase, in Lipoprotein Metabolism, ed. By Greten, Springer-Verlag, Berlin, New York pp. 28-41.

Glomset JA (1968). The plasma lecithin: cholesterol, acyltransferase reaction: Rev. J. Lipid Res., 9: 155-167.

Glomset JA, Norum KR (1973). Methods in enzymology. Adv. Lipid Res., 2: 1-65.

Hetzel BS, Charnock JS, Dwyer T, Maclennan PL (1989). Fall in coronary heartdisease mortality in USA and Australia due to sudden death: evidence for the role of polyunsaturated fat. J. Clin. Epidemiol., 42: 855-893.

Maroko PR, Maclean D, Ribeiro LGT, Braunwald E (1979). Pharmacological limitations of infarct size: Enzymatic, electrocardiographic and morphological studies in the experimental animal and man, In Enzymes in Cardiology, ed. Hearse J, de Leiris J, John Wiley and Sons, New York pp. 529-555.

Nordy G, Norum KR (1975). Substrate specificity of lecithin: cholesterol acyltransferase. Esterification of desmosterol, β-sitosterol and cholecalciferol in human plasma. Scand J. Clin. Lab. Invest., 35: 677-682.

Onyeneke EC, Oluba OM, Adeyemi O, Eriyamremu GE, Ojeaburu SI, Adebisi KE, Adeyemi O, Aboluwoye CO (2007). Effects of soy protein on serum level of cardiovascular disease diagnostic enzymes in cholesterol-fed rats. The Internet J. Nutr. Wellness, (4): 1.

Oluba OM, Adeyemi O, Ojieh GC, Isiosio IO (2008). Fatty acid composition of Citrullus lanatus (*egusi* melon) oil and its effect on serum lipids and some serum enzymes. The Internet J. Cardiovascular Res., (5): 2.

Pometta D, Suenram A, Sheybani E, Grab B, James R (1986). HDL cholesterol levels in patients with myocardial infarction and their families. Atherosclerosis, 59: 21-21.

Relimpio A, Iriarte AJ (1981). Re-evaluation of cholesterol homeostatis and atherosclerosis. J. Biol. Chem., 256: 4478-4488.

Schumaker VH, Adams GH (1970). Circulating lipoproteins. J. Theor. Biol., 26: 89-91.

Searcy RL, Bergquist LM (1960). A new colour reaction for the quantitation of serum cholesterol. Clin. Chim. Acta. 5: 192-1999.

Stamler J, Wentworth D, Neaton J (1986). Is the relationship between serum cholesterol risk of death from coronary heart disease continous and graded? J. Am. Med. Assoc., 256: 2823-2828.

Stewart JCM (1980). Colorimetric determination of phospholipids with ammonium ferrocyanate. Anal. Biochem., 104: 10-14.

Wijendran V, Hayes KC (2004). Dietary n-6 and n-3 fatty acid balance and cardiovascular health. Annu. Rev. Nutr., 24: 597-615.

Varma KG, Soloff LA (1976). A method for the purification of milligram quantities of stable human phosphatidyl choline cholesterol acyltransferase. Biochem. J., 155: 583-599.

Zilversmith DB, Hughes CB, Balmer I (1975). Stimulation of cholesterol ester exchange by lipoprotein-free rabbit plasma. Biochim. Biophys. Acta. 409: 393-398.

Permissions

All chapters in this book were first published in AJBR, by Academic Journals; hereby published with permission under the Creative Commons Attribution License or equivalent. Every chapter published in this book has been scrutinized by our experts. Their significance has been extensively debated. The topics covered herein carry significant findings which will fuel the growth of the discipline. They may even be implemented as practical applications or may be referred to as a beginning point for another development.

The contributors of this book come from diverse backgrounds, making this book a truly international effort. This book will bring forth new frontiers with its revolutionizing research information and detailed analysis of the nascent developments around the world.

We would like to thank all the contributing authors for lending their expertise to make the book truly unique. They have played a crucial role in the development of this book. Without their invaluable contributions this book wouldn't have been possible. They have made vital efforts to compile up to date information on the varied aspects of this subject to make this book a valuable addition to the collection of many professionals and students.

This book was conceptualized with the vision of imparting up-to-date information and advanced data in this field. To ensure the same, a matchless editorial board was set up. Every individual on the board went through rigorous rounds of assessment to prove their worth. After which they invested a large part of their time researching and compiling the most relevant data for our readers.

The editorial board has been involved in producing this book since its inception. They have spent rigorous hours researching and exploring the diverse topics which have resulted in the successful publishing of this book. They have passed on their knowledge of decades through this book. To expedite this challenging task, the publisher supported the team at every step. A small team of assistant editors was also appointed to further simplify the editing procedure and attain best results for the readers.

Apart from the editorial board, the designing team has also invested a significant amount of their time in understanding the subject and creating the most relevant covers. They scrutinized every image to scout for the most suitable representation of the subject and create an appropriate cover for the book.

The publishing team has been an ardent support to the editorial, designing and production team. Their endless efforts to recruit the best for this project, has resulted in the accomplishment of this book. They are a veteran in the field of academics and their pool of knowledge is as vast as their experience in printing. Their expertise and guidance has proved useful at every step. Their uncompromising quality standards have made this book an exceptional effort. Their encouragement from time to time has been an inspiration for everyone.

The publisher and the editorial board hope that this book will prove to be a valuable piece of knowledge for researchers, students, practitioners and scholars across the globe.

List of Contributors

E. O. Ajani
Department of Biochemistry, Faculty of Basic Medical Sciences, Olabisi Onabanjo University, Remo Campus, Ikenne, Ogun State, Nigeria

B. A. Salau
Department of Biochemistry, Faculty of Basic Medical Sciences, Olabisi Onabanjo University, Remo Campus, Ikenne, Ogun State, Nigeria

E. N. Ezima
Department of Biochemistry, Faculty of Basic Medical Sciences, Olabisi Onabanjo University, Remo Campus, Ikenne, Ogun State, Nigeria

O. H. Oyefuga
Department of Biochemistry, Faculty of Basic Medical Sciences, Olabisi Onabanjo University, Remo Campus, Ikenne, Ogun State, Nigeria

P. O Osineye
Department of Biochemistry, Faculty of Basic Medical Sciences, Olabisi Onabanjo University, Remo Campus, Ikenne, Ogun State, Nigeria

Faouzi Laachari
Laboratoire de Biotechnologie Microbienne. Université Sidi Mohamed Ben Abdellah, B.P. 2202 - Route d'Imouzzer, FES, Morocco

Fatimazahra El Bergadi
Laboratoire de Biotechnologie Microbienne. Université Sidi Mohamed Ben Abdellah, B.P. 2202 - Route d'Imouzzer, FES, Morocco

Moulay Sadiki
Laboratoire de Biotechnologie Microbienne. Université Sidi Mohamed Ben Abdellah, B.P. 2202 - Route d'Imouzzer, FES, Morocco

Adel Sayari
Laboratoire de biochimie et de génie enzymatique des lipases, ENIS, BPW 3038 Sfax, Tunisia

Wifak Bahafid
Laboratoire de Biotechnologie Microbienne. Université Sidi Mohamed Ben Abdellah, B.P. 2202 - Route d'Imouzzer, FES, Morocco

Soumya Elabed
Université Sidi Mohamed Ben Abdellah, Centre Universitaire Régional d'Interface, Fès, Morocco

Iraqui Mohammed
Laboratoire de Biotechnologie Microbienne. Université Sidi Mohamed Ben Abdellah, B.P. 2202 - Route d'Imouzzer, FES, Morocco

Saad Koraichi Ibnsouda
Laboratoire de Biotechnologie Microbienne. Université Sidi Mohamed Ben Abdellah, B.P. 2202 - Route d'Imouzzer, FES, Morocco

Tahar Nouadri
Research Laboratory in Enzymatic Engineering, Department of Biochemistry and Microbiology, Mentouri-University, Constantine- Algeria

Zahia Meraihi
Research Laboratory in Enzymatic Engineering, Department of Biochemistry and Microbiology, Mentouri-University, Constantine- Algeria

Djekrif-Dakhmouche Shahrazed
Research Laboratory in Enzymatic Engineering, Department of Biochemistry and Microbiology, Mentouri-University, Constantine- Algeria

Bennamoun Leila
Research Laboratory in Enzymatic Engineering, Department of Biochemistry and Microbiology, Mentouri-University, Constantine- Algeria

Ezzatollah Fathi
Department of Clinical Sciences, Faculty of Veterinary Medicine, University of Tabriz, Tabriz-Iran

Raheleh Farahzadi
Department of Clinical Biochemistry, School of Medical Sciences, Tarbiat Modarres University, Tehran-Iran

C. Egwim Evans
Biochemistry Department, Federal University of Technology, P.M.B 65, Minna, Niger State, Nigeria

J. Yisa
Chemistry Department, Federal University of Technology, P.M.B 65, Minna, Niger State, Nigeria

P. O. Egwim
Science Laboratory Technology, Federal Polytechnic, P.M.B 55, Bida, Niger State, Nigeria

A. Obayomi
Department of Biochemistry, Lagos State University, Ojo, Lagos, Nigeria

S. A. Adeola
Department of Biochemistry, Lagos State University, Ojo, Lagos, Nigeria

H. A. Bankole
Department of Biochemistry, Lagos State University, Ojo, Lagos, Nigeria

O. G. Raimi
Department of Biochemistry, Lagos State University, Ojo, Lagos, Nigeria
Protein phosphorylation unit, College of Life Sciences, University of Dundee, Scotland, United Kingdom

Jafar Maleki
Department of Biochemistry, Tehran University of Medical Sciences, Hemmat High Way, Tehran, Iran

Ladan Haghighi
Tehran University of Medical Sciences, Shahid Akbarabadi Hospital, Tehran, Iran

Mohsen Korani
Department of Biochemistry, Tehran University of Medical Sciences, Hemmat High Way, Tehran, Iran

Soudabeh Fallah
Department of Biochemistry, Tehran University of Medical Sciences, Hemmat High Way, Tehran, Iran

Mohsen Firozrai
Department of Biochemistry, Tehran University of Medical Sciences, Hemmat High Way, Tehran, Iran

Morteza Seifi
Department of Biochemistry, Tehran University of Medical Sciences, Hemmat High Way, Tehran, Iran

Noriaki Saigusa
Department of Applied Microbial Technology, Faculty of Biotechnology and Life Science, Sojo University; 4-22-1 Ikeda, Nishi Ward, Kumamoto 860-0082, Japan

Seika Imayama
Department of Applied Microbial Technology, Faculty of Biotechnology and Life Science, Sojo University; 4-22-1 Ikeda, Nishi Ward, Kumamoto 860-0082, Japan

Yuji Teramoto
Department of Applied Microbial Technology, Faculty of Biotechnology and Life Science, Sojo University; 4-22-1 Ikeda, Nishi Ward, Kumamoto 860-0082, Japan

Okonji Raphael Emuebie
Department of Biochemistry, Obafemi Awolowo University, Ile-Ife, Osun State, Nigeria

Popoola Michael Olaoluwa
Department of Zoology, Obafemi Awolowo University, Ile-Ife, Osun State, Nigeria

Kuku Adenike
Department of Biochemistry, Obafemi Awolowo University, Ile-Ife, Osun State, Nigeria

Aladesanmi Omolara Titilayo
Department of Biochemistry, Obafemi Awolowo University, Ile-Ife, Osun State, Nigeria

J. E. Olajide
Department of Biochemistry Kogi State University Anyigba, Nigeria

M. A. Akanji
Department of Biochemistry, University of Ilorin, Nigeria

J. Omale
Department of Biochemistry Kogi State University Anyigba, Nigeria

N. O. Mohammad
Department of Biochemistry, University of Ilorin, Nigeria

T. Okunwaye
Biochemistry Division, Nigerian Institute for Oil Palm Research, P.M.B. 1030, Benin City, Edo State, Nigeria

J.U. Obibuzor
Biochemistry Division, Nigerian Institute for Oil Palm Research, P.M.B. 1030, Benin City, Edo State, Nigeria

E.A. Okogbenin
Biochemistry Division, Nigerian Institute for Oil Palm Research, P.M.B. 1030, Benin City, Edo State, Nigeria

R. McEwan
Department of Biochemistry and Microbiology, University of Zululand, P/B X1001, KwaDlangezwa 3886, South Africa

R. P. Madivha
Department of Biochemistry and Microbiology, University of Zululand, P/B X1001, KwaDlangezwa 3886, South Africa

T. Djarova
Department of Biochemistry and Microbiology, University of Zululand, P/B X1001, KwaDlangezwa 3886, South Africa

O. A. Oyedeji
Department of Chemistry, University of Zululand, P/B X1001 KwaDlangezwa 3886, South Africa

A. R. Opoku
Department of Biochemistry and Microbiology, University of Zululand, P/B X1001, KwaDlangezwa 3886, South Africa

M. A. Emokpae
Department of Chemical Pathology, Aminu Kano Teaching Hospital, Kano, Kano State, Nigeria
Department of Medical Laboratory Science, School of Basic Medical Sciences, University of Benin, Benin City, Edo State, Nigeria

P. O. Uadia
Department of Biochemistry, University of Benin, Benin City, Edo State, Nigeria

H. B. Osadolor
Department of Medical Laboratory Science, School of Basic Medical Sciences, University of Benin, Benin City, Edo State, Nigeria

Sacha Tabosa-Vaz
Department of Biotechnology and Food Technology, Durban University of Technology, Box 1339, Durban 4001, South Africa

Ajit Kumar
Department of Biotechnology and Food Technology, Durban University of Technology, Box 1339, Durban 4001, South Africa

Kugen Permaul
Department of Biotechnology and Food Technology, Durban University of Technology, Box 1339, Durban 4001, South Africa

Suren Singh
Department of Biotechnology and Food Technology, Durban University of Technology, Box 1339, Durban 4001, South Africa

A. U. Wurochekke
Biochemistry Department, Federal University of Technology, Yola, Adamawa State, Nigeria

A. J. Nok
Biochemistry Department, Ahmadu Bello University, Zaria, Kaduna State, Nigeria

H. M. Inuwa
Biochemistry Department, Ahmadu Bello University, Zaria, Kaduna State, Nigeria

C. E. Gimba
Chemistry Department, Ahmadu Bello University, Zaria, Kaduna State, Nigeria

Priyanka Dhar and Gurvinder Kaur
Department of Biotechnology, Indian Institute of Technology Guwahati, Guwahati, Assam, 781039, India

Nathan Wymer
Pfizer Global Research and Development, Groton, Connecticut (CT), USA

Jeremy Steflik
Pfizer Global Research and Development, Groton, Connecticut (CT), USA

Maria Brown
Pfizer Global Research and Development, Groton, Connecticut (CT), USA

Alaa El-Din H. Sayed
Department of Zoology, Faculty of Science, Assiut University, Egypt

Imam A. A. Mekkawy
Department of Zoology, Faculty of Science, Assiut University, Egypt
Department of Biology, Faculty of Science, Taif University, Taif, Saudi Arabia

Usama M. Mahmoud
Department of Zoology, Faculty of Science, Assiut University, Egypt

Fabian Nde Fon
Department of Agriculture, University of Zululand, KwaDlangezwa, 3886, South Africa

Ignatius Verla Nsahlai
Animal and Poultry Science, University of KwaZulu-Natal, Pietermaritzburg, South Africa

Peter Frank Scogings
Department of Agriculture, University of Zululand, KwaDlangezwa, 3886, South Africa

R. K. Rajesh
Department of Biochemistry, Karnatak University, Dharwad 580 003, Karnataka, India

H. B. Manjunatha
Department of Sericulture, University of Mysore, Mysore 570 006, Karnataka, India

H. S. Aparna
Department of Biochemistry, Karnatak University, Dharwad 580 003, Karnataka, India
Department of Biotechnology, University of Mysore, Mysore 570 006, Karnataka, India

E. E. Edet
Biochemistry Department, Faculty of Basic Medical Sciences, College of Health Sciences, University of Uyo, P. M. B. 1017, Uyo - Nigeria

M. I. Akpanabiatu
Biochemistry Department, Faculty of Basic Medical Sciences, College of Health Sciences, University of Uyo, P. M. B. 1017, Uyo - Nigeria

A. E. Eno
Physiology Department, Faculty of Basic Medical Sciences, College of Basic Medical Sciences, University of Calabar, P. M. B. 1115, Calabar – Nigeria

I. B. Umoh
Biochemistry Department, Faculty of Basic Medical Sciences, College of Health Sciences, University of Uyo, P. M. B. 1017, Uyo - Nigeria

E. H. Itam
Biochemistry Department, Faculty of Basic Medical Sciences, College of Health Sciences, University of Uyo, P. M. B. 1017, Uyo - Nigeria

Trayana Djarova
Department of Biochemistry and Microbiology, University of Zululand, KwaDlangezwa 3886, South Africa

Abigail Ramakoaba
Department of Biochemistry and Microbiology, University of Zululand, KwaDlangezwa 3886, South Africa

Albert Basson
Department of Biochemistry and Microbiology, University of Zululand, KwaDlangezwa 3886, South Africa

Sam Mugandani
Department of Biokinetics and Sport Science, University of Zululand, KwaDlangezwa 3886, South Africa

Jeanne Grace
Department of Biokinetics and Sport Science, University of Zululand, KwaDlangezwa 3886, South Africa

Jacque Cloete
Department of Mathematical Sciences, University of Zululand, KwaDlangezwa 3886, South Africa

Gregory Watson
School of Biochemistry, Genetics and Microbiology, University of KwaZulu Natal, Scottsville 3209, South Africa

C. I. Iheme
Department of Biochemistry, Federal University of Technology, P.M.B. 1526, Owerri, Imo State, Nigeria

R. N. Nwaoguikpe
Department of Biochemistry, Federal University of Technology, P.M.B. 1526, Owerri, Imo State, Nigeria

S. E. Abanobi
Department of Biochemistry, Federal University of Technology, P.M.B. 1526, Owerri, Imo State, Nigeria

C. U. Igwe
Department of Biochemistry, Federal University of Technology, P.M.B. 1526, Owerri, Imo State, Nigeria

E. C. Udenze
Department of Biochemistry, Federal University of Technology, P.M.B. 1526, Owerri, Imo State, Nigeria

A. U. Ezirim
Department of Biochemistry, Federal University of Technology, P.M.B. 1526, Owerri, Imo State, Nigeria

C. P. Ihedimbu
Department of Biochemistry, Federal University of Technology, P.M.B. 1526, Owerri, Imo State, Nigeria

D. I. Ukairo
Department of Biochemistry, Federal University of Technology, P.M.B. 1526, Owerri, Imo State, Nigeria

K. Suresh Babu Naidu
Department of Biomedical and Clinical Technology, Durban University of Technology, Durban-4000, South Africa

K. Lakshmi Devi
Faculty of Biochemistry, SriKrishna Devaraya University, Anantapur-515 003, Andhra Pradesh, India

K. Jamila Adam
Department of Biomedical and Clinical Technology, Durban University of Technology, Durban-4000, South Africa

A. Abdulazeez Mansurah
Department of Biochemistry, Ahmadu Bello University, Zaria, Kaduna State, Nigeria
Natural Product Research and Development Centre, Faculty of Pharmacy, Chiang Mai University, Thailand

Ibrahim Kassim
Department of Biochemistry, Ahmadu Bello University, Zaria, Kaduna State, Nigeria

Bulus Kenpia
Department of Biochemistry, Ahmadu Bello University, Zaria, Kaduna State, Nigeria

B. Babvoshia Hope
Department of Biochemistry, Ahmadu Bello University, Zaria, Kaduna State, Nigeria

Yusuf Abdullahi
Department of Biochemistry, Ahmadu Bello University, Zaria, Kaduna State, Nigeria

U. U. Gabriel
Department of Fisheries and Aquatic Environment, Rivers State University of Science of Science and Technology, P.M.B. 5080, Port Harcourt, Nigeria

F. G Obomanu
Department of Chemistry, Rivers State University of Science of Science and Technology, P.M.B. 5080, Port Harcourt, Nigeria

O. S. Edori
School of Remedial and General Studies, Rivers State College of Health Science and Technology, P.M.B. 5039, Port Harcourt, Nigeria

N. Annamalai
Marine Sciences, CAS in Marine Biology, Annamalai University, Parangipettai 608502, Tamil Nadu, India

S. Elayaraja
Marine Sciences, CAS in Marine Biology, Annamalai University, Parangipettai 608502, Tamil Nadu, India

S. Vijayalakshmi
Marine Sciences, CAS in Marine Biology, Annamalai University, Parangipettai 608502, Tamil Nadu, India

T. Balasubramanian
Marine Sciences, CAS in Marine Biology, Annamalai University, Parangipettai 608502, Tamil Nadu, India

Naoual BOUSSOUALIM
Laboratory of Applied Biochemistry, Department Biochemistry, Faculty of Natural Sciences and Life, University Ferhat Abbas-Setif, Algeria

Djalal MEZIANE-CHERIF
Institut Pasteur, Plateforme 6, CNRS URA 2185, 25 rue du Dr Roux, 75724 Paris Cedex 15, France

Abderrahmane BAGHIANI
Laboratory of Applied Biochemistry, Department Biochemistry, Faculty of Natural Sciences and Life, University Ferhat Abbas-Setif, Algeria

Atere Tope Gafar
Department of Biochemistry, College of Health Sciences, Osun State University, Osogbo, Nigeria

Ademakinwa Adedeji Nelson
Department of Biochemistry, College of Health Sciences, Osun State University, Osogbo, Nigeria

Agboola Femi Kayode
Department of Biochemistry, Obafemi Awolowo University, Ile-Ife 220005, Nigeria

S. GOMETI Ajiroghene
Department of Biochemistry, University of Nigeria, Nsukka, Nigeria

N.OGUGUA Victor
Department of Biochemistry, University of Nigeria, Nsukka, Nigeria

E. ODO Christian
Department of Biochemistry, University of Nigeria, Nsukka, Nigeria

E. JOSHUA Parker
Department of Biochemistry, University of Nigeria, Nsukka, Nigeria

N. OBI Kelechi
Department of Biochemistry, University of Nigeria, Nsukka, Nigeria

Gloria A. Otunola
Department of Home Economics and Food Science, University of Ilorin, Ilorin, Kwara State, Nigeria

Oyelola B. Oloyede
Department of Biochemistry, University of Ilorin, Ilorin, Kwara State, Nigeria

Adenike T. Oladiji
Department of Biochemistry, University of Ilorin, Ilorin, Kwara State, Nigeria

Anthony A. Afolayan
Department of Botany, Phytomedicine Research Centre, University of Fort Hare, Alice 5700, South Africa

L. S. Yaqub
Department of Veterinary Physiology, Ahmadu Bello University, Zaria, Nigeria

M. U. Kawu
Department of Veterinary Physiology, Ahmadu Bello University, Zaria, Nigeria

J. O. Ayo
Department of Veterinary Physiology, Ahmadu Bello University, Zaria, Nigeria

S. F. Ambali
Department of Veterinary Pharmacology and Toxicology, Ahmadu Bello University, Zaria, Nigeria

B. Habibu
Department of Veterinary Physiology, Ahmadu Bello University, Zaria, Nigeria

Youssef Ali Abou Hamin Neto
Enzyme Technology Laboratory, Department of Pharmaceutical Sciences, School of Pharmaceutical Sciences, Ribeirão Preto-USP, Ribeirão Preto, Brazil

Cristina Maria de Souza Motta
Micology Department, Federal University of Pernambuco, Recife, Brazil

Hamilton Cabral
Enzyme Technology Laboratory, Department of Pharmaceutical Sciences, School of Pharmaceutical Sciences, Ribeirão Preto-USP, Ribeirão Preto, Brazil

Sangeeta Negi
Biotechnology Section, Applied Mechanics Department, Motilal Nehru National Institute of Technology, 221 004 Teliarganj, Allahabad, India

Rintu Banerjee
Agriculture and Food Engineering Department, Microbial Biotechnology, 721302 Kharagpur, India

G. C. Ojieh
Department of Medical Biochemistry, College of Medicine, Ambrose Alli University, Ekpoma, Nigeria

G. O. Eidangbe
Department of Medical Biochemistry, College of Medicine, Ambrose Alli University, Ekpoma, Nigeria

O. M. Oluba
Department of Biochemistry, University of Benin, Benin-City, Nigeria